Lecture Notes in Mathematics

Edited by A. Dold and B. Eckmann

T0184819

1126

Algebraic and Geometric Topology

Proceedings of a Conference
held at Rutgers University, New Brunswick, USA
July 6–13, 1983

Edited by A. Ranicki, N. Levitt and F. Quinn

Springer-Verlag
Berlin Heidelberg New York Tokyo

Editors

Andrew Ranicki
Department of Mathematics, Edinburgh University
James Clerk Maxwell Building, The King's Buildings
Edinburgh EH9 3JZ, Scotland, U.K.

Norman Levitt
Department of Mathematics, Rutgers University
New Brunswick, NJ 08903, USA

Frank Quinn
Department of Mathematics, Virginia Polytechnic Institute
Blacksburg, VA 24061, USA

Mathematics Subject Classification (1980): 55-02, 57-02

ISBN 3-540-15235-0 Springer-Verlag Berlin Heidelberg New York Tokyo
ISBN 0-387-15235-0 Springer-Verlag New York Heidelberg Berlin Tokyo

Printing and binding: Beltz Offsetdruck, Hemsbach/Bergstr.
2146/3140-543210

PREFACE

This volume was conceived as the proceedings of a conference on surgery theory held at Rutgers University in July, 1983. The editors have taken the opportunity to considerably expand the subject matter.

The articles in this volume present original research on a wide range of topics in modern topology. They include important new material on the algebraic K-theory of spaces (Waldhausen, Vogell), the algebraic obstructions to surgery and finiteness (Cappell and Shaneson, Milgram, Pedersen and Weibel, Ranicki, Sondow), geometric and chain complexes (Davis, Quinn, Smith, Weinberger), characteristic classes (Levitt), and transformation groups (Assadi and Vogel).

A paper of J.Levine on homotopy spheres, written in 1969 ~~as~~ the sequel to the classic work of Kervaire and Milnor but never published, is also included.

Andrew Ranicki
Norman Levitt
Frank Quinn

November, 1984

TABLE OF CONTENTS

SEMIFREE FINITE GROUPS ACTIONS ON COMPACT MANIFOLDS

A. H. Assadi(*)
Department of Mathematics
University of Virginia
Charlottesville, Virginia 22903
USA

P. Vogel
Institut de Mathematiques
et d'informatique
Université de Nantes
44072 Nantes
Cédex, FRANCE

INTRODUCTION

One of the classical problems in transformation groups has been to study the properties of the stationary point sets of actions on manifolds, and to characterize them whenever possible. P. A. Smith theory in combination with various other topological considerations provide a number of necessary conditions to be satisfied by the stationary point sets of some restricted classes of actions. In the case of smooth actions of a compact Lie group G on a manifold W, the stationary point set, say F, is a manifold and its normal bundle in W, say ν, is a G-bundle which determines the action in a (tubular) neighborhood of F.

For a complete characterization (of the diffeomorphism type) of F, one needs to show that the above mentioned necessary conditions are sufficient as well, in the following sense. Assuming that the submanifold F of the prescribed manifold W, and the G-bundle ν given, one tries to find an action on W which would restrict to the given action in the tubular neighborhood of F provided by the G-bundle ν. Special cases of such problems have been considered under various circumstances by various authors: [J1], [J2], [A1], [A2], [A3], [A-B 1], [A-B 1], [L], [D-R], [S] to mention a few. In these and other related contexts, a common hypothesis is that W is simply-connected and this assumption is indispensable for the techniques and the arguments to be applicable.

In the following, we consider this and some other relevant questions in the case of non-simply connected compact manifolds on which a finite group G has a "simple semifree action," i.e. where action is free outside of the stationary point set, and a certain localized Borel construction becomes fibre homotopy trivial. Although semifree actions comprise a restricted class, their understanding seems essential in developing general theories with more complicated isotropy group structures. The further restriction of "simplicity" of actions has been imposed to bring the homotopy-theoretic constructions and algebraic

---

(*) Partially supported by an NSF grant.

calculations within reach, as well as to provide a satisfactory answer to the above-mentioned questions in the form of less-complicated necessary and sufficient conditions.

In the presence of the fundamental group of the ambient manifold — on which the desired G-action is to be constructed — much of the methods and results of the simply-connected cases (in their various forms and contexts) are inapplicable. Thus, one is led to construct a new obstruction group and a new invariant (depending on both $\pi_1 W$ and G) whose vanishing is one of the necessary conditions for the existence of such actions. The obstruction group fits in a five-term exact sequence relating various Whitehead groups, and conceivably it can be defined as the fundamental group of the fibre of a transfer map between two Whitehead spaces involved in the problem, although its definition given below is in purely algebraic terms. The above-mentioned invariant is related to a certain Reidemeister torsion-type invariant.

If $\pi = 1$, Wh_1^T becomes simply $\tilde{K}_0$. This functor takes into account the interaction between $\tilde{K}_0$ (the finiteness obstruction in the presence of G-actions) and Wh_1 (the Whitehead torsion involving the fundamental group $\pi_1 = \pi$) in a way which is necessary to study the above mentioned problems. Thus, in the geometric context, Wh_1^T plays the same role in the study of finite group actions on non-simply connected compact manifolds that $\tilde{K}_0$ does in the simply-connected case.

The organization of the paper is as follows. In Section I we introduce Wh_1^T and state some of its algebraic properties which are used subsequently to detect the (combined) finiteness and Whitehead torsion type obstructions as the image of a Reidemeister torsion type invariant. Section II illustrates some computations of Wh_1^T. (The details of the results in these sections will appear elsewhere.) Section III considers semifree simple actions and gives necessary and sufficient conditions for existence of simple actions in this context. The problem of characterization of the stationary point sets of simple semi-free actions on compact bounded manifolds and an extension theorem for free simple actions are reduced to the homotopy theoretic problem of constructing appropriate Poincaré complexes, which are carried out using mixing the localizations of diagrams of spaces involved. Section IV gives an indication of the proofs of the theorems of Section III. Section V gives some useful theorems on constructing free simple actions either by extending a given action on a subspace or by pulling back actions from a given space, thus formalizing and generalizing the constructions needed in Section III. Although these are non-simply connected versions of analogous results in [A2] and [A3] where free actions are constructed from homotopy actions on simply-connected spaces (which are not simple in general), there is little overlap in scope or the methods.

There is somewhat of an overlap between some of the results obtained independently by S. Cappell and S. Weinberger [CW] as well as S. Weinberger [W], P. Löffler [L], P. Löffler and M. Raußen [LR]. The papers of L. Jones [J] and F. Quinn [Qu] also deal with related problems.

SECTION 1. Let Λ be a ring and $P(\Lambda)$ denote the category of finitely generated projective Λ-modules. In the sequel, G will denote a finite group, and π a discrete group which denotes as well the subgroup $\pi\times\{1\}\subset\pi\times G$ for simplicity of notation. Consider the set $A = \{(P,B)\,|\,P\in P(\mathbb{Z}(\pi\times G)),\ B = \mathbb{Z}\pi\text{-basis for } P\}$. The operation of direct sum of modules and disjoint union of $\mathbb{Z}\pi$-bases in the given order gives A the structure of a monoid with neutral element $(0,\emptyset)$. We introduce the equivalence relation $(P,B)\sim(P',B')$ among the elements of A if there exists a $\mathbb{Z}(\pi\times G)$-linear isomorphism $\alpha : P\xrightarrow{\cong}P'$ such that $\tau_\pi(\alpha) = 0$ with respect to B and B', where $\tau_\pi(\alpha)\in Wh_1(\pi)$ is the Whitehead torsion. The set of equivalence classes $A' = A/\sim$ inherits the monoid structure of A, and contains the submonoid "of trivial elements"; namely, (P,B) represents a trivial element in A' if $\underline{P}$ is $\mathbb{Z}(\pi\times G)$-free, and B is induced by a $\mathbb{Z}(\pi\times G)$-basis. The quotient monoid A' modulo the submonoid of trivial elements is seen to be an abelian group and is denoted by $Wh_1^T(\pi\subset\pi\times G)$. We have an obvious homomorphism $\alpha : Wh_1^T(\pi\subset\pi\times G)\longrightarrow\tilde{K}_0(\mathbb{Z}(\pi\times G))$ induced by the forgetful map $(P,B)\longrightarrow P\in\tilde{K}_0(\mathbb{Z}(\pi\times G))$. There is a further homomorphism $\beta : Wh_1(\pi)\longrightarrow Wh_1^T(\pi\subset\pi\times G)$ which is induced by the operation of "twisting the standard basis;" namely, let $x\in Wh_1(\pi)$ be represented by $\phi : (\mathbb{Z}\pi)^n\longrightarrow(\mathbb{Z}\pi)^n$. After stabilization, we have a π-linear homomorphism $\phi\oplus\mathrm{id} : (\mathbb{Z}(\pi\times G))^m\longrightarrow(\mathbb{Z}(\pi\times G))^m$. Let B be the image of the standard basis of $(\mathbb{Z}(\pi\times G))^m$ under the $\mathbb{Z}\pi$- linear map $\phi\oplus\mathrm{id}$. Then B is a $\mathbb{Z}\pi$-basis for $(\mathbb{Z}(\pi\times G))^m$ and $((\mathbb{Z}(\pi\times G))^m,B)$ represents $\beta(x)\in Wh_1^T(\pi\subset\pi\times G)$.

<u>1.1 Theorem</u>. There is an exact sequence

$$Wh_1(\pi\times G)\xrightarrow{Tr}Wh_1(\pi)\xrightarrow{\beta}Wh_1^T(\pi\subset\pi\times G)\xrightarrow{\alpha}Wh_0(\pi\times G)\xrightarrow{tr}Wh_0(\pi)$$

in which Tr and tr are transfer homomorphisms and $Wh_0\cong\tilde{K}_0$.

The homomorphism $\mathbb{Z}\pi \to \mathbb{Z}_q\pi$ induces a homomorphism $Wh_1(\pi) \longrightarrow Wh_1(\pi;\mathbb{Z}_q) \overset{def}{\equiv} K_1(\mathbb{Z}_q\pi)/\{\pm\pi\}$ where $\mathbb{Z}_q = \mathbb{Z}/q\mathbb{Z}$. One has a further map $\gamma : Wh_1(\pi;\mathbb{Z}_q) \longrightarrow Wh_1^T(\pi \subset \pi \times G)$ defined as follows. Let $GL_n'(\mathbb{Z}\pi)$ be the monoid of $(n\times n)$-matrices which have an inverse in $GL_n(\mathbb{Z}_q\pi)$. Given $\phi \in GL_n'(\mathbb{Z}\pi)$, one has an exact sequence

$$(C_*) \; : \; 0 \longrightarrow (\mathbb{Z}\pi)^n \overset{\phi}{\longrightarrow} (\mathbb{Z}\pi)^n \longrightarrow M \to 0$$

Thus $M_q = M \otimes \mathbb{Z}_q = 0$. It follows that $\text{proj dim}_{\mathbb{Z}(\pi\times G)} M \leq 1$, and we may take a short projective resolution over $\mathbb{Z}(\Pi \times G)$ for M, where order $(G) = q$:

$$(C_*') \; : \; 0 \longrightarrow C_1' \longrightarrow C_0' \longrightarrow M \longrightarrow 0$$

such that C_1' is free and C_0' is projective over $\pi \times G$. There is a $\mathbb{Z}\pi$-linear chain homotopy equivalence $\zeta : C_* \to C_*'$. Since the finiteness obstruction of C_* over $\mathbb{Z}\pi$ vanishes, C_0' is stably trivial over $\mathbb{Z}\pi$ also. After stabliization, we choose $Z\pi$-basis for C_1' and C_0', say B_1' and B_2'. If we choose the "standard bases B_1 and B_0 in the resolution (C_*) above for $C_1 \cong (\mathbb{Z}\pi)^n$ and $C_0 \cong (\mathbb{Z}\pi)^n$, then it is possible to arrange for the choices of B_1' and B_0' so that becomes a simple homotopy equivalence over $\mathbb{Z}\pi$. Let $\gamma(\phi) = [(C_1',B_1')] - [(C_1',B_1')]$ in $Wh_1^T(\pi \subset \pi \times G)$. In general, for $\phi \in GL_n(\mathbb{Z}_q\pi)$, we take $\phi = \frac{1}{s}\psi$, where $(s,q) = 1$. Then $s(Id) \in GL_n'(\mathbb{Z}\pi)$ and $\psi \in GL_n'(\mathbb{Z}\pi)$.
Let $\gamma(\phi) = \gamma(s(Id))$.

<u>I.2. Theorem.</u> γ induces a well-defined homomorphism such that the following diagram commutes

$$\begin{array}{ccc} Wh_1(\pi) & \longrightarrow & Wh_1^T(\pi \subset \pi \times G) \\ & \text{canon.} \searrow \quad \swarrow \gamma & \\ & Wh_1(\pi;\mathbb{Z}q) & \end{array}$$

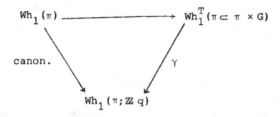

Suppose C_* is a chain complex over $\mathbb{Z}\pi$ such that $H_*(C_* \otimes \mathbb{Z}q) = 0$. Then the Reidemeister torsion of C_* is a well-defined element of $Wh_1(\pi;\mathbb{Z}q)$ and is denoted by $\tau(C_*)$. The main algebraic result of this section is the following:

<u>1.3. Theorem.</u> Let A_*' be a finite $\mathbb{Z}\pi$-based chain complex, and A_* be a finite $\mathbb{Z}(\pi \times G)$-based chain complex. Suppose there exists a $\mathbb{Z}\pi$-linear map $f:A_*' \longrightarrow A_*$ which is a $\mathbb{Z}\pi$-chain homotopy equivalence. Further, suppose $H_*(A \otimes \mathbb{Z}_q) = 0$ and that G acts trivially on $H_*(A)$, where order $(G) = q$. Then there is a finite $\mathbb{Z}(\pi \times G)$-based complex B_* and a $\mathbb{Z}(\pi \times G)$-chain homotopy equivalence $h:A_* \to B_*$ such that $hf:A_* \to B_*$ is π-simple if and only if $\gamma(\tau(A_*)) = 0$.

The above algebraic theory has the following application which is crucial in the construction of surgery problems of the next sections.

<u>1.4 Theorem.</u> Suppose we have a commutative diagram

$$\begin{array}{ccc} \tilde{Y} & \hookrightarrow & \tilde{X} \\ \beta \downarrow & & \downarrow \alpha \\ Y & \hookrightarrow & X \end{array}$$

with the following properties:

i) $\tilde{X}, \tilde{Y}$ and Y are finite connected CW complexes, and X is a connected CW-complex.

ii) $\pi_1(\tilde{X}) = \pi_1(\tilde{Y}) = \pi, \pi_1(X) = \pi_1(Y) = \pi \times G$.

iii) $\tilde{Y}$ is a covering space of Y and α induces a homotopy equivalence from $\tilde{X}$ to the covering space of X with the fundamental group π.

iv) $H_*(\tilde{X},\tilde{Y};\mathbb{Z}_q[\pi]) = 0$ and the Reidemeister torsion of $(\tilde{X},\tilde{Y})$ is $\tau(\tilde{X},\tilde{Y})$ in $Wh_1(\pi;\mathbb{Z}_q)$.

v) G acts trivially on $H_*(\tilde{X},\tilde{Y});\mathbb{Z}[\pi]) = H_*(X,Y;\mathbb{Z}[\pi \times G])$.

Then there exists a homotopy equivalence from X to a finite complex Z such that the composite map $\tilde{X} \xrightarrow{\alpha} X \longrightarrow Z$ induces a simple homotopy equivalence from $\tilde{X}$ to a covering space of Z, if and only if $\gamma(\tau(\tilde{X},\tilde{Y})) = 0$.

<u>Indication of Proof:</u> Let us denote by $C_*(-;M)$ the cellular chain complex with (twisted coefficients M. We have a π-linear homotopy equivalence $f : C_*(\tilde{X},\tilde{Y};\mathbb{Z}\pi) \longrightarrow (C_*(X,Y;\mathbb{Z}[\pi \times G])$. If there exists such a Z, then we have a π-simple homotopy equivalence

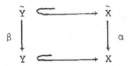

$$C_*(\tilde{X},\tilde{Y};\mathbb{Z}\pi) \longrightarrow C_*(Z,Y;\mathbb{Z}[\pi \times G])$$

from a finite π-based complex to a finite $\pi \times G$-based complex. Hence by Theorem 1.3, $\gamma(\tau(\tilde{X},\tilde{Y})) = 0$.

Conversely, suppose that $\gamma(\tau(\tilde{X},\tilde{Y}))$ vanishes. Then there exists a finite $\pi \times G$-based chain complex B_* and a $\pi \times G$-homotopy equivalence g from $C_*(X,Y;\mathbb{Z}[\pi \times G])$ to B_* such that $g \circ f$ is π-simple. This implies that the finiteness obstruction of X vanishes and there exists a homotopy equivalence from X to a finite complex $\tilde{Z}_1$. Moreover, we can add 2-cells and 3-cells to $\tilde{Z}_1$ in order to modify the simple type of Z_1 to obtain a finite complex Z such that the composite map

$B_* \xrightarrow{g^{-1}} C_*(X,Y;\mathbb{Z}[\pi \times G]) \longrightarrow C_*(Z,Y;\mathbb{Z}[\pi \times G])$ is a $\pi \times G$-simple homotopy equivalence. It is easy to see that the composite map $\tilde{X} \to X \to Z$ induces a simple homotopy equivalence from $\tilde{X}$ to the covering space of Z with fundamental group π.

<u>SECTION II</u>. Let $A = \mathbb{Z}\pi$ and $\omega = \sum_{g\ G} g$ be the norm of G. For simplicity of notation, let $A[G]/\omega A[G] \equiv A[G]/\omega, A/qA \equiv A_q$, and $\mathbb{Z}/2\mathbb{Z} \times M \cong \{+1,-1\} \times M \equiv \pm M$ for any group M. Consider the cartesian diagram:

$$\begin{array}{ccc} A[G] & \xrightarrow{\quad h \quad} & A[G]/\omega \\ \Big\downarrow f & & \Big\downarrow \\ A & \longrightarrow & A_q \end{array} \qquad \text{(C)}$$

where f is the augmentation and all other homomorphisms are cannonically defined quotient morphisms. The associated Mayer-Vietories sequence is:

$$K_1(A[G]) \longrightarrow K_1(A) \oplus K_1(A[G]/\omega) \longrightarrow K_1(A_q) \longrightarrow K_0(A[G]) \longrightarrow$$
$$K_0(A) \oplus K_0(A[G]/\omega) \longrightarrow K_0(A_q) \qquad \text{(MV)}$$

Corresponding to (MV), one has the following exact sequence if $G \neq \mathbb{Z}_2$

(U) $0 \longrightarrow \pm H_1(\pi) \times H_1(G) \longrightarrow \pm H_1(\pi) \oplus \pm H_1(\pi) \times H_1(G) \longrightarrow \pm H_1(\pi) \xrightarrow{0} \mathbb{Z} \longrightarrow \mathbb{Z} \oplus \mathbb{Z} \longrightarrow \mathbb{Z} \longrightarrow 0$

and if $G = \mathbb{Z}_2$, the sequence reads:

$$0 \longrightarrow \pm H_1(\pi) \times H_1(G) \longrightarrow \pm H_1(\pi) \oplus H_1(\pi) \times H_1(G) \longrightarrow H_1(\pi) \xrightarrow{0} \mathbb{Z} \longrightarrow \mathbb{Z} \oplus \mathbb{Z} \longrightarrow \mathbb{Z} \longrightarrow 0$$

The sequences (U) and the corresponding homomorphisms are also obtained from the diagram (C). The sequence (U) admits an injective homomorphism into the sequence (MV) and the quotient sequence is the exact sequence of the Whitehead groups below:

$$Wh_1(\pi \times G) \longrightarrow Wh_1(\pi) \oplus K_1(A[G]/\omega)/\pm H_1(\pi) \times H_1(G) \longrightarrow Wh_1(\pi;\mathbb{Z}_q) \xrightarrow{\partial}$$

$$\tilde{K}_0(A[G]) \longrightarrow \tilde{K}_0(A) \oplus \tilde{K}_0(A[G]/\omega) \longrightarrow \tilde{K}_0(A_q)$$

For simplicity of notation we write this sequence in terms of Whitehead groups (by a slight abuse of notation)

$$Wh_1(A[G]) \longrightarrow Wh_1(A) \oplus Wh_1(A[G]/\omega) \longrightarrow Wh_1(A_q) \xrightarrow{\partial} Wh_0(A[G])$$

$$\tag{W}$$

$$Wh_0(A) \oplus Wh_0(A[G]/\omega) \longrightarrow Wh_0(A_q)$$

The boundary map ∂ in the sequence is related to a generalization of the Swan homomorphism $(\mathbb{Z}_q)^* \xrightarrow{x} \tilde{K}_0(\mathbb{Z}G)$ in the case of $\pi = 1$, (cf. [Sw] or [M]). We continue to call ∂ the Swan homomorphism.

Let α and γ be as in Theorem I.1. Then the Swan homomorphism is $-\alpha\circ\gamma$. To see this, let $x \in Wh_1(A_q)$ correspond to the isomorphism $\phi : (A_q)^n \longrightarrow (A_q)^n$ induced by the (injective) homomorphism $\phi : A^n \longrightarrow A^n$. As in Section I, it follows that in the exact sequence

$$0 \longrightarrow A^n \longrightarrow A^n \longrightarrow M \longrightarrow 0$$

one has proj $\dim_{AG}(M) \leq 1$. Thus one has the commutative diagram:

$$\begin{array}{ccccccccc}
0 & \longrightarrow & A^n & \xrightarrow{\phi} & A^n & \longrightarrow & M & \longrightarrow & 0 \\
\uparrow & & \uparrow (f)^n & & \uparrow & & \uparrow 1 & & \uparrow \\
0 & \longrightarrow & K & \xrightarrow{\mu} & A[G]^n & \xrightarrow{\eta} & M & \longrightarrow & 0
\end{array}$$

where $(f)^n$ is induced by the augmentation f. Thus $\alpha\gamma([\phi]) = -[K]$ and the problem is reduced to show that the following diagram is cartesian:

$$\begin{array}{ccccc}
K & & \xrightarrow{\bar{\mu}} & & (A[G]/\omega)^n \\
\downarrow \lambda & & & & \downarrow (\nu)^n \\
A^n & \xrightarrow{(\rho)^n} & (A/q)^n & \xrightarrow{\bar{\phi}} & (A/q)^n
\end{array}$$

(Recall the definition of ∂ in the Mayer-Vietories sequence; cf. [M] e.g.). Since $\mathrm{Ker}\lambda \cong \mathrm{Ker}(f)^n \cong \mathrm{Ker}(\nu)^n$ and $(f)^n \circ \mu = \phi \circ \lambda$, one has the diagram:

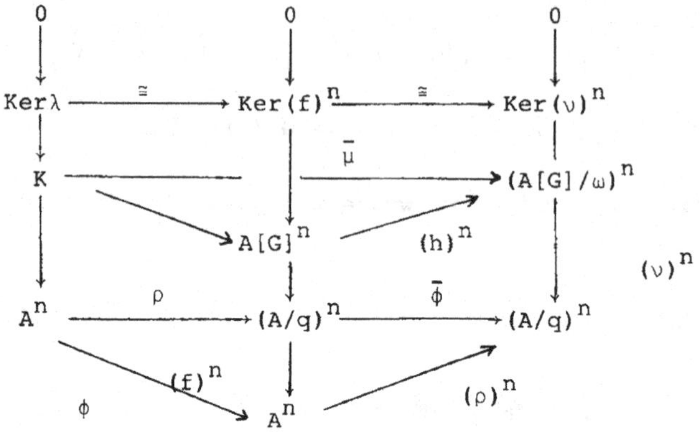

obtained from diagrams (B) and (C) above, and in which $\bar{\phi}\circ(\rho)^n\circ\lambda = (\rho)^n\circ\phi\circ\lambda = (\rho)^n\circ(f)^n\circ\mu = (\nu)^n\circ(h)^n\circ\mu = (\nu)^n\circ\bar{\mu}$. Thus (B) is cartesian.

Next, we identify the transfer $T_i : Wh_1(A[G]) \longrightarrow Wh_i(A)$, $i = 0,1$ in the 5-term exact sequence of Theorem I. Consider the diagram:

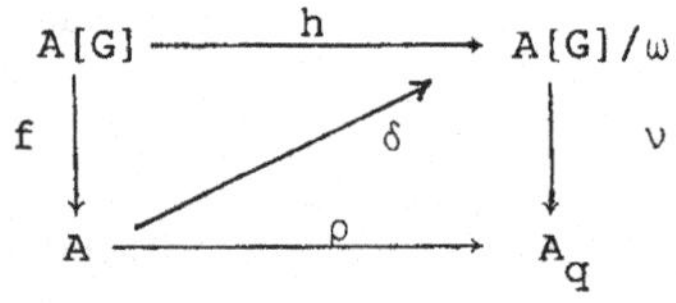

where δ is the composite $A \xrightarrow{\cong} A\times\{1\} \longrightarrow AG \longrightarrow AG/\omega$ so that $\nu\circ\delta = \rho$. Let $p \in P(AG)$ be given and tensor P over AG by the diagram (C) to obtain the cartesian diagram:

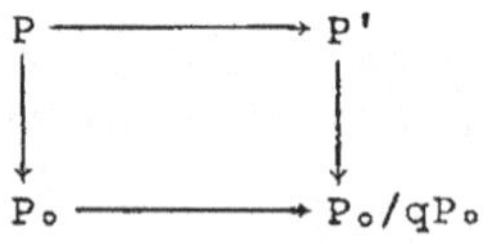

Thus one obtains four functors from $P(AG)$ to the categories $P(A[G])$, $P(A[G]/\omega)$, $P(A)$, and $P(A_q)$. The above cartesian diagram yields the commutative diagram:

$$\begin{array}{ccccccccc}
0 & \longrightarrow & P & \longrightarrow & P_\circ \oplus P' & \longrightarrow & P_\circ/qP_\circ & \longrightarrow & 0 \\
 & & \uparrow & & \uparrow 1+0 & & \uparrow 1 & & \\
0 & \longrightarrow & P_\circ & \xrightarrow[(\cdot)\times q]{} & P_\circ & \longrightarrow & P_\circ/qP_\circ & \longrightarrow & 0
\end{array}$$

which yields the exact sequence:

$$0 \longrightarrow P_\circ \longrightarrow P \oplus P_\circ \longrightarrow P_\circ \oplus P' \longrightarrow 0$$

The above sequence defines, in fact, a short exact sequence of the corresponding functors due to the functoriality of all the above constructions. It follows from Quillen's theorem on the additivity of functors (cf.[Q]) that the functor $P \longrightarrow P \oplus P_\circ$ is the sum of functors $P \longrightarrow P_\circ$ and $P \longrightarrow P_\circ \oplus P'$, which in turn implies that induced homomorphisms on K-theory satisfy $Tr = f_* \oplus tr\, h_*$, $Tr : K_*(A[G]) \longrightarrow K_*(A)$ and $tr : K_*(AG/\omega) \longrightarrow K_*(A)$ are transfer homomorphisms. Thus on the level of Whitehead groups, one has the following:

<u>II.1 Lemma</u>. Let $T_i : Wh_i(A[G]) \longrightarrow Wh_i(A)$ and $t_i : Wh_i(A[G]/\omega) \longrightarrow Wh_i(A)$ be transfer homomorphisms. Then $T_i = f_* \oplus t_i h_*$, where f_* and h_* are induced from diagram (C) above.

Let $\rho_i \equiv \rho_* : Wh_i(A) \longrightarrow Wh_i(A) \longrightarrow Wh_i(A_q)$. Specializing to the case $G = \mathbb{Z}_2$, the above calculation is continued to show

<u>II.2 Lemma</u>: The sequence $Wh_1(\pi) \xrightarrow{2\rho_1} Wh_1(\pi;\mathbb{Z}_2) \xrightarrow{\gamma} Wh_1^T(\pi \subset \pi \times \mathbb{Z}_2) \longrightarrow$

$\longrightarrow Ker\rho_\circ \longrightarrow 0$ is exact. In particular $Ker\gamma = Im 2\rho$.

This characterizes completely the obstructions which are discussed in Section I in terms of the $Wh_1(\pi)$ and the mod 2 reduction $Wh_1(\pi) \longrightarrow Wh_1(\pi;\mathbb{Z}_2)$. To obtain examples of nontrivial obstructions, let $\pi = \mathbb{Z}_8$. Then computations show

$$Wh_1(\mathbb{Z}_8;\mathbb{Z}_2) \cong \mathbb{Z}_2 \oplus \mathbb{Z}_2 \oplus \mathbb{Z}_4$$

<u>II.3 Corollary</u>: $Wh_1^T(\mathbb{Z}_8 \subset \mathbb{Z}_8 \times \mathbb{Z}_2) \cong \mathbb{Z}_2 \oplus \mathbb{Z}_2 \oplus \mathbb{Z}_2 \oplus \mathbb{Z}_2$ and $\ker\gamma$ consists of the 2-divisible elements of $Wh_1(\mathbb{Z}_8;\mathbb{Z}_2)$.

Although $Wh_1(\mathbb{Z}_5) \cong \mathbb{Z}$, one can show that in the case $\pi = \mathbb{Z}_5$, $G = \mathbb{Z}_2$, all the obstructions vanish.

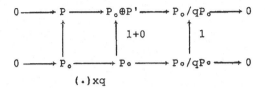

II.4 Corollary: $Wh_1(\mathbb{Z}_5;\mathbb{Z}_2) \cong \mathbb{Z}_3$ and $Im(\gamma) \equiv 0$.

Remark: In [Kw] Kwun has shown that the transfer $Wh_1(\mathbb{Z}_2 \times \mathbb{Z}_r) \longrightarrow Wh_1(\mathbb{Z}_r)$ is onto if and only if r = odd or $r = 2,4,6$. We thank the referee from bringing Kwun's result to our attention.

SECTION III. Let X be a finite dimensional CW complex with $\pi_1(X) = \pi$, and let G be a finite group of order q acting semifreely on X - i.e. the action is free outside of the stationary point set. In general, there is no explicit relationship between $H_*(X)$ and $H_*(X^G)$. The rather implicit information obtained using the localization theorems of Atiyah-Borel-Quillen-Segal type does not seem sufficient to yield a satisfactory characterization of the stationary point set X^G under general hypotheses. In the sequel, we will consider a class of actions which are encountered often in the geometric considerations, and to which it is possible to apply the present techniques of algebraic topology to obtain rather precise information and characterizations of X^G.

Given a connected space X and a subring of rational numbers Λ or $\Lambda = \mathbb{Z}_q$ we denote by X_Λ the localization of X which preserves $\pi_1 X$ and $\pi_i(X_\Lambda) \cong \pi_i(X) \otimes \Lambda$ for $i > 1$. For instance Bousfield-Kan's localization [B-K] applied to the universal covering space $\tilde{X}$ yields $\tilde{X}_\Lambda$ on which $\pi_1(X)$ operates freely and $\tilde{X} \longrightarrow \tilde{X}_\Lambda$ is equivariant. Then X_Λ can be defined as $\tilde{X}_\Lambda/\pi_1(X)$. For $\Lambda = \mathbb{Z}_q$, $\Lambda = \mathbb{Z}_{(q)}$ and $\Lambda = \mathbb{Z}[\frac{1}{q}]$ we can use the notations X_q, $X_{(q)}$ and $X(\frac{1}{q})$ respectively.

The key observation to reconstruct a space (respectively a diagram of spaces) from its localizations (respectively its diagrams of localizations) is the following:

III.1. Lemma. For any connected space X the following diagram is cartesian:

$$\begin{array}{ccc} X & \longrightarrow & X(\frac{1}{q}) \\ f \downarrow & & \downarrow f' \\ X_q & \longrightarrow & X_q(\frac{1}{q}) \end{array}$$

Proof: Since $H_*(X_q, X; \mathbb{Z}/q[\pi]) = 0$ it follows that $H_*(X_q, X; \mathbb{Z}\pi)$ is

$\mathbb{Z}[\frac{1}{q}]$-local. Hence the (homotopy) fibre of f is $\mathbb{Z}[\frac{1}{q}]$-local (Cf. [S]). Since the (homotopy) fibre of f' is also $\mathbb{Z}[\frac{1}{q}]$-local, f and f' has the same fibre (up to homotopy).

Definition. Let X be a connected G-CW complex, where G is a finite group of order q. X is called a simple G-space (and the action is called simple) if $(E_G \times_G X)_q$ is fibre homotopy equivalent to $(BG \times X)_q$.

For instance, if X has trivial mod q homology, then any G-action on X will be simple, or if X has the mod q homology of a sphere and $X^G \neq \emptyset$, the X-{point} has a simple action if we take out a point from X^G.

Proposition. Suppose G is a finite group of order q which has a simple semifree action on the finite dimensional complex X with $\pi_1 X = \pi$. Then $H_*(X, X^G; \mathbb{Z}_q \pi) = 0$, where the homology has local coefficients.

In the case of semifree simple actions on compact manifolds, one obtains further restrictions imposed on X^G. For simplicity, let us consider the case of a smooth semifree G-action on a compact manifold W^n with $\pi_1 W = \pi$. Then the stationary point set $W^G = F^k$ is a submanifold with normal bundle ν which is a G-bundle with a free G-representation at each fibre. Assume that $n-k > 2$. We identify the total space of the disk bundle $D(\nu)$ with a closed G-invariant tubular neighborhood of F. Let C^n = W-interior $D(\nu)$. One can choose an appropriate CW structure for W so that W, C, and $D(\nu)$ become G-CW complexes, and various cellular chain complexes have preferred bases. If the action is simple, then $H_*(W,F;\mathbb{Z}_q \pi) = 0$, and G acts trivially on $H_*(W,F;\mathbb{Z}\pi)$, as well as on $H_*(S(\nu);\mathbb{Z}[\frac{1}{q}](\pi)) \cong H_*(S(\nu)/G;\mathbb{Z}[\frac{1}{q}](\pi \times G))$. One further observation is that the geometry provides us with the dotted arrow in the following diagram in which $\pi = \pi_1(W)$:

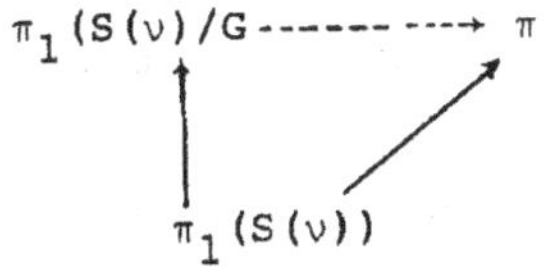

For a pair (W,F) as above, we define an element $\omega(W,F) \in Wh^T(\pi \subset \pi \times G)$ as follows. Given a free finite $\mathbb{Z}\pi$-based chain complex (A_*, A') and a free $\mathbb{Z}G$-resolution R_* of $\mathbb{Z}$, we form the $\mathbb{Z}(\pi \times G)$-complex $A_* = A'_* \otimes R_*$ which is $\mathbb{Z}\pi$-chain homotopy equivalent to A'_*. Suppose $H_*(A'_* \otimes \mathbb{Z}_q) = 0$.

Then by theorem I.3 there is a finite $\mathbb{Z}(\pi \times G)$-projective complex B_* with a π-basis $\mathcal{B}'$ such that $(B_*, \mathcal{B}')$ is π-simple homotopy equivalent to (A'_*, A'). Define $\omega(A'_*, A') = \sum(-1)^i [B_i, \mathcal{B}_i] \in Wh^T(\pi \subset \pi \times G)$ which is seen to be well-defined. Now let A'_* be the $\mathbb{Z}\pi$-chain complex of cellular chains of (W,F) with local $\mathbb{Z}\pi$-coefficients and let $\mathcal{R}'$ be the natural preferred bases provided by the cells. Then $\omega(W,F) = \omega(A'_*, \mathcal{R}')$ is well-defined. From section I, one can compute that $\omega(A'_*, \mathcal{R}') = \gamma\tau(A'_*)$.

<u>III.2. Theorem.</u> Let $\phi : G \times W^n \longrightarrow W^n$ be a smooth simple semifree action with $F^k = W^G$, $n-k > 2$, and ν = normal bundle of F in W, $\pi = \pi_1(W)$. Then:

1) $H_*(W,F;\mathbb{Z}_q\pi) = 0$,

2) G acts trivially on $H_*(S(\nu)/G;\mathbb{Z}[\frac{1}{q}](\pi \times G))$,

3) there is a homomorphism ι making the following diagram commute:

$$\begin{array}{ccc} \pi_1(S(\nu)/G & \xrightarrow{\iota} & \pi \\ \uparrow & \nearrow & \\ \pi_1(S(\nu)) & & \end{array}$$

4) $\omega(W,F) \in Wh_1^T(\pi \subset \pi \times G)$ vanishes.

Since $C_*(C^n, S(\nu);\mathbb{Z}\pi)$ is $\mathbb{Z}\pi$-chain homotopy equivalent to $C_*(W,F;\mathbb{Z}\pi)$, one verifies that $\omega(W,F)$ is defined under the following more general situation: $F^k \subset W^n$ is a submanifold with normal bundle ν, $n-k > 2$, and ν has G-bundle structure with a free representation on each fibre, and conditions (1) and (3) of Theorem II.2 are satisfied for (W,F).

The main results of this section are the following two theorems.

<u>III.3. Theorem (Characterization of stationary-point sets of simple actions).</u>

Let W^n be a compact manifold with connected boundary such that $\pi_1(\partial W) \cong \pi_1(W) = \pi$, and let $(F^k, \partial F^k) \subset (W, \partial W)$ be a smooth submanifold with normal bundle ν, $n-k > 2$, $n \geq 6$. Then there is a smooth simple semifree G-action on W^n with $(W^n)^{\overline{G}} = F$ if and only if F:

1) ν admits a G-bundle structure over F with a free representation on each fibre.

2) $H_*(W,F;\mathbb{Z}_q\pi) = 0$,

3) $\gamma\tau(W,F) \in Wh_1^T(\pi \subset \pi \times G)$ vanishes.

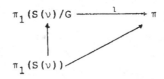

Remark: Condition (3) is equivalent to $\omega(W,F) = 0$. $\tau(W,F)$ is the Reidemeister torsion, and γ is the homomorphism of Theorem I.2.

The above theorem follows from the following extension theorem and III.2.

III.4. The Extension Theorem. Suppose C^n is a compact smooth manifold with $\partial C = \partial_+ C \cup_{\partial} \partial_- C$ where $\pi_1(\partial_- C) \cong \pi_1(C) = \pi$, $n \geq 6$.

Suppose that G is a finite group of order q and $\phi_+ : G \times \partial_+ C \longrightarrow \partial_+ C$ is a free smooth action such that:

1) $H_*(C, \partial_+ C; \mathbb{Z}_q \pi) = 0$,

2) there is a commutative diagram

$$\begin{array}{ccc} \pi_1(\partial_+ C/G) & \dashrightarrow & \pi \\ \uparrow & \nearrow & \\ \pi_1(\partial_+ C) & & \end{array}$$

3) G acts trivially on $H_*(\partial_+ C/G; \mathbb{Z}[\frac{1}{q}](\pi \times G))$.

Then there is a free G-action $\phi : G \times C \to C$ extending ϕ_+ with G acting trivially on $H_*(C/G; \mathbb{Z}[\frac{1}{q}](\pi \times G))$ if and only if $\gamma\tau(C, \partial_+ C) \in \widetilde{Wh}_1^T(\pi \subset \pi \times G)$ vanishes. Moreover, this action is unique up to concordance.

SECTION IV. We indicate an outline of proofs of the main theorems of Section III. Complete proofs and further applications will appear later.

Outline of the proof of Theorem III.3. The necessity of condition (2) follows from an application of the Atiyah-Segal-Quillen localization theorem for each prime order cyclic subgroup of G to the covering G-action on the universal covering space of W. Condition (3) is necessary due to Theorem I.3.

Given (W,F) satisfying (1) - (3) of III.3, we can apply The Extension Theorem III.4 to W-intD(ν) $\equiv$ C and the induced action of (1) to $S(\nu) \equiv \partial_+ C$ in order to construct a smooth simple semifree G-action on W with $W^G = F$.

An outline of the proof of III.4 is as follows. Theorems IV.1 and IV.2 allow us to construct an appropriate Poincare pair (X,Y) such that surgery problem provided by (X,Y) would yield the candidate for the

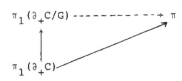

the orbit space $(C/G, C/G)$.

<u>IV.1. Theorem</u>. Let C^n be a compact manifold with $\pi_1(C) = \pi_1(\partial_- C) = \pi$, $\partial C^n = \partial_+ C^n \cup \partial_- C^n$, $\partial_+ C^n \cap \partial_- C^n = \partial_0 C = \partial(\partial_+ C) = \partial(\partial_- C)$. Suppose that $\phi : G \times \partial_+ C \to \partial_+ C$ is a free G-action such that:

1) $H_*(C, \partial_+ C; \mathbb{Z}_q[\pi]) = 0$

2) $\exists$ homomorphism $\int$ such that

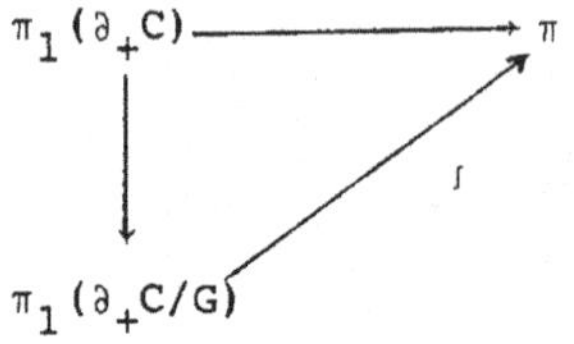

commutes.

3) G acts trivially on $H_*(\partial_+ C/G; \mathbb{Z}[\pi \times G])$, where $\mathbb{Z}[\pi \times G]$ is the local system for $\partial_+ C/G$ via $\int$.

Then there exists a Poincaré complex (X,Y) such that $Y = (\partial_+ C/G) \cup (\partial_- X)$, $(\partial_- X) \cap (\partial_+ C/G) = \partial_0 C/G$, and $(\tilde{X},\tilde{Y})$ is homotopy equivalent to $(C, \partial C)$ rel $\partial_+ C$, where $(\tilde{X},\tilde{Y})$ is the covering space with the covering transformation group G and fundamental group π.

<u>IV.2 Theorem</u>. Keep the notation and the hypotheses of IV.1. Then (X,Y) can be taken to be a finite Poincaré pair π-simple homotopy equivalent (rel $\partial_+ C$) to $(C, \partial C)$ if and only if $\gamma\tau(C, \partial_+ C) = 0$.

Assuming the proofs of IV.1 and IV.2, the proof of III.4 proceeds as follows. By IV.1, we have a Poincaré pair (X,Y) whose covering pair $(\tilde{X},\tilde{Y})$ with fundamental group π is homotopy equivalent to $(C, \partial C)$ rel $\partial_+ C$. By virtue of IV.2 and condition (3) of the hypotheses, we can choose (X,Y) so that $(\tilde{X},\tilde{Y})$ is simple-homotopy equivalent to $(C, \partial C)$. Next we show that the Spivak normal fibre space of (X,Y) has a linear structure which in turn shows that the set of normal invariants of (X,Y) is non-empty. Moreover, we can choose a normal invariant such that the corresponding normal map $(V, \partial V) \xrightarrow{f} (X,Y)$ is relative to $\partial_+ C/G$, i.e. $\partial_+ C/G \subset \partial V$ and $f|\partial_+ C/G : \partial_+ C/G \to \partial_+ C/G \subset Y$ is the inclusion (as a normal map). To see this, let $\lambda : X \to BG$ be the classifying map into Stasheff's classifying space for (stable) spherical fibrations. $\lambda|\partial_+ C/G$ has a life to BO induced by the given smooth structure of a $\partial_+ C/G$. Since G/O is an infinite loop space, the obstruction to extending this lift to X is an element $\alpha \in h^*(X,\partial_+C/G)$ where h^* is the generalized cohomology theory associated with $G/0$. We need the following lemma to show that this obstruction vanishes.

IV.3. Lemma. For any generalized cohomology theory $h^*(X,\partial_+C/G)$ is $(\frac{1}{q})$-local.

Let $\mu : \tilde{X} \to X$ be the covering with the covering transformation group . Then $\mu^*(\alpha) \in H^*(\tilde{X},\partial_+C)$ vanishes since $\tilde{X}$ is homotopy equivalent to C rel ∂_+C and the latter is a smooth manifold. The transfer $t : h^*(\tilde{X},\partial_+C) \to h^*(X,\partial_+C/G)$ is defined and $t\mu^*$ is multiplication by q. The above lemma implies that μ^* is a monomorphism and $\alpha = 0$ as a consequence. Hence we can choose a lift of λ to B0 which is compatible with the given life for $\lambda|\partial_+C/G$ so that the resulting normal map $f_1 : (V_1,\partial V_1) \to (X,Y)$ is rel ∂_+C/G as desired.

Let $K = \mathrm{Ker}(Wh_1(\pi\times G) \xrightarrow{\text{Transfer}} Wh_1(\pi))$. Although the Poincaré pair (X,Y) is not necessarily a simple Poincaré pair, the G-covering $(\tilde{X},\tilde{Y})$ is a simple Poincaré pair since it is π-simple homotopy equivalent to the manifold pair $(C,\partial C)$. Consequently the Whitehead torsion of the duality isomorphism lies in K. Denote by L^K_* the surgery obstruction groups of Wall where the homotopy equivalences are required to have Whitehead torsion belonging to K. Then the π-π theorem of Wall can be modified slightly to show that $L^K_*(\pi',\pi') = 0$ for a finitely presented group π'. Since the hypotheses imply that $\pi_1(X) \cong \pi_1(Y-\partial_+C/G)$, we may assume that f_1 is normally cobordant to a homotopy equivalence $f : (V,\partial V) \to (X,Y)$ rel ∂_+C/G with torsion in K.

Next, we can choose $(V,\partial V)$ such that the covering space $(V,\partial V)$ with group G is diffeomorphic to $(C,\partial C)$ rel ∂_+C. We have the commutative diagram of surgery exact sequences of Sullivan-Wall corresponding to (X,Y) rel ∂_+C/G and $(\tilde{X},\tilde{Y})$ rel ∂_+C and the maps induced by the covering projection:

$$\begin{array}{ccccccc}
L^s_{n+1}(\pi,\pi) & \longrightarrow & S^s(\tilde{X},\tilde{Y}) & \xrightarrow{\cong} & N(\tilde{X},\tilde{Y}) & \longrightarrow & L^s_n(\pi,\pi) \\
\uparrow & & \uparrow & & \uparrow & & \uparrow \\
L^K_{n+1}(\pi\times G,\pi\times G) & \to & S^K(X,Y) & \xrightarrow{\cong} & N(X,Y) & \longrightarrow & L^K_n(\pi\times G,\pi\times G)
\end{array}$$

The horizontal isomorphisms are due to Wall's (π-π) theorem. Now $N(X,Y) \cong h^0(X,\partial_+C/G)$ and $N(\tilde{X},\tilde{Y}) \cong h^0(\tilde{X},\partial_+C)$ where h^* is the cohomology theory associated with the G/O spectrum.

Consider the Carton-Leray-Serre spectral sequence for the G-covering $\mu : \tilde{X} \to X$ in which the E_2-term is $H^*(BG;h^*(X,\partial_+C))$ and it converges to $h^*(X,\partial_+C/G) \cong h^*(X,\partial_+C)^G$. The only nonvanishing terms are $H^0(BG;h^*(\tilde{X},\partial_+C) \cong h^*(\tilde{X},\partial_+C)^G$. Hence the spectral sequence collapses and $h^*(X,\partial_+C/G) \cong h^*(\tilde{X},\partial_+C)^G$. One can show that G acts trivially on $h^*(\tilde{X},\partial_+C)$. Hence $h^*(\tilde{X},\partial_+C) \cong h^*(X, \partial_+C/G)$ and μ^* induces the isomorphism. Thus we may choose the normal invariant in $N(X,Y)$ so that the corresponding homotopy smoothing $(V,\partial V)$ has the G-covering $(\tilde{V},\partial\tilde{V})$ diffeomorphic to $(C,\partial C)$ rel ∂_+C.

G acts freely on $(\tilde{V},\partial\tilde{V})$ as the group of covering transformations and this action extends the induced action on ∂_+C.

The idea of the proof of IV.1 is to construct a diagram

$$\begin{array}{ccc} \partial_0 C/G & \longrightarrow & \partial_+ C/G \\ \downarrow & & \downarrow \\ \partial_- X & \longrightarrow & X \end{array}$$

(In which ∂_-X, X and the dotted arrows are to be determined) with the property that the diagram $\tilde{\Delta}$ consisting of various G-covering spaces is (up to homotopy) the diagram D below:

$$\begin{array}{ccc} \partial_0 C & \longrightarrow & \partial_+ C \\ \downarrow & & \downarrow \\ \partial_- C & \longrightarrow & C \end{array}$$

In this vein, one constructs the diagrams Δ_q and $\Delta(\frac{1}{q})$ (of "localizations") such that there is a map $\Delta(\frac{1}{q}) \to \Delta_q(\frac{1}{q})$ which lifts to the appropriate maps of the G-coverings. The existence of such localized diagrams uses condition (1) and obstruction theory. A modified version of Lemma III.1 for diagrams yields the diagram Δ.

In Theorem IV.2, consider the diagram

$$\begin{array}{ccc} \partial_+ C & \longrightarrow & C \\ \downarrow & & \downarrow \\ \partial_+ C/G & \longrightarrow & X \end{array}$$

where X has been determined up to homotopy by Theorem IV.1. This diagram satisfies the hypothesis of Theorem 1.4, hence X is homotopy equivalent to a finite complex with $\tilde{X}$ being π-simple homotopy equivalent to C if and only if $\gamma\tau(C,\partial_+C) \in Wh_1^T(\pi \subset \pi \times G)$ vanishes. In particular if (X,Y) is π-simple homotopy equivalent to $(C,\partial C)$ rel ∂_+C then $\gamma\tau(C,\partial_+C) = 0$. (Observe that both X and ∂_-X in the above diagram are finitely dominated.)

To prove the converse, suppose that $\gamma\tau(C,\partial_+C) = 0$. Then by Theorem I.4, we can replace X in the above diagram by a finite complex whose covering with fundamental group π is π-simple homotopy equivalent to C. For simplicity of notation, assume that X is such a finite complex, so that $\tilde{X}$ is also finite. Consider the diagram

$$\begin{array}{ccc} \partial_-\tilde{X} & \longrightarrow & \tilde{X} \\ \downarrow & & \downarrow \\ \partial_-X & \xrightarrow{\eta} & X \end{array}$$

where ∂_-X and η are determined by the diagram Δ above. By Poincaré duality, $C^*(X,\partial_+C/G;\mathbb{Z}[\pi \times G])$ is chain homotopy equivalent to $C_*(X,\partial_-X;\mathbb{Z}[\pi \times G]) \equiv C_*(\eta;\mathbb{Z}[\pi \times G])$. Thus, there exists a $\pi \times G$-chain homotopy equivalence $f : C_*(X,\partial_-X;\mathbb{Z}[\pi \times G]) \to D_*$, where D_* is a finite $\pi \times G$-complex. Since $\pi_1(\partial_-X) \cong \pi_1(X) = \pi \times G$, by the additivity property of finiteness obstructions, it follows that the finiteness obstruction of $C_*(\partial_-X;\mathbb{Z}[\pi \times G])$ vanishes as well. Hence we may assume that ∂_-X and its covering $\partial_-\tilde{X}$ are finite complexes. At this point, the argument of Theorem I.4 goes through to show that we can choose ∂_-X to be a finite complex such that $\partial_-\tilde{X}$ is π-simple homotopy equivalent to ∂_-C. The additivity of Whitehead torsions shows that (X,Y) is a finite Poincaré pair such that $(\tilde{X},\tilde{Y})$ is π-simple homotopy equivalent to $(C,\partial C)$ rel ∂_+C.

<u>SECTION V.</u> To formalize some of the results of Section IV, in this section we prove some general results for constructing quasi-simple free actions on a given homotopy type (see below). The question of

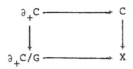

choosing a particular simple-type can be treated using the algebraic theory developed in Section I. The analogs of theorem I.4 which describes the obstructions for the choice of a simple type (lying in $Wh_1^T(\pi \subset \pi \times G)$) are valid and may be formulated in the context of Theorems V.1 and V.2 below.

<u>Definition</u>. A free action $G \times X \to X$ is called quasi-simple if $\pi_1(X/G) = \pi_1(X) \times G$ and G acts trivially on $H_*(X;\mathbb{Z}[\frac{1}{q}])$, $q = |G|$.

<u>V.1 Theorem</u>. (Pushing forward actions.) Suppose $\phi : G \times A \to A$ is a free quasi-simple action, and $f : A \longrightarrow X$ induces an isomorphism $f_* : H_*(A; f^*\mathbb{Z}_q[\pi] \longrightarrow H_*(X;\mathbb{Z}_q[\pi])$, where $\pi = \pi_1(X)$. Then there exists a free quasi-simple G-action on a space X', an equivariant map $\tilde{f} : A \longrightarrow X'$, and a homotopy equivalence $h : X \longrightarrow X'$ such that $h \circ f \longrightarrow f'$.

<u>Outline of Proof</u>: We need to construct a space Y and a map $A/G \xrightarrow{g} Y$ such that the G-covering $\tilde{Y}$ and the induced G-maps $\tilde{g} : A \to \tilde{Y}$ satisfy the property required for X' and f'. Let $g : (A/G)_q \longrightarrow Y$ be constructed as follows. Since $\pi_1(A) \longrightarrow \pi_1(X)$ is surjective with a q-perfect kernel, we can add free G-cells equivariantly to A to obtain $\tilde{Y}_q$ such that $\pi_1(\tilde{Y}_q) \cong \pi_1(X)$ and the inclusion $A \longrightarrow \tilde{Y}$ induces a $\mathbb{Z}_q[\pi]$-isomorphism (equivariant plus construction). Then define $Y_q \equiv \tilde{Y}_q/G$. Next, obstruction theory shows that $(A/G)(\frac{1}{q}) \simeq (A \times BG)(\frac{1}{q})$ since the action is quasisimple. Let $Y(\frac{1}{q}) = (X \times BG)(\frac{1}{q})$ and let $g(\frac{1}{q})$ be the composition $(A/G)(\frac{1}{q}) \xrightarrow{\sim} (A \times BG)(\frac{1}{q}) \longrightarrow (X \times BG)(\frac{1}{q})$. Then we have a map $Y(\frac{1}{q}) \xrightarrow{\alpha} Y_q(\frac{1}{q})$ by obstruction theory such that the pull-back diagram

$$\begin{array}{ccc} Y & \longrightarrow & Y(\frac{1}{q}) \\ \downarrow & & \downarrow \\ Y_q & \longrightarrow & Y_q(\frac{1}{q}) \end{array}$$

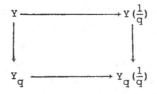

has the G-covering $\tilde{Y}$ homotopy equivalent to X via $h : X \longrightarrow \tilde{Y}$. Let $X' = \tilde{Y}$. The G-action on X' is quasi-simple by construction. The maps g and $g(\frac{1}{q})$ pull back to give the map $g : A/G \longrightarrow Y$ and we let the lift $\tilde{g} : A \longrightarrow \tilde{Y}$ be $f : A \longrightarrow X'$. One verifies that f' and X' satisfy the required properties.

V.2. Theorem (Pulling back actions). Let A be a free quasi-simple G-space with $\pi_1(A) = \pi$. Let $f : X \to A$ be such that $f_* : H_*(X; f^*\mathbb{Z}[\pi]) \to H_*(A;\mathbb{Z}[\pi])$ is an isomorphism. Then there exists a free quasi-simple G-space X' an equivariant map $f' : X' \to A'$ and a homotopy equivalence $h : X' \to X$ such that $f \circ h \simeq f'$.

Outline of Proof: As before, we need to construct the orbit space Y and $g : Y \to A/G$ satisfying the stated properties on the level of G-coverings. Let $Y_q = (A/G)_q$ with $g_q = \mathrm{id}$ and $Y(\frac{1}{q}) = X(\frac{1}{q}) \times BG$ with $g(\frac{1}{q}) = f(\frac{1}{q}) \times \mathrm{id}$. There exists a map $Y(\frac{1}{q}) \to Y_q(\frac{1}{q})$ which is up to homotopy the composition $X(\frac{1}{q}) \times BG \to X_q(\frac{1}{q}) \times BG \to A_q(\frac{1}{q}) \times BG \to (A/G)_q(\frac{1}{q})$. Let Y be pull-back of the diagram:

$$\begin{array}{ccc} Y & \longrightarrow & (A/G)_q \\ \big\downarrow & & \big\downarrow \\ X(\frac{1}{q}) \times BG & \longrightarrow & (A/G)_q(\frac{1}{q}) \end{array}$$

The Seifert-van Kampen theorem shows that $\pi_1(Y) \cong \pi_1(X) \times G$. Furthermore, functoriality of pull-backs and the Mayer-Vietoris theorem show that the G-covering Y is homotopy-equivalent to X. The maps g and $g(\frac{1}{q})$ yield $g : Y \to A/G$ (via the above pull-back) and we may define $X \to Y$ and $f' \equiv g : \tilde{Y} \to \tilde{A}$ the induced map on the covering spaces. One readily verifies that X' and f' satisfy the desired properties.

V.3. Theorem. (The relative version). Under the hypotheses of V.2 suppose that a subspace $X_0 \subset X$ is equipped with a quasi-simple free G-action such that $f|X_0$ is equivariant. Then it is possible to arrange for X_0 to be a G-invariant supspace of X', $f|X_0 = f|X_0$, and for h to be a homotopy equivalence rel X_0.

Proof: This is the quasi-simple analog of [A3] Proposotion 2.III with a similar obstruction theory argument.

Let A_1 and A_2 be the CW complexes. Call A_1 and A_2 "weakly $\mathbb{Z}_q$-

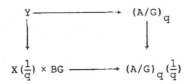

equivalent," if there exists a CW complex C and maps $f_1 : A_1 \to C$ such that $f_{i*} : (H_i(A_i; f_i^* \mathbb{Z}_q[\pi_1(C)]) \to H_i(C; \mathbb{Z}_q[\pi_1(C)])$ are isomorphisms. The equivalence relation generated by weak $\mathbb{Z}_q$-equivalence is simply called "$\mathbb{Z}_q$-equivalence".

V.4 Proposition. Suppose A_1 and A_2 are $\mathbb{Z}_q$-equivalent complexes. Then A_1 admits a free quasi-simple G-action if and only if A_2 does.

Proof: This follows from V.1, V.2, and the defintion of $\mathbb{Z}_q$-equivalence.

V.5 Remarks: The above results are valid for diagrams of spaces as it was needed in Section III.

REFERENCES

[A1] A. Assadi, "Finite Group actions on Simply-connected Manifolds and CW Complexes," Memoirs AMS, No. 257 (1982).

[A2] ________, "Extensions of Finite Group Actions From Submanifolds of a Disk," Proc. of London Top. Conf. (Current Trends in Algebraic Topology) AMS (1982).

[A3] ________, "Extensions Libres Des Actions Des Groupes Finis Dans les Variétés Simplement Connexes," (To appear in the Proc. Aarhus Top. Conf. Aug. 1982.)

[A-B1] A. Assadi and W. Browder, "On the Existence and Classification of Extenstions of Actions of Finite Groups on Submanifolds of Disks and Spheres" (to appear in Trans. AMS)

[A-B2] A. Assadi and W. Browder, "Construction of Free Finite Group Actions on Simply-Connected Bounded Manifolds." (In preparation.)

[B-K] A. K. Bousfield and D. M. Kan, Springer-Verlag LNM, No. 304 (1972).

[D-R] K. H. Dovermann and M. Rothenberg, "An Equivariant Surgery Sequence and Equivariant Diffeomorphism and Homomorphism Classification (A Survey)," Proc. Siegen Conf., Springer-Verlag Lecture Notes.

[J1] L. Jones, "The Converse to Fixed Point Theorem of P. A. Smith I," Ann. of Math. 94 (1971), 52-68.

[J2] ________, "Converse to Fixed Point Theorem Of P. A. Smith II," Indiana U. Math. J. 22 (1972), 309-325.

[L] P. Löffler, "$\mathbb{Z}_p$- Operationen auf Rationalen Homologiesphären mit Vorgeschriebener Fixpunktmenge," Manuscripta Math. 36. (1981) 187-221.

[Q] D. Quillen, "Higher Algebraic K-theory: I," Springer-Verlag LNM 341.

[SW] R. Swan, "Periodic Resolutions and Projective Modules," Ann. Math. 72 (1960), 552-578.

[M] J. Milnor, "Algebraic K-theory," Ann. Math. Studies, Princeton Press 1971.

[CW] S. Cappell - S. Weinberger, (to appear).

[W] S. Weinberger, "Homologically Trivial Actions I" and "II", (preprint, Princeton University 1983).

[LR] P. Löffler - M. Raußen, (to appear).

[J] L. Jones, "Construction of Surgery Problems," Proc. 1978 Georgia Topology Conference.

[Qu] F. Quinn, "Nilpotent Classifying spaces and actions of finite groups, " Houston J. of Math. 4, (1978) 239-248.

[Kw] K. Kwun, "Transfer Homomorphisms and Whitehead Groups of Some Cyclic Groups," Springer-Verlag LNM, 298 (1972).

Torsion in L-groups

by Sylvain E. Cappell[1] and Julius L. Shaneson[1]

Introduction.

Let $L_n^h(\pi,w)$ denote the Wall group for the homotopy equivalence problem, π a finitely presented group and $w: \pi \to \{\pm 1\}$ a homomorphism. These groups figure in many geometric problems, and their rank is the same as that of other related surgery groups which have been computed in many important cases. Their torsion, for π finite, reflects the subtle relation between signature and discriminant of quadratic forms, discussed below. This paper contributes two calculations and a sample application to manifolds with finite fundamental group.

Theorem A. *The torsion of* $L_{2k}^h(Z_{2^r})$ *is a vector space over* Z_2 *of dimension* $[2(2^{r-2}+2)/3]-[r/2]-\varepsilon$, $\varepsilon = 1$ *if* k *is even and* 0 *if* k *is odd.*

(In Theorem A, $[x]$ = greatest integer in x.)

Theorem B. $L_{2k+1}^h(Z_{2^r},-)$ *is a vector space over* Z_2 *of dimension* 2^{r-3}, $r \geq 3$, *and zero for* $r = 1,2$.

Since the rank of $L_{2k}^h(Z_{2^r})$ is well-known (see [W1]), Theorem A completely determines this group. Taylor and Oliver, and Milgram have obtained at least Theorem A independently. Their method involves use of the computation of projective L-groups and a study of the relation of L^h to these via an exact sequence whose third term is a cohomology group of Z_2 with coefficients in $K_0(\pi)$. In this paper, we apply some of the algebraic

[1]Both authors partially supported by NSF Grants.

results obtained in the course of our study of non-linear similarity. We use these to analyze the sequence (4.1) in [Sh] ("Rothenberg's sequence") relating L^s-groups (computed for cyclic groups in [W1]) to L^h-groups. We believe the independence lemma we use on units in the group ring $Z[Z_{2^r}]$ may be of independent interest.

Here is an application of the methods and results of this paper to the problem of providing a Poincaré Duality (PD) space with a manifold structure. Let X be a finite complex. Then X is called a PD space of dimension m if there exists a class $[X] \in H_m(X)$ such that for all i

$$\cap[X]:H^i(X;B) \to H_{n-i}(X,B)$$

is an isomorphism for every local coefficient system B over X. A closed manifold is a Poincaré duality space. If X is connected, then X will be a PD space if and only if there exists $[X] \in H_m(X)$ with $\cap[X]$ an isomorphism for the local coefficient system (i.e. the $Z\pi_1X$-module) $B = Z\pi_1X$.

Now suppose X is connected and $\pi_1(X) = G$ is a finite group and $m = 2k$. Then an invariant $\chi(X) \in R(G)$ is defined (in [W2, §13B], this invariant is denoted $\sigma(G,X)$.) To define it, (compare §3) let $\rho:G \to U(n)$ be an irreducible complex representation. Then $\mathbb{C}^n$ becomes a local coefficient system over X, and the isomorphism

$$(\cap[X])^{-1}:H_k(X;\mathbb{C}^n) \to H^k(X,\mathbb{C}^n) = H_k(X;\mathbb{C}^n)^*$$

is easily seen to be the adjoint of a unimodular $(-1)^k$ Hermitian form β over the complex numbers. Let

$$\sigma_\rho(X) = \text{signature of} \begin{cases} \beta & k \text{ even} \\ \sqrt{-1}\,\beta & k \text{ odd} \end{cases} .$$

Then let

$$\chi(X) = \sum_\rho \sigma_\rho(X)\rho,$$

the sum over irreducible representations. By a theorem of Atiyah-Bott, as extended by Wall to the PL and topological case, if X has the homotopy of a compact manifold, smooth, PL, or even topological, then $\chi(X)$ is a multiple of the regular representation.

A stable orientable linear, PL, or topological Euclidean space bundle (or block bundle) ξ over X is called reducible if its Thom class is spherical. If X has the homotopy type of manifold, such a bundle exists, namely, the stable normal bundle of the manifold.

In §6, the calculations of this paper will be applied to obtain a precise list of invariants for finding a manifold structure on a PD space X with $\pi_1 X = Z_{2^r}$. One consequence is:

Theorem C. Let X be a connected finite complex, and suppose that X is a PD space, of dimension $2k$ with $\pi_1 X = Z_{2^r}$, satisfying the following:

(i) $\chi(X)$ is a multiple of the regular representation,

(ii) there is a reducible (PL or TOP) bundle over X.

(iii) X has the homotopy type of the two-fold cover $\hat{Y}$ of a finite complex Y with $\pi_1 Y = Z_{2^{r+1}}$, and the image of $[X]$ in $H_{2k}(\hat{Y})$ is in the image of the transfer (from $H_{2k}(Y)$).

Then X has the homotopy type of a PL or TOP manifold (with the given reducible bundle as normal bundle).

We leave a smooth version to the reader. For k even, the usual condition on the L-polynomial of the bundle is needed to kill the simply-connected part of a surgery obstruction. For k odd, one has the usual Arf-invariant difficulties.

§1. A basis for $H^0(Wh(Z_{2^r}))$.

The group ring of Z_{2^r} is the ring $R = R_r = Z[T|T^{2^r} = 1]$, and this ring has the involution $p(T)^- = p(T^{-1})$. This involution induces one on the Whitehead group of Z_{2^r}, and $H^0(Wh(Z_{2^r}))$ denotes the cohomology of Z_2 with respect to this involution; i.e. elements satisfying $x = \bar{x}$, modulo elements of the form $x+\bar{x}$. When necessary for clarity, the generator T of Z_{2^r} will be denoted T_r, and we suppose $T_r = (T_{r+1})^2$. A unit $u(T) \in R^{\times}_{(r)}$, with $u(T) = u(T^{-1})$ represents an element of $H^0(Wh(Z_{2^r}))$. In this section we will give some units, $U^{(r)}_{m,s,i}$, $3 \leq s \leq m \leq r$, $i \equiv 1 (\text{mod } 4)$, $1 \leq i < 2^{s-1}$, which represent a basis of the Z_2-vector space $H^0(Wh(Z_{2^r}))$.

To define our units, we first set, for $r \geq 3$, $i \equiv 1(\text{mod } 4)$
$U^{(r)}_{r,r,i} = (\sum_{j=0}^{2^{r-1}} T^{ij})(\sum_{j=0}^{2^{r-1}} T^j) - (1+2^{r-2})(\sum_{j=0}^{2^r-1} T^j)$. To see that $U_{r,r,i}$ is a self-conjugate unit, consider the fibered square

$$
(1.1) \qquad \begin{array}{ccc} R_r & \xrightarrow{\gamma_r} & Z[T|1+T+\dots+T^{2^r-1} = 0] \\ \big\downarrow a_r & & \big\downarrow \bar{a}_r \\ Z & \longrightarrow & Z/2^rZ \end{array} ,
$$

$a_r(\sum a_i T^i) = \sum a_i$. Then $a_r(U_{r,r,i}) = 1$ and also

$$\gamma_r(U_{r,r,i}) = \left(\frac{T^{2^{r-1}+i}-1}{T^i - 1}\right)\left(\frac{T^{2^{r-1}+1}-1}{T - 1}\right), \text{ since } (T^i)^{2^{r-1}+1} = T^{2^{r-1}i+i} = T^{2^r+i}$$

if i is odd. An element in R_r is determined by its image under a_r and γ_r, and is a unit if and only if these images are. But

$\rho^{(r)}_{j,k} = \left(\frac{T^j-1}{T^k-1}\right)$ is always a unit in $Z[T|1+T+\dots+T^{2^r-1} = 0]$ if j is odd (a "cyclotomic unit") and $\overline{\rho}_{j,k} = T^{k-j}\rho_{j,k}$. It follows that $U_{r,r,i}$ is a self conjugate unit of R_r.

Now we define the units $U^r_{r,s,i}$, $3 \leq s < r$, $1 \leq i < 2^{s-1}$, $i \equiv 1 \pmod 4$. Assume by induction that $U^{(r-1)}_{r-1,s,i}$ has been defined, is self-conjugate, has image 1 under a_{r-1}, and that

$$\gamma_{r-1}(U^{(r-1)}_{r-1,s,i}) = T^{\delta}_{r-1} \prod_J \rho^{(r-1)}_{j,k}$$

is a product of T^{δ}_{r-1} and cyclotomic units, with $\sum_J (k-j) \equiv 2\delta \pmod{2^{r-1}}$. Note that

$$\overline{a}_r(\rho^{(r)}_{j,k}) \equiv \overline{a}_{r-1}(\rho^{(r-1)}_{j,k}) \bmod 2^{r-1}.$$

Hence $\overline{a}_r(T^{\delta}_r \prod_J \rho^{(r)}_{j,k}) \equiv 1 + \varepsilon \cdot 2^{r-1} \bmod 2^r$, where $\varepsilon = 0$ or 1. Hence there is a unique element $U^{(r)}_{r,s,i} \in R_r$ with $a_r(U^{(r)}_{r,s,i}) = 1$ and

$$\gamma_r(U^{(r)}_{r,s,i}) = \left(\frac{T^{2^{r-2}}(T^{2^{r-1}+1}-1)}{(T-1)}\right)^{\varepsilon} T^{\delta} \prod_J \rho^{(r)}_{j,k};$$

clearly $U_{r,s,i}$ will be a self-conjugate unit, with the appropriate image under γ_r to continue the inductive definition.

Finally, if $s \leq m < r$ write

$$U^{(m)}_{m,s,i} = \sum_{i=0}^{2^m-1} b_i T^i_m$$

and then let

$$U^{(r)}_{m,s,i} = \sum_{i=0}^{2^m-1} b_i T^{2^{r-m}i}.$$

This inductive definition was given for convenience only. An explicit formula can be given as follows: Let $\varepsilon_s(k)$, $k = 0,1,2,\ldots$ be the unique sequence of zeroes and ones with

$$(1+2^{s-1}) \prod_{t=s}^{m} (1+2^{t-1})^{\varepsilon_s(t-s)} \equiv 1 \pmod{2^m}.$$

Let

$$\lambda(m,s) = \frac{1}{2^m}\left[(1+2^{s-1}) \prod_{t=s}^{m} (1+2^{t-1})^{\varepsilon(t-s)} - 1\right].$$

Let

$$\omega(m,s,i) = -\sum_{t=s}^{m} \varepsilon(t-s)2^{t-2} - 2^{s-2}i.$$

Then

$$U_{m,s,i} = T^{2^{r-m}\omega(m,s,i)}\left(\sum_{j=0}^{2^{s-1}} T^{2^{r-m}ij}\right) \prod_{t=s}^{m} \left(\sum_{j=0}^{2^t-1} T^{2^{r-m}j}\right)^{\varepsilon(t-s)}$$

$$- \lambda(m,s)\left(\sum_{j=0}^{2^m-1} T^{2^{r-m}j}\right).$$

Let $I_r:R_{r-1} \to R_r$ and $\pi_r:R_r \to R_{r-1}$ be ring homorphisms (preserving 1) with $I_r(T_{r-1}) = T_r^2$ and $\pi_r(T_r) = T_{r-1}$. Let $\tau_r:R_r \to R_{r-1}$, the transfer, be defined as follows: if $u(T_r) \in R_r$, then $u(T_r)u(-T_r) = v(T_r^2) = v(T_{r-1})$. Set $\tau_r(u(T_r)) = v(T_r^2)$. Obviously, τ_r carries units to units.

(1.2) <u>Proposition</u>. Let $3 \leq s \leq m \leq r$. Then

$$I_r(U_{m,s,i}^{(r-1)}) = U_{m,s,i}^{(r)} \quad \text{if } m < r;$$

$$\pi_r(U_{m,s,i}^{(r)}) = T_{r-1}^{-\varepsilon_s(m-s)2^{r-2}} U_{m-1,s,i}^{(r-1)} \quad \text{if } s < m;$$

$$\pi_r(U_{m,m,i}^{(r)}) = 1;$$

$$\tau_r(U_{m,s,i}^{(r)}) = (U_{m,s,i}^{(r-1)})^2 \quad \text{if } m < r;$$

$$\tau_r(U_{r,r,i}^{(r)}) = 1; \text{ and}$$

$$\tau_r(U_{r,s,i}^{(r)}) = T_{r-1}^{-\varepsilon_s(r-s)2^{r-2}} U_{r-1,s,i}^{(r-1)} \quad \text{if } s < r.$$

<u>Proof</u>. The statement about I_r is clear, and so is that about π_r, from the inductive definition. To prove the 2nd & 3rd statement about τ_r, write

$$\gamma_r(U^{(r)}_{r,s,i}) = T^{\delta} \prod_{j} \rho^{(r)}_{j,k}$$

as above, with $2\delta = \sum_{j} (k-j) \bmod 2^r$ again, and since $i \equiv 1 (\bmod 4)$, j and k will always be $\equiv 1(\bmod 4)$ also; hence δ is even. It is not difficult to show that $\tilde{\tau}_r : Z[T_r | 1+T_r+\ldots+T_r^{2^{r}-1} = 0] \to Z[T_{r-1} | 1+T_{r-1}+\ldots+T_{r-1}^{2^{r-1}-1} = 0]$ is also defined, with $\tilde{\tau}_r\gamma_r = \gamma_{r-1}\tau_r$. A quick calculation gives $\tilde{\tau}_r(\rho^{(r)}_{j,k}) = \rho^{(r-1)}_{j,k}$; and $\tilde{\tau}_r(T_r^{\delta}) = T_{r-1}^{\delta}$. Hence (note $\rho^{(r-1)}_{2^{r-1}+i,i} = 1$)

$$\gamma_{r-1}(\tau_r U^{(r)}_{r,s,i}) = \begin{cases} \gamma_{r-1}(T_{r-1}^{-\varepsilon_s(r-s)2^{r-2}} U^{(r-1)}_{r-1,s,i}) & s < r \\ 1 & s = r . \end{cases}$$

From (1.1), it follows that on units γ_{r-1} is a monomorphism (with image those units having image $\equiv \pm 1$ in $Z/2^{r-1}$), for $r \geq 3$, which yields the result. Finally, the first statement concerning τ_r follows from the observation that $\tau_r I_r(x) = x^2$.

(1.3) Theorem. The elements $U^{(r)}_{m,s,i}$, $3 \leq s \leq m \leq r$, $i \equiv 1(\bmod 4)$, and $1 \leq i < 2^{s-1}$, represent a basis for the Z_2-vector space $H^0(Wh(Z_{2^r}))$.

Since $Wh(Z_{2^r}) \cong R_r^{x}/\{\pm T^i\}$ and the involution on $Wh(Z_{2^r})$ is actually trivial [B], (1.3) can be restated as follows:

(1.3)' The units $U^{(r)}_{m,s,i}$ and the trivial units -1 and T form a basis for $R_r^{x}/(R_r^{x})^2$.

To prove these results, we first recall that

$$\dim_{Z_2} H^0(Wh(Z_{2^r})) = 2^{r-1}-r.$$

Since there are exactly $2^{r-1}-r$ units $U_{m,s,i}$, $3 \leq s \leq m \leq r$, $i \equiv 1(\bmod 4)$, $1 \leq i < 2^{s-1}$, it suffices to check the independent of these units. The obvious inductive argument, together with Prop. 1.2 (the statements concerning π_r), shows that it suffices to prove that the units $U^{(r)}_{m,m,i}$, $3 \le m \le r$, $1 \le i < 2^{m-1}$, $i \equiv 1 (\mathrm{mod}\ 4)$ are independent modulo squares and trivial units. This will follow easily from the independence lemma of the next section.

§2. The independence lemma.

(2.1) *Suppose there exist units* $v \in Z[Z_{2^r}]^{\times}$ *and* $u \in Z[Z_{2^{m-1}}]^{\times}$ *and integers* δ_i, $i \equiv 1(\mathrm{mod}\ 4)$, $1 \le i < 2^{m-1}$ *and* ℓ, *so that*

$$(2.1.1) \qquad \prod_i (U^{(r)}_{m,m,i})^{\delta_i} = \pm T^{\ell} u v^2 .$$

Then $\delta_i \equiv 0(\mathrm{mod}\ 2)$ *for all* i.

In (2.1.1), u is identified with its image under the inclusion $I_r I_{r-1} \cdots I_m$ (recall $T_m = (T_r)^{2^{r-m}}$). The proof of (1.3) or (1.3)' is completed with (2.1) and decreasing induction on m.

To prove (2.1), assume (2.1.1) is satisfied and apply the composite $\pi_r \pi_{r-1} \cdots \pi_2$, projecting R_r to R_1, and then map to $R_1 \otimes Z_2 = Z_2[T_1 | T_1^2 = 0]$. Under this map, it follows from (1.2) that the left side of (2.1.1) maps to 1. Clearly $Z[Z_{2^{m-1}}]$ will map to Z_2, and so the unit u maps to 1 also. If v maps to $a+bT_1$, then v^2 maps to $(a+b) \in Z_2$ also. Hence ℓ must be even. Absorbing the sign into u and $T^{\ell/2}$ into v, we may therefore replace (2.1.1) with

$$(2.2) \qquad \prod_i (U^{(r)}_{m,m,i})^{\delta_i} = u v^2 .$$

Next, we wish to reduce to the case $m = r$. Suppose $m < r$, and let $\rho : R_r \to R_r$ be the involution $(T = T_r)$

$$\rho(\sum_i a_i T^i) = \sum_i (-1)^i a_i T^i.$$

Then ρ has the fixed point set R_{r-1} $(= I_r(R_{r-1}))$. Hence, if ρ is applied to (2.1.1), the same equation is obtained, except that v is replaced by $\rho(v)$. Hence

$$(v/\rho(v))^2 = 1.$$

But the torsion of $Z[Z_{2^r}]^x$ consists entirely of the trivial units $\{\pm T^i\}$ ([B], see also [M], [W2]). So

$$v/\rho(v) = \pm 1 \quad \text{or} \quad \pm T^{2^{r-1}}.$$

Now map to $R_r \otimes Z_2 = Z_2[T|T^{2^r} = 0]$. In this ring, v and $\rho(v)$ become equal, i.e. $v/\rho(v)$ maps to 1, where as $T^{2^{r-1}} \neq 1$ even mod 2. Thus $v/\rho(v) = \pm 1$; i.e. $v = \pm\rho(v)$. Hence either v or Tv is in $Z[Z_{2^{r-1}}]^x$, as $\rho(T) = -1$, so $\rho(Tv) = -T\rho(v)$. Replacing v with Tv if necessary, we therefore obtain an equation of the form (2.11), but with r replaced by $(r-1)$. (In fact, for the case $v = -\rho(v)$, in this equation in $Z[Z_{2^{r-1}}]$, $\ell = 1$. It follows from the argument preceding 2.2 that $\rho(v) = -v$ could not occur.)

Next we wish to derive from (2.2) an equation of the form of (5.1) of [CS1, §5]. First apply $\gamma = \gamma_r$ to both sides of (2.1); we then obtain, with $q = 2^{r-2}$,

$$(2.3) \qquad \prod_i \left(\frac{T^{2q+i}-1}{T^i - 1}\right)^{\delta_i} \left(\frac{T^{2q+1}-1}{T - 1}\right)^{\delta_i} = \gamma(uv^2);$$

the product is over i with $1 \leq i < 2q$ <u>and</u> $i \equiv 1 \pmod 4$. This is just an equation of the form of (5.1) of [CS1], with minor notational changes.

For a odd, let

$$f_a : \{0,1,\ldots 2^r-1\} \to \{0,1\} = Z/2Z$$

be defined by $f_a(x) = 1$ if the least non-negative residue of $ax \bmod 2^r$ is between 1 and 2^{r-1}, and $f_a(x) = 0$ otherwise. In §5 of [CS1] it is shown that (2.3) implies the vanishing of a certain cohomology class $\chi \in H^1(R_r^x) = H^1(Z_2, R_r^x)$, with respect to the involution ρ defined above. From §7 of [CS1], it follows that the vanishing of this class implies the following equation of functions to $Z/2Z$ (see also the last equation in the 2nd complete paragraph on page 341 of [CS1]):

$$\sum_i \delta_i f_i + (\sum_i \delta_i) f_1 + (\ell/2)(f_1 + f_{2q+1}) = 0.$$

In this case, $\ell = 2(\sum_i \delta_i)$. Again, all sums are over i with $1 \leq i < 2^{r-1}$ and $i \equiv 1 (\mathrm{mod}\ 4)$. Hence we obtain

$$(2.4) \qquad (\sum_i \delta_i f_i) + (\sum_i \delta_i) f_{2q+1} = 0 \quad (q = 2^{r-2}).$$

However, 2^r is a <u>tempered number</u> ([CS2], see also [CS1,3,4]). Hence all relations among the functions f_a are consequences of the "obvious" ones:

$$f_a + f_{2q+a} = f_1 + f_{2q+1}.$$

It follows that if $\delta_i \not\equiv 0 (\mathrm{mod}\ 2)$, for $1 < i < 2q$, (2.4) would have a term involving f_{2q+i} as well; since the sum is over i between 1 and $2q$, this does not occur. Hence $\delta_i \equiv 0 (\mathrm{mod}\ 2)$ for $1 < i < 2q$, and (2.4) becomes $\delta_1(f_1 + f_{2q+1}) = 0$. This obviously implies $\delta_1 \equiv 0 (\mathrm{mod}\ 2)$ also (as $f_1(1) + f_{2q+1}(1) = 1$). This completes the proof of (2.1).

§3. Signatures and determinants.

Let G be a finite group and let ρ be an irreducible complex representation, $\rho:G \to U(n)$. Let $\alpha = (H,\phi,\mu)$ be a $(-1)^k$ Hermitian unimodular[1] (quadratic) form over $Z[G]$, H a (stably) free $Z[G]$-module, representing an element $[\alpha]$ in $L^h_{2k}(G)$. Let $\alpha_{\mathbb{C}} = \alpha \otimes_{Z[G]} \mathbb{C}^n$, a $(-1)^k$-Hermitian form over the complex numbers, where $\mathbb{C}^n$ has a $Z[G]$-module structure via ρ. Then let

$$\sigma_\rho(\alpha) = \text{signature of} \begin{cases} \alpha_{\mathbb{C}} & k \text{ even} \\ \sqrt{-1}\,\alpha_{\mathbb{C}} & k \text{ odd} \end{cases}$$

Let $R(G)$ denote the complex representation ring of G; then a well-defined homomorphism (the multisignature)

$$\chi: L^h_{2k}(G) \to R(G)$$

is defined by

$$\chi([\alpha]) = \sum_\rho \sigma_\rho(\alpha)\rho,$$

where the sum is over irreducible representations. Let

$$\lambda: L^s_{2k}(G) \to L^h_{2k}(G)$$

be the natural map, $L^s_{2k}(G)$ the obstruction group for the surgery problem to obtain a simple homotopy equivalence. According to [Wl], the following holds:

[1] Throughout this paper "unimodular" is used to mean that the adjoint $\mathrm{Ad}\,\phi: H \to H^*$ is an isomorphism.

(3.1) Theorem. Let G be cyclic. Then the composite $\chi\lambda$ has the image $\{4(\rho+(-1)^k\bar{\rho}) \mid \rho \in R(G)\}$, is a monomorphism if k is even, and has kernel isomorphic to Z_2 for k odd (detected by the Arf invariant).

Now assume G is abelian (so that an irreducible representation is 1-dimensional). Let $(H,\phi,\mu) = \alpha$ be as above. Then, upon choice of a basis for H,

$$\det \alpha \in Z[G]^{\times}$$

is defined; a change of basis will multiply $\det \alpha$ by a unit of the form $x\bar{x}$.

(3.2) Proposition. Let G be abelian. Suppose α is a $(-1)^k$-Hermitian unimodular form over $Z[G]$ of rank $2r$. Let ρ be an irreducible representation of G. Then $\sigma_\rho(\alpha) \equiv 0 (\mathrm{mod}\ 2)$, and $\sigma_\rho(\alpha) \equiv 0 (\mathrm{mod}\ 4)$ if and only if $(-1)^{r(k+1)}\rho(\det \alpha) > 0$.

Remark. Since ρ is irreducible, $\rho: G \to S^1 \subset \mathbb{C}$, and extends to a homomorphism $Z[G] \to \mathbb{C}$, also denoted ρ. Since α is $(-1)^k$-Hermitian and H has even rank, $(\det \alpha)^- = \det \alpha$. Hence $\rho(\det \alpha)$ is real. Since $\rho(x\bar{x}) = \rho(x)\rho(x)^- > 0$, the sign of $\rho(\det \alpha)$ is unaffected by a change of basis.

This result is nothing more than a simple consequence of an old formula for computing the signature of a Hermitian form over $\mathbb{C}$; see e.g. [J]. Given a Hermitian form over $\mathbb{C}$, one can find a basis so that the (determinants of) the sequence of principle minors, ordered by size starting with zero, contains no successive zeroes. By convention, the determinant of the 0×0 minor is 1. The signature of the form is then given as $P-C$, where P is the number of permanences of sign and C the number of changes in the signs sequence of the principal minors. The sign of a zero is chosen arbitrarily.

Now apply this to $\alpha \otimes_{Z[G]} \mathbb{C} = \alpha_{\mathbb{C}}$ or $\sqrt{-T}\,\alpha_{\mathbb{C}}$ as above. Then $P+C = 2r$. Hence $\sigma_\rho(\alpha)$ is also even. For k even, the sign of the largest principal minor is just that of $\rho(\det \alpha)$, and for k odd it is $(-1)^r \rho(\det \alpha) = \det(\sqrt{-T}\,\alpha_{\mathbb{C}})$. If the sign of the largest minor is positive, then C must be even. Hence $P-C = P+C-2C = 2r-2C$ will be divisible by 4 if and only if r is even. Similarly, if the last sign is negative, $P-C$ will be divisible by 4 if and only if r is odd. The result follows.

A result similar to (3.1) for cyclic groups of odd order (and slightly misstated) is stated in [W2] and was used there in the classification of fake lens spaces (see also [BPW]).

(3.3) Proposition. *For $3 \leq s \leq m \leq r$, $1 \leq i < 2^{s-1}$, $i \equiv 1 (\text{mod } 4)$, there is a $(-1)^k$-Hermitian unimodular form $\alpha^{(r)}_{m,s,i}$, representing an element of $L^h_{2k}(Z_{2^r})$, with*

$$\det(\alpha_{m,s,i}) = T^{2^{r-1}\varepsilon_s(m+1-s)} U^{(r)}_{m,s,i},$$

with respect to a suitable basis. $(T = T_r)$.

Notes: 1. In particular,

$$\det(\alpha_{m,m,i}) = T^{2^{r-1}} U^{(r)}_{m,m,i}.$$

2. For k even, it follows that $\text{rank}(\alpha_{m,s,i}) \equiv 0 (\text{mod } 4)$, applying 3.2 to the trivial representation. Recall that a unimodular even form over Z has signature $\equiv 0(8)$.

(3.4) Lemma. *Let $x \in R_r^\times$, with $x = \bar{x}$ and $a(x) = 1$ for k odd. Then there is a $(-1)^k$ symmetric unimodular form β with*

$$\det \beta = x \quad \text{or} \quad T^{2^{r-1}} x,$$

with respect to a suitable choice of basis.

Proof: According to [W1, §], there is a short exact sequence

$$(3.5)\qquad 0 \to L^s_{2k}(Z_{2^r}) \xrightarrow{\lambda} L^h_{2k}(Z_{2^r}) \xrightarrow{d} Wh(Z_{2^r})\otimes Z_2 \to 0.$$

Of course, $Wh(Z_{2^r})\otimes Z_2 = H^0(Wh(Z_{2^r}))$. There is well known surjective determinant map (actually an isomorphism in this case [B])

$$Wh(Z_{2^r}) \to R_r^x/\{\pm T^i\} \to 1,$$

and that the composition of the map induced on H^0 with the appropriate map of the above exact sequence is a surjective homomorphism

$$d{:}L^h_{2k}(Z_{2^r}) \to R_r^x/\{\pm T^i y\bar{y} \mid y \in R_r^x\},$$

with $d[\alpha] = [\det \alpha]$.

The determinant of a form can be multiplied by $y\bar{y}$ merely by changing basis (even by multiplying a single basis element by y). Hence there exists a form β with $\det(\beta) = \pm T^i x$, with respect to a suitable choice of basis. Further, $\det \beta = (\det \beta)^-$, since β is Hermitian or skew-Hermitian of even rank. (To compute the rank, pass all the way to $Z/2Z$, to obtain a symmetric unimodular form with $x \cdot x \equiv 0$. Such a form always has even rank.) Hence $T^i = T^{-i}$, thus $i = 0$ or 2^{r-1}. So $\det \beta = \pm x$ or $\pm T^{2^{r-1}} x$.

If k is even and the minus sign appears, just replace β by its orthogonal sum with a kernel; i.e. with $\kappa = (k,\phi,\mu)$, where ϕ has the matrix

$$\begin{pmatrix} 0 & 1 \\ 1 & 0 \end{pmatrix}$$

with respect to some basis. Clearly this will change the sign.

Suppose k is odd. Then $\beta \otimes_{R_r} Z$ will be a unimodular skew form over the integers. It is well-known that such a form is a sum of kernels; in this case a kernel will have matrix $\begin{pmatrix} 0 & 1 \\ -1 & 0 \end{pmatrix}$. Hence $\det(\beta \otimes_{R^r} Z) = +1$.

It follows that under the augmentation $a_r : R_r \to Z$, $a_r(\det \beta) = +1$. Hence, since $a_r(x) = 1$, the minus sign is impossible.

Proof of 3.3. First apply (3.4), but with r replaced by $r+1$, to obtain a $(-1)^k$-Hermitian unimodular form β over R_{r+1} with (for a suitable basis)

$$\det \beta = U^{(r+1)}_{m+1,s,i} \quad \text{or} \quad T^{2^r}_{r+1} U^{(r+1)}_{m+1,s,i}.$$

The map $\pi_{r+1} : R_{r+1} \to R_r = Z[Z_{2^r}]$ provides an R_{r+1}-module structure on R_r, and it is not hard to see that

$$\det(\beta \otimes_{R_{r+1}} R_r) = \pi_{r+1}(\det \beta).$$

Let $\alpha_{m,s,i} = \beta \otimes_{R_{r+1}} R_r$. Since $\pi_r(T^{2^r}_{r+1}) = 1$, that $\alpha_{m,s,i}$ has the desired determinant now follows from (1.2).

§4. The image of the multisignature (Proof of Theorem A).

Let $R_{2k}(G)$, a Z_2 vector space, be the quotient of the group elements $2(\rho + (-1)^k \bar{\rho})$, $\rho \in R(G)$, by those of the form $4(\rho + (-1)^k \bar{\rho})$. Then by (3.1), (3.2) and the exact sequence (3.5), there is a diagram[1]

$$(4.1) \qquad \begin{array}{ccc} L^h_{2k}(Z_{2^r}) & \xrightarrow{\chi^{(r)}} & \{2(\rho + (-1)^k \bar{\rho}) \mid \rho \in R(Z_{2^r})\} \\ \downarrow d^{(r)} & \searrow \sigma^{(r)} & \downarrow \omega^{(r)} \\ H^0(Wh(Z_{2^r})) & \xrightarrow{\chi_2^{(r)}} & R_{2k}(Z_{2^r}) \end{array} .$$

[1] Recall $\overline{\chi(\alpha)} = (-1)^k \chi(\alpha)$.

Here $\chi^{(r)}$ is the multisignature, $d^{(r)}$ the determinant map d above with $H^0(Wh(Z_{2^r})$ identified with $R_r^{\times}/\{\pm T^i y^2 | y \in R_r^{\times}\}$; recall that $\bar{y} = T^j y$ some j, if $y \in R_r^{\times}$ [B][W2, §14]). Let $\sigma^{(r)} = \omega^{(r)}\chi^{(r)} = \chi_2^{(r)} d^{(r)}$.

(4.2) Theorem. *The elements* $\sigma^{(r)}(\alpha_{m,s,i})$ *(see 3.3), with* $3 \leq s \leq m \leq r$, $1 \leq i < 2^{s-1}$, $i \equiv 1 (\text{mod } 4)$, *and* $s \leq 2m-r$, *form a basis (over* Z_2*) for the image of* $\sigma^{(r)}$.

(4.3) Corollary. $\text{Dim}_{Z_2}(\text{Im } \sigma^{(r)}) = [2/3(2^{r-1}-1)]-[(r-1)/2]$.

The corollary follows by just computing the number of indices m,s,i with $s \leq 2m-r$. It can be restated as follows (see 4.1):

(4.4) Corollary. $\text{Dim}_{Z_2}(\chi^{(r)}(L_{2k}^h(Z_{2^r}))/\chi^{(r)}\lambda(L_{2k}^s(Z_{2^r}))) =$

$$= [2/3(2^{r-1}-1)]-[(r-1)/2].$$

Recall $\lambda = \lambda^{(r)}$ from (3.1). In view of (3.1), (3.5), and the fact that $\dim H^0(Wh(Z_{2^r})) = 2^{r-1}-r$, it follows easily that for k even

$$\dim \text{Torsion}(L_{2k}^h(Z_{2^r})) = (2^{r-1}-r) - \dim(\text{Im } \chi/\text{Im } \chi\lambda),$$

and one more than this for k odd.

Clearly the right side is just

$$[2/3(2^{r-2}+2]-[r/2]-1$$

which implies Theorem A.

The rest of this section is devoted to the proof of (4.2). Let t_r be the representation of Z_{2^r} to $\mathbb{C}$ determined by

$$t_r(T_r) = e^{2\pi i/2^r}.$$

Then $R(Z_{2^r}) = Z[t_r | t_r^{2^r} = 1]$. Let

$$(I_r)_* : L_{2k}^h(Z_{2^{r-1}}) \to L_{2k}^h(Z_{2^r}),$$

$$(\pi_r)_* : L_{2k}^h(Z_{2^r}) \to L_{2k}^h(Z_{2^{r-1}}),$$

$$(\tau_r)_* : L_{2k}^h(Z_{2^r}) \to L_{2k}^h(Z_{2^{r-1}}),$$

be the indicated induced maps. $(I_r)_*$ and $(\pi_r)_*$ are just induced by the maps I_r and π_r on group rings, and $(\tau_r)_*$ is by just the transfer map of surgery theory. The maps I_r, π_r and τ_r also induce maps between the quotients $H^0(Wh(Z_{2^r})) = R_r^x/\{\pm T^i y^2\}$ and $R_{r-1}^x/\{\pm T_{r-1}^i y^2\}$, and the obvious diagrams involving all these maps and d commute.

(4.5) <u>Proposition.</u> <u>Let</u> $x \in L_{2k}^h(Z_{2^{r-1}})$ <u>and</u> $y \in L_{2k}^h(Z_{2^r})$. <u>Assume that</u>

$$\sigma^{(r-1)}(x) = \sum_0^{2^{r-1}-1} \gamma_i t_{(r-1)}^i \quad \underline{\text{and}}$$

$$\sigma^{(r)}(y) = \sum_0^{2^r-1} \delta_i t_{(r)}^i \quad (\gamma_i, \delta_i \in 2Z/4Z).$$

<u>Then the following hold:</u>

(4.5.1) $$\sigma^{(r)}((I_r)_* x) = \sum_0^{2^{r-1}-1} \gamma_i (t_r^i + t_r^{2^{r-1}+i})$$

(4.5.2) $$\sigma^{(r-1)}((\pi_r)_* y) = \sum_0^{2^{r-1}-1} \delta_{2i} t_{r-1}^i; \quad \text{and}$$

(4.5.3) $$\sigma^{(r-1)}((\tau_r)_* y) = \sum_0^{2^{r-1}-1} (\delta_i + \delta_{2^{r-1}+i}) t_{(r-1)}^i.$$

These formulas follow from similar formulas for χ, whose proofs we leave to the reader. These formulas obviously provide maps

$$(\pi_r)_!:R_2(Z_{2^r}) \to R_2(Z_{2^{r-1}}),$$

$$(I_r)_!:R_2(Z_{2^{r-1}}) \to R_2(Z_{2^r}),$$

$$(\tau_r)_!:R_2(Z_{2^r}) \to R_2(Z_{2^{r-1}}),$$

with $(\pi_r)_! \sigma^{(r)} = \sigma^{(r-1)}(\pi_r)_*$, etc.

Let $W^{(r)}_{m,s} \subset R_2(Z_{2^r})$ be the span of the elements $\sigma^{(r)}(\alpha_{m,s,i})$, $1 \leq i < 2^{s-1}$, $i \equiv 1 (\mathrm{mod}\ 4)$.

(4.6) Proposition. $W^{(r)}_{r,r} = \ker(\pi_r)_! \cap \ker(\tau_r)_!$, and $\{\sigma^{(r)}(\alpha_{r,r,i}) | 1 \leq i < 2^{r-1}, i \equiv 1(4)\}$ is a basis for it.

(Note: We write $\sigma^{(r)}(\alpha_{r,r,i})$ for $\sigma^{(r)}$ applied to the equivalence class of it, and similarly for χ, χ_2, etc..)

Proof. From (3.3), $\det(\alpha_{r,r,i}) = T^{2^{r-1}} U_{r,r,i}$. By (1.2), $\tau_r(T^{2^{r-1}} U_{r,r,i}) = \pi_r(T^{2^{r-1}} U_{r,r,i}) = 1$. Hence $(\pi_r)_! \sigma(\alpha_{r,r,i}) =$ $= \sigma^{(r-1)}(\pi_r)_*(\alpha_{r,r,i}) = \chi_2^{(r)} \pi_r(\det \alpha_{r,r,i}) = \chi^{(r)}_{(2)}(1) = 0$. Similarly, $(\tau_r)_! \sigma(\alpha_{r,r,i}) = 0$. Hence $W_{r,r} \subset \ker(\pi_r)_! \cap \ker(\tau_r)_!$.

However, $\ker(\pi_r)_! \cap \ker(\tau_r)_!$ is precisely the elements of the form $\sum_{i \text{ odd}} \gamma_i t_r^i$ $(\gamma_i \in 2Z/4Z)$, $\gamma_i = \gamma_{2^r-i}$ and $\gamma_i = \gamma_{2^{r-1}+i}$. Hence this Z_2-vector space has dimension 2^{r-3}, $r \geq 3$ (and 0, $r = 1$ or 2) therefore it suffices to prove that the elements $\sigma(\alpha_{r,r,i})$ are linearly independent. This will be done using (3.2).

Let $\zeta = t_r(T)$, and let

$$\xi_{ij} = t^j(T^{2^{r-1}} U_{r,r,i}). \quad \text{Then}$$

$$\xi_{ij} = \frac{-(\zeta^{ij}+1)(\zeta^j+1)}{(\zeta^{ij}-1)(\zeta^j-1)} \quad ; \quad \text{hence}$$

$$\xi_{ij} = \frac{-(\zeta^{ij}-\zeta^{-ij})(\zeta^j-\zeta^{-j})}{(\zeta^{ij}-2+\zeta^{-ij})(\zeta^j-2+\zeta^{-j})} .$$

Clearly the denominator is a positive real number. Let σ_{ij} be the coefficient of t^j in $\sigma^{(r)}(\alpha_{r,r,i})$.

Now $\zeta^k-\zeta^{-k} = (\sqrt{-1})z$, where z is real and $z > 0$ if and only if the least positive residue of $k \bmod 2^r$ is less than 2^{r-1}. Hence, for $2^{r-1} \leq j < 2^r$, $j \equiv 1 (\mathrm{mod}\ 4)$,

$$\xi_{ij} < 0 \text{ if and only if } f_j(i) = 1.$$

Hence, by (3.2), $\sigma_{ij} = 2f_j(i) \pmod 4$ for $2^{r-1} \leq j < 2^r$, and in fact

$$\sigma(\alpha_{r,r,i}) = \sum_j 2f_j(i)(t^j+t^{-j}+t^{2^{r-1}+j}+t^{2^{r-1}-j})$$

the sum over $2^{r-1} \leq j < 2^r$, $j \equiv 1 (\mathrm{mod}\ 4)$.

(4.7) <u>Lemma</u> ([CS2]). <u>The matrix</u> $(f_j(i))$, $1 \leq i,j < 2^{r-1}$, $1 \equiv j \equiv 1 (\mathrm{mod}\ 4)$, <u>is non-singular over</u> Z_2.

This lemma clearly implies the independence of the elements $\sigma(\alpha_{r,r,i})$, and this completes the proof.

(4.8) <u>Proposition</u>. $\ker(\pi_r)_! \cap \mathrm{Image}(I_r)_! \subset W_{r,r}$.

<u>Proof</u>. It is obvious that $(\tau_r)_!(I_r)_! = 0$. Hence $(\ker \pi_r)_! \cap \mathrm{Im}(I_r)_! \subset (\ker \pi_r)_! \cap (\ker \tau_r)_! = W_{r,r}$, by 4.6.

(4.9) <u>Proposition</u>. <u>For</u> $k,t \geq 0$ <u>and</u> $2k+t \leq r-3$,

$$W^{(r)}_{r-k-t,r-2k-t} \subset \sum_{\ell=0}^{k} W^{(r)}_{r-\ell,r-2\ell}.$$

Proof. By induction on k. Suppose $k = 0$. Then we must show that $\sigma(\alpha_{m,m,i}) \in W^{(r)}_{r,r}$ for $m < r$. But

$$\begin{aligned}\sigma^{(r)}(\alpha_{m,m,i}) &= \chi_2^{(r)}(\det \alpha_{m,m,i}) = \chi_2^{(r)}(T^{2^{r-1}} U^{(r)}_{m,m,i}) \\ &= \chi_2^{(r)}(I_r(T^{2^{r-2}}_{r-1} U^{(r-1)}_{m,m,i}) = (I_r)_! \chi_2^{(r-1)}(\det \alpha^{(r-1)}_{m,m,i})) \\ &= (I_r)_! \sigma^{(r-1)}(\alpha^{(r-1)}_{m,m,i}).\end{aligned}$$

So $\sigma^{(r)}(\alpha^{(r)}_{m,m,i}) \in \mathrm{Im}(I_r)_!$.

Since $\pi_r(T^{2^{r-1}} U^{(r)}_{m,m,i}) = 1$, a similar argument implies that $\sigma(\alpha_{m,m,i}) \in \ker(\pi_r)_!$, and then the case $k = 0$ follows from (4.8).

Suppose $k > 0$. We claim that

$$(4.10) \qquad (\pi_r)_!(W^{(r)}_{m,s}) = \begin{cases} W^{(r-1)}_{m-1,s} & s < m \\ 0 & s = m \end{cases}.$$

In fact, if $s < m$, $(\pi_r)_!(\sigma^{(r)}_{m,s,i}) = (\pi_r)_!(\chi_2^{(r)}(T^{2^{r-1}} \varepsilon U^{(r)}_{m,s,i})) =$

$= \chi_2^{(r-1)}(\pi_r(T^{2^{r-1}} \varepsilon U^{(r)}_{m,s,i})) = \chi_2^{(r-1)}(T^{2^{r-2}} \varepsilon_s(m-s) U^{(r-1)}_{m-1,s,i}) =$

$= \chi_{(2)}^{(r-1)}(\det \alpha^{(r-1)}_{m-1,s,i}) = \sigma^{(r-1)}(\alpha^{(r-1)}_{m-1,s,i})$, and similarly one gets 0 if $m = s$.

Similarly, one shows that for $m < r$, $s < m$,

$$(4.11) \qquad W^{(r-1)}_{m-1,s} = (I_{r-1})_!(W^{(r-2)}_{m-1,s}).$$

Hence

$$(\pi_r)_!(W^{(r)}_{r-k-t,r-2k-t}) = (I_{r-1})_!(W^{(r-2)}_{(r-2)-h-t,(r-2)-2h-t}),$$

where $h = k-1$. By induction,

$$W^{(r-2)}_{(r-2)-h-t,(r-2)-2h-t} \subset \sum_{e=0}^{h} W^{(r-2)}_{(r-2)-e,(r-2)-2e}.$$

But, as we have just seen (with $h = e$ and $t = 0$),

$$(I_{r-1})_!(W^{(r-2)}_{(r-2)-e,(r-2)-2e}) = (\pi_r)_!(W^{(r)}_{r-(e+1),r-2(e+1)}).$$

Hence

$$(\pi_r)_!(W^{(r)}_{r-k-t,r-2k-t}) \subset (\pi_r)_!(\sum_{1}^{k} W^{(r)}_{r-\ell,r-2\ell}), \text{ i.e.}$$

(4.12) $$W^{(r)}_{r-k-t,r-2k-t} \subset \sum_{1}^{k} W^{(r)}_{r-\ell,r-2\ell} + \ker(\pi_r)_!.$$

For $1 \leq \ell$, $W^{(r)}_{r-\ell,r-2\ell} = (I_r)_!(W^{(r-1)}_{r-\ell,r-2\ell})$, and similarly $W^{(r)}_{r-k-t,r-2k-t} \subset \operatorname{Image}(I_r)_!$. Hence these all lie in $\ker(\tau_r)_!$, as $(\tau_r)_!(I_r)_! = 0$. Therefore in (4.12), $(\ker \pi_r)_!$ can be replaced by $(\ker \pi_r)_! \cap (\ker \tau_r)_!$, which equals $W^{(r)}_{r,r}$ by (4.6). This completes the proof of (4.9).

<u>Proof of 4.2</u>. By the previous proposition (4.9), the elements $\sigma(\alpha_{m,s,i})$, $3 \leq s \leq m \leq r$, $1 \leq i < 2^{s-1}$, $i \equiv 1 (\mathrm{mod}\ 4)$, and $s \leq 2m-r$ generate the image of σ. (To see this, note that if $m = r-k-t$, $s = m-2k-t$, then $s = 2m-r+t$.) Therefore it will suffice to prove their linear independence. For $r = 3$, this is a consequence of (4.6). We argue by induction on r.

As in the proof of (4.9), (compare (4.10) and (4.11)), we have

(4.13) $$(\pi_r)_!\sigma(\alpha^{(r)}_{m,s,i}) = \begin{cases} 0 & \text{if } m = s \\ \sigma(\alpha^{(r-1)}_{r-1,s,i}) & \text{if } m = r,\ s < m \\ (I_{r-1})_!\sigma(\alpha^{(r-2)}_{m-1,s,i}) & \text{if } m < r,\ s < m. \end{cases}$$

It is obvious that $(I_{r-1})_!$ is a monomorphism. It then follows easily from the inductive hypothesis, (4.6), and (4.13) that the elements $\sigma(\alpha^{(r)}_{m,s,i})$ with either $m = s = r$ or $s < m < r$ and $s \leq 2m-r$ $(= 2(m-1)-(r-2))$ are linearly independent.

On the other hand

$$(\tau_r)_! \sigma(\alpha^{(r)}_{m,s,i}) = \begin{cases} 0 & \text{if } r = m = s \text{ or } m < r \\ \sigma(\alpha^{(r-1)}_{r-1,s,i}) & \text{if } m = r \text{ and } s < m. \end{cases}$$

For example, for $s < r$, $(\tau_r)_! \sigma(\alpha^{(r)}_{r,s,i}) = (\tau_r)_! (\chi^{(r)}_2 (T^{2^{r-1}} \varepsilon U^{(r)}_{r,s,i})) =$

$= \chi^{(r-1)}_2 (\tau_r (T^{2^{r-1}} \varepsilon U^{(r)}_{r,s,i})) = \chi^{(r-1)}_2 (T^{2^{r-2}}_{r-1} \varepsilon s(r-s) U^{(r-1)}_{r-1,s,i})$

$= \chi^{(r-1)}_2 (\det \alpha^{(r-1)}_{r-1,s,i}) = \sigma(\alpha^{(r-1)}_{r-1,s,i})$, and the other cases are argued similarly, using (1.2).

So the elements $\sigma(\alpha^{(r)}_{m,s,i})$ with $m = s = r$ or with $s < m < r$ and $s \leq 2m-r$ map to 0 under $(\tau_r)_!$, whereas the elements $\sigma(\alpha^{(r)}_{r,s,i})$, $m < r$, map to the elements $\sigma(\alpha^{(r-1)}_{r-1,s,i})$, which, since $s \leq (r-1) = 2(r-1)-(r-1)$, are linearly independent by induction. This accounts for all the elements $\sigma(\alpha^{(r)}_{m,s,i})$ with $s \leq 2m-r$ and so completes the proof.

§5. Proof of Theorem B.

Consider the exact sequence [Sh, 4.1]

$$L^s_{4k+1}(Z_{2^r},-) \to L^h_{2k+1}(Z_{2^r},-) \to H^1(Wh(Z_{2^r})) \to$$

$$\to L^s_{4k}(Z_{2^r},-) \xrightarrow{\lambda} L^h_{4k}(Z_{2^r},-).$$

However, in this case the cohomology $H^1(Wh(Z_{2^r}))$ is taken with respect to the involution induced by

$$(\sum_i a_i T^i)^* = (\sum_i (-1)^i a_i T^{-i})$$

on the level of the group ring $R_r = Z[Z_{2^r}]$.

According to [W1], $L^s_{4k+1}(Z_{2^r},-) = 0$. Also from [W1], it follows that is a monomorphism, since $L^s_{4k}(Z_{2^r},-)$ can be detected by multisignatures and Arf invariants, whose definition extends to $L^h_{4k}(Z_{2^r})$ as well. Hence

$$L^h_{2k+1}(Z_{2^r},-) \cong H^1(Wh(Z_{2^r})),$$

the homology taken with respect to the above-mentioned involution. For $r = 1,2$, $Wh(Z_{2^r}) = 0$, so assume $r \geq 3$.

Let $S_r = \{u \in R_r^\times | u = \bar{u}$ and $a_r(u) = 1\}$. Then there is a short exact sequence

$$1 \to \{1,T^{2^{r-1}}\} \to S_r \to Wh(Z_{2^r}) \to 0.$$

To see this, just recall again [B] (compare [W2, §14]) that $Wh(Z_{2^r}) = R_r^\times/\{\pm T^i\}$ and the involution on $Wh(Z_{2^r})$ induced by $\bar{\ }$ is trivial. Hence every element of $Wh(Z_{2^r})$ is represented by $u \in R_r^\times$ with $a_r(u) = 1$ and

$$\bar{u} = \pm T^i u,$$

some i. Since $a(\bar{u}) = a(u) = 1$, the sign is positive. Project to $Z[Z_2]$; in this ring the involution $\bar{\ }$ maps to the identity. It follows that $i = 2j$. Clearly $T^j u \in S_r$ and represents the same element of $Wh(Z_{2^r})$. Hence the map from S_r is surjective, and the kernel is easily identified.

Passing to cohomology, we obtain a long exact sequence

$$\begin{aligned} H^0(S_r) &\to H^0(Wh(Z_{2^r})) \to \{1,T^{2^{r-1}}\} \to H^1(S_r) \to \\ &\to H^1(Wh(Z_{2^r})) \to \{1,T^{2^{r-1}}\} \\ &\to H^2(S_r) \to H^2(Wh(Z_{2^r})). \end{aligned}$$

As above, every element of $Wh(Z_{2^r})$ has a representative $u \in R_r^X$ with $u = \overline{u}$; hence

$$u(T)^* = u(-T^{-1}) = u(-T).$$

Hence an element of $H^0(Wh(Z_{2^r}))$ will be represented by a unit u with $u(T) = T^j u(-T)$. Pass to $Z_2 \otimes R_r$; this equation then implies $T^j = 1$; i.e. $u(T) = u(-T)$. Hence u represents an element of $H^0(S_r)$, and so the map from $H^0(S_r)$ to $H^0(Wh(Z_{2^r}))$ is surjective, and similarly for H^2 as $H^2 = H^0$ for cohomology of Z_2. Hence

$$(5.1) \qquad L^h_{2k+1}(Z_2,-) \cong H^1(S_r)/\{1, T^{2^{r-1}}\}.$$

By definition, $H^1(S_r)$ consists of elements u of S_r with $u(T) = u(-T)^{-1}$, modulo those of the form $v(T)v(-T)^{-1}$, for some $v \in S_r$. But $u(T) = u(-T)^{-1}$ if and only if $\tau_r(u(T)) = u(T)u(-T) = 1$. Hence the inclusion $K = (\ker \tau_r) \cap S_r \subset S_r$ induces a <u>surjective</u> map

$$K/K^2 = H^1(K) \to H^1(S_r).$$

Now suppose $u \in R_r$ and $u^2 \in K$. Then $\tau_r(u^2) = 1$. So $u(T)^2 u(-T)^2 =$ Hence, since the torsion of R_r^X, consists entirely of trivial units, $u(T)u(-T) = \pm 1$ or $\pm T^{2^{r-1}}$. Hence

$$u^2 = \pm T^{\varepsilon} u(T)/u(-T)^{-1}, \quad \varepsilon = 0 \text{ or } 2^{r-1}.$$

By application of a_r, it is clear that the sign must be positive. Hence the previous map induces a surjective map

$$\omega : K/(R_r^X)^2 \cap K \to H^1(S_r)/\{1, T^{2^{r-1}}\}.$$

The next step is to apply (1.3)' and (1.2). Since the torsion of R_r^X consists entirely of the trivial units, it follows from (1.3)' that the units

$U^{(r)}_{m,s,i}$ generate a free (abelian) subgroup V of R^{x}_{r} of rank $2^{r-1}-r$, i.e. they are linearly independent. From (1.2) it then follows that $V \cap K$ has the basis (as a free abelian group) $\{U^{(r)}_{r,r,i} | 1 \leq i < 2^{r-1},\ i \equiv 1 (\text{mod } 4)\}$. It then follows by (1.3)' that these units represent a basis for the image of K in $R^{x}_{r}/(R^{x}_{r})^2$ as a Z_2-vector space. Since $K/K \cap (R^{x}_{r})^2 \subset R^{x}_{r}/(R^{x}_{r})^2$, it finally follows that $K/K \cap (R^{x}_{r})^2$ has as a basis the elements represented the elements $U^{(r)}_{r,r,i}$, $1 \leq i < 2^{r-1}$, and in particular, has dimension 2^{r-3}.

Now suppose $\omega(x)$ is trivial. Let x be represented by a product

$$\prod_i (U_{r,r,i})^{\delta_i},$$

with $\delta_i = 0$ or 1, $1 \leq i < 2^{r-1}$. Then

$$\prod_i (U_{r,r,i})^{\delta_i} = T^{\varepsilon}(u(T)/u(-T)),$$

$\varepsilon = 0$ or 2^{r-1}. Let $v(T) = u(T)u(-T) = \tau_r(u(T))$; $v(T) \in R_{r-1}$ $(= I_r(R_{r-1})) \subset R_r$. Then

$$\prod_i (U_{r,r,i})^{\delta_i} = (T^{\varepsilon/2}/u(-T))^2 v(T).$$

Hence by (2.1), $\delta_i = 0$ for all i; i.e. x is trivial. Hence ω is an isomorphism. By (5.1), this proves Theorem B.

Finally here is an exercise for the reader:

<u>Prove that</u> $\text{Tor}(L^h_{2k}(Z_{2^r},-)) = \text{Tor}(L^h_{2k}(Z_{2^{r-1}}))$.

§6. Smoothing Poincaré Complexes

Let X be a connected Poincare Duality space of dimension $2k$, with $\pi_1 X = G$. Let $\xi \in C_{2k}(X)$ be a cycle representing $[X]$. Let $C_*(X;Z[G])$

be the chain complex $C_*(\tilde{X})$ of the universal covering space, and let $C^*(X;Z[G])$ be the corresponding co-chain complex (just the usual co-chains of $\tilde{X}$ if G is finite). Then

$$\cap\xi : C^*(X;Z[G]) \to C_{2k-*}(X;Z[G])$$

is a chain equivalence which, up to chain homotopy, depends only upon X. The cells of X determine preferred bases of these chain and co-chain complexes. Hence $\cap\xi$ has a torsion in $Wh(G)$ which depends only on X. Denote this element $\overline{\Delta}(X)$. Then $\overline{\Delta}(X) = \overline{\Delta}(X)^-$, and so $\overline{\Delta}(X)$ represents an element $\Delta(X) \in H^0(Wh(G))$.

Now suppose $G = Z_{2^r}$. Then, by (1.3), $\Delta(X)$ has a unique representative of the form

$$\prod_{m,s,i} (U_{m,s,i})^{\delta_{m,s,i}},$$

where the product is over $3 \leq s \leq m \leq r$, $i \equiv 1 (\mathrm{mod}\ 4)$, $1 \leq i < 2^{s-1}$, and $\delta_{m,s,i} = 0$ or 1. Let

$$\Delta_{m,s,i}(X) = \delta_{m,s,i}.$$

(6.1) <u>Theorem. The connected Poincaré duality space</u> X <u>of dimension</u> $2k \geq 6$ <u>with</u> $\pi_1 X = Z_{2^r}$ <u>has the homotopy type of a</u> PL (<u>or</u> TOP) <u>manifold if and only if all of the following hold:</u>

(i) <u>there is a reducible</u> PL (<u>or</u> TOP) <u>bundle over</u> X;

(ii) $\chi(X)$ <u>is a multiple of the regular representation;</u>

(iii) $\Delta_{m,s,i}(X) = 0$ <u>for</u> $3 \leq s \leq m \leq r$, $i \equiv 1 (\mathrm{mod}\ 4)$, $1 \leq i < 2^{s-1}$, <u>and</u> $s > 2m-r$.

<u>Proof.</u> Necessity of (i) and (ii) has already been explained. For (iii), if M is a manifold and $h: M \to X$ a homotopy equivalence, then one considers the diagram

$$\begin{array}{ccc} C^*(\tilde{X}) & \xrightarrow{\cap\xi} & C_*(\tilde{X}) \\ h^* \downarrow & & \downarrow h_* \\ C^*(\tilde{M}) & \xrightarrow{\cap\eta} & C_*(\tilde{M}) \end{array} ,$$

η representing $[M]$. Since $\cap\eta$ is a simple equivalence, it follows that $\overline{\Delta}(X) = \tau(h)+\tau(h)^{-}$; i.e. $\Delta(X)$ is trivial in $H^0(Wh(Z_{2^r}))$.

To prove the converse, suppose ξ is reducible. Then, by the well-known transversality arguments, there is a degree one normal map

$$\begin{array}{ccc} \nu_M & \xrightarrow{b} & \xi \\ \downarrow & & \downarrow \\ M & \xrightarrow{f} & X \end{array}$$

into X, with surgery obstruction $\sigma(f,b) \in L^h_{2k}(Z_{2^r})$. Further, the following formulas hold:

$$\chi(\sigma(f,b)) = \chi(M) - \chi(X) \quad \text{and}$$

$$d(\sigma(f,b)) = \Delta(X) \qquad \text{(see (4.1))}.$$

These can be proven by standard arguments of surgery theory.

Let ρ denote the regular representation. Then the first equation and (ii) imply that

$$\chi(\sigma(f,b)) = q\rho,$$

where q is an integer. On the other hand, if k is even the coefficient of the trivial representation in $\chi(\sigma(f,b))$ is just the difference of the signatures $I(M)-I(X) = 8t$; hence $q = 8t$. Hence we may replace M by its connected sum with $|t|$ copies of a P.L. manifold of signature $8t/|t|$, to kill $\chi(\sigma(f,b))$. If k is odd, since $\chi(\sigma(f,b)) = -\chi(\sigma(f,b))^{-}$, $q = 0$ automatically. Hence we may assume $\chi(\sigma(f,b)) = 0$.

It follows from (4.2) (see also (4.1)) that $\Delta_{m,s,i}(X) = 0$ for $s \leq 2m-r$. Hence, by (iii), $\Delta_{m,s,i}(X) = 0$ for all m,s,i; i.e. $\Delta(X)$ is trivial. Hence by (3.5), $\sigma(f,b)$ actually is in the image of $L^s_{2k}(Z_{2^r})$. Hence, by (3.1), if k is even $\sigma(f,b) = 0$. If k is odd, $\sigma(f,b)$ can be killed by replacing M with its connected sum with a Kervaire manifold. So a normal map (f,b) with $\sigma(f,b) = 0$ is obtained, hence (f,b) is normally cobordant to a homotopy equivalence, which completes the proof.

Proof of Theorem C. Let X be as in Theorem C, and let $h:X \to \hat{Y}$ be a homotopy equivalence. Let $[Y] \in H_{2k}(Y)$, with transfer $h_*[X] \in H_{2k}(\hat{Y})$. Then h induces a homotopy equivalence $\tilde{h}:\tilde{X} \to \tilde{Y}$, and it is not hard to check that $\tilde{h}_*([X] \cap z) = \tilde{h}^*([Y] \cap \tilde{h}_* z)$. It follows that

$$\cap[Y]:H^i(\tilde{Y}) \to H_{2k-i}(\tilde{Y})$$

is an isomorphism for all i, and hence Y is a Poincaré Duality space.

Hence the invariant $\Delta(Y) \in H^0(Wh(Z_{2^{r+1}}))$ is defined. It is not hard to see that

$$\tau_{r+1}\Delta(Y) = \Delta(X),$$

here τ_{r+1} denotes the map induced on H^0 by the transfer. By (1.3)

$$\Delta(Y) \equiv \Pi(U^{(r+1)}_{m,s,i})^{\lambda_{m,s,i}}.$$

Hence $\Delta(X) \equiv \Pi\tau_{r+1}(U^{(r+1)}_{m,s,i})^{\lambda_{m,s,i}}$. Note that squares of self-conjugate units are trivial in $H^0(Wh(Z_{2^r}))$. Hence it follows from 1.2 that $\Delta_{m,s,i}(X) = 0$ for $m \neq r$. In particular, $\Delta_{m,s,i}(X) = 0$ for $s > 2m-r$, $3 \leq s \leq m \leq r$. Hence Theorem (6.1) applies to conclude that X has the homotopy type of a manifold.

References

[B] H. Bass. Algebraic K-theory. Benjamin, 1968.

[BPW] W. Browder, T. Petrie, and C.T.C. Wall. The classification of actions of cyclic groups of odd order on homotopy spheres, Bull. A.M.S. 77 (1971), 455-459.

[CS1] S.E. Cappell and J.L. Shaneson. Non-linear similarity. Annals of Math. 113 (1981), 315-355.

[CS2] ________________. On tempered numbers, to appear.

[CS3] ________________. Fixed points of periodic differentiable maps. Inventiones Math. 68 (1982), 1-19.

[CS4] ________________. Class numbers and periodic smooth maps. Comm. Math. Helv. 58 (1983), 167-185.

[J] Burton W. Jones. The arithmetic theory of quadratic forms. Carus Mathematical Monographs 10, Mathematical Association of America, Wiley, New York, 1950.

[M] J.W. Milnor. Whitehead torsion. Bull. A.M.S. 72 (1966), 358-426.

[Sh] J.L. Shaneson. Wall's surgery groups for $Z \times G$. Annals of Math. 90 (1969), 296-334.

[W1] C.T.C. Wall. Classification of Hermitian forms: VI, Group rings. Ann. of Math. 103 (1976), 1-80.

[W2] __________. Surgery on compact manifolds. Academic Press, New York, 1970.

S. Cappel
Courant Institute of Mathematics
New York University
New York, NY 10012, USA

J. Shaneson
Department of Mathematics
University of Chicago
Chicago, IL 60637, USA

Higher Diagonal Approximations and Skeletons of $K(\pi,1)$'s

By

James F. Davis*
University of Notre Dame

The cup product is graded commutative on the cohomomology lvel, but not on the cochain level. The failure of commutativity is measured by the higher diagonal approximations underlying such invariants as the Steenrod squares [6] and the symmetric signature associated to a Poincare duality space [5].

An n-skeleton of a $K(\pi,1)$ is a CW complex X of dimension n with $\pi_i(X) = 0$ for $1 < i < n$ and $\pi_1 X = \pi$. For example, X could be a space form, a manifold whose universal cover is the sphere or Euclidean space. This paper shows how the geometric higher diagonal approximations of X can be calculated purely algebraically from the cellular chains of the universal cover $\tilde{X}$.

This work was motivated by certain questions of John Jones and R. James Milgram concerning the Cappell-Shaneson detection [1] of a non-zero element $\sigma(S^3/Q_8)$ in the symmetric L-group $L^3(\mathbb{Z}Q_8)$. I wish to thank Andrew Ranicki for repeatedly bringing these questions to my attention.

Using the results of this paper one can compute the symmetric signature $\sigma(S^n/G) \in L^n(\mathbb{Z}G)$ for any free action of a finite group G on S^n. The symmetric signature appears in Ranicki's product formula for surgery obstructions. However, algebraic quadratic surgery shows that the product formula depends only on the chain level Poincare duality map (depending on Δ_0 defined below) and not on the higher diagonal approximations.

1. Preliminaries.

Let W be the standard free $\mathbb{Z}[\mathbb{Z}/2]$ - resolution of $\mathbb{Z}$

$$W: \quad \cdots \longrightarrow \mathbb{Z}[\mathbb{Z}/2] \xrightarrow{1-T} \mathbb{Z}[\mathbb{Z}/2] \xrightarrow{1+2} \mathbb{Z}[\mathbb{Z}/2] \xrightarrow{1-T} \mathbb{Z}[\mathbb{Z}/2] .$$

Here $\mathbb{Z}/2 = \langle T \rangle$. Let e_i denote the generator of the i-chains of W. Then $\partial(e_i) = (1+(-1)^i T)e_{i-1}$.

---

*Partially supported by NSF grants.

Let C be a chain complex. Then $\mathbb{Z}/2$ acts on $C \otimes C$ (= $C \otimes_{\mathbb{Z}} C$) via the interchange map

$$T:\ C \otimes C \longrightarrow C \otimes C$$

$$T(a \otimes b) = (-1)^{\deg a \deg b}\, b \otimes a .$$

We will consider $\mathbb{Z}[\mathbb{Z}/2]$ - module chain maps

$$\Delta:\ W \otimes C \longrightarrow C \otimes C .$$

Define $\Delta_i:\ C \longrightarrow C \otimes C$ by $\Delta_i(c) = \Delta(e_i \otimes c)$. Since Δ is a chain map the Δ_i satisfy relations

1.1 $$\partial\Delta_i - (-1)^i \Delta_i \partial = \Delta_{i-1} + (-1)^i\, T\Delta_{i-1} .$$

Thus Δ_0 is a chain map, Δ_1 is a chain homotopy between Δ_0 and $T\Delta_0$, Δ_2 is a chain homotopy between Δ_1 and $T\Delta_1$, etc. Conversely given a sequence of maps $\{\Delta_i\}$ satisfying 1.1, they give rise to a $\mathbb{Z}[\mathbb{Z}/2]$-module chain map Δ .

Let $S(X)$ denote the singular chain complex of a topological space X .

Theorem 1.2.

There exist functorial $\mathbb{Z}[\mathbb{Z}/2]$ -module chain maps

$$\Delta:\ W \otimes S(X) \longrightarrow S(X) \otimes S(X)$$

such that $\Delta_0(c) = c \otimes c$ for any singular 0-simplex c .

Proof. Method of acyclic models. □

If a group π acts on a space X , then functoriality implies that Δ is a $\mathbb{Z}[\mathbb{Z}/2 \times \pi]$ -module chain map.

Preposition 1.3.

Let π act freely and cellularly on a connected CW complex X . There is a splitting of $\mathbb{Z}\pi$-module chain complexes $S(X) = A \oplus B$ where A is isomorphic to the cellular chain complex $C(X)$ and $H_*(B) = 0$.

Proof.

Following Wall [7] let

$$D_i(X) = \ker\,(\partial:\ S_i(X^i) \longrightarrow S_{i-1}(X^i)/S_{i-1}(X^{i-1}))$$

Let $E_i(X) = \ker(D_i(X) \longrightarrow C_i(X))$. Then we have an exact sequence of $\mathbb{Z}\pi$-chain complexes

$$0 \longrightarrow E(X) \longrightarrow D(X) \longrightarrow C(X) \longrightarrow 0 .$$

Since $D(X) \longrightarrow C(X)$ induces an isomorphism in homology, $H_*(E(X)) = 0$, and hence $E(X)$ is chain contractible. It then follows that $D(X) = C(X) \oplus E(X)$ as chain complexes. Likewise the inclusion $D(X) \longrightarrow S(X)$ splits as a map of chain complexes. □

Let $f: C(X) \longrightarrow S(X)$ and $g: S(X) \longrightarrow C(X)$ be splitting maps in 1.3. Any $\mathbb{Z}[\mathbb{Z}/2 \times \pi]$-module chain map

$$\psi: W \otimes C(X) \longrightarrow C(X) \otimes C(X)$$

chain homotopic to

$$(g\otimes g) \circ \Delta \circ (1\otimes f): W \otimes C(X) \longrightarrow C(X) \otimes C(X)$$

is called a geometric π-higher diagonal approximation. The mod π reduction of ψ gives a geometric 1-higher diagonal approximation on X/π . If $\pi = 1$, ψ can be used to compute cup products and Steenrod squares. If X is simply connected and X/π is a Poincare complex, then ψ can be used to compute the symmetric signature [5]

$$\sigma(X/\pi) \in L^n(\mathbb{Z}\pi)$$

occuring in the product formula for surgery obstructions.

Let $C = \{C_i, \partial\}_{i\geq 0}$ be a chain complex of $\mathbb{Z}$-modules with augmentation $\varepsilon: C_0 \longrightarrow \mathbb{Z}$. Let $\varepsilon \otimes 1: C \otimes C \longrightarrow \mathbb{Z} \otimes C = C$. Let

$$(C\otimes C)_k = \bigoplus_{i+j=k} C_i \otimes C_j$$

and

$$(C\otimes C)^k = \bigoplus_{\substack{i<k \\ j\leq k}} C_i \otimes C_j .$$

Consider $\mathbb{Z}[\mathbb{Z}/2]$-module chain maps $\Delta: W \otimes C \longrightarrow C \otimes C$ satisfying

(i) $\Delta(W\otimes C_i) \subset (C\otimes C)^i$ for all i .

(ii) $(\varepsilon\otimes 1) \circ \Delta_0 = 1$
$(1\otimes\varepsilon) \circ \Delta_0 = 1$.

(iii) For all i , for any $c \in C_i$, there is an $a \in C_i \otimes C_i$ such that $\Delta_i(c) - c\otimes c = a + (-1)^i Ta$.

These conditions are geometrically inspired. Condition (ii) corresponds to the fact that for any cohomology class α , $\alpha \cup 1 = 1 \cup \alpha = \alpha$. Note that (ii) is satisfied for the Alexander-Whitney diagonal approximation. Condition (iii) is related to the identity $Sq^0(\alpha) = \alpha$.

Proposition 1.4.

On the category of topological spaces there exist functorial $\mathbb{Z}[\mathbb{Z}/2]$ - module chain maps

$$\Delta: \quad W \otimes S(X) \longrightarrow S(X) \otimes S(X)$$

satisfying (i), (ii), (iii).

Proof.

Condition (i) will hold for any functorial map. Let $C(\Delta^n)$ be the simplicial complex of the standard n-simplex. Consider $C(\Delta^n)$ as a subcomplex of $S(\Delta^n)$. By acyclic model theory there exists a functorial Δ such that $\Delta(W \otimes C(\Delta^n)) \subset C(\Delta^n) \otimes C(\Delta^n)$ for all n. Induction on n shows that $(\varepsilon \otimes 1)(\Delta_0(\Delta^n)) = \Delta^n$, $(1 \otimes \varepsilon)(\Delta_0(\Delta^n)) = \Delta^n$. The proof that $Sq^0 = Id$ (see [6]) shows that condition (iii) holds for $c = \Delta^n$. Then linearity and functorality shows that (ii) and (iii) always hold. □

2. The Main Theorem.

Theorem 2.1.

Let $C = \{C_i, \partial\}_{0 \leq i \leq n}$ be a chain complex of free $\mathbb{Z}\pi$-modules such that $H_0(C) = Z$ and $H_i(C) = 0$ for $0 < i < n$. Then there exist $\mathbb{Z}[\mathbb{Z}/2 \times \pi]$ -module chain maps

$$\Delta: \quad W \otimes C \longrightarrow C \otimes C$$

satisfying conditions (i), (ii), and (iii). Given two such maps they are chain homotopic.

Here the action of π on $C \otimes C$ is given by $g(x \otimes y) = gx \otimes gy$.

Corollary 2.2.

If X is a skeleton of a $K(\pi,1)$ and $C = C(\tilde{X})$ then any map satisfying (i), (ii), and (iii) is a geometric π-higher diagonal approximation.

Before we embark on the proof of 2.1, we need a lemma.

Lemma 2.3.

Let $\varepsilon = \pm 1$. If $b \in (C \otimes C)_{2i+1}$ is ε-symmetric $(Tb = \varepsilon b)$ and a boundary, then it is the boundary of an ε-symmetric chain. If $b \in (C \otimes C)_{2i}$ is ε-even $(b = a + \varepsilon Ta$ for some $a)$ and a boundary, then it is the boundary of an ε-even chain.

Proof.

Suppose D and E are chain complexes such that $D \otimes D$ and

$E \otimes E$ satisfy the conclusion of lemma 2.3. Then $(D \oplus E) \otimes (D \oplus E) = D \otimes D \oplus (D \otimes E \oplus E \otimes D) \oplus E \otimes E$ also satisfies the conclusion.

Now as a $\mathbb{Z}$-module chain complex C is isomorphic to a direct sum of $H_n(C)$, $H_0(C)$ and elementary chain complexes of the form $\mathbb{Z} \xrightarrow{\sim} \mathbb{Z}$. If D is any one of these chain complexes it is easily verified that $D \otimes D$ satisfies the conclusion. The lemma then follows by induction. □

Proof of 2.1.

To construct the $\mathbb{Z}[\mathbb{Z}/2 \times \pi]$-module chain map $\Delta: W \otimes C \longrightarrow C \otimes C$ it suffices to construct of sequence $\Delta_i: C \longrightarrow C \otimes C$ of $\mathbb{Z}\pi$-maps satisfying the relations 1.1 as well as conditions (i), (ii), and (iii). Choose a $\mathbb{Z}\pi$ basis of C_0 of the form $\Lambda \cup \{\rho_0\}$ where $\varepsilon(\rho_0) = 1$ and $\varepsilon(f) = 0$ for $f \in \Lambda$. Define $\Delta_0(\rho_0) = \rho_0 \otimes \rho_0$. For $f \in \Lambda$, define $\Delta_0(f) = \rho_0 \otimes f + f \otimes f + f \otimes \rho_0$. Extend to a map of C_0 by linearity.

Fix $k > 0$ and $\ell < k-1$. Assume now that Δ has been defined on $W \otimes C_j$ for $j < k$ and that Δ_i has been defined on C_k for $i < \ell$. Let x be a basis element of C_k. We first consider the case $\ell=0$. Let $Z_i = \ker(\partial: C_i \longrightarrow C_{i-1})$. By the Kunneth theorem $H_i((C \otimes C)^i) = Z_i \otimes \mathbb{Z} \oplus \mathbb{Z} \otimes Z_i$. In particular if b is a i-cycle in $(C \otimes C)^i$ with $(\varepsilon \otimes 1)b = 0 = (1 \otimes \varepsilon)b$ then b is a boundary in $(C \otimes C)^i$. So $\Delta_0(\partial x) - \partial x \otimes \rho_0 - \rho_0 \otimes \partial x$ is a boundary in $(C \otimes C)^{k-1}$. Say it is ∂a. Then define $\Delta_0(x) = x \otimes \rho_0 + \rho_0 \otimes x + a$. For $\ell > 0$, $(-1)^\ell \Delta_\ell(\partial x) + \Delta_{\ell-1}(x) + (-1)^\ell \Delta_{\ell-1}(x)$ is a boundary in $(C \otimes C)^k$ say ∂a. Define $\Delta_\ell(x) = a$. Extend to a map $\Delta_\ell: C_k \longrightarrow C \otimes C$ by linearity.

Now fix $k > 0$ and assume that Δ has been defined on $W \otimes C_j$ for $j < k$ and that Δ_i has been defined on C_k for $i < k-1$. Let x be a basis element of C_k. Then $(-1)^{k-1}\Delta_{k-1}(\partial x) + \Delta_{k-2}(x) + (-1)^{k-1}T\Delta_{k-2}(x) - (-1)^{k-1} \partial x \otimes \partial x$ is a $(-1)^{k-1}$-even boundary in $(C \otimes C)^k \cap (C \otimes C)_{2k-2}$ and hence by lemma 2.3 lifts to $a + (-1)^{k-1}Ta$ with $a \in (C \otimes C)^k \cap (C \otimes C)_{2k-1}$. We define $\Delta_{k-1}(x) = a + (-1)^{k-1}Ta + \partial x \otimes x$. (For the case $k = 1$ we also have to guarantee that $(\varepsilon \otimes 1)\Delta_0(x) = x$ and that $(1 \otimes \varepsilon)\Delta_0(x) = x$, but this can be done by the proof of 2.3.). Extend Δ_{k-1} to a map of C_k by linearity. Now

$(-1)^k \Delta_k(\partial x) + \Delta_{k-1}(x) + (-1)^k T\Delta_{k-1}(x) = \partial x \otimes x + (-1)^k x \otimes \partial x$.

So define $\Delta_k(x) = x \otimes x$ for the basis element x . Extend by linearity. This completes the existence part of theorem 2.1.

For the uniqueness part of 2.1 consider $\mathbb{Z}[\mathbb{Z}/2]$ -module chain maps $\Delta: W \otimes C \longrightarrow C \otimes C$ satisfying

(i') $\Delta(W\otimes C_i) \subset (C\otimes C)^i$ for all i .

(ii') $(\varepsilon\otimes 1) \circ \Delta_0 = 0$

$(1\otimes\varepsilon) \circ \Delta_0 = 0$.

(iii') For all i , $\Delta_i(C_i) \subset \operatorname{im}(1+(-1)^i T)$.

Lemma 2.4.

A $\mathbb{Z}[\mathbb{Z}/2 \times \pi]$ -module chain map satisfying (i'), (ii'), and (iii') is of the form $\Delta = \partial\chi + \chi\partial$ for some degree one map χ .

Proof.

Define $\chi_i(c) = \chi(e_i \otimes c)$. Then $\Delta = \partial\chi + \chi\partial$ is equivalent to

2.5 $$\Delta_i = \partial\chi_i + (-1)^i\chi_i\partial + \chi_{i-1} + (-1)^i T\chi_{i-1} .$$

Let x be a basis element of C_0 . Then $\Delta_0(x)$ is even and a boundary. Thus there is an $a \in (C\otimes C)_1$ such that $\partial(a+Ta) = \Delta_0(x)$ and $(\varepsilon\otimes 1)a = 0 = (1\otimes\varepsilon)a$. Define $\chi_0(x) = a + Ta$. Extend to a map of C_0 by linearity. Replace Δ by $\Delta - \partial\chi_0 - \chi_0\partial$.

We now assume Δ satisfies (i'), (ii'), and (iii') and that $\Delta_0(C_0) = 0$. We will now only consider χ such that $\chi(C^j) \subset (C\otimes C)^j$.

Fix $k < 0$ and $\ell < k - 1$. Assume that χ has been defined on $W \otimes C_j$ for $j < k$ and that χ_i has been defined on C_k for $i < \ell$. Furthermore assume

$$\Delta_i(c) = \partial\chi_i(c) + (-1)^i\chi_i(\partial c) + \chi_{i-1}(c) + (-1)^i T\chi_{i-1}(c)$$

for $c \in C_j$, $j < k$ and for $c \in C_k$, $i < \ell$. Let x be a basis element of C_k . Choose an element $a \in (C\otimes C)^k$ such that

$$\partial a = \Delta_\ell(x) - (-1)^\ell\chi_\ell(\partial x) - \chi_{\ell-1}(x) - (-1)^\ell T\chi_{\ell-1}(x) .$$

Define $\chi_\ell(x) = a$. Extend to a map of C_k by linearity.

Now fix $k > 0$ and assume that χ has been defined on $W \otimes C_j$ for $j < k$ and that χ_i has been defined on C_k for $i < k - 1$. Furthermore assume that the relation 2.5 is satisfied for $c \in C_j$,

$j < k$ and for $c \in C_k$, $i < k-1$. Choose $b \in (C \otimes C)^k$ such that

$$\partial b = \Delta_{k-1}(x) - \chi_{k-2}(x) - (-1)^{k-1} T\chi_{k-2}(x) .$$

Now $\Delta_k(x) - (b+(-1)^k Tb)$ is a $(-1)^k$-even cycle so by lemma 2.3,

$$\Delta_k(x) = b + (-1)^k Tb + c + (-1)^k Tc$$

for some cycle $c \in (C \otimes C)^k$. Define $\chi_{k-1}(x) = b + c$. Extend by linearity. This completes the proof of 2.1. □

There are two cases where one can avoid some of the above homological algebra to calculate the geometric higher diagonal approximations. First, if X is a simplicial complex, one can apply acyclic model theory in the simplicial category to X directly. Second, if X is actually a $K(\pi,1)$, the construction of Δ follows from the "fundamental lemma" of homological algebra from a projective complex to an acyclic one. Indeed, if C is acyclic, <u>any</u> two $\mathbb{Z}[\mathbb{Z}/2 \times \pi]$ - module maps

$$W \otimes C \longrightarrow C \otimes C$$

commuting with the augmentation are chain homotopic.

If X is a n-skeleton of a $K(\pi,1)$, then the homotopy type of X is determined by $\pi_1 X$, $\pi_n X$, and the first Eilenberg-MacLane k-invariant $k^{n+1}(x) \in H^{n+1}(\pi_1 X; \pi_n X)$. (See, for example, Olum [4]). Now $k^{n+1}(x)$ can be defined algebraically as follows: Let $D = \{D_i, \partial\}_{i \geq 0}$ be a projective $\mathbb{Z}\pi$-resolution of $\mathbb{Z}$. Choose a chain map

$$\{D_i, \partial\}_{i \leq n} \longrightarrow C_*(\tilde{X})$$

commuting with augmentation. Induced is a map

$$D_{n+1} \longrightarrow \ker(C_n(\tilde{X}) \longrightarrow C_{n-1}(\tilde{X})) = \pi_n X .$$

This cocycle gives $k^{n+1}(X)$. Hence the homotopy type of X is determined by the chain homotopy type of $C(\tilde{X})$. Thus every homotopy invariant of X should be computable algebraically. This gives a philosophical justification for Corollary 2.2.

3. <u>Product Formulae</u>

Given a product map

$$\mathrm{Id} \times f: N^m \times M^n \longrightarrow N^m \times X^n$$

with N^m a closed manifold and $f: M^n \longrightarrow X^n$ a degree one normal map, the surgery obstruction $\sigma(\mathrm{Id} \times f) \in L_{m+n}(\mathbb{Z}[\pi_1(X)])$ is determined by the symmetric signature $\sigma(N) \in L^m(\mathbb{Z}[\pi_1(N)])$ and the surgery obstruction $\sigma(f) \in L_n(\mathbb{Z}[\pi_1(X)])$. Indeed $\sigma(N)$ can be represented by a symmetric Poincare complex (C,ϕ) with $C = C_*(N)$ and $\sigma(f)$ by a quadratic Poincare complex (D,ψ) with $H_*(D) = K_*(M)$. Then according to Ranicki's product formula [5,II.8.1]

$$\sigma(\mathrm{Id} \times f) = (C \otimes D,\ \phi \otimes \psi)\ ,$$

using the algebraically defined pairing

$$L^m(\mathbb{Z}[\pi_1 N]) \otimes L_n(\mathbb{Z}[\pi_1 X]) \longrightarrow L_{n+m}(\mathbb{Z}[\pi_1 N] \otimes \mathbb{Z}[\pi_1 X])\ .$$

$$(C,\phi) \otimes (D,\psi) \longmapsto (C \otimes D,\ \phi \times \psi)\ ,$$

and the identification $\mathbb{Z}[\pi_1 N] \otimes \mathbb{Z}[\pi_1 X] = \mathbb{Z}[\pi_1(N \times X)]$.

Here ϕ and ψ are represented by a sequence of maps

$$\{\phi_i \in \mathrm{Hom}_{\mathbb{Z}[\pi_1 N]}(C^{n-r+i}, C_r) \mid r \in \mathbb{Z},\ i \geq 0\}$$

$$\{\psi_i \in \mathrm{Hom}_{\mathbb{Z}[\pi_1 X]}(D^{n-r-i}, D_r) \mid r \in \mathbb{Z},\ i \geq 0\}$$

and $(\phi \otimes \psi)_i = \phi_i \otimes \psi_i$.

A geometric $\pi_1 N$-higher diagonal approximation

$$\Delta:\ W \otimes C(\tilde{N}) \longrightarrow C(\tilde{N}) \otimes C(\tilde{N})$$

determines the symmetric signature $\sigma(N) = (C,\phi)$ as follows:

Choose a representative $[N] \in C_m(N;\mathbb{Z}^t)$ for the fundamental class of N. Let $\pi = \pi_1 X$. Apply $\mathbb{Z}^t \otimes_{\mathbb{Z}\pi}$ to the Δ_i associated to Δ to obtain

$$\Delta_i' : C(N;\mathbb{Z}^t) \longrightarrow C(\tilde{N}) \otimes_{\mathbb{Z}\pi} C(\tilde{N})\ .$$

Then the ϕ_i are defined via the slant product

$$\phi_i : C^{n-r+i} \longrightarrow C_r$$

$$\beta \longmapsto \Delta_i'([N])/\beta\ .$$

<u>Lemma 3.1</u>: The class of a quadratic Poincare complex (C',ψ') in $L_i(A)$ depends only on ψ_0. □

Proof: This is an immediate consequence of the algebraic theory of surgery [5,I.4.3].

<u>Corollary 3.2</u>: The class of $(C \otimes D,\ \phi \otimes \psi)$ in the product formula depends only on ϕ_0 and ψ_0.

We now restrict our attention to ϕ_0. As a corollary of the proof of 2.1 we have:

Corollary 3.3: Let $C = \{C_i, \partial\}_{0 \le i \le n}$ be a chain complex of free $\mathbb{Z}\pi$-module such that $H_0(C) = \mathbb{Z}$ and $H_i(C) = 0$ for $0 < i < n$. Then there exists a $\mathbb{Z}[\pi]$ - module chain map

$$\Delta_0 : C \longrightarrow C \otimes C$$

satisfying conditions (i), (ii), and (iii), for $i = 0$.

Because of its importance in the product formula we make explicit on algorithm for computing ϕ_0. A contracting chain homotopy for C is given by $\mathbb{Z}$-module maps $\{s, \delta\}$

$$\delta : \mathbb{Z} \longrightarrow C_0$$

$$s : C_{i-1} \longrightarrow C_i$$

satisfying $\partial s + \delta\varepsilon = Id$ on C_0

and $\partial s + s\partial + Id$ on C_i for $0 < i < n$.

Choose a $\mathbb{Z}\pi$-basis of C_0 of the form $\Lambda \cup \{\rho_0\}$ where $\rho_0 = \delta(1)$ and $\varepsilon(f) = 0$ for $f \in \Lambda$.

Define

$$\Delta_0(\rho_0) = \rho_0 \otimes \rho_0 ,$$

$$\Delta_0(f) = \rho_0 \otimes f + f \otimes f + f \otimes \rho_0 \quad \text{for} \quad f \in \Lambda .$$

Extend to a map on C_0 by linearity. Now assume Δ_0 has been defined on C_j for $j < k$. For a $\mathbb{Z}\pi$-basis element x of C_k define

$$\Delta_0(x) = x \otimes \rho_0 + \rho_0 \otimes x + (s \otimes 1 + \delta\varepsilon \otimes s)(\Delta_0(\partial x) - \partial x \otimes \rho_0 - \rho_0 \otimes \partial x) .$$

Extend to a map of C_k by linearity. This Δ_0 satisfies the desired properties.

Let X be an n-skeleton of a $K(\pi,1)$. Let $Y = K(\pi,1)$. Naively, one might try to avoid the algebra in 2.1 by constructing a geometric π-diagonal approximation

$$\Delta_0 : \ C(\tilde{Y}) \longrightarrow C(\tilde{Y}) \otimes C(\tilde{Y})$$

(Using, for example, the contracting chain homotopy $\{s \otimes 1 + \delta\,\varepsilon \otimes s,\ \delta \otimes \delta\}$ on $C(\tilde{Y}) \otimes C(\tilde{Y}))$, and then restricting the map

$$\Delta_0\big|_{C(\tilde{X})} : C(\tilde{X}) \longrightarrow C(\tilde{X}) \otimes C(\tilde{X})$$

to obtain a π-diagonal approximation for X. However, this Δ_0 need not satisfy the hypothes of 3.3, so there is no guarantee that $\Delta_0\big|_{C(\tilde{X})}$ is the correct chain homotopy class. In fact, unpublished computations of Jones and Milgram show that the above procedure can lead to a geometrically incorrect diagonal approximation for X. We now describe this example in some detail as it was the motivation of this paper.

A fundamental problem in surgery theory is the oozing problem, the problem of determining which elements of $L_*(\mathbb{Z}\pi)$ arise from surgery problems over closed manifolds. A critical example is the Cappell-Shaneson example

$$\mathrm{Id} \times f\colon S^3/\pi \times T^2 \longrightarrow S^3/\pi \times S^2$$

where $\pi = Q(2^n)$ is the generalized quaternion group and $f\colon T^2 \longrightarrow S^2$ is the Kervaire problem" representing the non-trivial element of $L_2(\mathbb{Z}) = \mathbb{Z}/2$. Here $\sigma(f)$ is represented by $(D,\psi) \in L_0(\mathbb{Z},-1) = L_2(\mathbb{Z})$ where

$$\psi = \begin{pmatrix} 1 & 1 \\ 0 & 1 \end{pmatrix} = D^0 = \mathbb{Z} \oplus \mathbb{Z} \longrightarrow D_0 = \mathbb{Z} \oplus \mathbb{Z} .$$

Using geometric reasoning, S. Cappell and J. Shaneson showed $\sigma(\mathrm{Id} \times f) \neq 0 \in L_1^h(\mathbb{Z}\pi)$. However, the product formula hints at an algebraic derivation of this result. In a preliminary attempt at this problem, Jones and Milgram constructed a map

$$\Delta_0\colon C(K(\pi,1)) \longrightarrow C(K(\pi,1)) \otimes C(K(\pi,1))$$

and restricted to the 3-skeleton to obtain

$$\Delta_0\colon C \longrightarrow C \otimes C$$

where $C = C(S^3/\pi)$. (Cartan and Eilenberg [2] give an explicit periodic $\mathbb{Z}\pi$-resolution of $\mathbb{Z}$ corresponding to a cell decomposition of S^3/π). The above Δ_0 lead to a chain homotopy equivalence

$$\phi_0\colon C^{3-*} \longrightarrow C_* .$$

Applying the product formula $(C \otimes D, \phi \otimes \psi)$, they obtained a trivial element of $L_1^h(\mathbb{Z}\pi)$, seemingly contradicting the Cappell-Shaneson example.

The resolution of this dilemma is that the naive approach does not lead to a geometrically correct result. Unpublished computations of the author show that the methods of this paper give a formation representing $\sigma(\mathrm{Id} \times f) \in L_1^h(\mathbb{Z}\pi)$, and prove algebraically the Cappell-Shaneson result that $\sigma(\mathrm{Id} \times f) \neq 0 \in L_1^h(\mathbb{Z}\pi)$, and $\mathrm{im}(\sigma(\mathrm{Id} \times f)) = 0 \in L_1^p(\mathbb{Z}\pi)$.

For an alternate algebraic approach to this result, see the paper of R. James Milgram, "The Cappell-Shaneson example," appearing in these proceedings.

References

1. S.E. Cappell and J.L. Shaneson, "A counterexample of the oozing problem for closed manifolds," Algebraic Topology: Aarhus 1978, Lecture Notes in Math. 763, Springer-Verlag (1978), 627-634.

2. H. Cartan and S. Eilenberg, Homological Algebra, Princeton U. Press, (1956).

3. R. James Milgram, "The Cappell-Shaneson Example," these proceedings.

4. P. Olum, "Homotopy type and singular homotopy type," Ann. of Math. 60 (1954), 317-325.

5. A. Ranicki, "The algebraic theory of surgery I-II," Proc. Lon. Math. Soc. 40 (1980), 87-283.

6. E.H. Spanier, Algebraic Topology, McGraw-Hill (1966).

7. C.T.C. Wall, "Finiteness conditions for CW complexes II," Proc. Royal Soc. A295 (1966), 129-139.

Department of Mathematics
University of Notre Dame
Notre Dame, IN 46556, USA

LECTURES ON GROUPS OF HOMOTOPY SPHERES

J. P. Levine
Department of Mathematics
Brandeis University
Waltham, Massachusetts 02254

Kervaire and Milnor's germinal paper [15], in which they used the newly-discovered techniques of surgery to begin the classification of smooth closed manifolds homotopy equivalent to a sphere (*homotopy-spheres*), was intended to be the first of two papers in which this classification would be essentially completed (in dimensions ≥ 5). Unfortunately, the second part never appeared. As a result, in order to extract this classification from the published literature it is necessary to submerge oneself in more far-ranging and complicated works (e.g. [7], [16], [30]), which cannot help but obscure the beautiful ideas contained in the more direct earlier work of Kervaire and Milnor. This is especially true for the student who is encountering the subject for the first time.

In Fall, 1969, I gave several lectures to a graduate seminar at Brandeis University, in which I covered the material which I believe would have appeared in Groups of Homotopy Spheres, II. Two students, Allan Gottlieb and Clint McCrory, prepared mimeographed notes from these lectures, with some extra background material, which have been available from Brandeis University. The present article is almost identical with these notes. I hope it will serve to fill a pedagogical gap in the literature.

The reader is assumed to be familiar with [15], [20]. In these papers, Kervaire-Milnor define the group θ^n of h-cobordism classes of homotopy n-spheres and the subgroup bP^{n+1} defined by homotopy spheres which bound parallelizable manifolds. The goal is to compute bP^{n+1} and θ^n/bP^{n+1}.

Section 1 reviews some well known results on vector bundles over spheres and the homotopy of the classical groups, as well as some theorems of Whitney on embeddings and immersions. Since a homotopy n-sphere Σ^n is h-cobordant to S^n (the n-sphere with its standard differential structure) iff Σ^n bounds a contractible manifold, in order to calculate bP^{n+1} we are interested in finding and realizing "obstructions" to surgering parallelizable manifolds into contractible ones. Section 2 contains some general theorems for framed surgery and describes which "obstructions" exist for each n. In [15] it is shown that bP^{n+1} is zero for $n+1$ odd. Sections 3 and 4 perform the corresponding calculations for $n+1 = 4k$ and $n+1 = 4k+2$ respectively. In section 5, by use of the Thom-Pontryagin construction, the calculation of θ^n/bP^{n+1} is reduced to a question of framed cobordism which is answered by using results from sections 3 and 4. Many results of these notes are summarized in a long exact sequence

$$\cdots \to P^{n+1} \xrightarrow{b} \theta^n \to A^n \to P^n \xrightarrow{b} \theta^{n-1} \to \cdots$$

which is discussed in the appendix.

Throughout these notes all manifolds are assumed to be smooth, oriented, and of dimension greater than 4. In addition all manifolds with boundary are assumed to have dimension greater than 5 (so that the boundary manifold will have dimension greater than 4).

§1. Preliminaries

A) Oriented vector bundles over spheres.

In [28] Steenrod gives the following method for viewing oriented k-plane bundles over S^n as elements of $\pi_{n-1}(SO_k)$. Let ξ be such a bundle. By section 12.9 of [28] the group of ξ may be reduced from $GL(k,\mathbb{R})$ to O_k. Since ξ is oriented O_k may be further reduced to SO_k. Cover S^n by two overlapping "hemispheres". Since the bundle is trivial over each hemisphere, it is determined by the transition function at each point of the equator. This function, $\alpha\colon S^{n-1} \to SO_k$, is well defined up to homotopy class by the equivalence class of ξ and is the obstruction to framing ξ. In addition the map $[\xi] \rightsquigarrow [\alpha]$ sets up a one-to-one correspondence between (oriented isomorphism) equivalence classes of oriented k-plane bundles and elements of $\pi_{n-1}(SO_k)$. For details the reader should see section 18 [28]. By abuse of notation we refer to $[\xi] \in \pi_{n-1}(SO_k)$.

<u>Lemma 1.1</u>. Let $[\xi] \in \pi_{n-1}(SO_k)$ be an oriented k-plane bundle $/S^k$. Then $[\xi \oplus \varepsilon] = i_*[\xi] \in \pi_{n-1}(SO_{k+1})$ where we view $SO_k \xrightarrow{i} SO_{k+1}$ as acting trivially on the last component of $\mathbb{R}^{k+1}$ (i.e. the matrix M goes to

$$\begin{pmatrix} & & & 0 \\ & M & & \vdots \\ & & & 0 \\ 0 & \cdots & 0 & 1 \end{pmatrix}$$

Proof. Cover S^n by two hemispheres as above. At a point x_o on the equation, the transition function for $\xi \oplus \varepsilon^1$ is $T \times \mathrm{id}: \mathbb{R}^{k+1} \to \mathbb{R}^{k+1}$ where T is the transition function for ξ at x_o. But this characterizes the element i_* as well.

Corollary 1.2. Oriented stable bundles over S^n are in 1-1 correspondence with elements of $\pi_{n-1}(SO)$.

B) Homotopy of the Classical Groups.

Let $(0,\ldots,0,1) = e_k \in S^k \subset \mathbb{R}^{k+1}$. Then the projection $SO_{k+1} \xrightarrow{p_k} S^k$ given by $\sigma \rightsquigarrow \sigma(e_k)$ gives a fibre bundle $SO_k \xrightarrow{i_k} SO_{k+1} \xrightarrow{p_k} S^k$. If M is a manifold, let $\tau(M)$ denote the tangent bundle of M.

By weaving together the resulting exact sequences one obtains:

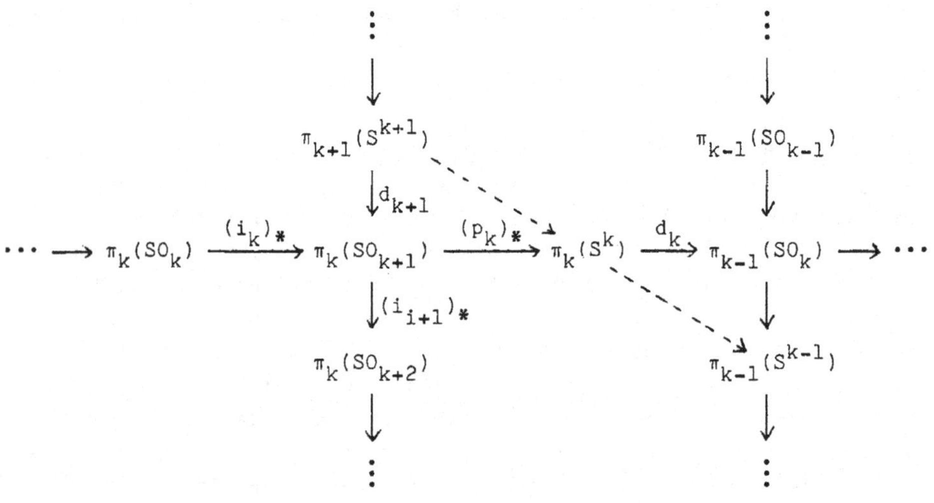

Diagram 1

where $d_k: \pi_k(S^k) \to \pi_{k-1}(SO_k)$ is the induced boundary map. By direct computation one checks that under d_k the generator is taken to $\tau(S^k) \in \pi_{k-1}(SO_k)$ and that, under $(p_{k-1})_*: \pi_{n-1}(SO_k) \to \pi_{n-1}(S^{k-1})$, a k-plane bundle ξ^k over S^n is taken to $O(\xi^k)$, the obstruction to finding a section (c.f. [28] §34.4). When $n = k$, $O(\xi^k) = \chi(\xi^k)$, the Euler class [28]. Since $\chi(\tau(S^k)) = \chi(S^k)$ generator where $\chi(S^k) = \begin{cases} 2 & k \text{ even} \\ 0 & k \text{ odd} \end{cases}$ is the Euler number, we have that the dashed maps are multiplication by 2 or 0 as indicated. This allows us to calculate the order of $\tau(S^k) \in \pi_{k-1}(SO_k)$. When k is even $(p_k)_*$ takes $\tau(S^k)$ to twice the generator and thus $\tau(S^k)$ has infinite order. When k is odd, twice the generator of $\pi_k S^k$ is in im p_{k*} so that $\tau(S^k)$ has order at most 2. Since S^k is parallelizable iff $k = 1, 3$ or 7 [3] we have

<u>Lemma 1.3</u>. $$\text{order } \tau(S^k) = \begin{cases} \infty & k \text{ even} \\ 1 & k = 1,3,7 \\ 2 & \text{otherwise} \end{cases}.$$

From the bundle exact sequence, we know that $(i_k)_*: \pi_j SO_k \to \pi_j SO_{k+1}$ is mono (resp. epi) unless $j = k-1$ (resp. $j = k$). Thus $\ker(\pi_{k-1}(SO_k) \to \pi_{k-1}(SO)) = \ker((i_k)_I: \pi_{k-1}(SO_k) \to \pi_{k-1}(SO_{k+1})) = \mathrm{im}(d_k: \pi_k S^k \to \pi_{k-1} SO_k)$. Applying Lemma 1.3 we obtain the first part of

<u>Theorem 1.4</u>. (1) $$\ker(\pi_{k-1}(SO_k) \to \pi_{k-1}(SO)) \cong \begin{cases} Z & k \text{ even} \\ 0 & k = 1,3,7 \\ Z_2 & \text{otherwise} \end{cases}$$

(2) $$\mathrm{coker}(\pi_k(SO_k) \to \pi_k(SO)) \cong \begin{cases} Z_2 & k = 1,3,7 \\ 0 & \text{otherwise} \end{cases}$$

(3) Let $V_{N,N-k}$ be the Steifel manifold of $N-k$ frames in N space. We have a bundle $SO_k \xrightarrow{i} SO_N \xrightarrow{p} V_{N,N-k}$. If N is large and $k = 3,7$, $\pi_k(SO_N) \xrightarrow{p_*} \pi_k(V_{N,N-k})$ is onto.

<u>Proof</u>. To prove (2) we need only investigate $\pi_k SO_k \xrightarrow{i_{k*}} \pi_k SO_{k+1} \xrightarrow{(i_{k+1})_*} \pi_k SO_{k+2}$ as the last group is also $\pi_k SO$. $(i_{k+1})_*$ is always epi. If k is even we see, from Diagram 1, that $d_k: \pi_k S^k \to \pi_{k-1}(SO_k)$ is mono and thus that i_{k*} is epi. If k is odd but unequal to 1, 3, or 7 the relevant part of Diagram 1 is

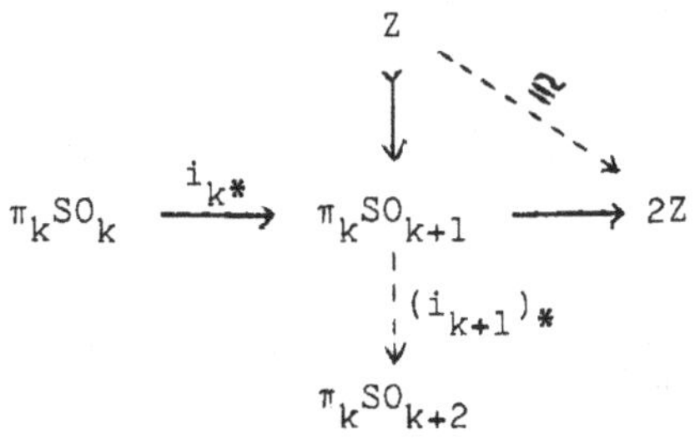

and a trivial diagram chase shows that $(i_{k+1} i_k)_*$ is epi. If $k = 1$, 3 or 7 we have

$$\begin{array}{ccccc}
 & & \mathbb{Z} & & \\
 & & \downarrow d_{k+1} & \overset{\times 2}{\searrow} & \\
\pi_k SO_k & \xrightarrow{i_{k*}} & \pi_k SO_{k+1} & \longrightarrow & Z \\
 & & \downarrow (i_{k+1})_* & & \\
 & & \pi_k SO_{k+2} & &
\end{array}$$

which concludes the proof.

(3) The bundle structure is given in [28] §7. This gives the sequences:

$$\pi_k(SO_N) \xrightarrow{p_*} \pi_k(V_{N,N-k}) \to \pi_{k-1}(SO_k) \to \pi_{k-1}(SO_N)$$

and the result now follows from (1).

We conclude this section by giving some results of Bott and Kervaire.

Theorem 1.5.

(1) $\pi_*(U)$ is periodic with period 2, $\pi_0 U = 0$, and $\pi_1 U = Z$

(2) $\pi_* 0$ is periodic with period 8 and the actual homotopy groups are

i mod 8	0	1	2	3	4	5	6	7
$\pi_i 0$	$\mathbb{Z}_2$	$\mathbb{Z}_2$	0	$\mathbb{Z}_2$	0	0	0	$\mathbb{Z}$

(3) For all j, $\pi_j(U/SO) \cong \pi_{j-2}(SO)$

(4) For all j, $\pi_{2j}(U_j) \cong Z_{j!}$.

Proof. (1) is proved in complete detail in [21] where a proof of (2) is also indicated. Both (2) and (3) can be found in [4] and (4) occurs in [5].

C) Some theorems of Whitney

Definition. An embedding $M \subset N$ of manifolds is proper if $\partial N \cap M = \partial M$ and M is transverse to ∂N.

Theorem 1.6. Let L^ℓ and M^m be compact proper submanifolds of N^n, $\ell + m = n$, such that L and M intersect transversely and the intersection number of L and M is zero. (The intersection number is an integer if L, M and N are oriented, and changing orientations changes its sign. If L, M or N is nonorientable, the intersection number is in Z_2.) If $\ell, m > 2$ and N is simply connected, then there is an ambient isotopy h_t of N such that $h_1(L) \cap M = \emptyset$.

Proof. Whitney's intersection removal technique is in [31]. See also Milnor [20].

The same technique yields

Theorem 1.7. Let $f: M^m \to N^{2m}$ be an immersion, M closed, with self-intersection number zero. (If m is even and M and N are oriented, the self intersection number is an integer. If M is odd, of if M or N is nonorientable, it is in Z_2.) If $m > 2$ and N is simply connected, then f is regularly homotopic to an embedding.

As a corollary of this theorem (and the approximation of continuous maps by immersions, and the fact that the self intersection number of an immersion can be changed arbitrarily without changing its homotopy type) we have:

Theorem 1.8. If N^{2k} is simply connected, $k > 2$, then any $\alpha \in \pi_k(N)$ can be represented by an embedded sphere.

Theorem 1.9. Let $f: (M^m, \partial M) \to (N^{2m-1}, \partial N)$ be a continuous map such that $f|\partial M$ is an embedding. Then f is homotopic to an immersion keeping $f|\partial M$ fixed.

Proof. See [32].

Definition. Let M and N be closed manifolds. Immersions $f_i: M \to N$, $i = 0,1$, are concordant if there is an immersion $f: M \times I \to N \times I$ such that $F^{-1}(N \times \{i\}) = M \times \{i\}$ and $F|M \times \{i\} = F_i$, $i = 0,1$.

Corollary 1.10. Let M^m and N^{2m} be closed manifolds. Two embeddings $f_i: M \to N$, $i = 0,1$, are homotopic if and only if they are concordant as immersions.

Proof. If $F: M \times I \to N \times I$ is a concordance, then $\pi \circ F: M \times I \to N$ is a homotopy, where $\pi: N \times I \to N$ is projection onto N. If $h: M \times I \to N$ is a homotopy from f_0 to f_1, let $H: M \times I \to N \times I$ be given by $H(x,t) = (h(t),t)$. Applying Theorem 1.9 to H, we obtain a concordance from f_0 to f_1.

§2. Some theorems on framed surgery

Let M be an oriented smooth manifold. Suppose that surgery is performed via the embedding $f\colon S^k \times D^{n-k} \to M^n$ to obtain a manifold $M' = (M - f(S^k \times D^{n-k})) \cup D^{k+1} \times S^{n-k-1}$. (We will always assume $f(S^k \times D^{n-k}) \subset \operatorname{Int} M$.) The "trace" W of the surgery is obtained by attaching the "handle" $D^{k+1} \times D^{n-k}$ to $M \times I$ by identifying $(\partial D^{k+1}) \times D^{n-k}$ with $f(S^k \times D^{n-k}) \times \{1\}$. Thus $\partial W = M' - M$. Let $\varepsilon^N(X)$ denote the trivial N-plane bundle over X. (We will write ε^N when the base space is clear from the context.)

Definition. A framed manifold (M,F) is a smooth manifold M together with a framing F of $\tau(M) \oplus \varepsilon^N(M)$ for some $N > 0$. A framed surgery of (M,F) is a surgery of M (as above) together with a framing G of $\tau(W) \oplus \varepsilon^k(W)$ $(k \geq N-1)$, where W is the trace of the surgery, satisfying $G|M = F \oplus t^{k-N+1}$, where t^{k-N-1} is the standard framing of ε^{k-N+1}. (Here M is identified with $M \times 0 \subset W$, and $\tau(W)|M$ is identified with $\tau(M) \oplus \varepsilon^1$ by using the inward normal vector field on $M \subset \partial W$.) Restricting G to $\partial W - M = M'$ we obtain a framed manifold (M',F'), the result of the framed surgery on (M,F). ($\tau(W)|M' = \tau(M') \oplus \varepsilon^1$ via outward normal field on M'.)

Remarks.

1) There is a corresponding definition of framed cobordism. Two closed framed manifolds (M,F) and (M',F') are framed cobordant if there is a compact framed manifold (W,G) such that $\partial W = M - M'$, $G|M = F$, and $G|M' = F'$. (More precisely, this means there exist integers $i,j,k \geq 0$ such that $G \oplus t^i|M = F \oplus t^j$ and $G \oplus t^i|M' = F' \oplus t^k$. Again we identify $\tau(W)|M$ with $\tau(M) \oplus \varepsilon^1$, and $\tau(W)|M'$ with $\tau(M') \oplus \varepsilon^1$.) It is easy to check that framed cobordism is an equivalence relation. Clearly if (M',F') is obtained from (M,F) by a finite sequence of framed surgeries, then (M',F') is framed cobordant to (M,F). Conversely (M',F') is framed cobordant to (M,F) implies that (M',F') can be obtained from (M,F) by a finite sequence of framed surgeries (compare Milnor [2]).

If (M_1,F_1) and (M_2,F_2) are framed manifolds, $(M_1,F_1) \# (M_2,F_2)$ denotes their framed connected sum. (See [10].) The set of framed cobordism classes of framed closed manifolds forms an abelian group under $\#$.

2) If F is homotopic to F', then clearly (M,F) is framed cobordant to (M,F'). By an easy obstruction argument, homotopy classes of framings of $\tau(M) \oplus \varepsilon^N$ for any fixed N are in one-to-one corre-

spondence with homotopy classes of framings of $\tau(M) \oplus \varepsilon^1$. Thus our definition of framed cobordism gives the same equivalence classes as the definition in Kervaire-Milnor [15].

3) The following conditions are equivalent:

i) M is s-parallelizable ("framable")--i.e. the bundle $\tau(M) \oplus \varepsilon^N$ is trivial for some N. (M is <u>parallelizable</u> if $\tau(M)$ is trivial: "s-parallelizable" means "stably parallelizable".)

ii) $\tau(M) \oplus \varepsilon^1$ is trivial. (This is the definition Kervaire-Milnor [15] gives for s-parallelizable.)

iii) M is a π-manifold (i.e. there is an N such that M embeds in R^N with trivial normal bundle). (See [15] and [20].)

(i) $\Leftrightarrow$ (iii) can be strengthened as follows: Let $i: M^n \to R^{n+k}$ be an embedding, k large. Then

$$\tau(M) \oplus \nu(i) = \tau(R^{n+k})|M = \varepsilon^{n+k} \qquad (\nu = \text{normal bundle})$$

so

$$\varepsilon^N \oplus \tau(M) \oplus \nu(i) = \varepsilon^{N+n+k} .$$

4) A manifold with boundary is s-parallelizable and only if it is parallelizable. (See [15].)

<u>Lemma</u>. Suppose N is large. Then if F is a framing of $\varepsilon^N \oplus \tau(M)$, there exists a framing F' of $\nu(i)$ such that $F \oplus F' \simeq t^{N+n+k}$, and any two such F' are homotopic. Conversely, if F' is a framing of $\nu(i)$, there exists a framing F of $\varepsilon^N \oplus \tau(M)$ such that $F \oplus F' \simeq t^{N+n+k}$, and any two such F are homotopic.

<u>Proof</u>. We will show that if ξ^k and ξ^ℓ are vector bundles over the manifold M^n with $\ell > n+1$, such that $\xi^k \oplus \eta^\ell \cong \varepsilon^{k+\ell}$, and F is a framing of ξ^k, then there exists a framing F' of η^ℓ, unique up to homotopy, such that $F \oplus F' = t^{k+\ell}$. F defines a map $\phi: M \to V_{k+\ell,k}$. Since $V_{k+\ell,\ell}$ is $\ell-1$ connected, $n < \ell$ implies that ϕ is null homotopy (by obstruction theory). Thus by the homotopy lifting property of $V_{k+\ell} \to V_{k+\ell,\ell}$, ϕ extends to a map $M \to V_{k+\ell,k+\ell}$. Thus F' exists. Suppose F'' is another framing of η^ℓ such that $F \oplus F'' \simeq t^{k+\ell}$. Then F' and F'' differ by a map $\alpha: M \to SO_\ell$, and if $i: SO_\ell \to SO_{k+\ell}$, $i \circ \alpha \simeq 0$. But $i_*: \pi_i SO_\ell \cong \pi_i SO_{k+\ell}$ for $i < \ell-1$, so since $n < \ell-1$, $i_*[\alpha] = 0 \to [\alpha] = 0$ (by obstruction theory). Thus $F'' \simeq F$.

<u>Definition</u>. Suppose that (M_1,F_1), (M_2,F_2) are normally framed manifolds

(i.e. F_i is a framing of an embedding $f_i: M_i \subset R^N$, N large). (M_1,F_1) and (M_2,F_2) are normally framed cobordant if there is a manifold W with $\partial W = M_1 \cup M_2$ and an embedding $g: W \to R^N \times I$ such that $\text{int } W \cap \partial(R^n \times I) = \emptyset$ and $g|M_i = f_i$, with a framing G of $\nu(g)$ such that $G|M_i = F_i$.

The set of normally framed cobordism classes of closed normally framed manifolds forms a group Ω_f^n under connected sum. By the lemma and remark 2) above, Ω_f^n is canonically isomorphic to the group of (tangentially) framed cobordims classes of (tangentially) framed manifolds. Pontryagin proved that Ω_f^n is isomorphic to the n-stem $\pi_n(S)$, the correspondence being the Thom-Pontryagin construction. For a proof, see [22]. In these notes, a "framed manifold" will usually mean a manifold with a framing of its stable tangent bundle. Normal framings are used only when the Thom-Pontryagin construction is needed.

Theorem 2.1. Let M be a compact framed manifold of dimension $n \geq 4$ such that ∂M is a homology sphere. By a finite sequence of framed surgeries M can be made $[\frac{n-1}{2}]$ connected.

Proof. This is 5.5 and 6.6 of Kervaire-Milnor [15].

This theorem says that for a compact framed manifold, surgery can be done to kill all homotopy groups "below the middle dimension." Therefore, by Poincaré duality, we have:

Corollary 2.2. Suppose that M^n is compact, framed, n odd ≥ 5, and ∂M is a homotopy sphere (resp. $\partial M = \emptyset$). By a finite sequence of framed surgeries M can be made contractible (resp. a homotopy sphere). Thus $bP^n = 0$ for n odd.

Surgery can be completed in the middle dimension of an even dimensional framed manifold if the middle homology group can be represented in a special way:

Theorem 2.3. Let M^{2k}, $k \geq 3$, be a compact framed $(k-1)$-connected manifold, ∂M a homotopy sphere (resp. $\partial M = \emptyset$). Suppose there is a basis $\alpha_1,\ldots,\alpha_r$, $\beta_1,\ldots,\beta_r$ of $H_k(M)$ such that

(1) $\alpha_i \cdot \alpha_j = 0$, $\beta_i \cdot \beta_j = \delta_{ij}$ for all i,j ("$\cdot$" is intersection number. Such a basis is called (weakly) symplectic).

(2) The α_i can be represented by disjoint embedded spheres with trivial normal bundles. (Note that the α_i are spherical by the Hurewicz theorem.)

Then M can be made contractible (resp. a homotopy sphere) by a finite sequence of unframed surgeries. The surgeries can be framed unless $k = 3$ or 7.

Proof. All but the last statement is included in the proof of Theorem 4 of Milnor [20]. As shown in §6 of Kervaire-Milnor [15] (see also the proof of 4.2b below), the obstruction to framing a surgery performed via an embedding $f: S^k \times D^k \to M^{2k}$ lies in $\pi_k(SO_{2k+N}) = \pi_k(SO)$, and this obstruction can be altered by any element in the image of the map $i_*: \pi_k(SO_k) \to \pi_k(SO)$. But i_* is surjective for $k \neq 1,3,7$ (1.4), so any surgery can be framed.

When can the hypotheses of this theorem be satisfied? If k is even (i.e. $n \equiv 0(4)$), we will see that $H_k(M)$ has a symplectic basis if and only if the signature (index) of M is zero. However, (2) always holds for k even, assuming (1) (see §3). If k is odd, $k \neq 3,7$, $H_k(M)$ has a symplectic basis, and the normal bundles of embedded spheres representing this basis are trivial if and only if the Kervaire (Arf) invariant of M is zero (§4). If $k = 3$ or 7, (1) and (2) both hold, but there is an obstruction to framing the surgery. In §4 this obstruction and the Kervaire invariant are shown to be manifestations of a single invariant which can be defined for all odd k.

Corollary 2.4. $bP^6 = bP^{14} = 0$.

§3. Computation of bP^{4k}

In this section we compute bP^{4k} by defining a surjective map from Z to bP^{4k}, and determining its kernel.

Let $\Sigma \in bP^{4k}$ say $\Sigma = \partial M^{4k}$ with M parallelizable. If Σ also bounds a contractible manifold, $\Sigma = 0$ in bP^{4k}, thus $\Sigma = 0$ if we can kill the homotopy of M by framed surgery. Theorem 2.1 allows us to assume that M is $(2k-1)$-connected, which places us in the situation described by Theorem 2.3, the hypotheses of which are satisfied iff the signature of M is zero.

Definition. Let M^{4k} be a compact oriented manifold with $H_{2k}M$ free (e.g. M $(2k-1)$-connected). The signature (index) of M $\sigma(M)$ is the situature of the quadratic (i.e. symmetric bilinear) form $< , >$: $H_{2k}M \otimes H_{2k}M \to Z$ given by the intersection pairing $<\alpha,\beta> = \alpha\cdot\beta$.

Remark. $\sigma(M \# M') = \sigma(M) + \sigma(M')$ where $\#$ is connected sum.

Proof. $< , >$ is dual to cup product, i.e.

$$\begin{array}{ccc} H_{2k}M \otimes H_{2k}M & \xrightarrow{\langle\, ,\, \rangle} & \mathbb{Z} \\ \cong \downarrow PD \otimes PD & & \downarrow \cong \quad \text{commutes} \\ H^{2k}(M, M) \otimes H^{2k}(M, \partial M) & \xrightarrow{\smile} & H^{4k}(M, \partial M) \end{array}$$

and (c.f. Milnor [19]) the signature of the cup product is additive with respect to connected sums.

Theorem 3.1. Let (M^{4k}, F) be a compact framed (2k-1)-connected manifold with ∂M a homotopy sphere (resp. $\partial M = \emptyset$). Then (M,F) can be framed surgered into a contractible manifold (resp. a homotopy sphere) iff $\sigma(M) = 0$.

Corollary 3.2. The Hirzebruch index theorem (below) implies that, if M^{4k} is framed and $\partial M = \emptyset$, then $\sigma(M) = 0$ and hence M is framed null cobordant.

Proof. ($\Rightarrow$) If a closed manifold N^{4k} bounds a compact manifold, then $\sigma(N) = 0$ (c.f. [17]). By the above remark, σ is thus an invariant of oriented cobordism. Therefore, if $\partial M = \emptyset$ and M can be surgered into a homotopy sphere Σ, $\sigma(M) = \sigma(\Sigma) = 0$.

Now suppose that N^{4k} is compact and $\partial N = \Sigma = \partial D$ with D contractible. We claim that $\sigma(N) = \sigma(N \underset{\Sigma}{\cup} D)$. Let $V = N \underset{\Sigma}{\cup} D$ and let $i: N \to V$ be the inclusion. Then we have the commuting diagram

$$\begin{array}{ccc} H^{2k}V \otimes H^{2k}V & \xrightarrow{\smile} & H^{4k}(V) \cong Z \\ \cong \downarrow i^* \otimes i^* & & \cong \downarrow i^* \\ H^{2k}(N, \partial N) \otimes H^{2k}(N, \partial N) & \xrightarrow{\smile} & H^{4k}(N, \partial N) \cong Z \end{array} .$$

As $\langle\, ,\, \rangle$ is dual to $\smile$, the claim follows. If $\partial M = \Sigma$ and M can be surgered to D, let W be the union of the traces of the surgeries. Then $\partial W \cong M \cup (D \cup \Sigma \times I)$. Thus, by our claim, $\sigma(M) = \sigma(\partial W) = 0$.

($\Leftarrow$) We will verify (1) and (2) of Theorem 2.3. Since $\sigma(M) = 0$, $\exists$ a symplectic basis, $\alpha_1, \ldots, \alpha_r, \beta_1, \ldots, \beta_r$ for $H_{2k}(M)$ (c.f. [26]). By the Hurewicz theorem, each α_i is spherical and can be represented by an embedding $f_i: S^{2k} \to M^{4k}$ (Theorem 1.8). Since $\alpha_i \cdot \alpha_j = 0$, the $f_i(S^{2k})$ can be isotoped so as to be disjoint (Theorem 1.6). Let $\nu(f_1)$ be the normal bundle. $[\nu(f_1)] \in \pi_{2k-1}(SO_{2k})$, and we have the commutative diagram (c.f. §1B)

$$\begin{array}{ccccc}
\pi_{2k}(S^{2k}) & \xrightarrow{d_{2k}} & \pi_{2k-1}(SO_{2k}) & \xrightarrow{(i_{2k-1})_*} & \pi_{2k-1}(SO_{2k+1}) = \pi_{2k-1}(SO) \\
& \searrow^{\times 2} & \downarrow{\scriptstyle p_{2k-1}} & & \\
& & \pi_{2k-1}(S^{2k-1}) & &
\end{array}$$

Now $\tau(S^{2k}) \oplus \nu(f_i) = f_i^*(\tau(M))$, so since $\tau(M)$ and $\tau(S^{2k})$ are stably trivial, so is $\nu(f_i)$, i.e. $i_*[\nu(f_i)] = 0 \in \pi_{2k-1}(SO)$. Thus $[\nu(f_i)] \in \operatorname{Im} \alpha$. But $p_{2k-1}[\nu(f_i)] = X(\nu(f_i))\cdot\text{gen} = (\alpha_i\cdot\alpha_i)\text{gen} = 0$, and $\operatorname{Im} d_{2k} \cap \operatorname{Ker} p_{2k-1} = 0$, since $p_{2k-1}d_{2k}$ is multiplication by 2, so $[\nu(f_i)] = 0$, i.e. $\nu(f_i)$ is trivial.

<u>Theorem 3.3</u>. Let M^{4k} be a framed $(2k-1)$ connected manifold whose boundary is empty or a homotopy sphere. Then $\sigma(M)$ is a multiple of 8.

<u>Proof</u>. Pick $\alpha \in H_{2k}M$ and let $\alpha' \in H^{2k}(M,\partial M)$ be its Poincaré dual. The mod 2 computation $\alpha' \cup \alpha' = Sq^{2k}\alpha' = V^{2k} \cup \alpha' = 0$ (V^{2k}, the $2k^{\text{th}}$ Wu class, is zero since $\tau(M)$ is stably trivial) shows that $\alpha \cup \alpha$ (and hence its dual $\langle\ ,\ \rangle$) is always even i.e. $\langle\ ,\ \rangle$ is an even quadratic form (c.f. [20] for a more geometric proof). Since the signature of an even unimodular integral quadratic form is a multiple of 8 (c.f. [26]), we need only show the:

<u>Assertion</u>. $\langle\ ,\ \rangle$ is unimodular.

<u>Proof</u>. We have the commuting diagram

$$\begin{array}{ccccccc}
H^{2k}(M,\partial M) & \xrightarrow[\cong]{i^*} & H^{2k}(M) & \xrightarrow[\cong]{PD} & H_{2k}(M,\partial M) & \xrightarrow[\cong]{i_*} & H_{2k}(M) \\
\alpha & \rightsquigarrow & \alpha & \rightsquigarrow & \alpha \cup \mu_M = \alpha & \leftarrow & \alpha \\
& & & \searrow \quad \swarrow & & & \\
& & & \operatorname{Hom}(H_{2k}M, Z) & \langle \alpha, \cdot \rangle & &
\end{array}$$

Where we have abused notation by not distinguishing between elements in isomorphic absolute and relative groups (μ_M is the fundamental class of M). $\langle\ ,\ \rangle$ is unimodular iff the map

$$\begin{array}{ccc}
H_{2k}M & \longrightarrow & \operatorname{Hom}(H_{2k}M, Z) \\
\alpha & \rightsquigarrow & \langle \alpha, \cdot \rangle
\end{array}$$

is an isomorphism. But the above diagram factors this map into the composition of three isomorphisms.

Theorem 3.4. Let $k > 1$ and $t \in Z$. Then $\exists$ a framed $4k$ manifold (M,F) with ∂M a homotopy sphere $\sigma(M) = 8t$.

A very complete proof can be found in [7] (see also [23]). The manifolds are constructed by plumbing disc bundles over spheres.

We now describe the map mentioned in the first paragraph of §3.

Definition. $b_k: Z \to bP^{4k}$ is defined as follows. Let $b_k(t) = [\partial M^{4k}]$ where M^{4k} is a framed manifold with signature $8t$ having boundary a homotopy sphere.

In the Appendix we will see that b_k can be thought of as a "boundary" map.

Lemma 3.5. (1) b_k is well defined, i.e. if M_1 and M_2 are as above, ∂M_1 is cobordant to ∂M_2.

(2) b_k is surjective.

Proof. For (1), it suffices to show that the connected sum $\partial M \# \partial M'$, a homotopy sphere, is cobordant to zero. From the boundary connected sum $W = M \# -M'$ (c.f. [1]). $\partial W = \partial M \# \partial M'$. But $\sigma(W) = 0$ so, by Theorem 3.1, W can be (interior) surgered into a contractible manifold. (1) follows from (2) is immediate from Theorem 3.4.

Corollary 3.6. $bp^{4k} \cong Z/\ker b_k$.

We now try to determine $\ker b_k$.

Suppose $t \in \ker b_k$. Then we have a framed manifold (M,F) with signature $8t$ whose boundary, Σ, is a homotopy sphere that bounds a contractible manifold D. Attaching D to M by identifying ∂M with ∂D gives an almost framed closed manifold N of dimension $4k$ with $\sigma(N) = 8t$. (An almost framed manifold is a pair (N,G) where G frames $\tau(N)|_{N-\{x\}}$ for some $x \in N$.) Conversely, given an almost framed closed manifold N^{4k} with $\sigma(N) = 8t$, let $D \subset N$ be any embedded disc. Then $N - \text{int}\, D$ is framed and has signature $8t$ and boundary S^{4k-1}. This gives:

<u>Theorem 3.7</u>. $t \in \ker b_k$ if $\exists$ an almost framed closed $4k$-manifold with signature $8t$.

This theorem leads us to investigate the signature of almost framed closed manifolds. Our tool is of course the:

<u>Hirzebruch signature (née index) theorem</u>. For any closed manifold M^{4k}, $\sigma(M)$ is the Kroneker index $\langle L_k(P_1(M),\dots,P_k(M)),\mu_M\rangle$ where L_k is a rational function and the P_i's are the Pontrjagin classes (see [19] or [12]). The only fact that we will use about L_k is that $L_k(x_1,\dots,x_k) = s_k x_k$ + terms not involving x_k where

$$s_k = \frac{2^{2k}(2^{k-1}-1)B_k}{(2k)!}$$

(B_k is the kth Bernoulli number.)

Let (M^{4k},F) be an almost framed closed manifold. Since $p_i(M)=0$ $i<k$, $\sigma(M) = s_k p_k(M)$. We will see that the obstruction to extending the almost framing to a stable framing of M (i.e. a framing of $\tau(M) \oplus \varepsilon^N$) actually determines $\sigma(M)$ and is thereby useful in calculating $\ker b_k$ and consequently bP^{4k}.

The obstruction $O(M,F) \in \pi_{4k-1}(SO) \cong Z$ (Theorem 1.5 (2)) can be defined as follows. Let $x \in M$ be the point where F is not defined. Next choose $x \in U \cong D^{4k}$ and let F' be the usual framing of D^{4k} (which orients D^{4k} consistently with M). $O(M,F) \in \pi_{4k-1}(SO)$ is the obstruction to forcing agreement of the stable framings F and F' on $U - \{x\} \cong S^{4k-1}$.

Let $\tau\colon M \to BSO$ be the classifying map of the stable tangent bundle of M. Since $M - \{x\}$ is stably parallelizable, $\tau|M - \{x\}$ is null-homotopy and thus factors (up to homotopy) as

$$\begin{array}{ccc} M & \longrightarrow & BSO \\ & {\scriptstyle\phi}\searrow \quad \nearrow & \\ & S^{2k} & \end{array}$$

where ϕ collapses to a point the complement of an open disk containing x. Hence $\exists\ \xi$, a stable oriented vector bundle $/S^{4k} \ni \phi^*\xi$ is the stable tangent bundle of M. As usual (c.f. §1A) we view $[\xi] \in \pi_{4k-1}(SO)$. one checks that $[\xi] = \pm\, O(M,F)$.

The above factorization of τ shows that the Pontryagin classes of almost framed $4k$-manifolds can be determined by examining the kth Pontryagin class of stable vector bundles $/S^{4k}$.

Theorem 3.8. (c.f. [18]). If ξ is a stable vector bundle over S^{4k}, then $p_k(\xi) = \pm a_k(2k-1)![\xi]$ where $a_k = \begin{cases} 1 & k \text{ even} \\ 2 & k \text{ odd} \end{cases}$.

Proof. One task is to make sense of the above equation as a priori the two sides lie in different groups. We will see that each group is isomorphic to Z and hence, up to sign, they are canonically isomorphic to each other.

By definition $p_k(\xi) = \pm C_{2k}(\xi^C)$ (the $2k^{th}$ Chern class of the complexification of ξ) and, just as $[\xi] \in \pi_{4k-1}(SO_N)$, $[\xi^C]$ $\pi_{4k-1}(U_N) \in$ (N large). In fact ξ^C is $i_k(\xi)$ where $i\colon SO_N \to U_N$ is the inclusion.

Let $W_{m,\ell}$ be the space of complex orthonormal ℓ frames in C^m (cf. [28]).

$C_{2k}((\xi^N)^C) \in H^{4k}(S^{3k}, \pi_{4k-1}(W_{N,N-2k+1})) \cong \pi_{4k-1}(W_{N,N-2k+1})$, is the obstruction to extending an N-2k+1 dimensional complex framing of ξ^C from the 4k-1 skeleton to S^{4k} itself. Equivalently it is the obstruction to extending an N-2k+1 dimensional framing of ξ^C from the southern hemisphere to S^{4k}. Since ξ^C is the obstruction to extending to complete framing from the southern hemisphere to S^{4k}, we see that $C_{2k}(\xi^C) = P_*(\xi^C)$ where $p\colon U_n \to U_n/U_{2k-1}$ is the usual projection. We have the exact sequence:

$$\pi_{4k-1}(U_N) \xrightarrow{p_*} \pi_{4k-1}(W_{N,N-2k+1}) \xrightarrow{\partial} \pi_{4k-2}(U_{2k-1}) \longrightarrow \pi_{4k-2}(U_N)$$

By (1.5) ($\pi_j(W_{m,\ell})$ is calculated in [3]) the above sequence becomes

$$Z \xrightarrow{p_*} Z \longrightarrow Z_{(2k-1)!} \longrightarrow 0 .$$

Hence p_* is multiplication by $(2k-1)!$. Since we have

$$\begin{array}{ccccc} Z & & Z & & Z \\ \wr\| & & \wr\| & & \wr\| \\ \pi_{4k-1}(SO_N) & \xrightarrow{i_*} & \pi_{4k-1}(U_N) & \xrightarrow{p_*} & \pi_{4k-1}(W_{N,N-2k+1}) \\ \xi & \rightsquigarrow & \xi^C & \rightsquigarrow & \pm C_{2k}(\xi^C) = \pm p_k(\xi) \end{array}$$

it remains to show that i_* is multiplication by $\pm a$. As N is large we may work with the stable map $i_*\colon \pi_{4k}(SO) \to \pi_{4k}(U)$. But we have the exact sequence

$$\pi_{4k}(U/SO) \longrightarrow \underset{\underset{Z}{\wr\|}}{\pi_{4k-1}(SO)} \xrightarrow{i_*} \underset{\underset{Z}{\wr\|}}{\pi_{4k-1}(U)} \longrightarrow \pi_{4k-1}(U/SO)$$

and in §1B we have also shown that $\pi_{4k}(U/SO) \cong \pi_{4k-2}(SO) \cong 0$ and that $\pi_{4k-1}(U/S) = \pi_{4k-3}(SO) = \begin{cases} 0 & k \text{ even} \\ 0 & k \text{ odd} \end{cases}$. The result follows.

Let M be an almost framed closed manifold. We have:

<u>Corollary 3.9</u>. $p_k(M) = \pm\, a_k(2k-1)!\,O(M,F)$.

<u>Corollary 3.10</u>. $O(M,F)$ is independent of F.

<u>Corollary 3.11</u>. $\sigma(M) = \dfrac{\pm\, a_k 2^{2k-1}(2^{2k-1}-1)B_k O(M,F)}{k}$.

<u>Corollary 3.12</u>. M is s-parallelizable if $\sigma(M) = 0$.

In order to completely determine bP^{4k} some basic properties of the J-homomorphism are needed (c.f. [13]).

<u>Definition</u>. Given n and ℓ we define $J = J_{n,\ell}\colon \pi_m(SO_\ell) \to \pi_{m+\ell}(S^\ell)$ as follows: Let $[\alpha] \in \pi_m(SO_\ell)$. $J(\alpha)\colon S^{m+\ell} \to S^\ell$ is constructed in two stages. We view $S^{m+\ell}$ as $(S^m \times D^\ell) \cup (D^{m+1} \times S^{\ell-1})$ and first define $J(\alpha)$ on $S^m \times D^\ell$ as the composition $S^m \times D^\ell \xrightarrow{\psi} D^\ell \xrightarrow{c} S^\ell$ where $\psi(x,y) = \alpha(x)(y)$ and c collapses ∂D^ℓ to a point. The second stage, extending $J(\alpha)$, is trivial as $c \circ \psi(\partial(S^m \times D^\ell))$ is just one point. $J([\alpha])$ is defined as $[J(\alpha)]$. One then verifies the

<u>Lemma 3.13</u>. View S^m as $S^m \times \{0\} \subset S^m \times D^\ell \subset S^{m+\ell}$ with F_0 the standard normal framing $S^m \subset S^m + D^\ell$. Given $[\alpha] \in \pi_m(SO_\ell)$, let F_α be the framing obtained by "twisting F_0 via α" (i.e. at $x \in S^m$, $F_\alpha(x) = \alpha(x)(F_0(x))$). The Thom-Pontryagin construction applied to $(S^m \subset S^\ell, F_\alpha)$ gives $\pm\, J([\alpha])$.

Since

$$\begin{array}{ccc} \pi_m(SO_\ell) & \xrightarrow{J} & \pi_{m+\ell}(S^\ell) \\ \downarrow{\scriptstyle i_*} & & \downarrow{\scriptstyle \Sigma} \\ \pi_m(SO_{\ell+1}) & \xrightarrow{J} & \pi_{m+\ell+1}(S^{\ell+1}) \end{array}$$

commutes (Σ is the suspension homomorphism), we obtain the stable J

homomorphism $J: \pi_m(SO) \to \pi_m(S)$, where $\pi_m(S) = \lim_{\overrightarrow{\ell}}\{\pi_{m+\ell}(S^\ell) \overset{\Sigma}{\to} \pi_{m+\ell+1}(S^{\ell+1})\}$ is the m-stem. The relevance of the J homomorphism to our work to the following:

<u>Theorem 3.14</u>. Given $\alpha \in \pi_{m-1}(SO)$, $\exists$ an almost framed closed manifold (M^m,F) with $O(M,F) = \alpha$ iff $J(\alpha) = 0$.

<u>Proof</u>. $\Rightarrow$ We may assume that F is a framing of $M^m - \operatorname{int} D^m$ with D^m a closed disc. Now imbed M^m in $\mathbb{R}^N$ (N large) so that D^m is the northern hemisphere of the standard m-sphere in $\mathbb{R}^N$. Let F_0 be the usual (outward) normal framing of $D^m \subset \mathbb{R}^N$. Let $F_\alpha = F_0|_{S^{m-1}}$ twisted via α. Hence the Thom-Pontryagin construction applied to (S^{m-1},F_α) gives $\pm J(\alpha)$. Since $\alpha = O(M,F), F = F|_{S^{m-1}}$. Thus $(S^{m-1},F_\alpha) = \partial(M^m - \operatorname{int} D^m, F)$ so (S^{m-1},F_α) is framed null cobordant. Therefore, the Thom-Pontraygin construction yields $\sigma \in \pi_{m-1}(S)$.

$\Leftarrow$ $S^{m-1} \subset D^m$. Let F_0 be the standard framing of $D^m \subset S^N$ (N large). Since $J(\alpha) = 0$, a framed manifold (N^m,F) such that $\partial(N^m,F) = (S^{m-1},F_\alpha)$. Let $M^n \subset N \underset{S^{m-1}}{} D^m$. Then (M^m,F) is an almost framed closed manifold and $O(M^m,F) = \alpha$.

If we let j_k be the order of the image of the stable J homomorphism $Z \cong \pi_{4k-1}(SO) \overset{J}{\to} \pi_{4k-1}(S)$ we get the following:

<u>Corollary 3.15</u>. The possible values for $O(M,F)$ are precisely the multiples of j_k.

<u>Corollary 3.16</u>. The possible values for $\sigma(M)$ are precisely the multiples of $\dfrac{a_k 2^{2k-1}(2^{2k-1}-1)B_k j_k}{k}$.

In order to (finally) get exact information about bP^{4k} we need a hard

<u>Theorem of Adams 3.17</u>. [1] [33] Let $J: \pi_m(SO) \to \pi_m(S)$.

1) If $m \not\equiv 3(4)$, J is injective.
2) j_k = denominator $(B_k/4k)$.

Although our primary interest in in 2), 1) also has important consequences.

Corollary 3.18. If M is an almost framed closed manifold and $\dim M \neq 0(4)$ then the almost framing of M extends to a complete framing.

Since homotopy $4k$ spheres have signature 0, we get:

Corollary 3.19. Any homotopy sphere is s-parallelizable.

We have already seen that bP^{4k} is a finite factor group of Z. Let t_k be the order of that group, we have (using 3.7 and 3.16) that

$$8t_k = \frac{a_k 2^{2k-1}(2^{2k-1}-1)B_k j_k}{k}.$$

Thus $t_k = a_k 2^{2k-2}(2^{2k-1}-1)(B_k/4k)j_k$ and, applying 3.17, this gives the final

Corollary 3.20. $bP^{4k} = Z_{t_k}$ where $t_k = a_k 2^{2k-2}(2^{2k-1}-1)$ numerator $\cdot$ $(B_k/4k)$.

§4. Computation of bP^n for $n = 2 \bmod 4$.

We proceed as in §3, computing bP^n by studying the kernel of a surjective map $Z_2 \to bP^n$.

Suppose that $\Sigma \in bP^n$, i.e. $\Sigma = \partial M^{2k}$, where k is odd, and M is a parallelizable manifold. By Theorem 2.1, M can be made $(k-1)$-connected by a finite sequence of framed surgeries. We wish to discuss the "obstruction" to a compact, framed $(k-1)$-connected manifold (M^{2k},F), k odd, satisfying the hypotheses of Theorem 2.3.

First notice that the intersection pairing $H_kM \otimes H_kM \to Z$ is skew-symmetric (since k is odd) and unimodular (by the proof of Theorem 3.3). Therefore [26] there is a symplectic basis for $H_k(M)$, i.e. there is a basis $\alpha_1,\ldots,\alpha_r,\beta_1,\ldots,\beta_r$ for $H_k(M)$ with intersection matrix

$$\begin{pmatrix} 0 & 1 \\ -I & 0 \end{pmatrix}$$

As in §3, each α_i is spherical by the Hurewicz theorem, and so if $k > 2$, the α_i are represented by disjoint embedded spheres (by 1.7 and 1.8). Furthermore any two embeddings $f\colon S^k \to M^{2k}$ representing an element $\alpha \in H_k(M^{2k})$ are concordant as immersions by 1.10. Now

$$f^*\tau(M) \cong \tau(S^k) \oplus \nu(f)$$

so

$$\varepsilon^N(s^k) \oplus f^*\tau(M) \cong \varepsilon^N(S^k) \oplus \tau(S^k) \oplus \nu(f)$$

$\varepsilon^N(S^k) \oplus f^*\tau(M) = f^*(\varepsilon^N \oplus \tau(M))$, so the framing F of $\phi^N \oplus \tau(M)$ pulls back to give a framing f^*F of $\varepsilon^N \oplus f^*\tau(M)$. If F_0 is the usual framing of $\varepsilon^{N-1} \oplus \tau(D^{k+1})$, $F_0|S^k$ gives a framing of $\varepsilon^{N-1} \oplus \tau(D^{k+1})|S^k = \varepsilon^N \quad \tau(S^k)$:

$$(*) \qquad \underbrace{\varepsilon^N \oplus f^*\tau(M)}_{f^*F} \cong \underbrace{\varepsilon^N \oplus \tau(S^k)}_{f_0|S^k} \oplus \nu(f)$$

Thus since f^*F gives a trivialization of $\varepsilon^N \oplus \tau(S^k) \oplus \nu(f)$, the framing $F_0|S^k$ assigns to each point in S^k an element of $V_{2k+N,k+N}$. Thus we get an element

$$\Phi(f) \in \pi_k(V_{2k+n,k+n}) \cong Z_2 \quad (k \text{ odd})$$

depending on M, F, and f. We will show that $\Phi(f)$ does not in fact depend on the choice of the embedding f representing α. Suppose f_0,f_1: $S^k \to M$ are embeddings representing α. Let H: $S^k \times I \to M \times I$ be an immersion concordance between them (1.10). Then we have the following bundles and framings over the space $S^k \times I$

$$(**) \qquad \underbrace{\varepsilon^{N-1} \oplus H^*\tau(M \times I)}_{H^*G} \cong \underbrace{\varepsilon^{N-1} \oplus \tau(S^k \times I)}_{G_0|S^k \times I} \oplus \nu(H) \cdot\cdot$$

where G is the framing of $\varepsilon^{N-1} \oplus \tau(M \times I)$ corresponding to F under the identification $\tau(M \times I) = \varepsilon^1 \oplus \tau(M)$, and G_0 corresponds to F_0 under $\tau(S^k \times I) = \varepsilon^1 + \tau(S^k)$. Thus $G_0|S^k \times I$ determines a map f: $S^k \times I \to V_{2k+N,k+N}$, which is a homotopy from $\Phi(f_0)$ to $\Phi(f_1)$ since $(**)$ restricted to $S^k \times \{i\}$ yields $(*)$, $i = 0,1$. Therefore we obtain a well-defined element $\Phi(\alpha) \in \pi_k(V_{2k+N,k+N}) \cong Z_2$.

<u>Remark 1</u>. In fact it is true that if the embeddings F_0,F_1: $S^k \to M^{2k}$ are homotopic, then they are regularly homotopic. This is an easy corollary of Smale-Hirsch immersion theory [27] [11]. (In fact for $M = R^{2k}$, Φ is identical with Smale's obstruction to homotoping an immersion of S^k to the standard embedding.) Thus if f_0,f_1 are

embeddings representing $\alpha \in H_k(M^{2k})$, $\nu(f_0) \cong \nu(f_1)$, and so (*) defines $\Phi(\alpha)$ indepdnently of the choice of embedding.

Remark 2. There is an alternate way to define Φ, using Smale-Hirsch theory. Given $\alpha \in H_k(M^{2k})$, M s-parallelizable, there is a certain regular homotopy class of immersions $f: S^k \times D^k \to M$ such that $F \circ i$ represents α, where $i: S^k \to S^k \times D^k$ by $i(x) = (x,0)$ (see [24]). $\Phi(\alpha)$ is defined to be the self-intersection number of the immersion $f \circ i$. For a presentation of this definition, see [24] and [30].

Theorem 4.2. (a) For $k \neq 3,7$, $\Phi(\alpha) = 0$ if and only if ν is trivial. (b) For $k = 3$ or 7 (i.e. $\dim M = 6$ or 14), $\nu(f)$ is trivial, and $\Phi(\alpha) = 0$ if and only if the surgery on M via $f: S^k \times D^k \to M^{2k}$ can be framed.

Proof of (a). Consider the long exact homotopy sequence of the bundle $SO_k \to SO_{2k+N} \to V_{2k+N,k+n}$:

$$\cdots \longrightarrow \pi_k SO_k \xrightarrow{i_*} \pi_k(SO_{2k+N}) \xrightarrow{p_*} \pi_k(V_{2k+N,k+N})$$
$$\xrightarrow{\partial_*} \pi_{k-1}(SO_k) \xrightarrow{i_*^!} \pi_{k-1}(SO_{2k+N}) \longrightarrow \cdots .$$

It is clear from the definitions that $\partial_*\Phi(\alpha) = [\nu(f)] \in \pi_{k-1}(SO_k)$. For $k \neq 3,7$, i_* is surjective (1.4), so p_* is 0, so ∂_* is injective. Thus $\Phi(\alpha) = 0 \Leftrightarrow [\nu(f)] = \partial_*\Phi(\alpha) = 0$.

Remark. Thus for $k \neq 3,7$, $\Phi(\alpha)$ can be defined directly as the obstruction to trivializing $\nu(f)$, i.e. $\Phi(\alpha) = [\nu(f)] \in \operatorname{Ker} i_*^! \cong Z_2$, and the two definitions correspond via the isomorphism $\partial_*: \pi_k(V_{2k+N,k+N}) \xrightarrow{\cong} \operatorname{Ker} i_*^!$.

Proof of (b). $\nu(f)$ is trivial because $\ker i_*^! = 0$ for $k = 3$ or 7 (1.4). As stated in the proof of Theorem 2.3, the obstruction to framing the surgery lies in $\operatorname{Coker} i_*$. For $k = 3$ or 7, $\operatorname{Im} i_*$ is a subgroup of $\pi_k(SO_{2k+N}) = \pi_k(SO)$ of index 2 (1.4), i.e. $\operatorname{Coker} i_* \cong Z_2$. Furthermore, since $\operatorname{Ker} i_*^! = 0$, p_* is surjective, i.e. $p_*: \operatorname{Coker} i_* \cong \pi_k(V_{2k+N,k+N})$. To see that $p_*(\mathcal{O}) = \Phi(\alpha)$, recall the definition of $\mathcal{O}$: A trivialization of $\nu(f)$ gives an embedding $S^k \times D^k \subset M$, and we would like to frame the trace $W = M \times I \bigcup_{S^k \times D^k} D^{k+1} \times D^k$ of the surgery of M via this embedding:

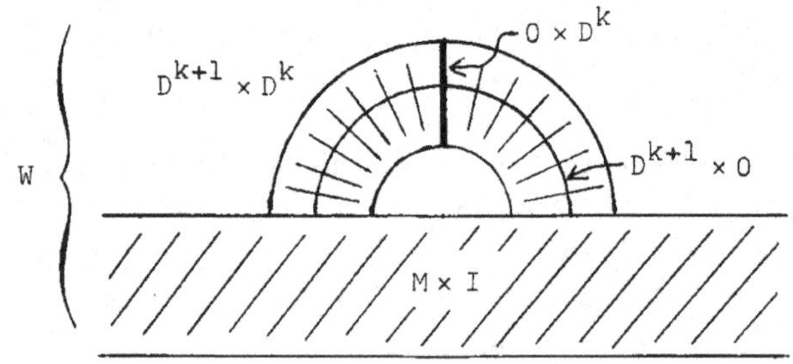

We have a framing of the stable tangent bundle of $M \times I$ which restricts to the given framing F of $\tau(M) \oplus \varepsilon^N \cong \tau(M \times I) \mid M \times \{1\}$. We also have a canonical framing $F_0 \times F_0'$ of $\tau(D^{k+1} \times D^k)$. Thus comparing $F_0 \times F_0'$ with F on $S^k \times 0$, we get a map $g: S^k \to SO_{2k+N}$. Changing the framing of $\nu(f)$ by an element of $\pi_k(SO_k)$ changes the homotopy class of g by an element in the image of $i_*: \pi_k(SO_k) \to \pi_k(SO_{2k+N})$. This defines $\mathcal{O} \in \text{Coker } i_*$. Now $p_*(\mathcal{O})$ is the homotopy class of $S^k \xrightarrow{g} V_{2k+N,2k+N} \xrightarrow{p} V_{2k+N,k+N}$. p forgets the last k vectors, so $p \circ g$ compares F_0 with F on $S^k \times 0$ and thus $[p \circ g] = \Phi(\alpha) \in \pi_k(V_{2k+N,k+N}) = Z_2$. This completes the proof of 4.2.

Let $\Phi_2: H_k(M;Z_2) \to Z_2$ be the map

$$H_k(M;Z_2 \xrightarrow{\cong} H_k(M;Z) \otimes Z_2 \xrightarrow{\Phi \otimes \text{id}} Z_2 .$$

We will show that Φ_2 is a "honsingular quadratic function."

Definition. Let V be a finite dimensional vector space over Z_2, $<\ ,\ >$ a symmetric bilinear form on V. A quadratic function with associated pairing $<\ ,\ >$ is a function $\psi: V \to Z_2$ such that

$$\psi(\alpha+\beta) = \psi(\alpha) + \psi(\beta) + <\alpha,\beta> .$$

ψ is called nonsingular if $<\ ,\ >$ is nonsingular. Let $\alpha_1,\ldots,\alpha_r$, $\beta_1,\ldots,\beta_r$ be a symplectic basis for $(V,<\ ,\ >)$. Define the Arf invariant of $(\psi,<\ ,\ >)$ by

$$A(\psi,<\ ,\ >) = \sum_i \psi(\alpha_i)\psi(\beta_i) .$$

Remark. It's not hard to show that A is independent of the choice of symplectic basis.

Proposition 4.3. A and rank V are complete invariants of the isomorphism class of $(V, \langle\ ,\ \rangle, \psi)$. (Isomorphism class means the obvious thing.)

Proof. See [2].

Proposition 4.4. Let M, Φ be as above. Then for $\alpha, \beta \in H_k(M)$ represented by embedding spheres,

$$\Phi(\alpha+\beta) = \Phi(\alpha) + \Phi(\beta) + (\alpha\cdot\beta)_2,$$

where $(\alpha\cdot\beta)_2$ is the intersection number of α and β reduced mod 2.

Proof. Let $f,g\colon S^k \to M$ be embeddings representing α,β respectively. Joining $f(S^k)$ and $g(S^k)$ by a tube gives an immersion $f \# g$ representing $\alpha+\beta$. Observe that the definition of Φ makes sense for an immersion (it is an invariant of regular homotopy) and it is not hard to see that

$$(*) \qquad \Phi(f \# g) \doteq \Phi(f) + \Phi(g)$$

The self-insersection number of the immersion $f \# g$ is just $(\alpha\cdot\beta)_2$. Thus if $(\alpha\cdot\beta)_2 = 0$, $f \# g$ is an embedding (after isotopy) representing $\alpha+\beta$, so the proposition is true by (*). If $(\alpha\cdot\beta)_2 = 1$, let $h\colon S^k \to M$ be a small null-homotopic immersion with self-intersection number $I(h) = 1$. Then by 1.7 $f \# g \# h$ is regularly homotopic to an embedding representing $\alpha+\beta$, so

$$\Phi(\alpha+\beta) = \Phi(f \# g \# h) = \Phi(f) + \Phi(g) + \Phi(h) = \Phi(\alpha) + \Phi(\beta) + \Phi(h)\ .$$

Thus we must show that $\Phi(h) = 1 = (\alpha\cdot\beta)_2$.

For a given manifold M, h is obtained by composing a fixed immersion $h_0\colon S^k \to R^{2k}$ having self-intersection number 1, with a coordinate embedding $R^{2k} \to M$. (For a definition of h_0, see [6].) Since $\Phi(h) = \Phi(h_0)$, it is enough to check that $\Phi(h) = 1$ for some particular choice of M. Let $M = S^k \times S^k$. $H_k(S^k \times S^k) \cong Z \oplus Z$, with generators α,β represented by the embeddings $a,b\colon S^k \to S^k \times S^k$ given by $a(x) = (x,x_0)$, $b(x) = (x_0,x)$. Clearly $(a,b)_2 = 1$. Let $d\colon S^k \to S^k \times S^k$ be the diagonal map $d(x) = (x,x)$. d is an embedding representing $\alpha+\beta$. Therefore $\Phi(d) = \Phi(a) + \Phi(b) + \Phi(h)$ for any framing F of $\varepsilon^N \oplus \tau(S^k \times S^k)$. Let F be the framing of $\varepsilon' \oplus \tau(S^k \times S^k)$ which is the restriction of the standard framing of

R^{2k+1} to the standard embedding $S^k \times S^k \subset R^{2k+1}$. Then it is clear that $\Phi(a) = \Phi(b) = 0$, so $\Phi(h) = \Phi(d)$. (Or one can produce a framing F such that $\Phi(a) = \Phi(b) = 0$ by the proof of Proposition 4.11 below.) For $k \neq 3,7$, $\Phi(d) = [\nu(d)] = [\tau(S^k)] = 1$. It remains to show that $\Phi(d) = 1$ for $k = 3$ or 7.

It should be possible to give a direct proof that $\Phi(d) = 1$. In lieu of such, here is an alternative proof of Proposition 4.4 for $k = 3$ or 7. It is sufficient to show that if $h: S^k \to R^{2k}$ is an immersion with self-intersection number 1, then $\phi(h) = 1$. But it is easily seen that for any immersion $f: S^\ell \to R^{2\ell}$, $\phi(f)$ is precisely Smale's obstruction to regularly homotoping f to the standard embedding of S^ℓ in $S^{2\ell}$ [27]. (It follows that $\phi(f)$ equals the self-intersection number of f --this is immediate when ℓ is odd, because $\phi(f)$ and the self-intersection number are in Z_2. (See [27].) Therefore $\phi(h) = 1$.

<u>Corollary 4.5</u>. $\Phi_2: H_k(M;Z_2) \to Z_2$ is a nonsingular quadratic function with associated pairing $\langle\alpha,\beta\rangle = (\alpha\cdot\beta)_2$.

<u>Definition</u>. Let (M^{2k},F), i odd be a compact framed $(k-1)$-connected manifold such that $H_k(M;Z)$ is free abelian. The <u>Kervaire</u> (Arf) <u>invariant</u> $c(M,F)$ is defined as

$$A(\Phi_2, (\ ,\ )_2) \in Z_2.$$

<u>Remark</u>. By a previous remark, for $k \neq 3,7$, $c(M,F)$ does not depend on F, so for $k \neq 3,7$ we let $c(M) = c(M,F)$.

<u>Theorem 4.6</u>. Let (M^{2k},F), k odd, be a compact framed $(k-1)$-connected manifold with ∂M a homotopy sphere (resp. empty). (M,F) can be made contractible (resp. a homotopy sphere) by a finite sequence of framed surgeries if and only if $c(M,F) = 0$.

<u>Proof</u>. ($\Leftarrow$) Let $\alpha_1,\ldots,\alpha_r,\beta_1,\ldots,\beta_r$ be a symplectic basis for $H_k(M;Z)$. Suppose $c(M,F) = 0$, i.e. $\sum_i \Phi(\alpha_i)\Phi(\beta_i) = 0 \in Z_2$.

<u>Claim</u>. We can find a new symplectic basis $\alpha_1',\ldots,\alpha_r',\beta_1',\ldots,\beta_r'$ for $H_k(M;Z)$ such that $\Phi(\alpha_i') = 0$ for all i. Assuming this, Theorem 4.2 (a) implies that the α_i' are represented by embedded spheres with trivial normal bundles. By Theorem 2.3, the homotopy groups of M can be killed by surgery. By 4.2 (b) the surgery can be framed even when $k = 3$ or 7.

<u>Proof of Claim</u>. If $\Phi(\alpha_i)\Phi(\beta_i) = 0$, take

$$\alpha_i' = \begin{cases} \alpha_i & \text{if } \Phi(\alpha_i) = 0 \\ \beta_i & \text{if } \Phi(\alpha_i) \neq 0 \quad (\text{and hence } \Phi(\beta_i) = 0) \end{cases}$$

$$\beta_i' = \begin{cases} \beta_i & \text{if } \Phi(\alpha_i) = 0 \\ \alpha_i & \text{if } \Phi(\alpha_i) \neq 0 \end{cases}$$

Since $\sum \Phi(\alpha_i)\Phi(\beta_i) = 0$, $\Phi(\alpha_i)\Phi(\beta_i) \neq 0$ for an even number of values of i. Suppose $\Phi(\alpha_i)\Phi(\beta_i) \neq 0$ and $\Phi(\alpha_2)\Phi(\beta_2) \neq 0$. Let

$$\alpha_1' = \alpha_1 + \alpha_2 \qquad \beta_1' = \beta_1$$

$$\alpha_2' = \beta_2 - \beta_1 \qquad \beta_2' = \alpha_1 .$$

It is easy to check that replacing $\alpha_1, \alpha_2, \beta_1, \beta_2$ by $\alpha_1', \alpha_2', \beta_1', \beta_2'$ gives a new symplectic basis with $\Phi(\alpha_1') = \Phi(\alpha_2') = 0$. Thus for each pair of values of i such that $\Phi(\alpha_i)\Phi(\beta_i) = 0$, we can replace the four basic elements involved with new ones such that $\Phi(\alpha_i) = 0$.

($\Rightarrow$). By an argument completely analogous to the one given in the proof of 3.1, it suffices to show that if (M,F) is as in the theorem and there is a framed manifold (V,G) with $\partial V = M$ and $G|\partial V = F$, then $c(M,F) = 0$. Let $i_*: H_k(M) \to H_k(V)$ be induced by inclusion.

<u>Assertion (1)</u>. $i_*(\alpha) = 0 \Rightarrow \Phi(\alpha) = 0$. Represent α by an embedding $f: S^k \to M$. Since V is framed, we can perform surgery to make V $(k-1)$-connected (without touching $M = V$) by Theorem 3.1. Now $i_*(\alpha) = 0 \Rightarrow i \circ f$ is a null-homologous singular sphere in V, and therefore $i \circ f$ is null-homotopic, since $H_k V \cong \pi_k V$ by the Hurewicz theorem. Therefore $i \circ F$ extends to a continuous map $g: D^{k+1} \to V$. By 1.9 g is homotopic rel S^k to an immersion. Consider the following commutative diagram of bundle isomorphisms and framings (where $i \circ f = f$ for simplicity):

$$\begin{array}{ccc}
\overbrace{\varepsilon^{N-1} \oplus h^*\tau(V)|S^k}^{h^*G} & \xrightarrow{\cong} & \overbrace{\varepsilon^{N-1} \oplus \tau(D^{k+1})|S^k}^{f_0|S^k} \oplus \nu(h)|S^k \\
\uparrow\cong & & \uparrow\cong \\
\underbrace{\varepsilon^{N} \oplus f^*\tau(V)}_{f^*F} & \xrightarrow{\cong} & \underbrace{\varepsilon^{N} \oplus \tau(S^k)}_{f_0|S^k} \quad \oplus \nu(f)
\end{array}$$

This diagram shows that the map $S^k \to V_{2k+N,k+N}$ representing $\Phi(\alpha)$ lifts to a map $D^{k+1} \to V_{2k+N,k+N}$, i.e. $\Phi(\alpha) = 0$, which proves assertion (1).

Assertion (2). There is a symplectic basis $\alpha_1,\ldots,\alpha_r,\beta_1,\ldots,\beta_r$ for $H_k(M;Z)$ such that $\alpha_i \in \operatorname{Ker} i_*$ for all i. Together with (1), this implies that $c(M,F) = \sum_i \Phi(\alpha_i)\Phi(\beta_i) = 0$, as desired.

For (2), consider the commutative diagram (Z coefficients)

$$\begin{array}{ccccccccc} \cdots \longrightarrow & H_{k+1}(V,M) & \xrightarrow{\partial} & H_k(M) & \xrightarrow{i_*} & H_k(V) & \xrightarrow{j_*} & H_k(V,M) & \longrightarrow \cdots \\ & \uparrow \cap \mu_V & & \uparrow \cap \mu_m & & \uparrow \cap \mu_V & & \uparrow \cap \mu_m & \\ \cdots \longrightarrow & H^i(V) & \xrightarrow{i^*} & H^k(M) & \xrightarrow{\delta_*} & H^{k+1}(V,M) & \xrightarrow{j_*} & H^{k+1}(V) & \longrightarrow \cdots \end{array}$$

Now $(u \cap \mu_M)\cdot(v \cap \mu_M) = (u \cup v) \cap \mu_M$ (intersection is dual to cup product), so it will suffice to find a symplectic basis $u_1,\ldots,u_r$, $v_1,\ldots,v_r$ for $H^k(M;Z)$ such that $u_i \in \operatorname{Ker} \delta_k$, $i = 1,\ldots,r$.

Lemma 4.7. $\operatorname{Ker} \delta_k$ is its own annihilator with respect to the cup product pairing, i.e. $u \cup v = 0$ for all $v \in \operatorname{Ker} \delta_k \Leftrightarrow u \in \operatorname{Ker} S_k$.

Proof. $\delta_{2k}(i^*\alpha \cup \beta) = \alpha \cup \delta_k\beta$ for every $\alpha \in H^k(V)$, $\beta \in H^k(M)$ [29]. Now $\delta_{2k}\colon H^{2k}(M) \overset{\cong}{\to} H^{2k+1}(V,M)$ (both groups are Z, and δ_{2k} is surjective by the diagram), so $u \in \operatorname{Ker} \delta_k = \operatorname{Im} i^* \Rightarrow u \cup v = 0$ for all $v \in \operatorname{Ker} \delta_k$. Conversely, if $u \in H^k(M)$ and $u \cup v = 0$ for all $v \in \operatorname{Ker} \delta_k = \operatorname{Im} i^*$, then $\alpha \cup \delta_k u = \delta_{2k}(i^*\alpha \cup u) = \delta_{2k}(0) = 0$ for all $\alpha \in H^k(M)$. Since the cup product pairing is nonsingular, $\delta_k u = 0$.

Remark. The proof of this lemma used only that V and $M^{2k} = \partial V$ are oriented manifolds and $H^k(M)$ is free.

Corollary 4.8. $\operatorname{Ker} \delta_k$ is a direct summand of half rank of $H^k(M)$.

Proof. Consider the following diagram with exact rows ($A^* = \operatorname{Hom}(A,Z)$):

$$\begin{array}{ccccccccc} 0 \longrightarrow & \operatorname{Ker} \delta_k & \xrightarrow{f} & H^k(M) & \longrightarrow & R & \longrightarrow & 0 & \quad (R = \operatorname{Coker} f) \\ & \wr\| & & \wr\| & & \wr\| & & & \\ 0 \longrightarrow & R^* & \longrightarrow & (H^k(M))^* & \longrightarrow & (\operatorname{Ker} \delta)^* & \longrightarrow & 0 & \end{array}$$

The middle arrow is an isomorphism since $\cup$ is a nonsingular pairing (by Poincaré duality). The left dotted arrow exists because $\operatorname{Ker} \delta_k$ annihilates itself under $\cup$. It is an isomorphism because $\operatorname{Ker} \delta_k$ equals its annihilator. A diagram chase then proves that the second dotted arrow is well-defined and injective. Therefore R is free, and so $\operatorname{Ker} \delta_k \cong R^* \cong R$. (Thus the right dotted arrow is an isomorphism and both sequences split.)

To complete the proof of assertion (2), let $u_1,\ldots,r_r$ be any basis of $\operatorname{Ker} \delta_k$, and let $v_1,\ldots,v_r$ be a "dual basis" of $R \cong (\operatorname{Ker} \delta_k)^*$, i.e. $u_i \cup v_j = \delta_{ij}$. $\operatorname{Ker} \delta_k$ annihilates itself under $\cup$, so $u_i \cup u_j = 0$. However, $v_i \cup v_i$ may be nonzero. Let $v_i' = v_i - (v_i \cup v_i)u_i$. Then it's easy to check that $u_1,\ldots,u_r,v_1',\ldots,r_r'$ is a symplectic basis for H^kM.

This completes the proof of Theorem 4.6.

Now we apply Theorem 4.6 to the computation of bP^n for $n = 2k$, k odd, $\neq 3$ or 7. (Recall that $bP^6 = bP^{14} = 0$ (Corollary 2.4).) We wish to define a map

$$b_k \colon Z_2 \to bP^{2k}$$

by $b_k(t) = \partial M$, where M is any compact framed $(k-1)$-connected $2k$-manifold with ∂M a homotopy sphere and $c(M) = t$. To show that b_k is well-defined, we must prove:

<u>Theorem 4.9</u>. (a) $c(M_1) = c(M_2) \Rightarrow \partial M_1$ is h-corbordant to ∂M_2.
(b) For each odd $k \neq 3$ or 7 and each $t \in Z_2$ there is a framed manifold M^{2k} such that ∂M is a homotopy sphere and $c(M) = 5$. Thus b_k is surjective.

<u>Proof</u>. (a) Let F_1 and F_2 be framings for M_1 and M_2, and let $(N,G) = (M_1,F_1) \# (M_2,F_2)$ be the framed boundary connected sum (cf. §6). $\partial N = \partial M_1 \# -\partial M_2$, and $c(N) = c(M_1) + c(M_2) = 0$. Therefore, by Theorem 4.6, N can be made contractible by framed surgeries. Thus $\partial M_1 \# - \partial M_2$ bounds a contractible manifold, i.e. ∂M_1 is h-cobordant to ∂M_2.

(b) If $t = 0$, take $M^{2k} = D^{2k}$. If $t = 1$, M^{2k} is constructed by plumbing two copies of the tangent disc bundle of S^k (see [7] or [23]).

This theorem shows that $bP^{2k} = 0$ if $b_k = 0$ and $bP^{2k} = Z_2$ if

$b_k \neq 0$, so we would like to determine when $b_k = 0$. This happens if and only if there exists a compact framed $(k-1)$-connected $2k$-manifold M with boundary the standard sphere S^{2k-1} such that $c(M) = 1$. Attaching a disc to the boundary of such an M we obtain an almost framed closed $2k$-manifold N with $c(N) = 1$. By Corollary 3.18, N is framed. Thus $b_k = 0 \Leftrightarrow$ there is a closed framed $(k-1)$-connected $2k$-manifold N with $c(N) = 1$. By the proof of Theorem 4.6, c is well-defined on framed cobordism classes. The framed cobordism group Ω_f^n is isomorphic to the n-stem $\pi_n(S) = \pi_{n+k}(S^k)$ for k large (§2). Thus for each odd k, the Kervaire invariant gives a map

$$c_k\colon \pi_{2k}(S) \to Z_2$$

and for $k \neq 3,7$, $b_k = 0 \Leftrightarrow c_k \neq 0$. According to Browder [6], $c_k = 0$ if $k \neq 2^{\ell} - 1$. (Kervaire [14] originally showed $c_6 = 0$ and $c_9 = 0$; then Brown and Peterson [8] showed $c_{4\ell+1} = 0$.) Mahowald and Tangora have shown that $c_{15} \neq 0$, and Barratt, Mahowald and Jones have shown $c_{31} \neq 0$ (see [6], [33], [34]). Therefore we have

<u>Theorem 4.10</u>. For k odd $bP^{2k} = \begin{cases} Z_2 & k \neq 2^{\ell} - 1 \\ 0 & k = 3,7,15, \text{ or } 31 \end{cases}$.

The following proposition, which extends our discussion of the existence of closed framed $2k$-manifolds with nonzero Kervaire invariant to the case $k = 3$ or 7, will be needed in §5:

<u>Proposition 4.11</u>. For $k = 3$ or 7 there is a framing F of $S^k \times S^k$ such that $c(S^k \times S^k, F) = 1$.

<u>Proof</u>. $H_k(S^k \times S^k) \cong Z \oplus Z$, with generators α, β represented by the embedded spheres $S^k \times *$, $* \times S^k$ respectively. Let G be any framing of $\tau(S^k \times S^k) \oplus \varepsilon^1$, $c(S^k \times S^k, G) = \Phi(\alpha)\Phi(\beta)$. <u>Claim (1)</u>: G can be altered so as to realize any values of $\Phi(\alpha)$ and $\Phi(\beta)$. This implies the proposition. Let $f\colon S^k \to SO_{2k+1}$. <u>Claim 2</u>: chainging G on $S^k \times *$ by f alters $\Phi(\alpha)$ by the map $S^k \xrightarrow{f} SO_{2k+1} \to V_{2k+1,k+1}$. Assuming this, we prove (1) as follows: For $k = 3$ or 7, $\pi_k(SO_N) \to \pi_k(V_{N,N-k})$ is surjective (1.4). Now $S^k \times S^k = (S^k \vee S^k) \cup D^{2k}$. Thus we can change G on $S^k \vee S^k$ to obtain a framing F on $S^k \vee S^k$ such that $\Phi(\alpha) = \Phi(\beta) = 0$. The obstruction to extending F over the $2k$-cell is an element of $\pi_{2k-1}(SO_N) = 0$ for $k = 3$ or 7 (1.5).

<u>Proof of (2)</u>. This follows from the definition of Φ. Suppose $g\colon S^k \to V_{2k+1,k+1}$ represents $\Phi(\alpha)$. It is clear that changing G by

f changes g to the map $\tilde{g}(x) = f(x)\,g(x)$, where SO_{2k+1} acts on $V_{2k+1,k+1}$ by rotation. But we can assume that $g(x)$ is the standard $(k+1)$-frame for x in the northern hemisphere, and that $f(x)$ is the identity element of SO_{2k+1} for x in the southern hemisphere, so $[\tilde{g}] = [g] + [h]$, where $h(x)$ is the standard $(k+1)$-frame in R^{2k+1} rotated by $f(x)$, i.e. $h(x) = \pi f(x)$, $\pi\colon SO_{2k+1} \to V_{2k+1,k+1}$. Thus $[\tilde{g}] = [g] = \pi_*[f]$, which proves (2).

Remark. We can define $c(M,F)$ for any compact framed $2k$-manifold, k odd, with ∂M empty or a homotopy sphere, as follows. Convert (M,F) to a $(k-1)$-connected framed manifold (N,G) by a finite sequence of framed surgeries, and let $c(M,F) = c(N,G)$. The proof of Theorem 4.6 shows that $c(M,F)$ is well-defined, and that it is an invariant of framed surgery. For $k = 3$ or 7, c is not an invariant of unframed surgery by Proposition 4.11, since $S^k \times S^k$ is null-corbordant. Theorem 4.10 implies that c is not an invariant of unframed surgery for some other values of k. Since $c_{15} \neq 0$, there is a closed framed 14-connected 30-manifold N with $c(N) = 1$. However, since N is framed it has zero Stiefel-Whitney and Pontryagin numbers, so N is unframed (oriented) null-corbordant. (In contrast, recall that the signature of a $4k$-manifold is an invariant of unframed surgery.)

Recall that if $k \neq 3, 7$ and M^{2k} is $(k-1)$-connected, then $c(M,F)$ does not depend on F. However, it is not known whether $c(M,F)$ depends on F for arbitrary M.

§5. Computation of θ^n/bP^{n+1}.

The results of this section are all in Kervaire-Milnor [15]. Suppose that the homotopy sphere Σ^n is embedded in R^{n+k} (k large) with a framing F of its normal bundle (recall that homotopy spheres are π-manifolds by Corollary 3.19). Then the Thom construction applied to (Σ,F) yields an element $T(\Sigma,F)$ of $\pi_{n+k}(S^k)$, which is an invariant of the normal cobordism class of (Σ,F). $T((\Sigma,F)\#(\Sigma',F')) = T(\Sigma,F) + T(\Sigma',F)$.

Lemma 5.1. Let $f\colon \Sigma^n \to SO_k$, and let $\alpha = [f] \in \pi_n SO_k$. Then if F is altered to F' via α,

$$T(\Sigma,F') = T(\Sigma,F) \pm J(\alpha)\ .$$

Proof. Recall (Lemma 3.13) that $T(S^n,F_\alpha) = \pm J(\alpha)$, where F_α is the standard framing F_0 of S^n altered by α. Thus

$$(\Sigma,F') = (\Sigma,F') \# (S^n,F_0) = (\Sigma,F) \# (S^n,F_\alpha) ,$$

and the lemma follows by applying T to both sides.

Corollary 5.2. $T(\Sigma) = \{T(\Sigma,F),\ F$ a framing of $\Sigma^n \subset R^{n+k}\}$ is a coset of $J(\pi_n SO_k)$ in $\pi_{n+k}S^k$.

Therefore we can define $T: \theta^n \to \text{Coker } J_n$, where $J_n: \pi_n SO \to \pi_n S$ is the J-homomorphism.

Proposition 5.3. $bP^{n+1} = \text{Ker } T$.

Proof. $\Sigma \in bP^{n+1} \Leftrightarrow \Sigma$ bounds a parallelizable manifold. $T\Sigma = 0 \Leftrightarrow$ there exists a normal framing, F, of Σ such that (Σ,F) bounds a normally framed manifold.

Thus we have an exact sequence

$$0 \longrightarrow bP^{n+1} \longrightarrow \theta^n \xrightarrow{T} \text{Coker } J_n .$$

Corollary 5.4. θ^n is a finite group $(n \geq 4)$.

Now $\theta^n/bP^{n+1} \cong \text{Im } T$. Suppose $\tilde{\alpha} \in \text{Coker } J_n$. $\tilde{\alpha} \in \text{Im } T$ if and only if $\tilde{\alpha}$ is represented by $\alpha \in \pi_{n+k}S^k$ (k large) such that $\alpha = T(\Sigma,F)$ for some (Σ,F). By the inverse Thom construction, and $\alpha \in \pi_{n+k}S^k$ equals $T(M,F')$ for some framed manifold (M,F'). $\alpha = T(\Sigma,F)$ is and only if (M,F') is framed cobordant to a homotopy sphere (Σ,F). Define

$$P^n = \begin{cases} 0 & n \text{ odd} \\ Z & n \equiv 0(4) \\ Z_2 & n \equiv 2(4) \end{cases}$$

(so that $bP^{n+1} = \text{Im}(b)$, $b: P^{n+1} \to \theta^n$ as in §3 and §4, and define $\phi: \Omega_n^f \to P^n$ by

$$\phi(M^n,F) = \begin{cases} 0 & n \text{ odd} \\ \sigma(M) & n \equiv 0(4) \\ c(M,F) & n \equiv 2(4) \end{cases}$$

ϕ is well-defined, since σ and c are invariants of framed cobordism, and $\phi(M,F) = 0 \Leftrightarrow (M,F)$ is framed cobordant to a homotopy sphere (Corollary 2.2 and Theorems 3.1 and 4.6). Let $\phi' = \phi T^{-1}$:

$$\begin{array}{ccc} \pi_n S & & \\ \cong\Big\uparrow T & \searrow^{\phi'} & \\ \Omega_n^f & \xrightarrow{\ \phi\ } & P^n \end{array}$$

Clearly $\phi'(\mathrm{Im}\ J_n) = 0$, so ϕ' induces a map

$$\phi'' \colon \mathrm{Coker}\ J_n \to P^n .$$

By the above analysis of $\mathrm{Coker}\ J_n$ we have:

Theorem 5.4. The sequence

$$P^{n+1} \xrightarrow{\ b\ } \theta^n \xrightarrow{\ T\ } \mathrm{Coker}\ J_n \xrightarrow{\ \phi''\ } P^n$$

is exact.

The new information here is that $\theta^n/bP^{n+1} \cong \mathrm{Ker}\ \phi''$. If n is odd, of course $\phi'' = 0$, since $P^n = 0$. If $n = 0(4)$, we have seen that $\phi'' = 0$ (by Corollary 3.12). If $n \equiv 2(4)$, $\phi'' = 0$ for $n \neq 2^i - 2$, and $\phi \neq 0$ for $n = 6, 14, 30$, or 62 (by the discussion preceding Theorem 4.10).

In summary, we have computed bP^{n+1} (except for $n+1 = 2^i - 2$, $i > 6$), and we have $\theta^n \cong \mathrm{Coker}\ J_n$ except when $n = 2^i - 2$. Then we have computed θ^n up to group extension.

Remark. Brumfield and Frank have then proved that for $n \neq 2^k - 1$ or $2^k - 2$

$$0 \longrightarrow bP^{n+1} \longrightarrow \theta^n \longrightarrow T\theta^n \longrightarrow 0$$

splits. (See e.g. [9].)

Appendix. The Kervaire-Milnor long exact sequence.

The results of these notes can be elegantly expressed by means of a long exact sequence

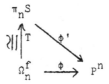

(i) $$\cdots \longrightarrow A^{n+1} \xrightarrow{p} P^{n+1} \xrightarrow{b} \theta^n \xrightarrow{i} A^n \xrightarrow{p} P^n \longrightarrow \cdots$$

θ^n is the group of homotopy n-spheres [15]. A^n is the group of "almost framed" cobordism classes of almost framed (i.e. framed except at a point) closed n-manifolds. (If M_1 is framed except at x_1 and M_2 is framed except at x_2, an almost framed cobordism between M_1 and M_2 is a corbordism W between M_1 and M_2 and a framing of $W - \alpha$, α an arc in int W from x_1 to x_2 which restricts to the given framings on the ends of the corbordism.) P^n is the group of framed cobordism classes of parallelizable n-manifolds with boundary a homotopy sphere. (A framed cobordism between M_1 and M_2 is a framed manifolds W with boundary $M_1 \underset{\partial M}{\cup} N \underset{\partial M_2}{\cup} M_2$, where N is an h-corbordism between ∂M_1 and ∂M_2, and the framing of W restricts to the given framings of M_1 and M_2.)

b is induced by the boundary map, and it is well defined by the definition of P^n. i is induced by "inclusion", i.e. any homotopy sphere is s-parallelizable, and so is almost framed. i is clearly well-defined. p is induced by "punching out a disc" containing the non-framed point to obtain a parallelizable manifold with boundary S^n. p is clearly well defined.

The discussions preceding Theorems 3.5 and 4.10 show that Ker(b) = Im(p). It is clear from the definition of A^n that Ker(i) = Im(b). It is also easy to see that Ker(p) = Im(i).

Corollary 2.2 implies that $P^n = 0$ for n odd. Theorems 3.1, 3.2, and 3.3 imply that $P^n \cong Z$ for $n \equiv 0(4)$. Theorems 4.6 and 4.9 imply that $P^n \cong Z_2$ for $n \equiv 2(4)$.

Now A^n lies in the exact sequence

(ii) $$\cdots \longrightarrow \pi_n(SO) \xrightarrow{J} \pi_n(S) \xrightarrow{t} A^n \xrightarrow{O} \pi_{n-1}(S) \xrightarrow{J} \pi_{n-1}(S) \longrightarrow \cdots$$

where J is the stable J-homomorphism, t is the inverse Thom construction which takes $\pi_n(S) \cong \Omega_f^n$ (the framed cobordism group), followed by the inclusion of Ω_f^n in A^n. Theorem 3.14 says that Ker(J) = Im(O). Ker(O) = Im(t) is clear. Ker(t) = Im(J) is easy to show (cf. Lemma 3.13).

Corollary 3.16 determines $\mathrm{Im}[A^n \xrightarrow{p} P^n \cong Z]$, $n \equiv 0(4)$. (This map assigns to an almost framed closed manifold its signature divided by 8.) Theorem 4.10 determines $\mathrm{Im}[A^n \xrightarrow{p} P^n \cong Z_2]$ for almost all $n \equiv 2(4)$. (This map assigns to a manifold its Kervaire invariant.)

The results of §5 can be interpreted as follows. By the exact sequence (ii), Coker (J) $\cong$ Im(t) $\subset A^n$. The discussion following 5.3

shows that $\mathrm{Im}(i) \subset \mathrm{Im}(t)$, so we have the exact sequence

$$P^{n+1} \xrightarrow{b} \theta^n \xrightarrow{T} \mathrm{Coker}\,(J) \xrightarrow{\phi} P^n$$

where $T(\Sigma) = i(\Sigma)$ and ϕ: Coker $(J) \subset A^n \xrightarrow{p} P^n$.

Remark. Let θ_f^n be the group of framed h-corbordism classes of framed homotopy n-spheres. Then we have the exact sequences

$$\text{(iii)} \qquad \cdots \longrightarrow P^{n+1} \longrightarrow \Omega_f^n \longrightarrow P^n \longrightarrow \theta_f^{n-1} \longrightarrow \cdots$$

($\Omega_f^n \to P^n$ is "punching out a disc"; $P^n \to \theta_f^{n-1}$ is "taking the boundary"), and

$$\text{(iv)} \qquad \cdots \longrightarrow \pi_n SO \longrightarrow \theta_f^n \longrightarrow \theta^n \xrightarrow{0} \pi_{n-1} SO \longrightarrow \cdots$$

($\pi_n SO \to \theta_f^n$ sends α to S^n with its standard framing changed by α; $\theta_f^n \to \theta^n$ forgets the framing).

Combining the long exact sequences (i), (ii), (iii), (iv) (replacing Ω_f^n by $\pi_n S$ in (iii)), we obtain the Kervaire-Milnor "braid":

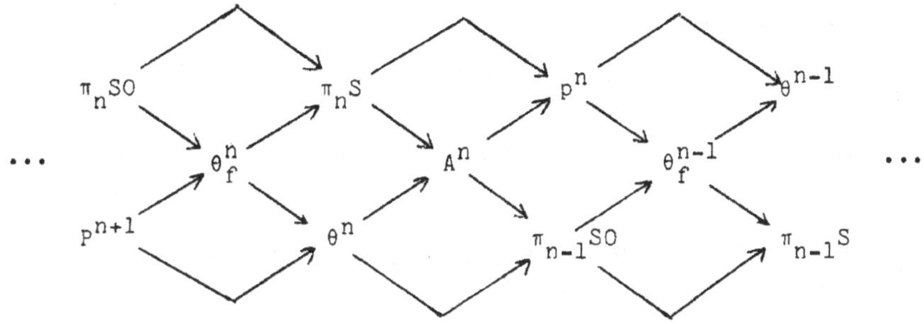

This braid is isomorphic to a braid of the homotopy groups of G, PL, and O (see e.g. [24]). Levine [16] has a nonstable version of the Kervaire-Milnor braid.

References

1. Adams, J.F., On the groups J(X) IV, Topology 5 (1966), 21-71.

2. Arf, Untersuchung über quadratische Formen in Körpern der Characteristik 2, Crelles Math. Journal, 183 (1941), 148-167.

3. Bott, R. and Milnor, J. W., On the parallelizability of the spheres, Bull. Amer. Math. Soc. 64 (1958), 87-89.

4. Bott, R., The stable homotopy of the classical groups, Annals of Math. 70 (1959), 313-337.

5. Bott, R., A report on the unitary group, <u>Proceedings of Symposia in Pure Math</u>. Volume 3, Differential Geometry (1961).

6. Browder, W. The Kervaire invariant of framed manifolds and its generalizations, <u>Annals of Math</u>. 90 (1969), 157-186.

7. Browder, W. Surgery on simply connected manifolds, <u>Ergebnisse</u>, Springer-Verlag, N.Y.

8. Brown, E.M. and Peterson, F.P., The Kervaire invariant of (8k+2)-manifolds, <u>Bull. Amer. Math. Soc</u>. 71 (1965), 190-193.

9. Brumfiel, G., Homotopy equivalences of almost smooth manifolds, Wisconsin Summer Institute on Algebraic Topology, 1970.

10. Haefliger, A., Knotted (4k-1)-spheres in 6k-space, <u>Ann. of Math</u>. 75 (1962), 452-466.

11. Hirsch, M.W., Immersions of manifolds, <u>Trans. Amer. Math. Soc.</u>, 93 (1959), 242-276.

12. Hirzebruch, F., <u>Topological Methods in Algebraic Geometry</u>, Springer Verlag (1966).

13. Kervaire, M.A., An interpretation of G. Whitehead's generalization of Hopf's invariant, <u>Annals of Math</u>. 69 (1959), 345-365.

14. Kervaire, M.A., A manifold which does not admit any differentiable structure, <u>Comment. Math. Helv</u>. 34 (1960), 257-270.

15. Kervaire, M.A. and Milnor, J.W., Groups of homotopy spheres, <u>Annals of Math</u>. 77 (1963), 504-537.

16. Levine, J., A classification of differentiable knots, <u>Annals of Math</u>. 82 (1965) 15-50.

17. Milnor, J.W., On simply connected 4-manifolds, <u>Symposium internacional de topologia algebraica</u>, Mexico (1958).

18. Milnor, J.W. and Kervaire, M.A., Bernoulli numbers, homotopy groups, and a theorem of Rochlin, <u>Proc. of the International Congress of Mathematicians</u>, Edinburgh (1958).

19. Milnor, J.W. and Stasheff, J., Characteristic classes, <u>Annals of Math. Studies</u> No. 76, Princeton U. Press, 1975.

20. Milnor, J.W., A procedure for killing homotopy groups of differentiable manifolds, <u>Proceedings of Symposia in Pure Mathematics</u>, Vol. 3, Differential Geometry (1961).

21. Milnor, J.W., <u>Morse Theory</u>, Princeton (1963).

22. Milnor, J.W., <u>Topology from the Differentiable Viewpoint</u>, The University Press of Virginia, (1965).

23. Milnor, J.W., Differential Topology, <u>Lectures in Modern Mathematics</u>, Vol. II, Saaty ed., Wiley (196).

24. Orlik, P., <u>Seminar notes on simply connected surgery</u>, Institute for Advanced Study, 1968.

25. Rourke, C.P. and Sanderson, B.A., Blockbundles III, Annals of Math. 87 (1968), 431-483.

26. Serre, J.P., Formes bilinéaires symétriques entières à discriminant $\pm$ 1, Séminaire Cartan, 14 (1961-62).

27. Smale, S., The classification of immersions of spheres in euclidean spaces, Annals of Math. 69 (1959), 327-344.

28. Steenrod, N.A., The Topology of Fibre Bundles, Princeton University Press (1951).

29. Spanier, E.M., Algebraic Topology, McGraw-Hill (1966).

30. Wall, C.T.C., Surgery of non-simply-connected manifolds, Annals of Math. 84 (1966), 217-276.

31. Whitney, H., The self-intersections of a smooth n-manifold in 2n-space, Annals of Math. 45 (1944), 220-246.

32. Whitney, H., Singularities of a smooth n-manifold in (2n-1)-space, Annals of Math. 45 (1944), 247-293.

33. Barratt, M.G., Jones, J.D.S., and Mahowald, M.E., The Kervaire Invariant in dimension 62 (to appear).

34. Mahowald, M.E., and Tangora, M., Some differentials in the Adams spectral sequence, Topology 6 (1967), 349-370.

Some Remarks on Local Formulae for p_1

by Norman Levitt

§0. Introduction

A well-known paper [GGL1] of Gabrielov, Gelfand and Losik, which was further explicated by MacPherson [M] and Stone [S_1,S_2], shows how a rational cocycle representing the first rational Pontrjagin class p_1 of a manifold may be computed directly from a combinatorial triangulation of the manifold, provided that certain other data, viz, "configuration" and "hyper-simplicial" data, are given as well. There has been, in addition, a further attempt by Gabrielov [G] to extend these ideas to the higher Pontrjagin classes. But this is only partially successful in that there is a conceptual obstruction to carrying out the suggested procedure which resides in the fact that the topology of certain configuration spaces, in particular the rational homology thereof, is not at all well understood.

The point of the present paper is that much of the apparatus of the original Gabrielov-Gelfand-Losik work is needlessly complicated and obscures what is, at base, a relatively straight-forward geometric concept. The essentials of the methodology can, in fact, be transcribed into a framework that has been in the literature for forty years: the Cairns proof [Ca 1] [Ca 2] of the smoothability of PL 4-manifolds. For that matter it is not too much to say that a prescription for determining local formulae for p_1 is already implicit in Cairns' foundational work on smoothing theory. Recall that this work [which contained some minor lacunae subsequently repaired by J.H.C. Whitehead [W] considered combinatorially triangulated manifolds simplex-wise convex-linearly embedded in general position in a high-dimensional Euclidean space. The main question posed by Cairns was whether a transverse hyperplane field can be found; for the existence of such a field was shown to imply smoothability. (In more modern language, Cairns was essentially demonstrating that a vector-bundle reduction of the PL normal bundle of a PL manifold yields a smoothing.) For our purposes, the most salient fact is Cairns' discovery that there is no obstruction to obtaining a normal field over the 4-skeleton of the dual cell-structure. This fact reduces to a theorem about the path-connectedness of certain configuration spaces which was proved in [Ca 2].

We shall show in what follows that the rather explicit construction of transverse plane fields readily allows the local calculation of a real 4-cocycle representing the real Pontrjagin class p_1. In fact, by slightly amplifying the combinatorial data, we may in fact obtain an integral cocycle representing the integral p_1.

It should be emphasized that the formulae we obtain are quite similar in spirit to those of [GGL1].

Perhaps one should raise, at this point, the question of explicitness. The methods developed below for evaluating a cocycle representing p_1 are formulated in such a way as to involve appeal to an explicit transverse field over the dual 4-skeleton. While it is shown that the value of the cocycle on a typical 4-cell depends only on the restriction of the field on the boundary 3-sphere (where it can easily be constructed explicitly), the actual computation would seem to depend on having a specific field on the 4-cell itself. Cairns' construction is, essentially, an "existence proof" rather than an explicit algorithm. However, we show in the concluding section how this difficulty can be avoided in principle so that the computation can go forward quite constructively in the presence of the appropriate data. Moreover, our derivation is far more transparent than that of [GGL1] and, in particular, avoids the "hypersimplicial" formalism of that paper. The superfluous complexity of the Gabrielov-Gelfand-Losik approach is, as we have intimated, an artifact of its failure to exploit directly the work of Cairns on transverse fields. Finally, this outline, read in conjunction with [G], makes it clear why the attempt to extend this approach to higher Pontrjagin classes runs into difficulty.

§1. Transverse fields

Let M^n be a topological manifold embedded in R^{n+k}. Recall [Ca 1, Wh] that a linear k-plane P in R^{n+k} is said to be transverse to M^n at $x \in M^n$ provided that there exists an open neighborhood U of x in M^n such that for any two distinct points $u_1, u_2 \in U$, $u_1 - u_2 \not\in P$. Here, subtraction denotes ordinary vector-subtraction in R^{n+k}. We let $G_{k,n}$ denote the usual Grassmannian of k-planes in R^{n+k}. A continuous assignment $F: M^n \to G_{k,n}$ such that $F(x)$ is transverse to M^n at x is called a <u>transverse field</u>. Clearly, if we view F as the classifying map of a k-dimensional vector bundle ν over M_n, ν is then a vector-bundle reduction of the stable TOP normal bundle of M^n. Cairns and Whitehead showed that the existence of such a transverse field implies that M^n is smoothable and, in fact, that it allows an isotopy of the embedding, pointwise arbitrarily small, to an embedding whose image is a smooth submanifold of R^{n+k}.

We next consider a closed n-form $\omega \in \Omega(G_{k,n})$ representing a stable real characteristic class for vector bundles (so, implicitly, $n \equiv 0(4)$). Consider, now, a combinatorially-triangulated oriented n-sphere Σ^n together with a distinguished equatorial (n-1)-sphere Σ^{n-1}, assumed to be a subcomplex. We also assume a PL embedding of Σ^n in R^{n+k} and a transverse k-plane field $\phi:\Sigma^n \to G_{k,n}$ (which is piecewise-smooth on the n-simplices of Σ^n). We orient Σ^n and distinguish one of the hemispheres into which Σ^{n-1} divides it as D^n_+, so that $\partial D^n_+ = \Sigma^{n-1}$.

We now define a real number $\eta = \eta(\Sigma^n, \Sigma^{n-1}, \omega, \phi)$ by:

$$\eta = \int_{D^n_+} \phi^*\omega \ .$$

<u>Lemma 1</u>. For a given embedding of Σ^n in R^{n+k}, η depends only on $\phi|D^n_-$, where D^n_- is the hemisphere of Σ^n opposite to D^n_+.

<u>Proof</u>. Since ϕ classifies a stably-trivial vector bundle (the normal bundle of a sphere in Euclidean space), it follows that $\int_{\Sigma^n} \phi^*\omega = 0$. Consequently, if ϕ and ϕ_1 are transverse fields which agree on D^n_-, we have:

$$\int_{D^n_+} \phi^*\omega = -\int_{D^n_-} \phi^*\omega = -\int_{D^n_-} \phi_1^*\omega = \int_{D^n_+} \phi_1^*\omega .$$

<u>Lemma 2</u>. For a given embedding of Σ^n, η depends only on $\phi|\Sigma^{n-1}$.

Proof. By the same reasoning as in Lemma 1 above, η depends only on $\phi|D^n_+$. Let ϕ,ψ now be transverse fields which agree on Σ^{n-1}. Let ϕ_+,ϕ_- denote restriction to D^n_+, D^n_- respectively, and similarly for ψ_+,ψ_-. Let $\eta(\alpha)$ abbreviate $\eta(\Sigma^n, \Sigma^{n-1}, \alpha,\omega)$ for any transverse field α. Then, by the preceding observations:

$$\eta(\phi) = \eta(\phi_+ \cup \psi_-) = \eta(\psi_+ \cup \psi_-) = \eta(\psi).$$

In like manner, we have the following:

Lemma 3. η depends only on the restriction to a neighborhood of Σ^{n-1} of the embedding and the transverse field ϕ.

Proof. Let $f_1\colon \Sigma^n \subset R^{n+k}$, ϕ_1 and $f_2\colon \Sigma^n \subset R^{n+k}$, ϕ_2 both be pairs consisting of an embedding and a transverse field. Assume first that f_1,ϕ_1 coincides with f_2,ϕ_2 on a neighborhood of D^n_-. By the reasoning of Lemma 1, the respective η's agree. Then, extending the reasoning of Lemma 2, we see that the η's still agree on the weaker assumption that (f_1,ϕ_1), (f_2,ϕ_2) merely agree on a neighborhood of Σ^{n-1} in S^n. Note that for this last part of the calculation, we may have to refer to a field transverse to an immersed, rather than embedded Σ^n. But this makes no difference, practically speaking.

Corollary. η depends only on the restriction of the embedding to a neighborhood of Σ^{n-1} and on $\phi|\Sigma^{n-1}$.

Proof. Given an embedding and two fields ϕ_1,ϕ_2 which agree on Σ^{n-1} it is easy to deform both ϕ_1,ϕ_2 to transverse fields ϕ_1', ϕ_2' such that on a collar neighborhood $\Sigma^{n-1} \times I$, $\phi_i'(x,t) = \phi_i(x)$, $x \in \Sigma^{n-1}$, and so that $\eta(\phi_i') = \eta(\phi_i)$ (for $i = 1,2$). Therefore $\eta(\phi_1) = \eta(\phi_2)$ since, by the lemma $\eta(\phi_1') = \eta(\phi_2')$.

From the discussion above, it follows that an invariant η may be defined by data consisting of:

an embedding i of $\Sigma^{n-1} \times I$;

a transverse field ϕ to the embedded $\Sigma^{n-1} \times I$;

a closed form ω on $G_{k,n}$

(Provided, of course, that it is understood that the embedding i and the transverse field ϕ extend, in some fashion, to an embedding and a transverse field on Σ^n. We also assume, obviously, that $\Sigma^{n-1} \times I$ is oriented.)

Thus, we shall revise our notation and speak of $\eta(i,\phi,\omega)$.

We may then note the following, understanding that in so doing, we identify Σ^{n-1} with $\Sigma^{n-1} \times \{\frac{1}{2}\} \subset \Sigma^{n-1} \times I$.

Let M^n be a compact oriented n-manifold with boundary $\partial M^n = \Sigma^{n-1}$; let $f\colon M^n \to R^{n+k}$ be an embedding (or even an immersion) of M^n which agrees with i on $\Sigma^{n-1} \times \{0,\frac{1}{2}\}$ (thought of as a collar of Σ^{n-1} in M^n. Let ψ be a transverse field to the embedded M^n which agrees with ϕ on $\Sigma^{n-1} \times \{0,\frac{1}{2}\}$. Let γ denote the real characteristic class represented by ω. Finally, let M_+^n denote the closed manifold $M^n \; c\Sigma^{n-1}$.

<u>Lemma 4</u>. The characteristic number $\gamma[M_+^n]$ is given by the formula $\gamma[M_+^n] = \int_{M_+^n} \psi^* \omega - \eta(i,\phi,\omega)$.

This is almost self-evident. The reader should note the formal analogy to the η-invariant of oriented Riemannian

(4j-1)-manifolds defined by Atiyah, Patodi, and Singer [APS]. The idea is that Riemannian data has now been replaced by geometric data consisting of the embedding i and the transverse field.

Finally, we develop a slight extension of these ideas. Rather than an $(n-1)$-sphere Σ^{n-1} and an embedding $\Sigma^{n-1} \times I \subset R^{n+k}$, consider a $(j-1)$-sphere Σ^{j-1} and an embedding $i:\Sigma^{j-1} \times I \times D^{n-j} \subset R^{n+k}$. ϕ will now be a transverse field to this embedding, and ω becomes a j-form on $G_{k,n}$. It is understood that i and ϕ extend, in some fashion, to $\Sigma^{j} \times D^{n-j}$.

Then $\eta(i,\phi,\omega)$ is defined as $\int_{D^j_+} \psi^*\omega$, where D^j_+ is identified with $D^j_+ \times \{0\}$ in $\Sigma^j \times D^{n-j}$. The statements analogous to Lemmas 1,2, and 3 are then easily proved.

§2. Real cocycles representing p_1

We now consider explicitly the case of 4-dimensional characteristic classes, which reduces, in essence, to a discussion of p_1.

Let M^n be a PL triangulated manifold embedded in R^{n+k} so that each simplex is convex-linearly embedded. Suppose a k-plane field ϕ is defined on $M_*^{(4)}$ = 4-skeleton of the cell complex Poincaré dual to the given triangulation. We assume that for every $x \in M_*^{(4)}$, $\phi(x)$ is transverse to M^n at x. Let ω be a closed differential 4-form on the Grassmannian $G_{k,n}$ representing some real characteristic class γ (e.g. p_1 or p_1 of the complementary bundle). Define an oriented real 4-cochain c on M_* by stipulating for any oriented 4-cell e of M_*

$$c(e) = \int_e \phi^*\omega$$

Lemma 5. c is a cocycle representing $\gamma(M) \in H^4(M^n;R)$.

Proof. Let ν denote the PL normal (block) bundle of M^n. Quite clearly, the vector bundle over $M_*^{(4)}$ defined by ϕ constitutes a vector-bundle reduction of $\nu|M_*^{(4)}$ so that regarded as a cochain, and hence a cocycle, of $M_*^{(4)}$, c certainly represents $\gamma(\nu|M_*^{(4)})$. Since $H^4(M^n;R) \to H^4(M_*^{(4)};R)$ is monic it follows that we need merely show that c is a cocycle of M_* itself, for then its cohomology class must coincide with $\gamma(M)$. To see this, we need merely consider an arbitrary oriented 5-cell d of M_*. Since $\phi|\partial d$ is easily seen to represent the stable normal bundle of $\partial d \cong S^4$, it follows that $\int_{\partial d}\phi^*\omega = 0$. But $\int_{\partial d}\phi^*\omega = \sum_{e^4 \subset \partial d} \int_{e^4}\phi^*\omega = \sum_{e^4 \subset \partial d} c(e^4)$. Hence, $\delta c = 0$ and we are done.

We now recall the work of Cairns [Ca 1] and Whitehead [W] to remind the reader of how a transverse k-plane field ϕ on $M_*^{(4)}$ can always be constructed under the very weak assumptions

(1) The embedding of M^n in R^{n+k} is in general position, i.e., the images of the vertices of any star form a linearly independent set of $(n+k)$-vectors.

(2) Every star of an $(n-4)$-simplex is a Brouwer star, i.e., $st(\sigma^{n-4})$ embeds in R^n with the embedding convex linear on each simplex.

(It is well known that any combinatorially triangulated manifold admits a subdivision wherein all stars are Brouwer stars.)

If we consider an abstract Brouwer star of the form $\Delta^p * \Sigma^{n-p-1}$, where Σ^{n-p-1} denotes a triangulated $(n-p-1)$-sphere, we note that the property of being a Brouwer star is equivalent to the fact that the complex $c\,\Sigma^{n-p-1}$ embeds in

R^{n-p} so that the embedding is convex-linear on each simplex. We define the <u>configuration space</u> $CF(\Sigma^{n-p-1})$ to be the space (with the obvious topology) of all such embeddings (normalized so that the cone-point $* \to 0$) modulo the action of $GL(n-p,R)$. The role of configuration spaces in analyzing transverse fields is revealed by the following:

<u>Lemma 6 (see [W])</u>. Let the Brouwer star $\Delta^{p}*\Sigma^{n-p-1}$ be embedded in R^{n+k} in general position convex-linearly on simplices. Let N $G_{k,n}$ denote the set of k-planes P such that P is transverse to $\Delta^{p}*\Sigma^{n-p-1}$ at b = barycenter of Δ^{p}. Then N is homeomorphic to $CF(\Sigma^{n-p-1}) \times R^{i}$. where $i = nk - q(n-p)$ and q = number of vertices of Σ^{n-p-1}.

(By convention, if $p = n$, i.e., $\Sigma^{n-p-1} = \emptyset$ then $CF(\Sigma^{n-p-1}) = *$.)

Let us review Cairns' proof of the smoothability of 4-manifolds, assuming the crucial result that the existence of a transverse field implies smoothability. We need only analyze, in a rather straightforward way, the obstructions to obtaining a field transverse to a simplex-wise convex-linear, general position embedding of M^4_* in R^{4+k}.

First, to each simplex σ assign, in arbitrary fashion, a k-plane P_σ transverse to M^4 at the barycenter b_σ. Next, we try to extend this to a transverse field defined everywhere. An obvious fact is that if $x \in M^4$ and $\sigma(x)$ denotes the unique simplex such that $x \in \text{int}\,\sigma$, and if we let $N(x)$ denote the set of k-planes transverse to M^4 at x, then

$N(x) \cong N(b_{\sigma(x)}) \sim CF(\ell k\sigma)$.

Now, since $\ell k\sigma$ is of dimension ≤ 3, there are only a few cases we need analyze.

Lemma 7. If $\dim \ell k\sigma < 2$, i.e $\dim \sigma = 2,3$ or 4, then $CF(\ell k\sigma)$ is contractible; if $\dim(\ell k\sigma) = 2$, i.e., $\dim \sigma = 1$, then $CF(\ell k\sigma)$ is path connected.

The first part of the lemma is a triviality. The second part, however, is far from trivial; it represents the substance of a separate paper of Cairns [Ca 2].

Remark: As of this writing, it remains an open question whether $CF(\Sigma^2)$ is, in fact, contractible for a triangulated 2-sphere Σ^2. We may reformulate this question slightly by characterizing $CF(\Sigma^2)$ as the space of geodesic triangulations of the standard S^2 realizing the simplicial complex Σ^2 such that each simplex is contained in an open hemisphere and such that one particular 2-simplex, σ^2, is realized in a fixed way. Block, Conolly and Henderson [BCH] have proved the following: Let K^2 be a subdivision of the standard 2-simplex such that K triangulates the boundary S^1 in the standard way. Let C denote the set of simplex-wise convex linear homeomorphisms $K^2 \to \Delta^2$ which are the identity on the boundary. Then C is contractible. This can be read as strong evidence for a positive answer to the stated open question.

Returning to Cairns' proof of smoothability for 4-manifolds, we exploit Lemma 7 in the following way. Consider the first barycentric subdivision K of the given triangulation of M^4.

We wish to construct a transverse field ϕ on M such that $\phi(b_\sigma) = P_\sigma$ for simplices σ of the original triangulation. The assignment $b_\sigma \longmapsto P_\sigma$ defines ϕ on the 0-skeleton of K. Now consider a 1-simplex τ of K; extending ϕ to

τ is tantamount to finding a path between two points in $N(b_\sigma) \sim CF(\ell k\sigma)$, where σ is the smallest simplex of the original triangulation such that $\tau \subset \sigma$. Lemma 7 guarantees that we can do this, the most difficult case occurring when $\dim \sigma = 1$. Proceeding to the s-skeleton we must, for any 2-simplex τ of K, find a way of extending ϕ, now defined on $\dot{\tau}$ to all of τ. Again, with σ the smallest original simplex containing τ, this is a question of contracting a loop in a space homotopy equivalent to $CF(\ell k\sigma)$; but since $\dim \sigma \geq 2$, Lemma 7 tells us that $CF(\ell k\sigma)$ is contractible. We continue in like manner to define ϕ on the 3-skeleton and then the 4-skeleton of K, which is to say, all of M. Hence as asserted, a transverse field does exist.

Transposing this argument to the more general context of triangulated n-manifolds (for arbitrary n), we see that exactly the same procedure works to construct a transverse field over the union of all simplices τ of the first subdivision such that the smallest original simplex σ containing τ satisfies $\dim \sigma \geq n-4$. In other words, the method works to construct a transverse k-plane field ϕ over the 4-skeleton $M_*^{(4)}$ of the cell-complex M_* Poincaré dual to the original triangulation.

Given now a closed 4-form $\omega \in \Omega(G_{k,n})$ whose deRham class is the real characteristic class γ, it follows from Lemma 5 that the real 4-cochain C defined on oriented 4-cells e of M_* by:

$$c(e) = \int_e \phi^* \omega$$

is a cocycle representing $\gamma(M)$.

We may now, without loss of generality, make the following

assumptions about the plane-field ϕ. Let e be a 4-cell of M_* with boundary ∂e. Then, typically, e is Poincaré dual to an $(n-4)$-simplex σ and is identified with a particular subspace of $b_\sigma * \ell k(\sigma) \subset st(\sigma)$. Thus, e naturally has the structure of a cone, viz., $c\partial e$ where b_σ corresponds to the cone point. Let C_e^+ denote a collar neighborhood of ∂e in $c\partial e = e$. We can assume that, for $x \in C_e^+$, $\phi(x) = \phi(px)$, where p denotes projection of C_e^+ upon ∂e.

Now, let C_e^- denote a collar neighborhood of ∂e in $(b_\sigma * \ell k\sigma)$ - int e. It is then easily seen that ϕ may be extended to a field ϕ_1 defined on $e \cup C_e^-$, again by letting $\phi(x) = \phi(px)$ for $x \in C_e^-$. Now it is quite obvious that the embedding of the 4-ball $e \cup C_e^-$ into R^{n+k} extends to an embedding of a sphere $\Sigma^4 = e \; C_e^- \cup c\partial_1$ (where ∂_1 denotes the copy of ∂e bounding $e \cup C_e^-$, at least if k is large enough. If we consider further the product neighborhood of $e \cup C_e^-$ in M^n, it is clear that this embedding of a 4-ball $\times D^{n-4}$ extends to an embedding $i: \Sigma^4 \times D^{n-4} \subset R^{n+k}$. As for ϕ_1, this extends to a k-plane field $\overline{\phi}$ on Σ^4 transverse to the embedded $\Sigma^4 \times D^{n-4}$. Thus, we are in the situation alluded to at the end of §1, and, reverting to the notation of that section, we have $c(e) = n(i, \overline{\phi}, \omega)$. It follows that $c(e)$ depends only the embedding of M in a neighborhood of ∂e^4 and on $\phi|\partial e$.

Our main objective, be it recalled, is to characterize a local formula for γ. Thus, taking ω, for the moment, as a given, we need to look a bit more closely at the actual construction of ϕ on ∂e. Taking e, as usual, to be the cell dual to a simplex σ, we see that $\phi|\partial e$ has been specified in the following way according to the procedure devised by Cairns:

(1) For each simplex ρ with $\rho > \sigma$ we have picked a point in $CF(\ell k\rho)$ (which yields a particular k-plane P transverse to M^n at b_ρ lying within the $(n+k-\dim \rho)$-plane comprised of all vectors perpendicular to τ.)

(2) For $\dim \rho = n, n-1, n-2$ P can, in fact, be chosen canonically. That is, for ρ of dimension n, the obvious choice is to make P the k-plane perpendicular to ρ; for $\dim \rho = n-1$ we take P to be the k-plane determined by the "obvious" configuration of the cone on the 0-sphere in R^1, viz., the two points of S^0 at $-1,1$ respectively with the cone point at 0. Finally, for $\dim \rho = n-2$, the choice of configuration of $c\ell k\rho$ is almost equally obvious; we embed $c\ell k\rho$ in R^2 as a regular polygon inscribed in the unit 2-disk.

(3) Note, in contrast to the foregoing, that, for $\dim \rho = n-3$, the choice of P, i.e. of an element in $CF(\ell k\rho)$, is not, in any obvious way, canonical. We must therefore rest content with an arbitrary choice.

(4) To complete the construction of ϕ over ∂e we must now choose contractions of the configuration spaces $CF(\ell k\rho)$ to the aforementioned canonical points for ρ of dimension n, $n-1$, $n-2$. Fortunately, these choices are also canonical, or very nearly so. To remove lingering ambiguities, it is helpful to make use of the notation of local ordering (introduced in [L-R]).

Definition. A local ordering on a locally-finite simplicial complex K is a partial ordering on the vertices of K such that the vertices of $st(\sigma)$ are linearly ordered for any simplex σ.

We will assume, henceforth, that the triangulations we work with are locally ordered in this sense. I.e., the linear order on each star will be assumed as part of the local data, in addition to the underlying combinatorial data, per se.

With the assumption that $st(\rho)$ is linearly ordered, it is easy to construct canonical contractions of $CF(\ell k\rho)$ for $\dim \rho = n, n-1, n-2$. This may be done trivially in the first two cases. For the case $\dim \rho = n-2$, we may view $CF(\ell k\rho)$ in the following way: Think of $CF(\ell k\rho)$ as the space of all simplex-wise linear embeddings of the triangulated disc $c\ell k\rho$ in R^2 which put the cone point at 0 and which take the "earliest" simplex of $\ell k\rho$ to an edge of the standard regular j-gon (j = # of vertices of $\ell k\rho$). Specifically, "earliest" means with respect to the induced lexicographic ordering on pairs of vertices. We choose the "standard" regular j-gon to have one edge lying in the right half plane and with the ususal x-axis as perpendicular bisector. If (v_0,v_1) is the earliest edge of $\ell k\rho$ ($v_0 < v_1$, in the given ordering), v_0 is assigned to the endpoint of the standard edge lying below the x-axis, and v_1 to the endpoint above the x-axis. The point is [see W] that any element $a \in CF(\ell k\rho)$ is represented by one and only one embedding with this property; thus the space of all embeddings with this property is, in fact, identical to $CF(\ell k\sigma)$. Thus, the "canonical" element of $CF(\ell k\rho)$ is the one which, subject to this condition, embeds $\ell k\rho$ as the standard regular j-gon. Now, if f is some arbitrary embedding satisfying the given constraint, then we form a one-parameter family of embeddings connecting f and the canonical embedding

Clearly we can slide $f(v)$ to $s(v)$ by first sliding it along a circle of radius $|f(v)|$ in the proper angular direction until it is radially in line with $s(v)$ and then sliding radially until it coincides. Do the "angular slide at uniform angular velocity for $0 \leq t \leq \frac{1}{2}$ and the "radial" slide at uniform linear velocity for $\frac{1}{2} \leq t \leq 1$. Doing this simultaneously for all v deforms f through configurations to the standard configuration.

This is jointly continuous in f and t and thus yields the desired contraction of $CF(\ell k\rho)$.

We may conclude, on the basis of (1)-(4) above that the only data needed to specify a cocycle c representing γ are

(1) The local ordering of M

(2) The form ω

(3) The choice of an element of $CF(\rho)$ for $(n-3)$-dimensional simplices ρ.

Clearly, ω being assumed, $c(e)$ will only depend on the data on $st(\sigma)$, σ dual to e. That is, we need only know the linear order on $st(\sigma)$, the embedding of $st(\sigma)$ in R^{n+k}, and the choice of an element in $CF(\ell k\rho)$ for $\rho^{n-3} \subset st(\sigma)$.

In order to obtain a purely local formula, i.e. one depending on the structure of $st(\sigma)$ as a simplicial complex, and on that alone, there are a number of simplifications available. First, we can take the form ω to be invariant under the action of $O(n+k)$ on $G_{k,n}$. Moreover, we might as well assume that we have chosen $\omega \in \Omega^4(G_{k,n})$ consistently for all n,k. This means that in the natural double sequence

$$
\begin{array}{ccccc}
 & \downarrow & & \downarrow & \\
\cdots \dashrightarrow & G_{k,n} & \dashrightarrow & G_{k+1,k} & \dashrightarrow \\
 & \downarrow & & \downarrow & \\
\cdots \dashrightarrow & G_{k,n} & \dashrightarrow & G_{k+1,n+1} & \dashrightarrow \cdots \\
 & \downarrow & & \downarrow & \\
 & \vdots & & \vdots &
\end{array}
$$

the choice of ω is consistent with pullback on Ω^4.

Next, given M^n merely as a triangulated manifold, we may pick an embedding $M^n \subset R^{n+k}$ by embedding M^n as a subcomplex of the standard simplex $\Delta^{n+k-1} \subset R^{n+k}$. This amounts to assigning vertices v of M^n to standard basis vectors of R^{n+k} in arbitrary fashion and then extending convex-linearly to a map on M^n.

Having done this, we obtain a transverse k-plane field on $M_*^{(4)}$ upon choosing a local order for M and choosing, for each σ^{n-3}, an element in $CF(\ell k\sigma)$.

If we have proceeded as above it is clear that we obtain a 4-cocycle c such that, for a typical dual 4-cell e, $c(e)$ depends only on the combinatorial structure of $\ell k\sigma$ (σ dual to e), on the linear ordering of $st\ \sigma$, and on the choice of an element in $CF(\ell k\tau)$ for all τ^{n-3} with $\sigma < \tau$. Thus we may write $c(e)$ as a function of such data without reference to M^n per se.

Now we may write down a formula:

$$\overline{c}(e) = E(c(e))$$

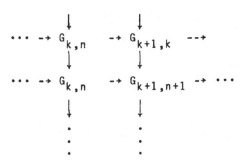

where E denotes expected value over all choices of a linear order on $st\ \sigma$ and all choices of configurations $\tau^{n-3} \dashrightarrow \alpha \in CF(\ell k\tau)$. Here we are implicitly assuming that the manifold $CF(\ell k\tau)$ has the natural structure of a measure space; but this is not at all hard to justify.

The following result is immediate:

Theorem 1. Let M^n be a combinatorially-triangulated manifold. Then the assignment $e \to \overline{c}(e)n$ is well-defined on oriented 4-cells and the resulting oriented cochcain with real coefficients, $\overline{c}(M)$, is a cocycle representing the characteristic class $\gamma(M) \in H^4(M;R)$.

So, in particular, if γ designates the first (tangential or normal) Pontrjagin class, we see that Theorem 1 gives us a local formula, one which is, the author trusts, less obscure than that in [GGL1].

§3. A local-ordered formula for the integral p_1.

In the foregoing section, we constructed a reasonably explicit local formula for the real first Pontrjagin class. On the other hand, the work of the author and C. Rourke [L-R] gives theoretical grounds for asserting the existence of a local formula for the integral p_1, provided that the local data is now understood to encompass a linear order of stars. I.e., we should expect to find, for any triangulated locally-ordered manifold M^n, an oriented integral 4-cocycle g representing the integral (tangential or normal) p_1 such that, given an an oriented dual 4-cell e, $g(e) \in Z$ depends only on the combinatorial type of the complex $st(\sigma^{n-4})$ (σ^{n-4} dual to e) and on the linear order on $st(\sigma)$.

The claim is that the ideas of the previous sections can be somewhat extended in order to create just such a formula, which we call a local-ordered formula. It is slightly unfortunate that in order to create it, we must restore some of the arbitrary choices that were "averaged away" at the end of the previous section. That is, given an oriented dual 4-cell e, in order to know the value of the formula on e one would have to know not only the a linear order on $st(\sigma)$ but, as well, a specific choice of configuration $\alpha(\tau) \in CF(\ell k\tau)$ for every $\tau^{n-3} < \sigma$. Therefore, in a strict sense, we only have a local ordered formula for p_1 once we have made, a priori, a choice of $\alpha(\Sigma^2) \in CF(\Sigma^2)$ for every possible triangulated 2-sphere. Thus we create not one but rather a multiplicity of local-ordered formulae, one for each such assignment.

The basic idea, once more, is to call upon Cairns' work on construction of transverse fields over $M_*^{(4)}$. But rather than using such a field to pull back a 4-form from $G_{k,n}$ to be integrated over 4-cells, we use it to define certain intersection numbers which are, perforce, integers.

As a preliminary step, we recall the work of Thom [T] on dual characteristic cycles in Grassmannians. Consider once more the Grassmannian $G_{k,n}$ of linear k-planes in R^{n+k}, $k > n$. Let Q be an arbitrarily chosen linear n-plane in R^{n+k}. We then have defined a certain subset $V(Q) \subset G_{k,n}$ by

$$V(Q) = \{P \in G_{k,n} | \dim(P \cap Q) \geq 2\}.$$

$V(Q)$ is a manifold of dimension $nk-4$ (i.e., of codimension 4) away from certain low-dimensional singularities. Moreover $V(Q)$ is a naturally-co-oriented cycle. That is to say, the normal bundle in $G_{k,n}$ of the non-singular part of $V(Q)$ has a natural 4-dimensional integral Thom class p. For a given k-plane bundle ξ over an arbitrary C·W complex K classified by $f:K \to G_{k,n}$, $p_1(\xi)$ may be computed as follows: assume f is in general position, so, in particular, $f(K^{(3)}) \cap V(Q) = \emptyset$. Then, for any oriented 4-cell e of K, the intersection number $d(e) = f(e) \cdot V(Q)$ is well-defined, using the natural co-orientation of $V(Q)$ as well as the orientation of e. The assignment $e \longmapsto d(e)$ is an oriented cocycle of M^n, and $p_1(\xi)$ is its integral-cohomology class.

We now update some ideas from §1. Let $\Sigma^4 \times D^{n-4}$ be embedded in R^{n+k}; Σ^3 is taken to be the equatorial sphere; i.e., $\Sigma^3 = D^4_- \cap D^4_+$, $\Sigma^4 = D^4_- \cup D^4_+$. As before, we assume that $\Sigma^4 \times D^{n-4}$ admits a transverse k-plane field ϕ. We make a choice of reference n-plane $Q \subset R^{n+k}$ and assume ϕ is in general position, viz, $\phi(\Sigma^3)\ \ V(Q) = 0$. We then obtain an integer

$$\eta = \phi(D^4_+) \cdot V(Q).$$

In analogy to the work of §1 where integrals of forms, rather than intersection numbers, were used to define η, we have:

<u>Lemma 8</u>. η depends only on the embedding and the field ϕ on a neighborhood of Σ^3.

In view of Thom's result, we may easily obtain the following consequence.

Let M^n be a triangulated manifold embedded simplex-wise convex-linearly in R^{n+k} in general position. For each simplex σ of codimension ≤ 4, pick an element $\alpha(\sigma) \in CF(\ell k\sigma)$. Assuming a local ordering of M^n, follow the procedure of §2 to obtain, over $M^{(4)}_*$, a transverse field ϕ to M^n. Now let Q be a generic n-plane in R^{n+k}, i.e., one chosen such that $\phi(\partial e) \cap V(Q) = \emptyset$, for any dual 4-cell e. We then define

$$d(e) = \phi(e) \cdot V(Q).$$

Essentially, $d(e)$ is the value of η determined by the embedding and the field ϕ in a neighborhood of ∂e, as well as by the reference plane Q.

Lemma 9. d is an integral 4-cocycle whose cohomology class is the first integral Pontrjagin class p_1 of the stable normal bundle of M^n.

The proof proceeds much as in the case of Lemma 5.

Remark: At this point, we could, effectively, claim to have obtained another local formula for the real first Pontrjagin class. Again, the formula would emerge through an averaging procedure, i.e., by choosing a "standard" embedding of M^n and by averaging over the set of reference planes Q as well as the ordering and configuration data which lead to the explicit construction of ϕ. Details are omitted.

Comparison with [GGL2] suggests that this formula might be replacable by one which involves only a finite averaging procedure, rather than one expected value over a measure space. This we leave as a conjecture, observing only that the key point seems to be this: Pick a reference plane Q and configuration data at the simplices $\tau > \sigma$, (σ^{n-4} dual to e^4). We see that $d(e)$, provided that it is well defined (i.e., provided that the field $\phi|\partial e$ determined by configuration data has image disjoint from $V(Q)$) remains constant under small perturbations of Q and of configuration data. This would seem to suggest that averages (perhaps weighted) should be taken over connected components of the set of all possible choices for Q and for configuration data, rather than an expected value whose determination involves an integral. Whether this would, in practice, represent an actual computational improvement is not clear.

It is now our purpose to sharpen Lemma 9 to the extent of obtaining a local formula for the integral (normal) first Pontrjagin class which disregards all data save the local ordering, the combinatorics of links of $(n-4)$-simplices and the choice of configuration for links of $(n-3)$-simplices.

Our first observation is directed towards eliminating the role of the arbitrarily-chosen reference plane Q from our existing formula.

<u>Lemma 10</u>. Let ϕ be a transverse k-plane field to M^n R^{n+k}. Let $q\colon M^n \to G_{n,k}$ be a homotopically trivial map. Let $V(q) \subset M^n \times G_{k,n}$ denote the set $\{(x,P) \mid \dim(P \cap q(x)) \geq 2\}$. Let Φ be the section of $M^n \times G_{k,n} \xrightarrow{\text{proj.}} M^n$ induced by ϕ, i.e., $\Phi(x) = (x,\phi(x))$. Finally, assume Φ is in general position with respect to $V(q)$. Then $p_1(\nu M^n)$ is the primary obstruction in $H^4(M^n;Z)$ to deforming Φ off $V(q)$.

The proof is trivial. For the case of q = the constant map $q(x) = Q \in G_{n,k}$, the lemma is merely a slightly roundabout statement of Thom's result. But if q_1, q_2 are homotopic maps then $V(q_1)$ and $V(q_2)$ are homologous cycles in $M^n \times G_{k,n}$ from which it readily follows that the primary obstructions to deforming Φ off $V(q_1)$ and $V(q_2)$ respectively must coincide. The lemma is then immediate.

We may paraphrase Lemma 10 as follows: Assume that Φ is in general position with respect to $V(q)$ in the sense that $\Phi(M_*^{(3)})\ V(q) = \emptyset$ and $\Phi|e^4$ is transverse to $V(q)$, for any oriented dual 4-cell e^4. Then the intersection number $d(e) = \Phi(e) \cdot V(q)$ is well-defined if e be oriented. (In fact, strictly speaking, we need not assume $\Phi|e^4$ is transverse to $V(q)$ to have this intersection number well defined.) The assignment $e \longrightarrow d(e)$ is then an oriented integral cocycle representing the first Pontrjagin class $p_1(\nu M^n)$.

We may even go further: the observation above remains valid provided merely that Φ arises from a transverse field ϕ defined over $M_*^{(4)}$.

We now construct a locally-determined map $q\colon M_n \to G_{n,k}$, using only data provided by the local ordering, so that q will turn out to be globally trivial. In fact, we define a frame-field over M^n. First consider the barycentric subdivision of M^n, and a typical vertex b_σ (= barycenter of σ) where σ is a simplex of the initial triangulation. We let $q(b_\sigma)$ be defined as the n-frame $(v_1(\sigma)\cdots v_n(\sigma))$ where $v_1(\sigma)\ldots v_n(\sigma)$ represent the earliest vertices of $st(\sigma)$ in the presumed ordering. To define q more generally, recall that the generic point x of M^n may be uniquely denoted

$$x = \alpha_1 b_{\sigma_1} + \cdots + \alpha_s b_{\sigma_s}$$

where $\alpha_i > 0$, $\Sigma\alpha_i = 1$. (I.e., $b_{\sigma 1}\ldots b_{\sigma s}$ are the vertices of the unique simplex of the barycentric subdivision of M^n which contains x as an interior point; thus $\sigma_1 < \sigma_2 \cdots < \sigma_s$).

Let $v_j(x) = \Sigma\alpha_i v_j(\sigma_i) \in R^{n+k}$ for $j = 1,2 \ldots n$.

<u>Lemma 11</u>. For any x, $\{v_1(x),\cdots,v_n(x)\}$ is a linearly independent set.

<u>Proof</u>. Suppose there is a dependence relation:

(i) $$c_1 \cdot v_1(x) + \ldots + c_n \cdot v_n(x) = 0$$

for some x and some set of coefficients $c_1 \cdots c_n$ not all 0,

Let us remember that $v_i(x)$ is a linear combination of vertices of $st(\sigma_i)$. Since $st(\sigma_1) \supset st(\sigma_2) \cdots \supset st(\sigma_s)$ it follows that the left side of (i) may be rewritten as a linear combination of the vertices of $st(\sigma_1)$. But since this set, by the general position assumption concerning the embedding $M^n \subset R^{n+k}$, is linearly independent, we see that in this reformulation all coefficients must be 0.

Consider the earliest coefficient in (i) which is nonzero; call it c_j. Consider now the vertex $v_j(\sigma_1)$. In general, for any p, $v_p(\sigma_i) > v_p(\sigma_1)$ in the order on $st(\sigma_1)$. We use this fact to examine the coefficient of $v_j(\sigma_1)$ in the reformulation of the right-hand side of (i) in terms of the vertices of $st(\sigma_1)$.

First of all,

$$c_j v_j(x) = c_j \cdot \alpha_1 v_j(\sigma_1) + c_j \alpha_2 v_j(\sigma_2) \cdots$$

Since $c_j \neq 0$, $\alpha_i > 0$, it follows that $c_j \cdot v_j(x) = A \cdot v_j(\sigma_1) +$ {other terms} where the additional terms do not involve $v_j(\sigma_1)$ and $A \neq 0$. On the other hand, consider the term $c_p \cdot v_p(x) = c_p \cdot \Sigma \alpha_i \cdot v_p(\sigma_i)$ for $p > j$. Now $v_p(\sigma_i) \geq v_p(\sigma_1) > v_j(\sigma_1)$. So $c_p \cdot v_p(x)$, rewritten as a linear combination of vertices of $st(\sigma_1)$, does not involve $v_j(\sigma_1)$. Thus $c_1 v_1(x) \cdots c_n \cdot v_n(x) = A \cdot v_j(\sigma_1) + B$ where B does not involve $v_j(\sigma_1)$, $A \neq 0$. But this contradicts the presumed linear independence of the vertices of $st(\sigma_1)$. Hence the lemma follows.

Thus we may define $q(x)$ for any x as the n-frame $(v_1(x) \cdots v_n(x))$. By slight abuse of notation, we also use q to denote the map $M^n \to G_{n,k}$ defined by $x \longmapsto \mathrm{span}(v_1(x), \cdots, v_n(x))$. Clearly, q represents a trivialized bundle and is thus null-homotopic in $G_{n,k}$.

Consider once more the k-plane field ϕ transverse to M^n defined over $M_*^{(4)}$ by data consisting of a local order on M^n and configuration data for simplices of $\dim \geq n-4$. We have, simultaneously, the n-frame field q. Thus, by the remarks following Lemma 10, we shall have a well-defined cocycle d(e) (depending on the embedding as well as the afore-mentioned data) simply provided that $\Phi(M_*^{(3)}) \cap V(q) = \emptyset$. We claim that this last condition will hold for a generic choice of configuration data, though the proof will be omitted.

As to dependency on the embedding, this may be eliminated simply by picking any "standard" embedding of M^n as a sub-complex of Δ^{n+k-1}, i.e., by assigning each vertex of M to an element of the standard basis for R^{n+k} and then extending convex linearly. It turns out that d(e) is independent of the particular embedding, and so d(e) is now seen to be intrinsic, depending only on the local ordering and the configuration data.

We may emphasize the intrinsic nature of the formula we have by freeing it entirely from the notion of embeddings and transverse fields.

Consider the typical dual 4-cell e^4 dual to σ^{n-4}. Decompose it into sub-cells $\{d_\tau\}_{\sigma<\tau}$, defined by $d_\tau = st(b_\tau, M'') \cap e^4$ where $st(\ ,M'')$ refers to the combinatorial star in the second barycentric subdivision of M.

Omitting details, we note that we may slightly modify the definition of the frame-field q so that for $x \in d_\tau$, $v_i(x)$ will lie in the vector subspace spanned by the vertices of $st(\tau)$. We also assume: $\phi(x)$ is transverse to M at b_τ if $x \in d_\tau$. These modifications in the definitions of ϕ and q are relatively simple to make and leave d(e) unaffected.

Concentrating on $d_\tau \cap e$, we let $\overline{\phi}(x)$ denote the $CF(\ell k\tau)$-coordinate of $\phi(x)$. For convenience, we shall think of $\overline{\phi}(x)$ as an embedding $\overline{\phi}(x): st\tau, b_\tau \subset R^n, 0$ where the embedding is convex linear on simplices and is the "standard embedding" on δ^n where δ^n denotes the n-simplex of $st\ \tau$ smallest with respect to the lexicographic ordering induced by the linear ordering of the vertices of $st\ \tau$. In fact, rather than using the standard R^n, we think of the n-plane P parallel to δ^n passing through the origin. So therefore, if we consider the direct-sum decomposition of R^{n+k} as $P \oplus \phi(x)$, the configuration of $st\ \tau$ in P comes merely by projection π_P onto the P-factor of $t - b_\tau$ for all vertices t.

Having oriented e and therefore d_τ we may define an integer $g(e,\tau)$ as the algebraic number of points $x \in d\tau$ such that

(ii) $$\dim \text{span } (\pi_P v_1(x) \ldots \pi_P v_n(x)) < n-2.$$

The point is that given the configuration of $st\ \tau$ in P we can determine π_P in purely algebraic terms (without reference to the embedding or the normal field). For these purposes, we may, without loss of generality assume that R^{n+k} is spanned by the vertices $\{t\}$ of $st\ \tau$. Now, from δ^n pick the earliest vertices among the t's (label them $u_1,\ldots,u_n$) so that $\{y_i\} = \{u_i - b_\tau\}$ spans P. Label the remaining vertices $w_1,\ldots,w_k$. Consider $z_i = w_i - \overline{w}_i - b_\tau$ where $\overline{w}_i$, a linear combination of y's, is the image of w_i under the configuration embedding $\overline{\phi}(x)$. Let B be the matrix $(y_1,\ldots,y_n, z_1 \ldots z_k)$ (writing vectors as columns with respect to the ordered basis $t_1,\ldots,t_{n+k}$). Let $Q = B^{-1}$, and set $=$ first n rows of Q. Write $R_1 = \binom{R}{0}$ an $n\times n$ matrix so R_1

is a projection expressed in y,z coordinates and $BR_1 = \Pi$ is projection π_p expressed in t-coordinates. Π, of course, depends on x. So $g(e,\tau)$ is the algebraic number of points x such that

(iii) $\quad \text{rank } \Pi(x)(v_1(x),\ldots,v_n(x)) \leq n-2.$

Now define $n(e)$ by

$$n(e) = \sum_{\tau} g(e,\tau)$$

Theorem. The assignment $e \mapsto n(e)$ is an integral oriented cocycle representing the first integral normal Pontrjagin class $p_1(\nu M)$.

We omit an explicit proof, but we do note that the formula for $n(e)$ represents, essentially, a computation of what we previously denoted as $d(e)$. This computation is a rather routine working out of consequences of the characterizations given in [Ca 1] and [W] of the space of k-planes transverse to M at a point in terms of configuration spaces.

This formula should be compared with the formula of [GGL].

There remains the slight problem of deriving a formula for the integral "tangential" first Pontrjagin class. Suffice it to say that the same general approach will work. I.e., in constructing a "normal" field over $M_*^{(4)}$--the transverse k-plane field ϕ--the Cairns procedure simultaneously constructs a "tangent" plane field $\phi^\perp$. By the same token, q must be replaced by a locally determined trivial k-plane bundle.

§4. Computational considerations

In the various local formulae we have exhibited above, there remains the problem of explicitness. That is, we have shown that given, say, an embedding of M^n in R^{n+k}, together with a local ordering a choice of configuration data at simplices of dimension $\geq$ n-4, we may construct a field transverse to M^n over $M_*^{(4)}$, from which a cocycle representing p_1 (with real or integral coefficients) may be computed. Note, however, these contrasting facts: We have shown that the value of the cocycle on a dual 4-cell e depends only on the given data on ∂e. (Strictly speaking, it depends on data for simplices τ of dimension $\geq$ n-3 to which σ, the dual simplex to τ, is incident.) Yet, in point of actual computation our formulas refer to data defined over all of e. Thus, by way of concrete example, if we adopt the approach of §2 and make use of a differential 4-form ω on $G_{k,n}$ we see that the value of the associated Pontrjagin cocycle on a typical e^4 is given as $\int_{e^4}\phi^*\omega$, even though this number is an invariant of ϕ restricted to ∂e^4. So, to compute this number, we should have to know ϕ explicitly on e^4, i.e., we should have to carry out the Cairns construction in detail using the detailed methods, as well as the actual results, of [Ca 2]. In some ways, the situation is analogous to what one finds for the Atiyah-Patodi-Singer η-invariant of an oriented Riemannian 4j-1 manifold M. $\eta(M)$ is, of course, dependent only upon M but is in general quite difficult to compute merely in terms of M. However, if M is known to be the boundary of the Riemannian 4j-manifold W (with the product metric near the boundary) then the value of $\eta(M)$ may be computed with relative ease.

In what follows, we indicate a computational alternative to

specifying the transverse field ϕ explicitly on the interior of e^4. Again, for simplicity's sake, we work in the context of real coefficients and differential forms.

The general idea is that one wants to avoid the difficulties of making Cairns' construction of the field ϕ over the 4-cells explicit. What is proposed, instead, is to use configuration data at (n-4)-simplices, together with the explicitly-constructible transverse field ϕ over the dual 3-skeleton to construct a certain field F over 4 cells $\overline{e}$ (one for each dual 4-cell e) so that integration can be done over $\overline{e}$ in place of $\int_e \phi^*\omega$ with essentially the same result, albeit F is <u>not</u> to be understood as a transverse field to e. Rather, F arises from constructing a 1-parameter family of embedded 3-spheres bounding a one-parameter family of 4-cells so that, for each 3-sphere in the family, we have an explicit field extending, in principle, to a field over the corresponding 4-cell. The process begins with ϕ and ends with a constant field so we can, for computations sake, regard it as a field over a 4-cell. The integral we get therefore represents the difference between the invariant $\int_e \phi^*\omega$ defined by ϕ itself and the invariant one gets from a constant field over a "flat" 3-sphere, i.e. one which bounds a "flat" 4-cell. Since the latter is obviously 0, we will have computed the former. The point is that F is, in principle, directly constructible without appeal to Cairns' theorem on connectivity of configuration spaces of cones on α-spheres.

Once more, assume that the combinatorially triangulated manifold M^n is simplex-wise linearly embedded in R^{n+k} in general position. We assume further data consisting of a local ordering on the triangulation and a choice, for every simplex σ of dimension $\geq$ n-4, of an element in $CF(\ell k\sigma)$.

With this data in hand, as we have noted continually, we obtain, by a perfectly straightforward construction, a k-plane field ϕ transverse to M^n over $M_*^{(3)}$. This, of course, makes no use of the configuration data at (n-4)-simplices. Rather than concerning ourselves with extending the transverse field to $M_*^{(4)}$, however, we now make use of the remaining data in a different way.

Given an (n-4)-simplex σ, the choice of a configuration of $\ell k \sigma$ is essentially a choice of a k-plane P transverse to M^n at b_σ. Choose an n-simplex ρ of $\mathrm{st}\ \sigma$, e.g., the smallest in the lexicographic order induced by the linear order on the vertices of $\mathrm{st}\ \sigma$. It follows, then, that every vertex v of $\mathrm{st}\ \sigma$ may be represented uniquely (in the conventional (n+k)-vector notation) as: $v = b_\sigma + r_v + n_v$ where r_v is a vector in the linear n-plane Q parallel to ρ and $n_v \in P$. Since P is, as assumed, transverse to M^n at b_σ, it follows that the projection map $v \mapsto b_\sigma + r_v$ is not only 1-1 but extends, in fact, by convex-linear extension, to an embedding $\mathrm{st}\ \sigma \subset b_\sigma + Q$. Furthermore, if we let u_t be defined as the convex-linear extension of the map $v \mapsto b_\sigma + r_v + (1-t)\cdot n_v$ for $0 \leq t \leq 1$, we obtain a simplex-wise linear embedding $\mathrm{st}\ \sigma \subset R^{n+k}$. If $t < 1$, u_t is in general position; therefore by Lemma 6 for such t we may use the configuration data on simplices of dimension $\geq$ n-3 to construct a k-plane field ϕ_t on $u_t(\partial e^4) - u_t(\mathrm{st}\ \sigma)$, where e^4 is the 4-cell dual to σ, such that $\phi_t(x)$ is transverse to $u_t(\mathrm{st}\ \sigma)$ at $u_t(x)$. This process is continuous on $\partial e^4 \times [0,1)$. Moreover, $\lim_{t \to 1} \phi_t(x) = \phi_1(x)$ exists for $x \in \partial e^4$ and we obtain thereby a vector bundle on $\partial e^4 \times I$, in fact a map $\Phi: \partial e^4 \times I \to G_{k,n}$.

Finally, $\phi_1(x)$ is transverse to the n-plane $b_\sigma + Q$.

Now we may deform ϕ_1 further (regarding it as a map $\partial e^4 \to G_{k,n}$) to a trivial map, i.e., the constant map $x \mapsto Q^\perp$. We think of this deformation (which may, of course, be constructed explicitly in terms of ϕ_1 without difficulty) as defining a map $\Psi\colon c\partial e^4 \to G_{k,n}$ with $\psi|\partial e^4 = \phi_1$. Thus concatenating with Φ, i.e., taking $F = \Phi \cup \Psi\colon \partial e^4 \times I \cup c\partial e^4 \cong c\partial e^4 \to G_{k,n}$, we get a map from a topological 4-cell to $G_{k,n}$. Call this 4-cell $\overline{e}$. Now let $\omega \in \Omega^4(G_{k,n})$ be, as in §2, a closed form representing the real Pontrjagin class p_1 in DeRham cohomology. We then have the following.

Theorem: Let $\overline{\phi}$ be any extension of ϕ to e^4 with $\overline{\phi}$ transverse to M. Then

$$\int_{e^4} \overline{\phi}^*\omega = \int_{\overline{e}'} F^*\omega.$$

We omit a proof. Suffice it to remark that what is being exploited here is a principle cognate to, though much simpler than, the well-known fact that, although the Atiyah-Patodi-Singer η-invariant of a $4j-3$ dimensional Riemannian manifold is not the integral of a locally-determined form, nonetheless the difference between the η-invariants of two Riemannian structures which differ by deformation can be represented as such an integral over the manifold $\times$ I.

As a final note, we observe that this method for computing a cocycle representing p_1 may easily be adapted to the integer coefficient case studied in §3 above, where intersection numbers are used in place of integrals.

References

[APS] Atiyah, M.F., Patodi, V. and Singer, I. Spectral Assymmetry and Riemannian Geometry, Bull. London Math. Soc. 5 (1973), 229-234.

[BCH] Bloch, E.D., Connelly, R., and Henderson, D.W., The space of simplexwise linear homeomorphisms of a convex 2-disc, Topology.

[Ca 1] Cairns, S., Homeomorphisms between topological manifolds and analytic manifolds, Ann. of Math. 41 (1940), 796-808.

[Ca 2] __________, Isotopic deformations of geodesic complexes on the 2-sphere and plane, Ann. of Math., 45 (1944), 207-17.

[G] Gabrielov, A.M., Combinatorial formulas for Pontrjagin classes and GL-invarient chains, Funkts. Anal. Prilozhen, 12, No. 2 (1978), 1-7.

[GGL1] Gabrielov, A.M., Gelfand, I.M. and Losik, M.V., A combinatorial calculation of characteristic classes, Funkts. Anal. Prilozhen, 9, No. 2, 12-28 and No. 3 5-26, 1975.

[GGL2] ______________________________, A local combinatorial formula for the first Pontrjagin class. Funkts. Anal. Prilozhen 10, No. 1 (1976).

[M] MacPherson, R., The combinatorial formula of Gabrielov, Gelfand and Losik for the first Pontrjagin class; Seminaire Bourbaki, 29, No. 497 (1976-77).

[LR] Levitt, N. and Rourke, C., The existence of combinatorial formulae for characteristic classes, Trans. A.M.S. 239 (1978), 391-397.

[S1] Stone, D., Notes on "A combinatorial formulae for $p_1(X)$," Advances in Math. 32 (1979), 36-97.

[S2] __________, On the combinatorial Gauss map for C^2-submanifolds of Euclidean space, preprint.

[T] Thom, R., Les singularités des applications differentiables, Ann. Inst. Fourier, Grenoble 6 (1955-56), 43-87.

[W] Whitehead, J.H.C., Manifolds with transverse fields in Euclidean space, Ann. of Math. 73 (1961), 154-212.

Department of Mathematics, Rutgers University
New Brunswick, NJ 08903, USA

EVALUATING THE SWAN FINITENESS OBSTRUCTION FOR PERIODIC GROUPS

by

R. James Milgram*

In [17] R. Swan introduced the finiteness obstruction $\sigma_n(G)$ for free actions of a periodic group G on finite complexes having the homotopy type of the sphere S^{n-1}. It takes its value in a certain quotient of $\tilde{K}_0(Z(G))$, $\tilde{K}_0(Z(G))/T$, and was one of the main motivations in the development of algebraic K-theory.

More recently Ib Madsen, C. Thomas, and C.T.C. Wall [22], [20], [10] proved a sharpened version of one of Swan's theorems, roughly that if G has period n then G acts freely on a homotopy S^{n-1} or S^{2n-1}, and no examples were known for which 2n-1 was actually necessary. Indeed it was somewhat hesitantly suggested that n-1 is always correct.

In [4], [5], [6], [8], [14], [15], a subgroup $D(G) \subset \tilde{K}_0(Z(G))$ was studied and shown to be computable in terms of determinants or reduced norms and the structure of units in certain algebraic number fields. D(G) contains T and we prove

<u>Theorem</u> 2.B.1: $\sigma_n(G) \in D(G)/T$.

In particular, we relate $\sigma_n(G)$ to the behaviour of these groups $Tor^i_{Z(G)}(M,Z)$ where M is a maximal order containing Z(G) in $\mathbb{Q}(G)$. In §3 we calculate these Tor groups for hyperelementary groups (this section may have independent interest), and in §4 we study D(G) for a class of periodic groups of period 4, Q(4p,q,1).

The Swan obstructions for these groups are written down in Theorem 4.B.6, and we have

<u>Theorem A</u> (4.C.2, 4.C.5, and 4.C.8): <u>The Swan obstruction</u> $\sigma_4(G) \neq 0$ <u>for</u> G = Q(12,5,1), Q(12,7,1), <u>or</u> Q(12,11,1). (The groups Q(a,b,c) are defined in §3.B, but the notation is standard.)

In fact we have

<u>Theorem B</u>: <u>Among periodic groups of period n and order</u> < 280 <u>only one group of each order</u> 120, 168, 240, <u>and</u> 264 <u>fails to act freely on a</u>

* Research supported in part by NSF MCS76-0146-A01

finite complex having the homotopy type of S^{n-1}.

However, for the next group in the series of theorem A we have

Proposition 4.C.1: $\sigma_3(Q(12,13,1)) = 0$.
Hence there is a 3-complex homotopic to S^3 on which Q(12,13,1) acts freely.

The number theory involved in these questions is subtle since Theorem A shows that Q(12,5,1), Q(12,7,1) and Q(12,11,1) can't even act freely on a homology 3-sphere.

Remark D: The proof that $\sigma_n(G) \in D(G)/T$ came out of several conversations with R. Oliver, and I also profited from a conversation with H. Bass. The initial question which led to this work came up in discussions with I. Hambleton.

This paper was originally written in 1978. For various reasons it has not been previously published but it has been circulated privately. It initiated a vigorous attack on the space form problem for the last and demonstrably most interesting class of groups, the Q(8a,b,c) with a,b,c odd coprime integers.

The initial theorem A and B above were quickly extended in [25] by constructing large numbers of odd index subgroups of the units in the cyclotomic fields $\mathbb{Q}(\lambda_p)$, $\mathbb{Q}(\lambda_{4p})$, $\mathbb{Q}(\lambda_p,\lambda_q)$ studied here. Indeed the proofs of the result in [25] are direct extensions of the proofs here. The only thing preventing their being given in this original paper was the fact that the unit theorems had not yet been proved.

The main result of [25] as slightly extended in [26] is

Theorem C: Let K_n be the maximal 2-abelian extension of $\mathbb{Q}$ contained in the cyclotomic field $Q(\zeta_n)$. If K is K_p, K_{pq} with the quadratic symbol $\left(\frac{p}{q}\right) = -1$, or the maximal 2 extension in $\mathbb{Q}(\zeta_{2^n}+\zeta_{2^n}^{-1}, \zeta_p+\zeta_p^{-1})$ $p \not\equiv 1(8)$, then the cyclotomic units have odd index in the units of K.

Next, by the techniques developed here and in [25] we obtain

Theorem D: Let p be a prime and

a) suppose $p \equiv 3(4)$ then $\sigma_4(q(8p,q,1)) = 0$ for q prime if and only if

(i) $q \equiv 1(8)$ or

(ii) $q \equiv 5(8)$ but $p^{\nu} \equiv \pm 1(q)$ for some odd ν.

b) $\sigma_4(Q(8p,q,1)) = 0$ if $p \equiv q = 1(4)$

but $\left[\frac{p}{q}\right] = -1$.

Using this as a starting place a spirited attack by the author and Ib Madsen on the actual surgery problems took place. (Preliminary results had already been indicated in [25]).

Again the results depended on finding sufficient units, hence had to be restricted to the cases covered by theorem C. The results [26] are

Theorem E: Let p,q be distinct odd primes and suppose $p \equiv -1 \pmod 8$; then

a) $Q(8p,q,1)$ acts freely on $\mathbb{R}^{8k+4}-(pt)$ $(k \geq 1)$ if $q \equiv 1 \pmod 4$ and p has odd order (mod q).

b) $Q(8p,q,1)$ acts freely on $\mathbb{R}^{8k+4}-(pt)$ but not on S^{8k+3} $(k \geq 1)$ if $q \equiv 5(8)$ and p has odd order (mod q).

c) $Q(8p,q,1)$ acts freely on S^{8k+3} $(k \geq 1)$ if $q \equiv 1(8)$ and p has odd order (mod q).

For special classes of numbers there are further results [27]. But all these results rest on the idea of identifying the Swan obstruction studied here as the image under ∂ of an element Θ where

$$\Theta \in \coprod_{\nu=2,p,q} \operatorname{im}\left(K_1(\hat{Q}_\nu(G)) \xrightarrow{\partial} \tilde{K}_0(Z(G))\right)$$

which is developed here. This class Θ rather than its image is shown to correspond to the surgery obstruction for certain surgery problems over the Poincaré complex constructed here, and using recent results on the calculation of these groups, if (as usual) one has sufficient information about units, Theorem E follows.

I would like to thank A. Ranicki and the other editors of this Proceedings for giving me the opportuniry to finally publish this work.

Edinburgh
September, 1984

§1. $K_0(Z(G))$ and $D(G)$ for G a finite group.

A. Preliminary remarks on $K_0(Z(G))$

Let $M_n(D)$ be a matrix ring over a division algebra D whose center F is a finite extension of the rationals $\mathbb{Q}$ or the complete local field $\hat{\mathbb{Q}}_p$. The reduced norm homomorphism

$$\tilde{N}: GL_n(D) \to F$$

is given by linear embedding $M_n(D) \subset M_n(K)$ for some extension K of F and taking the determinants of the images. Then Wang's theorem [21] identifies

1.A.1: $$K_1(M_n(D)) = K_1(D) \underset{\tilde{N}}{\hookrightarrow} F$$

where $\text{im}\,(K(D))$ is all those elements of F which are positive at all ∞ places at which D is a quaternion algebra.

Next, let M be a maximal order in $M_n(D)$ so $M = M_n(N)$ where N is a maximal order in D. The reduced $\tilde{K}_0(M) = \tilde{K}_0(N)$ is

1.A.2: $$\ker i: K_0(M) \to K_0(D) = Z$$

and Swan has shown in [18] that $\tilde{K}_0(N)$ is the ray class group of the center $\mathcal{O}$ of N consisting of all ideals modulo those principal ideals which are positive at all finite primes at which D is a quaternion algebra.

B. $D(G)$ and $E(G)$.

Let M be a maximal order in $\mathbb{Q}(G)$ containing $Z(G)$. Then the induced map

1.B.1: $$j: \tilde{K}_0(Z(G)) \to \tilde{K}_0(M)$$

does not depend on the particular M chosen and j is surjective (see e.g. [5]).

Definition 1.B.2: $D(G)$ is $\ker(j)$ in 1.B.1. so we have an exact sequence

$$0 \to D(G) \to \tilde{K}_0(Z(G)) \to \tilde{K}_0(M) \to 0.$$

A second group E(G) can be introduced which depends only on reduced norms. Let M_p be a maximal order in $\hat{\mathbb{Q}}_p(G)$ containing $\hat{Z}_p(G)$, and define

1.B.3: $$\mathcal{U}_p \subset K_1(\hat{\mathbb{Q}}_p(G))$$

as the image of $K_1(M_p)$. Writing

$$\hat{\mathbb{Q}}_p(G) = \coprod M_{n_i}(D_i)$$

we have $$M_p = \coprod_i M_{n_i}(N_i)$$

and $K_1(\hat{\mathbb{Q}}_p(G)) = \coprod_i (\dot{F}_i)$ where F_i is the center of D_i. Thus (one could use for example Quillen's localization sequence [12] to show this)

1.B.4: $$\mathcal{U}_p = \coprod_i \mathcal{U}(F_i)$$

the product of the units of the F_i's.

The composite maps

1.B.5:
$$\alpha_p\colon K_1(\hat{Z}_p(G)) \to K_1(M_p) \to \mathcal{U}_p$$
$$\beta_p\colon K_1(M) \to K_1(M_p) \to \mathcal{U}_p$$

then give

<u>Definition</u> 1.B.6: <u>The local defect $E_p(G)$ at p is</u> $\mathcal{U}_p/\mathrm{im}(\alpha_p)$.

<u>Remark</u> 1.B.7: If $p \nmid |G|$ then $M_p = \hat{Z}_p(G)$ so $E_p(G) = 1$. In any case $E_p(G)$ is a partial measure of the deviation of $\hat{Z}_p(G)$ from M_p and is finite for every finite group G.

$E_p(G)$ consists of a p primary part which is difficult to analyze and a somewhat easier part of order prime to p. To obtain this second part we can use the map

1.B.8: $$K_1(\hat{Z}_p(G)/J) \to \coprod_i \mathcal{U}(F_i)/(1+\mathfrak{m}_i)$$

where J is the Jacobson radical in $\hat{Z}_p(G)$ and $\mathfrak{m}_i$ is the maximal ideal in $\mathcal{O}(F_i)$.

Returning to 1.B.5 the map,

1.B.9: $$\beta = \coprod_{p \mid |G|} \beta_p : K_1(\mathcal{M}) \to \coprod_{p \mid |G|} E_p(G)$$

defines the quotient

1.B.10: $$E(G) = \coprod_{p \mid |G|} E_p(G)/\mathrm{im}(\beta).$$

Thus $E(G)$ is the product of the local defects factored out by the image of global units coming from $K_1(\mathcal{M})$. Its calculation is difficult but in specific cases presents no insuperable obstacles.

<u>Theorem</u> 1.B.11: $E(G) = D(G)$
(See for example [5].)

<u>Remark</u> 1.B.12: When the author initiated this work he was unaware of 1.B.11 and so derived his own proof, which we present in outline from here, as it may be easier for some people to reconstruct. Note, to begin, that there is a k so $|G|^k . \mathcal{M} \subset Z(G)$. So set $L_s(G) = \mathrm{im}(Z(G) \subset \mathcal{M}/|G|^{ks}\mathcal{M}$, and we have the pullback diagram

1.B.13: $$\begin{array}{ccc} Z(G) & \longrightarrow & \mathcal{M} \\ \downarrow & & \downarrow \pi_s \\ L_s(G) & \longrightarrow & \mathcal{M}/|G|^{ks}\mathcal{M} \end{array} .$$

Moreover, π_s is onto so Milnor's Mayer-Vietoris sequence can be applied obtaining

1.B.14: $$K_1(Z(G)) \to K_1(L_s(G)) \oplus K_1(\mathcal{M}) \to K_1(\mathcal{M}/|G|^{ks}\mathcal{M}) \xrightarrow{\partial_s} \tilde{K}_0(Z(G)) \to \tilde{K}_0(\mathcal{M}) \to 0$$

so $\mathrm{im}\,\partial_s = D(G)$. Now let s become large and pass to limits. In the limit we need some information about $K_1(\mathcal{M})$, $K_1(\mathcal{M}_p)$. This may be supplied using Quillen's exact sequence of a localization to show that the kernels of the <u>local</u> reduced norm maps are in the image of the elements in $K_1(\mathcal{M})$.

§2. <u>Swan's finiteness obstruction.</u>

A. <u>The definition and basic properties</u>

Let $T(G) \subset D(G) \subset \tilde{K}_0(Z(G))$ be defined as the image of $\coprod_{p \mid |G|} U(\hat{Z}_p^+)$

where $\hat{Z}_p^+ \subset M_p$ corresponds to the trivial representation. Now suppose G is periodic of period n. That is, there is an exact sequence

2.A.1: $$0 \to Z \to C_{n-2} \to \dots \to C_1 \to Z(G) \to Z \overset{\varepsilon}{\to} 0$$

where the C_j are projective Z(G) modules. In [17] Swan defined an invariant $\sigma_n(G) \in \tilde{K}_0(Z(G))/T$ which is zero if and only if G acts freely on an n-1 dimensional complex having the homology type of S^{n-1}. In fact, $\sigma_n(G)$ is the Euler class in $\tilde{K}_0(Z(G))/T$ of 2.A.1.

2.A.2: $$\sigma_n(G) = [C_0] - [C_1] + [C_2] \dots \pm [C_{n-1}].$$

If n is even then sequences 2.A.1 may be spliced either together to give $\sigma_{sn}(G) = s\sigma_N(G)$. Also if $H \subset G$ is a subgroup then by restriction Z(G) projectives become Z(H) projective so $r_H: \tilde{K}_0(Z(G)) \to \tilde{K}_0(Z(H))$ induces $\bar{r}_H: \tilde{K}_0(Z(G))/T \to \tilde{K}_0(Z(H))/T$ and

2.A.3: $$\bar{r}_H(\sigma_n(G)) = \sigma_n(H)$$

Swan's induction theorem [16] valid for arbitrary finite groups is

2.A.4: $$\coprod_{H \text{ hyperelementary}} \bar{r}_H: \tilde{K}_0(Z(G)) \longrightarrow \coprod_{\substack{H \subset G \\ H \text{ hyperelementary}}} \tilde{K}_0(Z(H))$$

is injective. This relates $\sigma_n(G)$ with the $\sigma_n(H)$ for all H. It must be handled carefully, however, since $\coprod_{H\ G} T(H)$ may not be equal in T(G), see e.g. [23], [24].

Finally, the types of hyperelementary subgroups of G are very restricted. For a discussion see e.g. [22].

B. $\sigma_n(G) \in D(G)/T(G)$.

Theorem 2.B.1 represents the fruit of several discussions with R. Oliver. It resulted from Oliver's attemps to provide counterexamples to the author's initially naive feeling that $\sigma_n(G)$ should lie in D(G)/T.

Theorem 2.B.1: $\sigma_n(G) \in D(G)/T$.

Proof: It suffices to consider hyperelementary subgroups since $\coprod r_H|$ in 2.B.2 is injective.

2.B.2:

$$\begin{array}{ccc} T(G) & \longrightarrow & \coprod_H T(H) \\ \downarrow & & \downarrow \\ D(G) & \longrightarrow & \coprod_H D(H) \\ \downarrow & & \downarrow \\ \tilde{K}_0(Z(G)) & \xrightarrow{\coprod r_H} & \coprod_H \tilde{K}_0(Z(H)) \\ {\scriptstyle p_G}\downarrow & & \downarrow{\scriptstyle \coprod p_H} \\ \tilde{K}_0(\mathcal{M}) & \xrightarrow{\coprod r_H|} & \coprod_H \tilde{K}_0(\mathcal{M}_H) \end{array}$$

Let N be a finitely generated torsion M module. By [13] N has projective length 1 so there is a short exact sequence

$$0 \to P_1 \to P_0 \to N \to 0$$

with P_0, P_1 finitely generated projective, and

$$\chi(N) = [P_0] - [P_1] \in \tilde{K}_0(M)$$

is well defined. We now have (and this is the heart of the matter)

Lemma 2.B.3: $p_G(\sigma_n(G)) = \sum_{i=0}^{n-1} (-1)^i \chi(\mathrm{Tor}^i_{Z(G)}(M(G), Z))$.

Proof: Tensor 2.A.1 over $Z(G)$ with M obtaining

2.B.4: $$0 \to C_{n-1} \otimes M \to \ldots \to M \xrightarrow{\varepsilon} Z \to 0 .$$

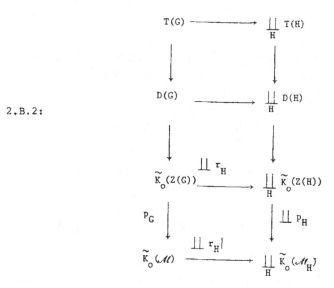

The homology of 2.B.4 is $\sum_0^{n-1} \widetilde{\mathrm{Tor}}^i_{Z(G)}(M, Z)$ which differs from the usual Tor group only in the part of involving Z^+ which is, in any case, a P.I.D. Clearly, $P_G(\sigma_n(G)) = \Sigma\ (-1)^j (C_j \otimes M)$, but this is given by the Tor formula in 2.B.3.

By results in [22], $\sigma_n(H)$ is zero except for H a 2-hyperelementary subgroup with 2-Sylow subgroup a generalized quaternion group.

For this group write $M = \coprod_i M_i$ where the M_i correspond to the irreducible representations of M_i so $\mathrm{Tor}^j_{Z(G)}(M,Z) = \coprod_i \mathrm{Tor}^j_{Z(G)}(M_i,Z)$.

In $3.6 we prove

<u>Lemma</u> 2.B.5 a. <u>$\mathrm{Tor}^j(M_i,Z)$ is a q-torsion abelian group with</u> $q \mid |G|$.

b. <u>If</u> q <u>is not prime then</u> $M_i = Z$.

c. <u>If</u> q <u>is an odd prime then</u> $M_i = M_2(\mathcal{O}_{q^j})$ with $\mathcal{O}_{q^j} = Z(\rho_{q^j} + \rho_{q^j}^{-1})$ <u>or</u> $M_i = \hat{Z}_q(\rho_{q^j})$.

d. <u>If</u> q <u>is</u> 2 <u>then</u> $M_i = M_2(\mathcal{O}_{2^j})$ <u>or</u> L <u>where L is the maximal order in the usual quaternion algebra over</u> $\mathbb{Q}(\rho_{2^j} + \rho_{2^j}^{-1})$.

(Here ρ_ℓ is a primitive ℓ^{th} root of unity, and (a) is well known.)

Now the proof of 2.B.1 follows directly since $\tilde{K}_0(Z) = 0$ and the prime ideal over p in $\mathcal{O}(\rho_{p^j} + \rho_{p^j}^{-1})$ is principal. The only remaining case is 2.B.5 (d), but there the results of [18] identify $K_0(L)$ with $K_0(\mathcal{O}(\rho_{2^j} + \rho_{2^j}^{-1}))$ (using Weber's theorem see e.g. [7]). This completes the proof of 2.B.1.

C. <u>The value of $\sigma_n(G)$ in $D(G)/T(G)$</u>

We suppose the groups $\mathrm{Tor}^j{}_{Z(G)}(M,Z)$ are known for the periodic group G, and local solutions

2.C.1: $$0 \to \hat{Z}_p \to C_{p,n-1} \xrightarrow{\partial p,\ n-1} C_{p,n-2} \to \cdots \to C_{p,1} \xrightarrow{p,1} Z_p(G) \xrightarrow{\xi} \hat{Z}_p \to 0.$$

of 2.A.1, with the $C_{p,j}$ free are also given.

Next, let the complex of free M-modules

2.C.2: $$0 \to Z \to M_{n-1} \to M_{n-2} \to \cdots \to M \to Z \to 0$$

be given with homology the $\mathrm{Tor}^j{}_{Z(G)}(M,Z)$. Localizing we have

<u>Lemma</u> 2.C.3: <u>Let N_p be a maximal order in a simple algebra over</u> $\hat{Z}_p$. <u>Then an isomorphism classification of finite chain complexes of free N_p modules</u> $C = \{C_i \to C_{i-1} \to \cdots \to C_0 \to 0\}$ <u>is given by the groups</u> $H_*(C)$

and the ranks $\dim_{N_p}(C_j)$.

Proof: This follows in the same way as the analogous theorem for free Z-chain complexes proved using the diagonalization theorem for matrices with coefficients in N_p. (See e.g. [13], p. 173, Theorem 17.7)

In particular, we assume 2.C.2 is a direct sum of complexes one for each simple order in M, and $\operatorname{rank}_M M_j = \operatorname{rank}_{\hat{Z}_p(G)}(C_{p,j}) = k_j$ for all $p \mid |G|$. Then we have

Lemma 2.C.4: There are elements $\alpha_i \in \coprod_{p \mid |G|} GL_{k_i}(M_p)$ which make 2.C.5 commute for each i

2.C.5:

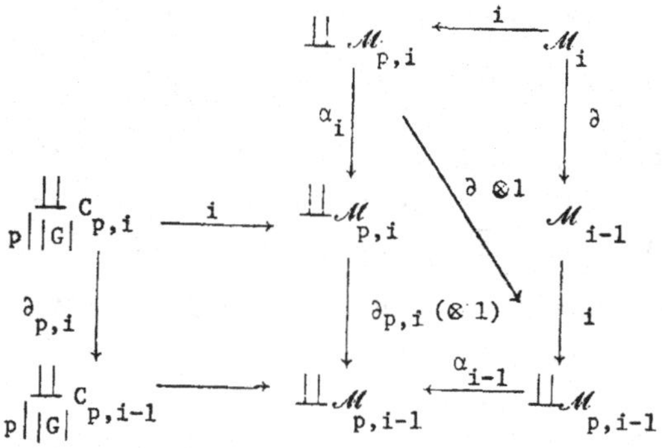

and

$$\sigma_n(G) = \{\alpha_1 \alpha_2^{-1} \alpha_3 \alpha_4^{-1} \ldots \alpha_{n-1}^{\varepsilon}\}^{-1}$$

Proof: The existence of the α_i is given by 2.C.3. Now let $C_i(Z(G))$ be obtained as the pullback diagram

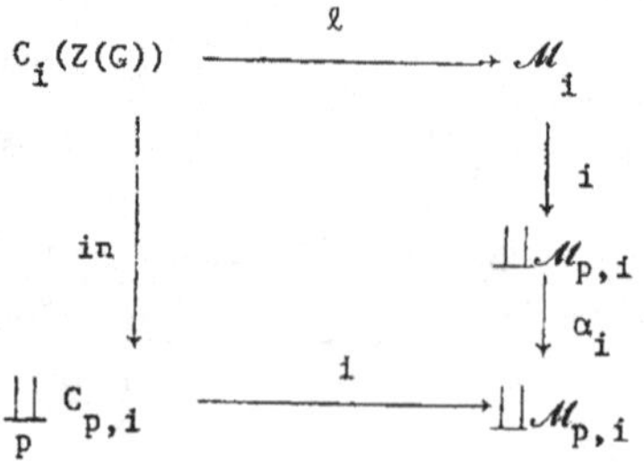

Then $C_i(Z(G))$ is projective and represented by $\{\alpha_i\}$ in $D(G)$. Moreover the following sequence of complexes is exact

$$0 \to C_*(Z(G)) \xrightarrow{(\ell,\ -in)} M_* \oplus \coprod (C_{p,*}) \xrightarrow{\alpha_{*i}\ \oplus\ i} \coprod M_{p,*} \to 0$$

and passing to homology, since α_{*i} induces isomorphisms (as $\mathrm{Tor}^i{}_{Z(G)}(M,Z)$ is torsion) in positive degrees, and in 0 $H_0(C_*(Z(G))) = Z$ we have $H_j(C_*(Z(G))) = 0$ $j > 0$ and the sequence obtained is a periodic resolution

Remark 2.C.6: We should be more explicit about the classes $\alpha_i \varepsilon \coprod \mathrm{Aut}(M_{p,i})$. Note that since $M_i = M \oplus \dots \oplus M$ we have an evident map $C_{p,i} = \hat{Z}_p(G) \oplus \dots \oplus \hat{Z}_p(G) \to M_{p,i}$. This identifies $M_{p,i}$ with $C_{p,i} \otimes M_p$. It is with respect to this that α_i is defined.

D. Wall's finiteness obstruction.

In [23, especially pp. 64-68] a finiteness obstruction $f_n(X)$ for complexes in a homotopy type satisfying "$D_n: H_i(\tilde{X}) = 0$ $i > n$, $H^{n+1}(X,B) = 0$ for all abelian coefficient bundles B $(n>2)$", was defined which genralizes Swan's obstruction. It makes its values in $\tilde{K}_0(Z(\pi_1(X))$, and for $\pi_1(X)$ finite we have the result corresponding to 2.B.3.

Lemma 2.D.1: $\quad p_{\pi_1(X)}(f_n(X)) = \pm \sum_{i=0}^{n} (-1)^i \chi(H_i(X,M))$.

(The proof is completely analogous to 2.B.3.)

Similarly, in case $p_{\pi_1(X)}(f_n(X)) = 0$, analysis of the Wall obstruction in terms of local piecing analogous to that in 2.C. may be casried through.

§3. Calculations of $\mathrm{Tor}^i{}_{Z(H)}(M(H),Z)$ for H a Hyperelementary Periodic Group.

A. Reduction to local cases.

The formal reduction

3.A.1: $$\mathrm{Tor}^i{}_{\hat{Z}_p(H)}(M \otimes \hat{Z}_p, \hat{Z}_p) = \mathrm{Tor}^i{}_{Z(H)}(M,Z) \otimes \hat{Z}_p$$

allows us to calculate the Tor^i locally. Thus, to begin we study the structure of the local group rings.

Lemma 3.A.2 Suppose p prime and $(p,\ell) = 1$, then

$$\hat{Z}_p(Z/\ell) = \hat{Z}_p \oplus \coprod_{j|\ell} \coprod_{i=1}^{r_j} \hat{Z}_p(\rho_j), \text{ where}$$

$r_j = \ell-1/\varphi_j(p)$ and $\varphi_j(p)$ is the least positive integer for which $p^{\varphi_\ell(p)} \equiv 1(j)$.

Proof: This is standard. Indeed since $(\ell,p) = 1$ $\hat{Z}_p(Z/\ell)$ is a maximal $\hat{Z}_p$ order. Also, the global maximal order is $\coprod_{j|\ell} Z_p(\rho_j)$ and

3.A.3: $$\hat{Z}_p \otimes Z(\rho_j) = \coprod_{i=1}^{r_j} \hat{Z}_p(\rho_j).$$

Compare with [9, p. 39].

More generally

Corollary 3.A.4: Write $n = p^m\ell$ with $(\ell,p) = 1$ then

$$\hat{Z}_p(Z/n) = \hat{Z}_p(Z/_pm) \oplus \coprod_{j|\ell} \coprod_1^{r_j} \hat{Z}_p(\rho_j)(Z/_pm).$$

Now suppose we are given a q-hyperelementary subgroup H. That is, H has a normal cyclic subgroup $Z/n \triangleleft H$ with quotient the Sylow q subgroup H_q of H. Then H is determined by a homomorphism

3.A.5: $$\lambda : H_q \to \mathrm{Aut}(Z/n) = (\dot{Z}/n)$$

and we can write H as an extension

3.A.6: $$0 \to Z/n \times \ker\lambda \to H \to (\mathrm{im}\lambda) \to 0.$$

Thus we can write $\hat{Z}_p(H) = \hat{Z}_p(Z/n \times \ker\lambda) \times_T (\mathrm{im}\lambda)$ and the splitting in 3.A.4 of $\hat{Z}_p(Z/n)$ implies an analogous splitting

3.A.7: $$\hat{Z}_p(H) = \left[(\hat{Z}_p(Z/_pm) \oplus \coprod_{j|\ell} \coprod_1^{r_j} \hat{Z}_p(\rho_j)(Z/_pm) \otimes \hat{Z}_p(\ker\lambda) \right] \times_T [\mathrm{im}\lambda]$$

In particular, assuming $p \neq q$ the argument of 3.A.2 shows

$$\hat{Z}_p(\ker\lambda) = \hat{Z}_p^+ \oplus \tilde{M}_p(\ker\lambda)$$

and

3.A.8: $\hat{Z}_p(H) = \hat{Z}_p(Z/_p m)[\mathrm{im}\lambda] \oplus N$

Factoring still further, we have the restriction map

$$r_p : \mathrm{Aut}(Z/n) \longrightarrow \mathrm{Aut}(Z/_p n)$$

and applying the above argument to the first factor in 3.A.8 (Which is $\hat{Z}_p(Z/_p m \times {}_T(\mathrm{im}\lambda))$ we obtain

3.A.9: $\hat{Z}_p(H) = \hat{Z}_p(Z/_p m \times {}_T r_p(\mathrm{im}\lambda)) \oplus N_1 \oplus N.$

Of course, when p = q, 3.A.8 takes the form

3.A.10: $\hat{Z}_q(\mathbb{H}) = \hat{Z}_q(\mathbb{H}_q) \oplus N.$

We call the various summands in 3.A.9, respectively 3.A.10, the p-blocks, respectively q-blocks of $\mathbb{H}$. Further, on tensoring with $\hat{\mathbb{Q}}_p$ the p-blocks each become direct sums of simple algebras, and we distinguish the block containing the trivial representation (the left-most block in 3.A.9 or 3.A.10) and write it

$$B_p(\mathbb{H}).$$

<u>Proposition</u> 3.A.10: $\mathrm{Tor}^*_{Z(H)}(M_i, Z)$ <u>contains</u> p-<u>torsion only if</u>

$$M_i \cap B_p(\mathbb{H}) \neq 0$$

<u>Proof</u>: We check locally. Now $B_p(\mathbb{H})$ is a direct summand of $\hat{Z}_p(\mathbb{H})$, hence projective. Moreover, the augmentation factors as

$$\begin{array}{ccc} \hat{Z}_p(\mathbb{H}) & \xrightarrow{\ \varepsilon\ } & \hat{Z}_p \\ & \searrow & \uparrow {\scriptstyle \varepsilon|} \\ & & B_p(\mathbb{H}). \end{array}$$

Thus, a suitable resolution of the augmentation is obtained by resolving $\varepsilon|$ over $B_p(\mathbb{H})$. Finally, tensoring this resolution with M_i gives zero identically unless $M_i \cap B_p(\mathbb{H}) \neq 0$.

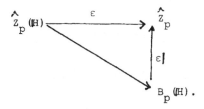

Corollary 3.A.11: $\mathrm{Tor}^i_{Z(H)}(M_i,Z)_p = \mathrm{Tor}^i_{B_p(H)}(M_i \otimes \hat{Z}_p, \hat{Z}_p)$.

This reduces tor calculations for these $\mathbb{H}$ to those for $Z/p^m \times_T V$ $V \subset \mathrm{Aut}\,(Z/p)$, for which calculations are easy, or in case $p = q$, for $\mathbb{H}_p$.

B. Calculations for $B_p(\mathbb{H})$ where $\mathbb{H}$ is a 2-hyperelementary group with H_2 generalized quaternion, $p \neq 2$.

In Milnor's notation $\mathbb{H} = Z/m \times \mathbb{Q}(2^i n, k, \ell)$. This means $\mathbb{Q}(2^i n, k, \ell)$ has a presentation $\{x,y,z,w,v \mid x^2 = y^{2^{i-1}}, y^{2^i} = 1, xyx^{-1} = y^{-1}, z^n = w^k = v^\ell = 1, xzvx^{-1} = z^{-1}v^{-1}, y(wv)y^{-1} = w^{-1}v^{-1}, yzy^{-1} = z, xwx^{-1} = w$ and z, w, v commute$\}$. Moreover, in order that $\mathbb{H}$ be periodic, m, n, k, ℓ must be coprime odd integers.

$$\text{The block } B_p(\mathbb{H}) = \begin{cases} \hat{Z}_p(Z/p^\ell) & \text{if } p/m \\ \hat{Z}_p(Z/p^\ell \times {}_T Z/2) & \text{if } p \nmid m. \end{cases}$$

The first case is well understood so we concentrate on the second case. (Note though that the first case gives the $\hat{Z}_p(\rho_{p^i})$ in 2.B.5 (c).) We have

3.B.1: $$\hat{\mathbb{Q}}_p(Z/p^\ell \times {}_t Z/2) = \coprod_{1 \le j \le \ell} M_2(\hat{\mathbb{Q}}_p(\rho_{p^j} + \rho_{p^j}^{-1})) \oplus \hat{\mathbb{Q}}_p^+ \oplus \hat{\mathbb{Q}}_p^-$$

so

3.B.2 $$M_p \cap B_p(\mathbb{H}) \otimes \hat{\mathbb{Q}}_p = \coprod_{1 \le j \le \ell} M_2(\hat{Z}_p(\rho_{p^j} + \rho_{p^j}^{-1})) \oplus \hat{Z}_p^+ \oplus \hat{Z}_p^- .$$

Note that 3.B.2 is already sufficient to prove all of 2.B.5 but (d). However, for our calculations in §4, we will need complete evaluations of the $\mathrm{Tor}_{\hat{Z}_p(\mathbb{H})}(M_p, \hat{Z}_p)$ so we give those results now.

Let $\lambda_{p^j} = \rho_{p^j} + \rho_{p^j}^{-1}$, then the representation

$$r_j : B_p(\mathbb{H}) \longrightarrow M_2(\hat{Z}_p(\lambda_{p^j}))$$

is given by

$$r_j(g) = \begin{pmatrix} 0 & 1 \\ -1 & \lambda_{p^j} \end{pmatrix}$$

3.B.3
$$r_j(t) = \begin{pmatrix} 0 & 1 \\ 1 & 0 \end{pmatrix}$$

Similarly

$$r_{\pm} : B_p(\mathbb{H}) \longrightarrow \hat{Z}_p^{\pm}$$

is given by $r_{\pm}(g) = 1$, $r_{\pm}(t) = \pm 1$. These representations define actions of $Z/p^{\ell} \tilde{\times}_T Z/2$ on $M_2(\hat{Z}_p(\lambda_{p^j}))$ or $\hat{Z}_p$, and we have

<u>Lemma</u> 3.B.4: $H_i^{r_j}(Z/_{p^{\ell}}, M_2(\hat{Z}_p(\lambda_{p^j}))) = \begin{cases} (Z/p)^2 & i \text{ even} \\ 0 & i \text{ odd} \end{cases}$

<u>Moreover the action of</u> t <u>on the groups above is via</u> $t(x) = (-1)^{i/2}x$.

<u>Proof</u>: Take the usual resolution of $Z(Z/_{p^{\ell}})$ with one copy of $Z(Z/_{p^{\ell}})$ in each degree, $\partial_{2i+1} = g-1$, $\partial_{2i} = 1+g+g^2+ \ldots +g^{p^{\ell}-1} = \Sigma_g$.
Then, tensor with $M_2(\hat{Z}_p(\lambda_{p^j}))$ over r_j and we have

$$\tilde{\partial}_{2i+1} = \begin{pmatrix} -1 & 1 \\ -1 & \lambda-1 \end{pmatrix}, \quad \tilde{\partial}_{2i} = 0$$

This means

$$\partial_{2i+1}(\alpha) = \alpha \cdot \begin{pmatrix} -1 & 1 \\ -1 & \lambda-1 \end{pmatrix} = \begin{pmatrix} a, & -a+b(\lambda-2) \\ c, & -c+d(\lambda-2) \end{pmatrix}$$

and the cokernel is $Z/p \times Z/p$ since $\lambda - 2$ is a uniformizing parameter for $\hat{Z}_p(\lambda)$.

Representing generators can be chosen as $\begin{pmatrix} 0 & 1 \\ 0 & 0 \end{pmatrix}, \begin{pmatrix} 0 & 0 \\ 0 & 1 \end{pmatrix}$,

and $A \rightarrow A(t) = A\begin{pmatrix} 0 & 1 \\ 1 & 0 \end{pmatrix}$ fixes these generators mod $\mathrm{im}(\partial_{2i+1})$.

Next note that the periodicity is given by capping with a generator e of $H^2(Z/_{p^\ell}, \hat{Z}_p) = \hat{Z}/_p$ [1, p. 115]. Then $e \cap H^{(r_j)}_{i+2}(Z/_{p^\ell}, M_2(\hat{Z}_p(\lambda_{p^j})))$

$= H^{(r_j)}_i(Z/_{p^\ell}, M_2(\hat{Z}_p(\lambda_{p^j})))$, $t(\alpha \cap \beta) = t(\alpha) \cap t(\beta)$ and $t(e) = -1(e)$,

so 3.B.4. follows.

Corollary: 3.B.5: $\mathrm{Tor}^i_{B_p(\mathbb{H})}(M_2(Z_p(\lambda_{p^\ell})), \hat{Z}_p) = \begin{cases} (Z/p)^2 & i \equiv 0(4) \\ 0 & \text{otherwise.} \end{cases}$

Proof: In the Hochschild-Serre spectral sequence

$$E^2_{i,k} = H_i(Z/2, H^{(r_j)}_k(Z/p^j; M_2(\hat{Z}_p(\lambda_{p^j}))))$$

which collapses to $E^2_{0\,*}$ since $p \neq 2$. But

$$E^2_{0,k} = H_k(Z/p^j, M_2(\hat{Z}_p(\lambda_{p^j})))/\mathrm{im}\,(t-1)$$

and $\mathrm{im}(t-1) = 0$, $k \equiv 0(4)$ but $\mathrm{im}(t-1) = \mathrm{im}(-2)$ if $k \equiv 2(4)$ so 3.B.5. follows.

Similarly, for $r_\pm$ we have

Corollary 3.B.6: (a) $\mathrm{Tor}^i_{B_p(\mathbb{H})}(\hat{Z}^+_p, \hat{Z}_p^+) = \begin{cases} \hat{Z}_p & i = 0 \\ Z/p^\ell & i \equiv 3(4) \\ 0 & \underline{\text{otherwise}}. \end{cases}$

(b) $\mathrm{Tor}^i_{B_p(\mathbb{H})}(\hat{Z}^-_p, \hat{Z}^+_p) = \begin{cases} Z/p^\ell & i \equiv 1(4) \\ 0 & \underline{\text{otherwise}}. \end{cases}$

Remark 3.B.7: Exactly the same chain of ideas works to calculate the Tor's in the case of $Z/p^\ell \times_T Z/_{q^s}$ where $q^s \mid (p-1)$. Here

3.B.8: $$M_p = \coprod_{1 \le j \le \ell} M_{q^s}(\hat{Z}_p(\rho_{p^j})^{(\mathrm{inv})}) \oplus \coprod_1^{q^s} \hat{Z}_p$$

the latter terms corresponding to embeddings $Z/q^s \hookrightarrow (\hat{Z}_p)$; taking the generator to the varions q^s'th roots of unity.

C. <u>The situation at</u> 2 <u>and the completion of the proof of</u> 2.B.5

$$B_2(\mathbb{H}) = \hat{Z}_2(H_2) = \hat{Z}_2(\{x,y \mid x^2 = y^{2^i},\ xyx^{-1} = y^{-1}\})$$

and we have

<u>Lemma</u> 3.C.1: <u>The maximal order of</u> $B_2(\mathbb{H})$ is

(a) $$\coprod_{i \geq j \geq 2} M_2(\mathbb{Z}_2(\rho_{2^j} + \rho_{2^j}^{-1})) \oplus \hat{Z}_2^{(4)}$$

<u>if</u> $i > 2$

(b) $\mathfrak{Q} \oplus Z_2^{(4)}$, <u>where</u> $\mathfrak{Q}$ <u>is the maximal order</u> $\hat{Z}_2(\rho_3) \oplus \hat{Z}_2(\rho_3)\pi$, $\pi^2 = -2$, <u>and</u> $\pi\lambda = \Psi(\lambda)\pi$ <u>for</u> $\lambda \in \hat{Z}_2(\rho_3)$ <u>where</u> Ψ <u>is the Galois automorphism of</u> $\hat{Z}_2(\rho_3)$.

(This follows from the results of [11]. However, for $i > 2$ the explicit representations are given easily. In particular the faithful representation ($i=j$) is given by

3.C.2: $$g \to \begin{pmatrix} 0 & 1 \\ -1 & \lambda_{2^i} \end{pmatrix}, \quad t \to \begin{pmatrix} 1 & C-\lambda \\ C & -1 \end{pmatrix}$$

where C is a root of

3.C.3: $$C^2 - \lambda_{2^i} C + 2 = 0.$$

Indeed 3.C.3 splits inside $\hat{\mathbb{Q}}_2(\lambda_{2^i})$ for $i > 2$ since its discriminant

$$d = \lambda^2 - 8 \equiv \lambda^2 (4\pi)$$

and is thus a square. The remaining representations are given by the same formulae as 3.B.3.)

This completes the proof of 2.B.5 since the only representations which restrict non-trivially at both 2 and at least one other prime are the 4 copies of Z.

D. Calculations for $B_2(\mathbb{H})$

A minimal resolution of $\mathbb{H}$ can be found on page 253 of [2]. It is periodic and has the form

$$\xrightarrow{\partial} Z(H_2)e \xrightarrow{\partial_3} Z(\mathbb{H}_2)c \oplus Z(\mathbb{H}_2)c' \xrightarrow{\partial_2} Z(\mathbb{H}_2)b \oplus Z(\mathbb{H}_2)b' \xrightarrow{\partial_1} Z(H_2)a \xrightarrow{\varepsilon} Z$$

where

$$\partial_3(e) = (y-1)c - (yx-1)c'$$

$$\partial_2(c') = (yx+1)b + (y-1)b'$$

3.D.1:

$$\partial_2(c) = (1+y+y^2+\dots+y^2+\dots+y^{2^{i-1}-1})b - (x+1)b'$$

$$\partial_1(b') = (x-1)a$$

$$\partial_1(b) = (y-1)a.$$

We can use 3.D.1 to calculate Tor groups explicitly in this case. The only ones we need are for the group of order 8 where the results are given by the table.

3.D.2:

	Tor^0	Tor^1	Tor^2	Tor^3
Z^{++}	$\hat{Z}_2$	$Z/2 + Z/2$	0	$Z/8$
Z^{+-}	$Z/2$	$Z/2$	$Z/2$	0
Z^{-+}	$Z/2$	$Z/2$	$Z/2$	0
Z^{--}	$Z/2$	$Z/2$	$Z/2$	0
$\mathcal{Q}$	F_4	F_4	F_4	0

§4. Evaluating the Swan obstruction for the groups $Q(4p,q,1)$, p,q odd primes.

In this section we calculate the Swan obstruction for some of the groups $Q(4p,q,1)$. We begin in 4.A. with partial calculations of the local defects. The method is to do it separately for each p-block and the maximal orders associated with it. As the methods and the results are rather technical they are summarized in Table 4.A.11. Using 4.A.11 the reader can skip to 4.B. where the Swan obstruction is calculated for these groups. Finally, in 4.C we give examples of specific calculations chosen mainly to illustrate the types of complexities encountered. In particular, these examples imply all the results mentioned in the introduction.

A. <u>The local defects.</u>

The p-blocks have the form

4.A.1: (a) $C(Z/p) \times {}_T Z/2 \times Z/2$

(b) $L(Z/p) \times {}_T Z/2$

where $C = \hat{Z}_p$, $\hat{Z}_p(\rho_q)$ or $(\hat{Z}_p(\rho_q))^2$ and $t^2 = -1$, while $L = \hat{Z}_p$, $M_2(\hat{Z}_p(\rho_q+\rho_q^{-1})$ or $M_2(\hat{Z}(\rho_q))$.

The second case occurs in C and L if $p^\nu \equiv -1(q)$ for some ν, and the third case occurs otherwise.

<u>Lemma</u> 4.A.2: <u>The non p-primary part of the local defect of a block of type</u> 4.A.1. <u>is</u>

$$\dot{F}_p(\rho_q+\rho_q^{-1})$$

<u>in cases</u> 2 <u>and</u> 3 <u>in</u> C <u>or</u> L <u>and</u> $\dot{F}_p$ <u>in the first case.</u>

<u>Proof:</u> Let R be the maximal order for 4.A.1 in $\hat{\mathbb{Q}}_p \otimes (\ )$. Then R contains $N = C \otimes_T Z/2 \times {}_T Z/2$ or $L(Z/2)$ as a direct summand, and

$$R = N \oplus M$$

where M is the maximal order in $\hat{\mathbb{Q}}_p \otimes C(\rho_p) \times {}_T Z/2 \times Z/2$ or $\hat{\mathbb{Q}}_p \otimes L(\rho_p) \times {}_T Z/2$.

Let J be the Jacobson radical of the block B, then

$$B|J \cong N/J$$

and hence the local defect C away from (p) for B is obtained from the diagram

4.A.3:
$$\begin{array}{ccccc} \mathcal{N}/J' & \xrightarrow{i} & B/J & \xrightarrow{\ell} & (\mathcal{N} \oplus \mathcal{M})/J'' \\ & & \downarrow \cong & & \downarrow p_2 \\ & & \mathcal{N}/J' & \xrightarrow{p_2 \circ \ell \circ i} & \mathcal{M}/J''' \end{array}$$

Thus, factoring out by imB/J simply identifies N/J with its image under $p_2 \circ \ell \circ i$ in M/J. Finally, observe that any element in one of the Jacobson radicals is p-adic so contributes only p-torsion to the local defect.

At 2 the situation is slightly more involved.

There are 3 types of blocks.

(a) $C_{p,q} \times_T \mathbb{H}_2$

(b) $C_p \times_T \mathbb{H}_2$

(c) $\hat{Z}_2(\mathbb{H}_2)$

Here $C_{p,q}$ is 2 or 4 copies of $\hat{Z}_2(\rho_p,\rho_q)$ and C_p is one or two copies of $\hat{Z}_p(\rho_p)$.

<u>Lemma</u> 4.A.5: <u>The local defect for the block</u> $\hat{Z}_2(\mathbb{H}_2)$ <u>is</u> $Z/2$ <u>and the generator corresponds to</u> $\langle 3\rangle$ <u>at the trivial representation (hence is in the image of</u> T).

(This may be directly obtained from the calculations of [6], [8].)

<u>Lemma</u> 4.A.6: <u>The</u> 2 <u>primary part of the local defect for a block of type</u> 4.A.4 (a) <u>is zero.</u>

<u>Proof:</u> A block of type 4.A.4 (a) has a maximal order $M_4(\mathfrak{O}) \oplus M_4(\mathfrak{O})$ where $\mathfrak{O}$ is one of $\hat{Z}_2(\rho_p,\rho_q)$, $\hat{Z}_2(\rho_p+\rho_p^{-1},\rho_q)$, $\hat{Z}_2(\rho_p,\rho_q+\rho_q^{-1})$ or $\hat{Z}_2(\rho_{pq}+\rho_{pq}^{-1})$. The two idempotents for these representations are

$\frac{1}{2}(y^2-1) = e_-$, $\frac{1}{2}(y^2+1) = e_+$. Suppose $C_{p,q} = (\hat{Z}_2(\rho_p,\rho_q))$, and take $\alpha = (a,0)$, then

$$2e_-\alpha + 1 \longmapsto \left(\begin{pmatrix} 1+2a & 0 & 0 & 0 \\ 0 & 1+2\bar{a} & 0 & 0 \\ 0 & 0 & 1 & 0 \\ 0 & 0 & 0 & 1 \end{pmatrix}, \quad I\right)$$

which has the image $(N(1+2a),1)$. But since the 4 possible extensions of $\hat{Z}_2$ which form the centers are all unramified, the units of the center are all obtained as norms of units in $\hat{Z}_2(\rho_p,\rho_q)$, and 4.A.6 follows. A similar argument holds in case $C = (\hat{Z}_2(\rho_p,\rho_q))^4$.

Finally we evaluate the local defect for 4.A.4(b).

Lemma 4.A.7: The 2 primary part of the defect for 4.A.4(b) is $(\hat{Z}_2(\rho_p+\rho_p^{-1})/m)^+$. The generators come from the representation $C_p \times_T Z/2$ $(t^2 = 1)$, and are represented by units of the form $1 + 2\nu$. The relation with the maximal representation in the block is, in case $C = \hat{Z}_2(\rho_p)$

$$1 + (\alpha+\bar{\alpha})(\iota - 1) \longleftrightarrow 1 + 2(\alpha+\bar{\alpha})$$

and in case $C = \hat{Z}_2(\rho_p)^2$,

$$1 + (\alpha + i\,\beta)(\iota - 1) \longleftrightarrow 1 + 2(\alpha + \iota).$$

Remark 4.A.8: The non-triviality of the local defect in this case is crucial to the calculations, as without it it would be routine to evaluate the Swan obstruction. Thus the reader is advised to check the proof of 4.A.7 most carefully.

Proof: The maximal order for 4.A.4(b) is

4.A.9: $$M_2(\mathfrak{G})^+ \oplus M_2(\mathfrak{G})^- \oplus M_2(\mathfrak{G})$$

when, in case $C_p = \hat{Z}_2(\rho_p)$, $\vartheta = \hat{Z}_2(\rho_p+\rho_p^{-1})$

$$\vartheta' = \hat{Z}_2(\rho_{4p}+\rho_{4p}^{-1})$$

while if $\quad C_p = \hat{Z}_2(\rho_p)^2 \qquad \sigma = \hat{Z}_2(\rho_p).$

$$\sigma = \hat{Z}_2(\rho_{4p}).$$

Also

$$C_p \times_T (\mathbb{H}_2)/J = M_2(Z/2(\rho_p + \rho_p^{-1})) \text{ or } M_2(Z/2(\rho_p)).$$

Thus

$$C_p \times_T \mathbb{H}_2 = M_2(F) \oplus M_2(F)[(y-1),2] \oplus M_2(F)[(g-1)^2,4,\ 2(y-1)] \oplus .$$

writing it in filtered form.

The 2 primary component of $SL_2(F) = \begin{pmatrix} 1 & \alpha \\ 0 & 1 \end{pmatrix} = W$ and since we are only interested in 2 torsion it suffices to analyze the image of

$$\begin{pmatrix} 1 & \alpha \\ 0 & 1 \end{pmatrix}[1 + \theta(y-1) + \theta'_2 + \theta''(y-1)^2 + \theta'''2(y-1) + \theta^{(iv)}4 + \dots].$$

Next, using the idempotents $\frac{1}{2}(y^2-1)$, $\frac{1}{2}(y^2-1)$ we find that

4.A.10 $\qquad (\dot{\sigma}/_{m^2} \times \dot{\sigma}/_{m^2} \times \dot{\sigma}/_{m^2})$

surjects only the 2 primary defect. Then we use in turn each of the 5 remaining generators $(y-1),2$, etc., to obtain alltthe relations in 4.A.10 and complete the proof of 4.A.7.

We summarize the results of §4.A in the following table

4.A.11: TABLE

Block	Defect	Representations
$\hat{Z}_p(Z/p \times_T Z/2)^+$	$\dot{Z}/p$	Z^{++}, Z^{-+}, $M_2(Z(\rho_p+(\rho_p^{-1}))^+$
$\hat{Z}_p(Z/p \times_T Z/2)^-$	$\dot{Z}/p$	Z^{+-}, Z^{--}, $M_2(Z(\rho_p+\rho_p^{-1}))^-$
$\hat{Z}_g(Z/q \times_T Z/2)^+$	$\dot{Z}/q$	Z^{++}, Z^{-+}, $M_2(Z(\rho_q+\rho_q^{-1}))^+$
$\hat{Z}_g(Z/q \times_T Z/2)^-$	$\dot{Z}/q$	Z^{-+}, Z^{--}, $M_2(Z(\rho_q+\rho_q^{+}))^-$
$M_2(\hat{Z}_p(\rho_q+\rho_q^{-1}))(Z/p \times_T Z/2)$	$\dot{F}_p(\rho_q+\rho_q^{-1})$	$M_2(Z(\rho_q+\rho_q^{-1}))^+$ $M_2(Z(\rho_q+\rho_q^{-1}))^-$ $M_4(Z(\rho_p+\rho_p^{-1},\rho_q+\rho_q^{-1}))$
$M_2(\hat{Z}_q(\rho_p+\rho_p^{-1}))(Z/q \times_T Z/2)$	$\dot{F}_q(\rho_p+\rho_p^{-1})$,	$M_2(Z(\rho_p+\rho_p^{-1}))^+$ $M_2(Z(\rho_p+\rho_p^{-1}))^-$ $M_4(Z(\rho_p+\rho_p^{-1}, \rho_q+\rho_q^{-1}))$
$\hat{Z}_2(\rho_p) \times_T H_8$	$\dot{F}_2(\rho_p+\rho_p^{-1}) \times F_2(\rho_p+\rho_p^{-1})^+$,	$M_2(Z(\rho_p+\rho_p^{-1}))^+$ $M_2(Z(\rho_p+\rho_p^{-1}))^-$ W_p
$\hat{Z}_2(\rho_q) \times_T H_8$	$\dot{F}_2(\rho_q+\rho_q^{-1}) \times F_2(\rho_q+\rho_q^{-1})^+$	$M_2(Z(\rho_q+\rho_q^{-1}))^+$ $M_2(Z(\rho_q+\rho_q^{-1}))^-$ W_q
$\hat{Z}_2(H_8)$	$Z/2$	Z^{++}, Z^{+-}, Z^{-+}, Z^{--}, $\mathcal{Q}$
$\hat{Z}_p(\rho_q)(Z/p) \times_T \mathbb{H}_{8_{t^2=-1}}$	$\dot{F}_p(\rho_q+\rho_q^{-1})$	W_q $W_{p,q}$
$\hat{Z}_q(\rho_p)(Z/q) \times_T \mathbb{H}_{8_{t^2=-1}}$	$\dot{F}_q(\rho_p+\rho_p^{-1})$	W_p $W_{p,q}$
$\hat{Z}_2(\rho_p, \rho q) \times_T \mathbb{H}_8$	0	$M_4(Z(\rho_p+\rho_p^{-1}, \rho_q+\rho_q^{-1}))$ $W_{p,q}$

Remark 4.A.12: W_r is the maximal order in

$$\mathbb{Q}(\rho_r) \times_T \mathbb{H}_{8_{t^2=-1}}$$

and $W_{p,q}$ is the maximal order in

$$\mathbb{Q}(\rho_{pq}) \times_T \mathbb{H}_{8_{t^2=-1}}$$

B The Swan Obstruction.

Let us consider $\mathbb{Q}(12,5,1)$. The local defects may be presented in an array as follows

4.B.1:

$$\begin{array}{lll}
\dot{Z}/3 = E(B_3(\mathbb{H})) & \dot{Z}/2 = E(B_2(\mathbb{H})) & \dot{Z}/5 = E(B_5(\mathbb{H})) \\
\dot{Z}/3 = E(B_3(\mathbb{H})^-) & & \dot{Z}/5 = E(B_5(\mathbb{H})^-) \\
 & Z/2 = E(B_2(\hat{Z}_2(\rho_3)(\mathbb{H}_2)) & \\
 & \dot{F}_4 \times (Z/2)^2 = E(\hat{Z}_2(\rho_5)(\mathbb{H}_2)) & \\
\dot{F}_9 & & \dot{Z}/5 \\
\dot{Z}/3(y^2 = -1) & & \dot{Z}/5(t^2 = -1) \\
\dot{F}_9(y^2 = -1) & & \dot{Z}/5(t^2 = -1)
\end{array}$$

The last 2 rows correspond to the blocks of type 4.A.4(b) and 4.A.4(a) respectively, which at odd primes are obtained from the blocks of type 4.A.1(b) with $Y^2 = -1$ or 4.A.1(a) respectively.

Remark 4.B.2: Lemmas 4.A.2 and 4.A.5 together imply that im T is precisely the first row of 4.B.1, and we have

Lemma 4.B.3 : The units in row 2 of 4.B.1 may be identified with the corresponding units in row 1 on factoring out by global units.

Proof: Take the global representation

$M_2(Z(\rho_5 + \rho_5^{-1}))^-$. Its units go to the generators of $\dot{Z}/5$ in row 2, to some elements in $\dot{F}_4 \times (Z/2)^2$ and to $\dot{F}_9$. But the corresponding units of $M_2(Z(\rho_5 + \rho_5^{-1}))^+$ hit F5 on row one and are identified with the images of $M_2(Z(\rho_5 + \rho_5^{-1}))^-$ in rows 3 and 4. The same argument works for $M_2(Z(\rho_3 + \rho_3^{-1}))$.

Thus, since the Swan obstruction is in $D(\mathbb{H})/T$, we may ignore the first two rows in 4.B.1. But these rows contain all the information from the 4 copies of Z. Now, by 2.B.5, the only remaining M_i which give torsion give it only for 2,3, or 5 <u>separately</u>.

We now describe the Swan obstruction. Choose periodic resolutions of $B_3(\mathbb{H})$, $B_5(\mathbb{H})$, $B_2(\mathbb{H})$, which can all be chosen of the form $\hat{Z}_p \to \Lambda \to \Lambda^2 \to \Lambda^2 \to \Lambda \xrightarrow{\varepsilon} \hat{Z}_p$ and otherwise, on the higher blocks the resolution locally are

$$\begin{array}{ccccccc} B & \xrightarrow{id} & B & & B & \xrightarrow{id} & B \\ & & \oplus & & \oplus & & \\ & & B & \xrightarrow{id} & B & & \end{array} .$$

Now, the complex of maximal order is

4.B.4

(a)
$$\begin{array}{ccccccc} \mathcal{Q} & \xrightarrow{(i-1)} & \mathcal{Q} & & \mathcal{Q} & \xrightarrow{(i-1)} & \mathcal{Q} \\ & & \oplus & & \oplus & & \\ & & \mathcal{Q} & \xrightarrow{(i-1)} & \mathcal{Q} & & \end{array}$$

(b)
$$M_2(Z(\rho_p + \rho_p^{-1})) = \Lambda\ ,\ \theta = \begin{pmatrix} 2-(\rho_p + \rho_p^{-1}) & 0 \\ 0 & 1 \end{pmatrix}$$

and

$$\begin{array}{ccccccc} \Lambda & \xrightarrow{id} & \Lambda & & \Lambda & \xrightarrow{\theta} & \Lambda \\ & & \oplus & & \oplus & & \\ & & \Lambda & \xrightarrow{id} & \Lambda & & \end{array} .$$

Then, using §2.C. we have $\sigma_4(Q(12,5,1))$ given by

4.B.5

$$\begin{array}{ccc} * & * & * \\ * & & * \\ 2-\rho_5-\rho_5^{-1} & & 2-\rho_3-\rho_3^{-1} \\ & 2-\rho_3-\rho_3^{-1} & \\ & 2-\rho_5-\rho_5^{-1} & \\ 2 & & 2 \\ 1 & & 1 \end{array}$$

Additionally, $\rho_3+\rho_3^{-1}= -1$ and on F_3, $2 = -1$ so $2-\rho_5-\rho_5^{-1}=-\rho_5^{-1}(\rho_5^3-1)/(\rho_5-1)$.

Also, at level 4, $\rho_5^2+\iota\rho_5^{-2}$ at $Z(\rho_{20}) \times_T Z/2 | t^2 = -1$ has norm

$$2 + \iota\rho_5 - \iota\rho_5^{-1}$$

which is congruent to 2 mod the maximal ideal over 5.

At level 5 note that $\rho_5^2 + \rho_5^{-2}$ is the image of $z^2 + yz^{-2}$ (in the notation of 3.B) which, at the maximal representation goes to

$$\begin{pmatrix} \rho_5^2 & 0 & \rho_5^{-2} & 0 \\ 0 & \rho_5^{-2} & 0 & \rho_5^2 \\ -\rho_5^{-2} & 0 & \rho_5^2 & 0 \\ 0 & -\rho_5^2 & 0 & \rho_5^{-2} \end{pmatrix}$$

and the determinant of this is $(\rho_5^2 + \rho_5^{-2})^2 = 2 + \rho_5 + \rho_5^{-1}$.

At level 3 and in the 2 local part

$$z^2 + yz^{-2} \to \left(\begin{pmatrix} \rho_5^2 + \rho_5^{-2} & 0 \\ 0 & \rho_5^{-2} + \rho_5^2 \end{pmatrix}, \begin{pmatrix} \rho_5^2 - \rho_5^{-2} & 0 \\ 0 & \rho_5^{-2} - \rho_5^2 \end{pmatrix}, * \right)$$

and so using $(\rho_5^2 + \iota\rho_5^{-2})$ we change the obstruction to

$$\begin{matrix} * & * & * \\ * & & * \\ -2+\rho_5+\rho_5^{-1} & & -3 \\ & 1 & \\ & 1 & \\ 1 & & 1 \\ (\rho_5+\rho_5^{-1})^2 & & 1 \end{matrix} .$$

Actually, the procedure outlined above is general and we have

<u>Theorem</u> 4.B.6: <u>In</u> $E_p \oplus E_q \oplus E_2$ <u>the lift of the Swan obstruction</u> (σ_4) <u>for</u> $\mathbb{Q}(4q,p,1)$ <u>can be chosen to be</u>

$$\begin{array}{ccc}
* & * & * \\
* & & * \\
\mathrm{im}(-2+\rho_q+\rho_q^{-1}) & & \mathrm{im}(-2+\rho_p+\rho_p^{-1}) \\
 & 1 & \\
 & 1 & \\
1 & & 1 \\
(\rho_q+\rho_q^{-1})^2 & & (\rho_p+\rho_p^{-1})^2
\end{array} .$$

C. <u>Examples.</u>

<u>Proposition</u> 4.C.1: <u>The Swan obstruction</u> $\sigma_4(\mathbb{H})$ <u>for</u> $\mathbb{H} = \mathbb{Q}(12,13,1)$ <u>is zero.</u>

<u>Proof</u>: In $Z(\rho_{13})$ the ideal (3) splits as $(3) = P_1 P_2 P_3 P_4$ and these are interchanged in pairs by conjugation. Hence the image of $-2+\rho_{13}+\rho_{13}^{-1}$ at the 2 primes over (3) in $Z(\rho_{13}+\rho_{13}^{-1})$ are respectively

$$-2 + \rho_{13} + \rho_{13}^{-1} \quad \text{and} \quad -2 + \rho_{13}^{2} + \rho_{13}^{-2}$$

but $Z(\rho_{13} + \rho_{13}^{-1})/P_1 = F_{27}$ and we have

$$N(1 + \rho_{13}+ \rho_{13}^{-1}) = \rho_{13}\left(\frac{\rho_{13}^{3}-1}{\rho_{13}-1}\right)\left(\frac{\rho_{13}^{9}-1}{\rho_{13}^{3}-1}\right)\left(\frac{\rho_{13}-1}{\rho_{13}^{9}-1}\right) = 1 ,$$

hence $1 + \rho_{13} + \rho_{13}^{-1}$ is a square in F_{27} and similarly for $1+\rho_{13}^{2}+\rho_{13}^{-2}$.

Again $3^3 \equiv -3(13)$ so -3 is a square but not a 4th power mod (13). Note in particular that 2 is a primitive generator mod (13) and

$(\rho_{13}+\rho_{13}^{-1})^{10} \equiv -3(\rho_{13}-1)$. Now use the unit $(\rho_{13}+\rho_{13}^{-1})^2$ at $M_2(Z(\rho_{13}+\rho_{13}^{-1}))$. Since it is a square it leaves $(1+\rho_{13}^{\varepsilon}+\rho_{13}^{-\varepsilon})$ equal to a square and sets $-3 \equiv 1$. The only remaining obstruction is $(\rho_{13}+\rho_{13}^{-1})^{-2}$ at level 5. But this is a square in each F_{27}.

Hence, an odd multiple of $\sigma_3(\mathbb{H})$ is trivial. But there is an orthogonal, free representation of $\mathbb{Q}(2^i a,b,c)$ in $O(8)$ (see e.g. [22]), hence $2\sigma_3(\mathbb{H}) = 0$ and 4.C.1 follows.

In marked contrast to 4.C.1 we have

<u>Proposition</u> 4.C.2: <u>The Swan obstruction</u> $\sigma_3(\mathbb{H})$ <u>for the group</u> $\mathbb{Q}(12,5,1)$ <u>is non-zero</u>.

<u>Proof</u>: We begin by noting that all global representations at levels 4 and 5are quaternionic at ∞. This is evident for $\mathbb{Q}(\rho_{4p}) \times_T Z/2, t^2 = -1$, and proved in [11] for the faithful representation. Hence, the global units which occur for them must be positive at all infinite places.

The centers in question are

$$K_1 = \mathbb{Q}(\rho_{12}+\rho_{12}^{-1}),\ K_2 = \mathbb{Q}(\rho_{20}+\rho_{20}^{-1}),\ K_3 = \mathbb{Q}(\rho_5+\rho_5^{-1})$$

$$U(K_1) = Z \times Z/2,\ U(K_2) = Z^3 \times Z/2,\ U(K_3) = Z \times Z/2.$$

<u>Lemma</u> 4.C.3: a. <u>The generating unit of</u> $U(K_1)$ <u>is positive at all infinite places, and</u> $N(\rho_3 + i\rho_3^{-1})$ <u>is an odd power of this generator.</u>

b. The positive elements of $U(K_2)$ are generated by $U(K_2)^2$, $N(\rho 5 + i\rho 5^{-1})$.

c. <u>The positive elements of</u> $U(K_3)$ <u>are</u> $U(K_3)^2$.

<u>Proof</u>: Consider the units of $\mathbb{Q}(\rho_n)$ as compared with those in the maximal real subfield $\mathbb{Q}(\rho_n+\rho_n^{-1})$. By the Dirichlet unit theorem the ranks of the torsion free parts are equal, and H. Hasse [7] has proved index (torsion free part $U(\mathbb{Q}(\rho_n+\rho_n^{-1}))$) in (torsion free part $U(\mathbb{Q}(\rho_n))$) is 2 if n is composite, and one if n is a prime power. In particular, for $n = 4p$ the extra unit ν has the property $\bar{\nu} = i\nu$. Hence, $\rho_p + i\rho_p^{-1}$ represents this extra unit. Clearly, its norm is positive, and evidently, a non-square. This proves (a).

To show (b) we check the signs of the quotients $\rho_p + \iota\rho_p^{-1}/\rho_p^j + \iota\rho_p^{-j} = \lambda_j$. These are invariant under conjugation, so contained in the real subfield, and we easily check that the signs are independent for a suitable subclass of them. Thus, the λ_j and $N(\nu)$ generate $\mathbf{U}(K_2)$ up to odd index, so (b) follows.

(c) is similar, but easier. $\rho_5 + \rho_5^{-1} = \frac{-1+\sqrt{5}}{2}$ has norm -1. Hence, the signs of its infinite embeddings are (+,-).

We now return to the proof of 4.C.2. Look at the level 4 and 5 part of 4.B.6. An element λ_j^2 changes

$$(\rho_5 + \rho_5^{-1})^{-2} \quad \text{to} \quad (\rho_5 + \rho_5^{-1})^{-2}(\tilde{\lambda})^4.$$

But in $\dot{F}_9$ $(\rho_5 + \rho_5^{-1})$ has order 8 since taking norms gives

4.C.4: $$N(\rho_5 + \rho_5^{-1}) = N\left(\rho_5^{-1}\left(\frac{\rho_5 - 1}{\rho_5 - 1}\right)\right) = -1$$

Thus, the only possible element for removing $(\rho_5 + \rho_5^{-1})^{-2}$ is $(\rho_5 + \rho_5^{-1})^2$ in $\mathbf{U}(K_3)$. But at 5 this is 4 $\equiv -1(5)$ and these are no remaining global units to convert this -1 to a 1. 4.C.2. follows.

For $\mathbf{Q}(12,5,1)$ the Swan obstruction was non-zero in level 5. Our next example shows the obstruction can also be non-trivial in level 3.

<u>Proposition</u> 4.C.5: <u>The Swan obstruction</u> $\sigma_3(\mathbf{H})$ <u>for</u> $\mathbf{Q}(12,7,1)$ <u>is non-zero.</u>

<u>Proof:</u> Referring to 4.B.6, and noting that $\hat{\mathbf{Q}}_3(\rho_7)$ has degree 6 over $\hat{\mathbf{Q}}_3$, we apply a calculation analogous to 4.C.4 to show that an odd multiple of $\sigma_3(\mathbf{H})$ is represented by

4.C.6: $$\begin{bmatrix} * & * & * \\ * & & * \\ -1 & & 1 \\ & 1 & \\ & 1 & \\ 1 & & 1 \\ 1 & & 1 \end{bmatrix}$$

We must, as before, study units, this time for $K_4 = \mathbb{Q}(\rho_7 + \rho_7^{-1})$ and $K_5 = \mathbb{Q}(\rho_{28} + \rho_{28}^{-1})$. We have

Lemma 4.C.7: a. The units of K_4 are generated up to odd order by $\rho_7 + \rho_7^{-1}$, $\rho_7^3 + \rho_7^{-3}$, -1.

b. The positive units in K_5 are generated by $(\mathbf{U}(K_5)^2)$, $N(\rho_7 + \iota\rho_7^{-1})$)

(a) is well known, see e.g. [3], and (b) follows as in 4.C.3.

Now, we cannot use -1 to cancel the -1 in 4.C.6, as we easily check. Moreover, the effect of $N(\rho_7 + \iota\rho_7^{-1})$ at E_2 has already been noted, and the remaining positive units of K_5, being squares, have no effect. Hence, the only remaining candidates are $\rho_7 + \rho_7^{-1}$, $\rho_7^3 + \rho_7^{-3}$. However, a calculation analogous to 4.C.4 shows

$$N(\rho_7 + \rho_7^{-1}) = N(\rho_7^3 + \rho_7^{-3}) = 1$$

from F_{27} to F_3 and 4.C.5 follows.

Remark 4.C.8: A similar calculation in the case of $\mathbb{Q}(12,11,1)$ shows $\sigma_3(\mathbb{H})$ is non-zero in this case as well. As the details are similar to those in 4.C.3, 4.C.5 we omit them.

Remark 3.C.9: In view of Wall's result that all remaining p-hyperelementary periodic groups of period n have $\sigma_n(\mathbb{H}) = 0$, we see that for hyperelementary groups of order less than $280 = |Q(20,7,1)|$ the only groups with $\sigma_n(\mathbb{H}) \neq 0$ are $Q(12,5,1)$, $Q(12,7,1)$, $Q(24,5,1)$ and $Q(12,11,1)$. (Note that $Q(24,5,1) \supset Q(12,5,1)$ as a subgroup so restriction shows its Swan invariant is non-zero.)

In addition, we obtain an infinite number os composite groups with non-trivial Swan obstruction, for example $\mathbb{Q}(12,5,q)$, $\mathbb{Q}(12,5q,1)$ etc. But at present we don't have an infinite number of groups $Q(4p,q,1)$ for which the obstruction $\sigma_3(\mathbb{H})$ vanishes.

Remark 4.C.10: The theorem in the introduction follows from 4.C.9 and Wall's results [22].

Remark 4.C.11: Further progress in these questions would seem to involve identifying some odd index subgroups of $\mathbf{U}(\mathbb{Q}(\rho_p))$ and $\mathbf{U}(\mathbb{Q}(\rho_{4p}))$, but these appear to be very difficult problems.

Stanford University

Bibliography

[1] J.W. Cassels - A. Fröhlich, Algebraic Number Theory, Academic Press (1967)

[2] Cartan-Eilenberg, Homological Algebra, Princeton U. Press (1956)

[3] H. M. Edwards, Fermat's Last Theorem, A Geometric Introduction to Algebraic Number Theory, Springer Verlag (1977).

[4] A. Fröhlich, "On the classgroup of integral group rings of finite Abelian groups", Mathematika 16 (1969), 143-152.

[5] ———, Locally free modules over arithmetic orders", J. Reine Angew. Math. 274/75 (1975) 112-138.

[6] A. Fröhlich, M. E. Keating, and S. M. J. Wilson "The classgroup of quaternion and dihedral 2-groups", Mathematika 21 (1974) 64-71.

[7] H. Hasse, Über die Klassenzahl abelscher Zahlenkörper, Berlin, Akademie-Verlag.(1952).

[8] M. E. Keating, "On the K-theory of the quaternion group", Mathematika 20 (1973) 59-62.

[9] S. Lang, Algebraic Number Theory, Addison-Wesley (1968).

[10] Ib Madsen, C. B. Thomas, and C. T. C. Wall, "The topological spherical space form problem II, existence of free actions", Topology 15 (1976) 375-382.

[11] R. J. Milgram, "Determination of the Schur subgroup", mimeo Stanford (1977).

[12] D. Quillen, "Higher algebraic K-theory I", Algebraic K-theory-I, Lecture Notes in Methematics, Vol. 341, Springer-Verlag (1973) 85-147.

[13] I. Reiner, Maximal Orders, Academic Press (1975).

[14] I. Reiner, S. Ullom, "A Mayer-Vietoris sequence for class groups", J. Alg. 31 (1974), 305-342.

[15] ———, ———, "Class groups of integral group rings", Trans. A. M. S. 170 (1962) 1-30.

[16] R. G. Swan, "Induced representations and projective modules", Ann. of Math. (2) 71 (1960) 552-578.

[17] ________, "Periodic resolutions for finite groups", Ann. of Math. (2) 72 (1960) 267-291.

[18] ________, K-theory of finite groups and orders, Notes by E. G. Evans, Lecture Notes in Mathematics #149, Springer-Verlag (1970).

[19] C. B. Thomas, "Free actions by finite groups on S^3", Proceedings of Symposia in Pure Mathematics, Vol. 32 (1977) A. M. S.

[20] C. B. Thomas and C. T. C. Wall, "The topological space form problem I", Compositio Math 23 (1971) 101-114.

[21] S. Wang, "On the commutator group of a simple algebra", Amer. J. of Math 72 (1950) 323-334.

[22] C.T.C. Wall, "Free actions of finite groups on spheres", Proceedings of Symposia in Pure Mathematics", Vol. 32 (1977) A.M.S.

[23] ________, "Finiteness conditions for CW-complexes", Ann. of Math. 81 (1965) 56-69.

[24] ________, "Periodic projective resolutions", Proc. L. M. S. 39 (1979), 509-553.

[25] R. J. Milgram, "Odd-index subgroups of units in cyclotomic fields and applications", Lecture Notes in Mathematics, vol. 854, 269-298.

[26] J. F. Davis and R. J. Milgram, "The spherical space form problem", Harwood (1984).

[27] S. Bentzen, Thesis, Aarhus (1983)

Department of Mathematics
Stanford University
Stanford, CA 94305, USA

THE CAPPELL-SHANESON EXAMPLE

by R. James Milgram,

Stanford University

Introduction

In this note we shall be considering the quaternion group $Q_8 = \{x,y \mid x^2 = y^2 = (xy)^2\}$, denoting it by π. From [6] we have that $\tilde{K}_0(\mathbb{Z}(\pi)) = \mathbb{Z}/2$ with generator the Euler characteristic of the trivial $\mathbb{Z}(\pi)$-module $\mathbb{Z}/3$, which we denote $<3^{++}>$.

From e.g. [5] we have that $L_1^h(\mathbb{Z}(\pi)) = \mathbb{Z}/2 \oplus \mathbb{Z}/2$ with a canonical non-zero class [A] given by the image of the non-trivial class in $\hat{H}_{ev}(\mathbb{Z}/2; \tilde{K}_0(\mathbb{Z}(\pi)))$ in the Ranicki-Rothenberg exact sequence.

Now, π acts freely on S^3, and in [2] Cappell and Shaneson prove that the surgery obstruction in $L_1^h(\mathbb{Z}(\pi))$ of the map

I.1 $$1 \times f : S^3/\pi \times K^{4j+2} \longrightarrow S^3/\pi \times S^{4j+2}$$

is non-trivial, and given by [A], with f representing the simply-connected Kervaire problem. However, their proof proceeds by an intricate "peeling" argument, and it has seemed desirable for a number of reasons to have a purely algebraic proof of their result.

In the current volume Jim Davis' paper [3] is concerned with this question, and provides a general recognition principle by which one can decide if a symmetric Poincaré structure on a chain complex (necessary for the application of the surgery product formula of [7]) is "geometric". Also Hambleton and Ranicki in as yet unpublished joint work have obtained other algebraic proofs based on "peeling".

In this note we first construct a 3-dimensional chain complex C_* of f.g. free $Z(\pi)$-modules and a chain equivalence $\phi : C^{3-*} \longrightarrow C_*$. We then analyze the class of ϕ in $\mathbb{Z} \otimes_{\mathbb{Z}(\pi)} (C_* \otimes C_*)$ and note that this class determines ϕ up to chain homotopy. Comparing this class with that of the diagonal chain map gives that ϕ is the base map of a suitable symmetric Poincaré structure on C_*, so taking the product of (C_*, ϕ) and the algebraic Kervaire problem gives an explicit quadratic Poincaré complex whose surgery obstruction is that of I.1.

This problem is then quickly evaluated (the procedures used here may have independent interest) and the Cappell-Shaneson example is the result. Indeed, by way of illustrating this last comment the final section indicates how to extend these results to the remaining compact space forms.

A. The complex C_* and the map ϕ

There is exactly one chain homotopy class of finitely generated free $\mathbb{Z}(\pi)$-module chain complexes with the homology of S^3 (since the set of such classes is given by $U(\mathbb{Z}/8)/\langle\pm 1, \mathrm{im} T\rangle = \{1\}$, with T the Swan subgroup, see e.g. [4]). A representative is given in [1] and is specified as follows

A.1

i	C_i	Generators	∂
0	$\mathbb{Z}(\pi)$	a	0
1	$\mathbb{Z}(\pi)\oplus\mathbb{Z}(\pi)$	b	$(x-1)a$
		b'	$(y-1)a$
2	$\mathbb{Z}(\pi)\oplus\mathbb{Z}(\pi)$	c	$(1+x)b-(y+1)b'$
		c'	$(xy+1)b+(y+1)b'$
3	$\mathbb{Z}(\pi)$	e	$(x-1)c-(xy-1)c'$.

Then C^* is specified by the formulae

A.2
$$\delta(c^*) = (x^3-1)e^*$$
$$\delta(c'^*) = -(yx-1)e^*$$
$$\delta(b^*) = (1+x^3)c^* + (yx+1)c'^*$$
$$\delta(b'^*) = -(y^3+1)c^* + (x^3-1)c'^*$$
$$\delta(a^*) = (x^3-1)b^* + (y^3-1)b'^* \quad .$$

The chain equivalence $\phi: C^{3-*} \longrightarrow C_*$ is given by the equations

A.3
$$\phi(e^*) = a$$
$$\phi(c^*) = -x^3 b$$
$$\phi(c'^*) = -(yb+b')$$
$$\phi(b^*) = (yx-(y^3+1)(1+x))c' + (2y^3-y)c$$
$$\phi(b'^*) = x^{-1}c'$$
$$\phi(a^*) = (2y^3-y)e \quad .$$

B. <u>ϕ is the beginning map in the Mishchenko-Ranicki symmetric structure on S^3/π</u>

The set of chain homotopy classes of $\mathbb{Z}(\pi)$-module chain maps

$$\phi : C^{3-*} \longrightarrow C_*$$

is in 1-1 correspondence with $H_3(\mathbb{Z}\otimes_{\mathbb{Z}(\pi)}(C_*\otimes C_*))$, where $\mathbb{Z}(\pi)$ acts on $C_*\otimes C_*$ via the diagonal map

B.1 $$\Delta : \mathbb{Z}(\pi) \longrightarrow \mathbb{Z}(\pi)\otimes\mathbb{Z}(\pi) = \mathbb{Z}(\pi\times\pi) \ ; \ g \longmapsto g\otimes g \quad .$$

This is well known, see e.g. [7].

<u>Proposition B.2</u> $H_3(\mathbb{Z}\otimes_{\mathbb{Z}(\pi)}(C_*\otimes C_*)) = \mathbb{Z}\oplus\mathbb{Z}$ <u>with generators A,B (say) and the projection</u> $p: C_*\otimes C_* \longrightarrow \mathbb{Z}\otimes_{\mathbb{Z}(\pi)}(C_*\otimes C_*)$ <u>induces an injection</u>

$$p_* : H_3(C_*\otimes C_*) \longrightarrow H_3(\mathbb{Z}\otimes_{\mathbb{Z}(\pi)}(C_*\otimes C_*)) \ ;$$
$$e\otimes 1 + 1\otimes e \longmapsto 8A \ , \ e\otimes 1 \longmapsto B \ .$$

<u>Proof</u>: There is a spectral sequence converging to $H_*(\mathbb{Z}\otimes_{\mathbb{Z}(\pi)}(C_*\otimes C_*))$ with

$$E^2_{i,j} = H_i(\pi, H_j(C_*\otimes C_*))$$

so $E^2_{i,j} \neq 0$ only for $j = 0,3,6$. Moreover $H_4(\pi,\mathbb{Z}) = 0$, so $d_4 = 0 : E_{4,0} \longrightarrow E_{0,3}$ and $H_3(\mathbb{Z}\otimes_{\mathbb{Z}(\pi)}(C_*\otimes C_*))$ is given as an extension

B.3 $$0 \longrightarrow E_{0,3} \xrightarrow{\ i\ } H_3(\mathbb{Z}\otimes_{\mathbb{Z}(\pi)}(C_*\otimes C_*)) \longrightarrow E_{3,0} \longrightarrow 0$$

where $E_{0,3} = \mathbb{Z}\oplus\mathbb{Z}$ and $E_{3,0} = \mathbb{Z}/8$. Moreover i in B.3 is the map p_*.

To determine the extension in B.3 note that the geometric diagonal $d: S^3 \longrightarrow S^3\times S^3$ is π-equivariant, so that there is an algebraic chain approximation such that the diagram below commutes

B.4 $$\begin{array}{ccc} C_* & \xrightarrow{\ d_\#\ } & C_*\otimes C_* \\ \Big\downarrow{\scriptstyle p_0} & & \Big\downarrow{\scriptstyle p} \\ \mathbb{Z}\otimes_{\mathbb{Z}(\pi)} C_* & \xrightarrow{\ \hat{d}_\#\ } & \mathbb{Z}\otimes_{\mathbb{Z}(\pi)}(C_*\otimes C_*) \end{array} \quad .$$

But $H_3(C_*) = \mathbb{Z}$ and $H_3(\mathbb{Z}\otimes_{\mathbb{Z}(\pi)} C_*) = \mathbb{Z}$, with p_{0*} multiplication by 8. On the other hand $d_*: H_3(C_*) \longrightarrow H_3(C_*\otimes C_*)$ sends the generator e to $e\otimes 1 + 1\otimes e$, and B.2 follows.

[]

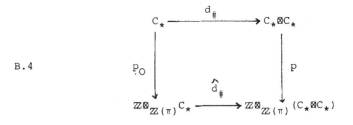

Corollary B.5 The map ϕ defined in A.3 is chain homotopic to the map corresponding to the diagonal in B.4. More exactly $[\phi] = \hat{d}_*(f)$ where f is a generator of $H_3(\mathbb{Z}\otimes_{\mathbb{Z}(\pi)}C_*)$.

Proof: It suffices to show that $[\phi] \in H_3(C_*\otimes C_*)$ is $d_*(e)$ from B.2. But this is the case if and only if

$$(\phi)_* \ : H^3(C) \longrightarrow H_0(C)$$

is dual to $(\phi)_* : H^0(C) \longrightarrow H_3(C)$ and both are isomorphisms. This is easily checked and the result follows since the desired symmetric structure on C_* restricts to the class of the diagonal map in degree 0.

[]

Remark B.6 In this case the class of the lifting in B.4 determined the class. In general this is not true as there may be many classes in $H_n(\mathbb{Z}\otimes_{\mathbb{Z}(\pi)}(C_*\otimes C_*))$ which lift to the same class in $H_n(C_*\otimes C_*)$. (Here "class" means $d_\#, \hat{d}_\#$ and "lift" multiply by the order of π).

[]

C. The evaluation of the surgery obstruction for I.1

Proposition C.1 Let τ be a finite 2-group, and suppose C_*, D_* are finitely generated free $\mathbb{Z}(\tau)$-module chain complexes, with a chain equivalence $\lambda_\#: C_* \longrightarrow D_*$. If $\mathbb{Z}/2\otimes_{\mathbb{Z}(\tau)}C_*$ and $\mathbb{Z}/2\otimes_{\mathbb{Z}(\tau)}D_*$ both have trivial boundary maps then $\lambda_\#$ is an injection. Moreover, for each i $D_i/\mathrm{im}\lambda_i$ is a finite odd torsion module.

Proof: Since $\lambda_\#$ is a chain isomorphism

$$\lambda_* \ : H_*(\mathbb{Z}/2\otimes_{\mathbb{Z}(\tau)}C_*) \longrightarrow H_*(\mathbb{Z}/2\otimes_{\mathbb{Z}(\tau)}D_*)$$

is an isomorphism, but since $\mathbb{Z}/2\otimes_{\mathbb{Z}(\tau)}\partial = 0$ in both complexes it follows that

$$\mathbb{Z}/2\otimes\lambda_\# \ : \mathbb{Z}/2\otimes_{\mathbb{Z}(\tau)}C_i \longrightarrow \mathbb{Z}/2\otimes_{\mathbb{Z}(\tau)}D_i$$

is an isomorphism for each i. Now apply Nakayama's Lemma and C.1 follows.

[]

Corollary C.2 Let (D,ψ) be the $(4i+2)$-dimensional quadratic Poincaré complex over $\mathbb{Z}$ with Kervaire invariant 1,

$$D_{2i+1} = \mathbb{Z}\oplus\mathbb{Z} \quad , \quad D_j = 0 \text{ for } j \neq 2i+1$$

$$\psi_0 = \begin{pmatrix} 1 & 1 \\ 0 & 1 \end{pmatrix} : D^{2i+1} = \mathbb{Z}\oplus\mathbb{Z} \longrightarrow D_{2i+1} = \mathbb{Z}\oplus\mathbb{Z} \ ,$$

and consider the product $(4i+5)$-dimensional quadratic complex

C.3 $\qquad (C_*\otimes D, \phi\otimes\psi)$.

The chain equivalence

C.4 $$\phi\otimes\psi - \phi^*\otimes\psi^* : (C_*\otimes D)^{4i+5-*} \longrightarrow C_*\otimes D$$

is an injection in each degree, with odd torsion cokernel.

Proof: A direct application of C.1.

[]

In dimension 0

$$\lambda_0 = \phi_0\otimes\psi - \phi_0^3\otimes\psi^* = \begin{pmatrix} -1+2y-y^3 & -1 \\ 2y-y^3 & -1+2y-y^3 \end{pmatrix} ,$$

while in dimension 1

$$\lambda_1 = \phi_1\otimes\psi - \phi^2\otimes\psi^* = \begin{pmatrix} x^3+2y-y^3 & x^3 & * & * \\ 2y-y^3 & x^3+2y-y^3 & * & * \\ 0 & 0 & x+1 & 1 \\ 0 & 0 & x & x+1 \end{pmatrix} .$$

Hence the order of an odd torsion quadratic form representing the surgery obstruction of the product $(C_*\otimes D, \phi\otimes\psi)$ is

$$\det(\phi_1\otimes\psi - \phi^2\otimes\psi^*)/\det(\phi_0\otimes\psi - \phi^3\otimes\psi^*) ,$$

where $\det(\theta)$ means the class of θ in $K_1(\mathbb{Q}(\pi))$. Restricting to the five irreducible representations of π we have the table

	Representation	++	+-	-+	--	Q
C.5	λ_0	1	3	1	3	73
	λ_1	9	-3	1	-3	73

Hence the form in question is represented by a torsion module of order 9 at the trivial representation and 0 at all other representations. Since the form is SKEW SYMMETRIC this must be $\mathbb{Z}/3\oplus\mathbb{Z}/3$ with each $\mathbb{Z}/3$ a torsion lagrangian.

But this torsion class exactly represents the image of the class $<3^{++}>$ from $\hat{H}_{ev}(\mathbb{Z}/2,\tilde{K}_0(\mathbb{Z}(\pi)))$ in $L_1^h(\mathbb{Z}(\pi))$ and we have proved that the surgery obstruction for I.1 is non-trivial.

D. The algebraic evaluation of the surgery obstruction for other space forms

Let τ be one of the groups $\mathbb{Z}/2^m$ or $Q_{2^n} = \{x,y \mid x^{2^{n-2}} = y^2 = (xy)^2\}$, so that τ acts freely preserving orientation on the sphere S^{2m+1} in the first case or S^{4n+1} in the second. Each such action corresponds to a finite chain complex with the chain homotopy type unambiguously specified by the first k-invariant κ_{i+1} of the resulting quotient S^i/τ. (See e.g. [4] for discussion and references).

For these complexes $C(\tau,\kappa_{m+1})$ Proposition B.2 generalizes, and the only change in the statement is

D.1 $$p_*(e\otimes 1 + 1\otimes e) = |\tau|A \quad .$$

D.1 together with C.1 provide an effective method for determining the obstruction. When $\tau = \mathbb{Z}/2^m$ here is the result.

Proposition D.2 Let $\tau = \mathbb{Z}/2^m$ and suppose $C(\tau,u_{2n+2})$ is the $\mathbb{Z}(\tau)$-module chain complex of dimension $2n+1$

$$\mathbb{Z}(\tau) \xrightarrow{x^u-1} \mathbb{Z}(\tau) \xrightarrow{\Sigma} \cdots \xrightarrow{\Sigma} \mathbb{Z}(\tau) \xrightarrow{x-1} \mathbb{Z}(\tau) \xrightarrow{\Sigma} \mathbb{Z}(\tau) \xrightarrow{x-1} \mathbb{Z}(\tau)$$

with Poincaré duality chain equivalence $\phi: C(\tau,u_{2n+2})^{2n+1-*} \longrightarrow C(\tau,u_{2n+2})$, for some unit u in the ring $\mathbb{Z}/2^n$. Then

a. For n+1 odd $\sigma((C(\tau,u_{2n+2}),\phi)\otimes\text{Kervaire}) = 0 \in L^h_{2n+3}(\mathbb{Z}(\tau))$,

b. For n+1 even $\sigma((C(\tau,u_{2n+2}),\phi)\otimes\text{Kervaire}) \neq 0 \in L^h_{2n+3}(\mathbb{Z}(\tau))$ and has non-trivial image in $L^p_3(\mathbb{Z}(\tau)) = \mathbb{Z}/2$.

Proof: The geometrically induced ϕ is such that $\phi_0 = \text{id.}$ and $\phi_{2n+1} = (-)^{n+1}.\text{id.}$ Hence a suitable ϕ is given by the table

D.3

dimension	ϕ
0	id
1	$1+x+\ldots+x^{m-u-1}$
2	$(m-u)$
3	$-(m-u)x^{-1}$
4	$-(m-u)$
5	$(m-u)x^{-1}$
6	$(m-u)$
7	$\vdots$
$\vdots$	
2n	$(-1)^{n+1}(1+x^{-1}+\ldots+x^{-(m-u-1)})$
2n+1	$(-1)^{n+1}$

.

Moreover (C_*,ϕ) satisfies the conditions of C.1, so

D.4 $$(C_*\otimes \text{Kervaire}, \phi\otimes\psi - (-1)^{n+1}\phi^*\otimes\psi^*)$$

satisfies the hypothesis of C.1 as well. Hence the answers come from evaluating the alternating product of the images in K_1 of the $(\phi_i\otimes\psi - (-1)^{n+1}\phi^*_{2n+1-i}\otimes\psi^*)$ through dimension $i = n$.

This calculation is direct. The matrices which appear are $\begin{pmatrix} 0 & 1 \\ -1 & 0 \end{pmatrix}$ in degree 0, $\begin{pmatrix} 0 & 1+x+..+x^{u-1} \\ -(1+x+..+x^{u-1}) & 0 \end{pmatrix}$ in 1, and $u\begin{pmatrix} 1-x^{-1} & 1 \\ -x^{-1} & 1-x^{-1} \end{pmatrix}$ otherwise. The result is alternately

D.5 $$\left\{ \frac{u^2(1-x^{-1}+x^{-2})}{(1+x+..+x^{u-1})^2} \right\} \qquad n+1 \text{ even}$$

$$\{(1+x+..+x^{u-1})^2\} \qquad n+1 \text{ odd} \quad .$$

The odd case clearly gives 0. For the even case we check at the trivial representation and the -1 representation $r_-(x^i) = (-1)^i$ obtaining

D.6 $$\begin{pmatrix} + & - \\ u^2 & 3u^2 \end{pmatrix}$$

which represents the non-trivial element in $L_3^p(\mathbb{Z}/2^m)$.

Bibliography

[1] H.Cartan and S.Eilenberg
Homological algebra
Princeton University Press (1956)

[2] S.Cappell and J.Shaneson
A counterexample to the oozing conjecture
Algebraic Topology, Arhus '78, Springer Lecture Notes 763 (1979), 627 - 634

[3] J.F.Davis Higher diagonal approximations and skeletons of $K(\pi,1)$'s
these proceedings

[4] ______ and R.J.Milgram
The spherical space form problem
Harwood (1984)

[5] I.Hambleton and R.J.Milgram
The surgery obstruction groups for finite 2-groups
Invent. Math. 61 (1980), 33 - 52

[6] M.E.Keating On the K-theory of the quaternion group
Mathematika 20 (1973), 59 - 62

[7] A.A.Ranicki The algebraic theory of surgery
Proc. L.M.S. 40 (1980), 87 - 283

A Nonconnective Delooping of Algebraic K-theory

by

Erik K. Pedersen and Charles A. Weibel*

Abstract: Given a ring R, it is known that the topological space $BGl(R)^+$ is an infinite loop space. One way to construct an infinite loop structure is to consider the category $\underline{F}$ of free R-modules, or rather its classifying space $B\underline{F}$, as food for suitable infinite loop space machines. These machines produce connective spectra whose zeroth space is $(B\underline{F})^+ = \mathbb{Z}\times BGl(R)^+$. In this paper we consider categories $\underline{C}_0(\underline{F}) = \underline{F}$, $\underline{C}_1(\underline{F}),\ldots$ of parametrized free modules and bounded homomorphisms and show that the spaces $(B\underline{C}_0)^+ = (B\underline{F})^+$, $(B\underline{C}_1)^+,\ldots$ are the connected components of a nonconnective Ω-spectrum $B\underline{C}(F)$ with $\pi_i B\underline{C}(F) = K_i(R)$ even for negative i.

0. Introduction.

Given a ring R, let $\underline{F}$ be the category of finitely generated free R-modules and isomorphisms. Form the "group completion" category $\underline{F}^{-1}\underline{F}$ of $\underline{F}$ (see [G]); it is known that its classifying space $B\underline{F}^{-1}\underline{F}$ is the algebraic K-theory space $BGl(R)^+\times\mathbb{Z}$. The purpose of this paper is to produce a nonconnective delooping of $BGl(R)^+ \times K_0(R)$ by using the parametrized versions $\underline{C}_0(\underline{F}) = \underline{F}$, $\underline{C}_1(\underline{F}),\ldots$ of $\underline{F}$ given in [P]. Our main result is this:

Theorem A. Write B_i for the classifying space of the category $\underline{C}^{-1}\underline{C}$, except that $B_0 = BGl(R)^+$. Then the spaces B_i are connected, and for $i \geq 0$ we have

$$\Omega B_{i+1} = B_i \times K_{-i}(R).$$

* Partially supported by an NSF/grant

hus the sequence of spaces $\hat{B}_i = B_i \times K_{-i}(R)$ forms a nonconnective -spectrum $\hat{\underline{B}}$ with homotopy groups

$$\pi_i(\hat{\underline{B}}) = K_i(R) \ , \quad i \text{ any integer.}$$

n particular, the negative homotopy groups of $\hat{\underline{B}}$ are the negative -groups of Bass [B].

Actually, we work in the generality of a small additive ategory $\mathcal{A}$, rather than just with the additive category $\mathcal{F}$ of initely generated free R-modules. For example, one could take $\mathcal{P}$, he category of finitely generated projective R-modules. The ategory $\mathcal{P}$ is the idempotent completion of $\mathcal{F}$, and we recover the ame spectrum $\hat{\underline{B}}$ if we replace $\mathcal{F}$ by $\mathcal{P}$. Note that $B\underline{P}^{-1}\underline{P}$ is $BGl(R)^+$ $K_o(R)$, where $\underline{P}$ is the category of isomorphisms in $\mathcal{P}$.

Given $\mathcal{A}$, we consider the additive categories $\mathcal{C}_i(\mathcal{A})$ of $\mathbf{Z}^i$-graded bjects and bounded homomorphisms (see section 1 for details). If $= \mathcal{F}$ this definition specializes to the groups $\mathcal{C}_i$ of [P]. Let $\hat{\mathcal{C}}_i$ be he idempotent completion of $\mathcal{C}_i(\mathcal{A})$, and let $\underline{A}$, $\underline{C}_i$, $\hat{\underline{C}}$ be the ubcategories of isomorphisms in $\mathcal{A}$, $\mathcal{C}_i$ and $\hat{\mathcal{C}}_i$, respectively. Our econd result is this:

heorem B. Write $\hat{B}_i$ for the classifying space of the category $_i^{-1}\hat{\underline{C}}_i$ and B_i for the classifying space of $\underline{C}_i^{-1}\underline{C}_i$. Then

$$\Omega\hat{B}_{i+1} = \hat{B}_i$$

$$\Omega^i \hat{B}_i = \hat{B}_o = \text{"group completion" } (B\underline{A})^+ \text{ of } B\underline{A} \ .$$

he connected component of $\hat{B}_i$ is B_i (except for $i=0$), and the equence of spaces $\hat{B}_o, \hat{B}_1, \ldots$ is a nonconnective Ω-spectrum. In articular, $\hat{B}_i$ is an i-fold delooping of $(B\underline{A})^+$.

The outline of this paper is as follows. In section 1 we give he definitions of the $\mathbf{Z}^i$-graded categori $\mathcal{C}_i(\mathcal{A})$. In section 2, we ecall the passage from categories to spectra, and review the main oints of Thomason's paper [T] that we need. In section 3, we prove heorems A and B.

The authors would like to thank Bob Thomason for his lucid xposition in [T], which clarified a number of technical points.

The second author would also like to thank the Danish Natural Science Research Council and Odense University for its hospitality during the writing stage.

1. The categories $\mathcal{C}_i$.

In this section we give the definition of the categories $\mathcal{C}_i(\mathcal{A})$ associated to a small additive category $\mathcal{A}$. We also review the notions of filtered additive categories and of the idempotent completion of $\mathcal{A}$ for the convenience of the reader.

Definition 1.1. An additive category $\mathcal{A}$ is said to be filtered if there is an increasing filtration

$$F_o(A,B) \subseteq F_1(A,B) \subseteq \ldots \subseteq F_n(A,B) \subseteq \ldots$$

on Hom(A,B) for every pair of objects A,B of $\mathcal{A}$. Each $F_n(A,B)$ is to be a subgroup of Hom(A,B) and we must have $\cup F_n(A,B) = \text{Hom}(A,B)$. We require 0_A and 1_A to be in $F_o(A,B)$, and assume that the composition of morphisms in $F_m(A,B)$ and $F_n(B,C)$ belongs to $F_{m+n}(B,C)$. We also assume that the projections $A\oplus B \to A$, and inclusions $A \to A\oplus B$ and coherence isomorphisms all belongs to F_o. If ϕ is in $F_d(A,B)$ we say that ϕ has filtration degree d.

The reason for concerning ourselves with filtered categories is that the categories $\mathcal{C}_i$ come with a natural filtration. Of course every additive category has a trivial filtration, obtained by setting $F_o(A,B) = \text{Hom}(A,B)$.

Example (1.1.1). Given a $\mathbb{Z}$-graded ring A such as $R[t,t^{-1}]$, let $\mathcal{A}$ be the category of graded A-modules. We can filter $\mathcal{A}$ by legislating that homogeneous maps of degree $\pm d$ have filtration degree d.

We now give our definition of the filtered category $\mathcal{C}_i$. Let the stance between points $J = (j_1,\dots,j_i)$ and $K = (k_1,\dots,k_i)$ in $\mathbf{Z}^i$ be iven by

$$\|J-K\| = \max_s |j_s-k_s| \quad .$$

efinition (1.2). Let $\mathcal{A}$ be a (filtered) additive category. We efine $\mathcal{C}_i(\mathcal{A})$ to be the category of $\mathbf{Z}^i$-graded objects and bounded omomorphisms. This means that an object A of $\mathcal{C}_i$ is a collection of jects $A(J)$ in $\mathcal{A}$, one for each J in $\mathbf{Z}^i$. A morphism $\phi : A \to B$ in $\mathcal{C}_i$ filtration degree d is a collection

$$\phi(J,K) : A(J) \to B(K)$$

$\mathcal{A}$-morphisms, where we require $\phi(J,K) = 0$ unless $\|J-K\| \leqslant d$. If $\mathcal{A}$ filtered, we also require each $\phi(J,K)$ to have filtration degree d. Composition of $\phi = A \to B$ with $\psi : B \to C$ is defined by

$$(\psi \circ \phi)(J,L) = \sum_K \psi(K,L)\circ\phi(J,K) \quad .$$

te that composition is well-defined because only finitely many lements in this sum are different from 0. It is easily seen that $_0(\mathcal{A}) = \mathcal{A}$.

ample (1.2.1). If $\mathcal{F}$ is the category of finitely generated free -modules (with trivial filtration), the category $\mathcal{C}_i(\mathcal{F})$ is the same the category $\mathcal{C}_i(R)$ constructed in [P]. In that paper it was oven that

$$K_1(\mathcal{C}_{i+1}(R)) = K_{-i}(R) \ , \quad i \geqslant 0 \ .$$

is indicated that $\mathcal{C}_{i+1}$ might be a delooping of K-theory, and was e original motivation for this paper. That it cannot be exactly e case follows from (1.3.1) below.

Example (1.2.2). Since $\mathcal{C}_i(\mathcal{A})$ is filtered, we can iterate the onstruction. It is easy to see that

$$\mathcal{C}_i(\mathcal{C}_j(\mathcal{A})) = \mathcal{C}_{i+j}(\mathcal{A}) \ .$$

owever, if we forget the filtration on $\mathcal{C}_j(\mathcal{A})$ this is no longer the case.

Remark (1.2.3). If V is any metric space, we can define a category $\mathcal{C}_V(\mathcal{A})$ in a way generalizing the case $V = \mathbf{Z}^i$. An object A of $\mathcal{C}_V$ is a collection of objects A(v), one for each v in V, subject to the following constraint: for every $d > 0$ and v, $A(w) \neq 0$ for only finitely many w of distance less than d from v. Morphisms are defined as for $\mathcal{C}_i$. It is easy to see that if $V = \mathbb{R}^i$ then $\mathcal{C}_V$ is naturally equivalent to its subcategory $\mathcal{C}_i$. This shows that the difference between $\mathcal{C}_i$ and $\mathcal{C}_{i+1}$ is the rate of growth in d of the number n(d,J) of points K within a distance of d from J.

Example (1.2.4). If we take $V = \langle 0,1,2,\ldots\rangle$ then we will let $\mathcal{C}_+(\mathcal{A})$ denote $\mathcal{C}_V(\mathcal{A})$. This is the full subcategory of $\mathcal{C}_1(\mathcal{A})$ whose objects satisfy $A(j) = 0$ for $j < 0$. Similarly, if we take $V = \langle 0,-1,-2,\ldots\rangle$, we will write $\mathcal{C}_-(\mathcal{A})$ for $\mathcal{C}_V(\mathcal{A})$. We can identify $\mathcal{C}_+(\mathcal{A}) \cap \mathcal{C}_-(\mathcal{A})$ with $\mathcal{A}$ in the obvious way.

There is a shift functor $T : \mathcal{C}_1(\mathcal{A}) \to \mathcal{C}_1(\mathcal{A})$ sending A to TA with $TA(j) = A(j-1)$, and T restricts to an endofunctor of $\mathcal{C}_+(\mathcal{A})$. There is an obvious natural isomorphism t from A to TA in both $\mathcal{C}_1$ and $\mathcal{C}_+$. We include the following result here for expositional purposes, and will generalize it in section 3 below.

Lemma (1.3). Every object of $\mathcal{C}_+(\mathcal{A})$ is stably isomorphic to 0. In particular, the Grothendieck group $K_o(\mathcal{C}_+)$ is zero.

Proof. Given A in $\mathcal{C}_+$, let $B = \Sigma\, T^n A$. That is, $B(j) = A(j)\oplus A(j-1)\oplus\ldots\oplus A(0)$. It is clear that $A\oplus TB = B$. The result follows from the observation that $t : B \cong TB$ is an isomorphism in $\mathcal{C}_+(\mathcal{A})$.

Corollary (1.3.1). If $i \neq 0$ then every object of $\mathcal{C}_i(\mathcal{A})$ is stably isomorphic to 0. In particular, $K_o(\mathcal{C}_i) = 0$.

Proof. By (1.2.2) we can assume that $i = 1$. But every object of

1 can be written $A_+ \oplus A_-$ with A_+ in $\mathcal{C}_+$ and A_- in $\mathcal{C}_-$. Hence $K_o(\mathcal{C}_1)$ is quotient of $K_o(\mathcal{C}_+) \oplus K_o(\mathcal{C}_-) = 0$.

Here is a quick discussion of idempotent completion, as applied o the $\mathcal{C}_i$ construction.

efinition (1.4) (see, e.g., [F, p.61]). Let $\mathcal{A}$ be an additive ategory. The idempotent completion $\hat{\mathcal{A}}$ of $\mathcal{A}$ has as objects all orphisms $p : A \to A$ of $\mathcal{A}$ satisfying $p^2 = p$. An $\hat{\mathcal{A}}$-morphism from p_1 to 2 is an $\mathcal{A}$-morphism ϕ from the domain A_1 of p_1 to the domain A_2 of 2 satisfying $\phi = p_2 \phi p_1$. It is easily seen that $\hat{\mathcal{A}}$ is an additive ategory and that $\mathrm{Hom}(p_1, p_2)$ is a subgroup of $\mathrm{Hom}(A_1, A_2)$. Hence $\hat{\mathcal{A}}$ nherits any filtered structure that $\mathcal{A}$ might have. There is a full mbedding of $\mathcal{A}$ in $\hat{\mathcal{A}}$ sending A to 1_A; if this is an equivalence ogf ategories, we say that $\mathcal{A}$ is idempotent complete.

xample (1.4.1). The idempotent completion of the category $\mathcal{F}$ of ree R-modules is equivalent to the category $\mathcal{P}$ of projective -modules.

emma (1.4.2). The categories $\mathcal{A}$ and $\mathcal{C}_i(\mathcal{A})$ are cofinal in their dempotent completions $\hat{\mathcal{A}}$ and $\hat{\mathcal{C}}_i(\mathcal{A})$. Moreover, $\mathcal{C}_i(\mathcal{A})$ is cofinal in $_i(\hat{\mathcal{A}})$.

roof. This is an easy computation. For example, if p is an object f $\mathcal{C}_i(\hat{\mathcal{A}})$, define q by $q(J) = 1 - p(J)$. Then $p \oplus q$ belongs to $\mathcal{C}_i(\mathcal{A})$.

To compute the K-theory of $\mathcal{A}$, we need to know which sequences re "exact": a different embedding of $\mathcal{A}$ in an ambient abelian ategory will result in a different family of short exact sequences see [Q]). In particular, we cannot talk about $K_1\mathcal{C}_i(\mathcal{A})$ unless we now which sequences in $\mathcal{C}_i$ are "exact". It is not clear what the otion of "exact" should be, unless either (a) all exact sequences n $\mathcal{A}$ split (we insist the same is true of $\mathcal{C}_i$), or (b) $\mathcal{A}$ is embedded n an abelian category $\tilde{\mathcal{A}}$ closed under countably infinite direct sum for then $\mathcal{C}_i$ is embeddable in $\tilde{\mathcal{A}}$). In either case, it follows from

(1.4.2) and Theorem 1.1 of [Gr] that

$$K_n\mathcal{C}_i(\mathcal{A}) = K_n\mathcal{C}_i(\hat{\mathcal{A}}) = K_n\hat{\mathcal{C}}_i(\mathcal{A}) , \quad n \geq 1 .$$

Note that our proofs of theorem A and B only apply to situation (a).

Example (1.5). Let p_- be the idempotent natural transformation in $\mathcal{C}_1(\mathcal{A})$ given by

$$(p_-)_A : A \to A , \quad p_-(j,k) = \begin{cases} 1 & \text{if } j = k \leq 0 \\ 0 & \text{otherwise} \end{cases}$$

Given an object A of $\mathcal{A}$. let A_- denote the image of p_- on the constant object $A(j) = A$ of $\mathcal{C}_1(\mathcal{A})$. Thus $A_-(j) = 0$ if $j > 0$ and $A_-(j) = A$ if $j \leq 0$. The map t is an endomorphism of the constant object $A \cong TA$; write s for the restriction of p_-t to A_-. Then $1-s : A_- \to A_-$ is both a monomorphism and an epimorphism in $\mathcal{C}_1(\mathcal{A})$, but not an isomorphism. This is because its "inverse" Σs^n is not bounded. In particular, $\mathcal{C}_1(\mathcal{A})$ can never be an abelian category, even if $\mathcal{A}$ is.

We conclude this section with the following result, which provides motivation for our Theorem B. It is also a consequence of Theorem B. Since we will not use this result, we merely sketch the proof.

Proposition (1.6). If all short exact sequences in $\mathcal{A}$ split, then $K_1(\mathcal{C}_{i+1}(\mathcal{A})) = K_0(\hat{\mathcal{C}}_i(\mathcal{A}))$. In particular, $K_1\mathcal{C}_1(\mathcal{A}) = K_o(\hat{\mathcal{A}})$.

Sketch of proof. This is proven in section 1 of [P], modulo terminology.

First of all, we can assume that $\mathcal{A}$ is idempotent complete and that $i = 0$ by (1.4.2) and (1.2.2). The map from $K_o(\mathcal{A})$ to $K_1\mathcal{C}_1(\mathcal{A})$ sends the object A of $\mathcal{A}$ to the shift automorphismt of the constant object $A(j) = A$ of $\mathcal{C}_1(\mathcal{A})$. The map $\phi : K_1(\mathcal{C}_1) \to K_o(\mathcal{A})$ is defined by sending the class of $\alpha \in \mathrm{Aut}(A)$ to the difference (for $d \gg 0$) in $K_0(\mathcal{A})$:

$$\phi(\alpha) = [(\alpha p_-\alpha^{-1})(\bigoplus_{j=-2d}^{2d} A(j))] - [p_-(\bigoplus_{j=-2d}^{2d} A(j))] .$$

If α has filtration degree less than d, one shows as in [P,(1.11)] at this map ϕ is well-defined and independant of d. Clearly the omposition is the identity on $K_o(\mathcal{A})$. The proof of [P, (1.20)] plies to show that ϕ is monic, which proves the proposition.

xample (1.6.1). Again, let $\mathcal{F}$ be the category of finitely enerated free R-modules. Then for $i \geq 1$ we have
$\mathcal{C}_i(R) = 0$ but $K_o\hat{\mathcal{C}}_i(R) = K_1\mathcal{C}_{i+1}(R) = K_{-i}(R)$.

ote: Example (1.6.1) follows from [P], not from (1.6).

. The pasage to topology.

In this section we recall various results on the passage from e categories $\mathcal{A}$, $\mathcal{C}_i$ etc. to infinite loop spaces and spectra. We lso recall Thomason's simplified double mapping cylinder from ection 5 of [T]. We urge the reader to consult [T] for more etails.

A symmetric monoidal category $\underline{S}$ is a category together with a unctor $\oplus : \underline{S}\times\underline{S} \to \underline{S}$ and natural isomorphisms

$$\alpha : (A\oplus B)\oplus C \cong A\oplus(B\oplus C)$$

$$\gamma : A\oplus B \cong B\oplus A .$$

hese natural isomorphisms are subject to coherence conditions that ertain diagrams commute. We refer the reader to [Mac] for a more etailed definition, contending ourselves with:

xample (2.1). If $\mathcal{A}$ is an additive category then $\mathcal{A}$ is a symmetric onoidal category under $\oplus$ = direct sum. The subcategory $\underline{A}$ of the somorphisms in $\mathcal{A}$ is also symmetric monoidal under $\oplus$ = direct sum. t follows that $\mathcal{C}_i(\mathcal{A})$ and its category $\underline{C}_i(\mathcal{A})$ of isomorphisms are lso symmetric monoidal.

There is a functor Spt from the category of small symmetric monoidal categories to the category of connective Ω-spectra (i.e., sequences of spaces X_n with X_n being (n-1)-connected and with $X_n = \Omega X_{n+1}$). This functor satisfies

(a) A functor $\underline{A} \to \underline{B}$ preserving $\oplus$ up to coherent natural transformation , a "lax" functor, induces a map $Spt(\underline{A}) \longrightarrow Spt(\underline{B})$ of infinite loop spectra.

(b) The zeroth space $Spt_o(\underline{A})$ is the "group completion" of $B\underline{A}$, the classifying space of the category $\underline{A}$.

The construction of Spt is basically due to May and to Segal, and Spt is unique up to homotopy equivalence. See [A]. One description of Spt may be found in the Appendix of [T].

Lemma (2.2). Suppose that $\underline{A} \to \underline{B}$ is a lax functor of small symmetric monoidal categories, and that $B\underline{A} \to B\underline{B}$ is a homotopy equivalence of topological spaces. Then $Spt_o(\underline{A}) \to Spt_o(\underline{B})$ is a homotopy equivalence.

Proof. See (2.3) of [T].

Lemma (2.3). Suppose that $\underline{A}$ is a full, cofinal subcategory of the small symmetric monoidal category $\underline{B}$. Then the connected components of $Spt_o(\underline{A})$ and $Spt(\underline{B})$ are homotopy equivalent.

Proof. This is wellknown. The point is that

$$H_*[Spt_o(\underline{A})_o] = \underset{A\in\underline{A}}{\underrightarrow{colim}}\ H_*\ B\ Aut(A)$$

$$= \underset{B\in\underline{B}}{\underrightarrow{colim}}\ H_*\ B\ Aut(B)$$

$$= H_*[Spt_o(\underline{B})_o]\ .$$

Lemma (2.4) (Quillen). Let $\underline{S}$ be a small symmetric monoidal category in which all morphisms are isomorphisms, and assume that all translations $S\oplus : \underline{S} \to \underline{S}$ are faithful. Then there is a category $\underline{S}^{-1}\underline{S}$

ose objects are pairs (S_1,S_2) of objects in $\underline{S}$, such that $B\underline{S}^{-1}\underline{S}$ is
omotopy equivalent to $Spt_o(\underline{S})$.

roof. See [G, p.221] or p. 1657 of [T].

orollary (2.4.1). If $\mathcal{A}$ is a small additive category, let $\underline{A}$ denote
e category of isomorphisms in $\mathcal{A}$. Then $B\underline{A}^{-1}\underline{A}$ is homotopy equivalent
o $Spt_o(\underline{A})$.

xample (2.4.2). Let R be a ring for which $R^m \cong R^n$ implies that
= n, and let $\underline{F}$ be the category of finitely generated free
-modules and isomorphisms. The basepoint component of $\underline{F}^{-1}\underline{F}$ has
ojects $R^m = (R^m,R^m)$ and

$$Hom(R^m,R^{m+n}) = Gl_{m+n}(R) \underset{Gl_n(R)}{\times} Gl_{m+n}(R).$$

n particular, $Hom(0,R^m)$ is $Gl_m(R)$. The family of the $Hom(0,R^m)$
ives a map from $BGl(R)$ to the basepoint component $BGl^+(R)$ of $B\underline{F}^{-1}\underline{F}$

The main ingredient in the proof of Theorem B is the simplified
ouble mapping cylinder construction of R.W. Thomason, described in
5.1) of [T]. Let $\underline{A}$ be a symmetric monoidal category with all
orphisms isomorphisms and $u : \underline{A} \to \underline{B}$, $v : \underline{A} \to \underline{C}$ strong functors of
ymmetric monoidal categories (i.e. functors preserving direct sum
p to natural isomorphism). Define $\underline{P} = \underline{P}(\underline{A},\underline{B},\underline{C},u,v)$ to be the
ategory with objects triples (B,A,C) with A an object of $\underline{A}$, B of $\underline{B}$,
nd C of $\underline{C}$. A morphism $(B,A,C) \to (B',A',C')$ is a 5-tuple
$\psi,\psi_1,\psi_2,U,V)$ where U,V are objects of $\underline{A}$, $\psi : A \cong U\oplus A'\oplus V$,
$_1 : B\oplus uU \to B'$ and $\psi_2 : C\oplus vV \to C'$. U and V may be varied up to
somorphism. Composition of $(\psi,\psi_1,\psi_2,U,V) : (B,A,C) \to (B',A',C')$
ith $(\bar\psi,\bar\psi_1,\bar\psi_2,\bar U,\bar V) : (B',A',C') \to (B'',A'',C'')$ is given by

$$A \cong U\oplus A'\oplus V \cong (U\oplus\bar U)\oplus A''\oplus(\bar V\oplus V)$$

$$B\oplus u(U\oplus\bar U) \cong (B\oplus uU)\oplus u\bar U) \to B'\oplus u\bar U \to B''$$

$$v(\bar V\oplus V)\oplus C \cong v\bar V\oplus vV\oplus C \to v\bar V\oplus C' \to C''$$

nd direct sum in $\underline{P}$ is induced by direct sum in $\underline{A}$, $\underline{B}$ and $\underline{C}$. We then
ave

<u>Theorem 2.5</u> (R.W. Thomason [T,(5.2)]. Up to homotopy the diagram

$$\begin{array}{ccc} Spt_0\underline{\underline{A}} & \longrightarrow & Spt_0\underline{\underline{B}} \\ \big\downarrow & & \big\downarrow \\ Spt_0\underline{\underline{C}} & \longrightarrow & Spt_0\underline{\underline{P}} \end{array}$$

is a pullback diagram.

3. The proof of Theorem A and B.

In this section we prove Theorems A and B. We make the standing assumption that $\mathcal{A}$ is a small filtered additive category and that $\underline{\underline{A}}$ is the (symmetric monoidal) category of isomorphisms of $\mathcal{A}$. Similarly we write $\underline{\underline{C}}_i$, $\underline{\underline{C}}_+$ and $\underline{\underline{C}}_-$ for the categories of isomorphisms of $\mathcal{C}_i(\mathcal{A})$, $\mathcal{C}_+(\mathcal{A})$ and $\mathcal{C}_-(\mathcal{A})$. The idea is to show that the diagram

$$\begin{array}{ccc} \underline{\underline{A}} & \longrightarrow & \underline{\underline{C}}_+ \\ \big\downarrow & & \big\downarrow \\ \underline{\underline{C}}_- & \longrightarrow & \underline{\underline{C}}_1 \end{array}$$

induces a pullback diagram of spectra, and to use the following result:

<u>Proposition (3.1)</u>. $Spt_o(\underline{\underline{C}}_+)$ and $Spt_o(\underline{\underline{C}}_-)$ are contractible.

<u>Proof</u>. By symmetry it is enough to consider $\underline{\underline{C}}_+$. Recall from the discussion before (1.3) that there is a shift functor $T : \underline{\underline{C}}_+ \to \underline{\underline{C}}_+$ and a natural transformation t from A to TA. The category $\underline{\underline{C}}_+$ has an endofunctor $\sum_{n=0}^{\infty} T^n$ with

$$\Big(\sum_{n=0}^{\infty} T^n\Big)A(j) = \bigoplus_{n=0}^{j} A(j-n) .$$

ecall that $A(j) = 0$ for $j < 0$.)We can define $\sum_{n=1}^{\infty} T^n$ similarly. The atural isomorphism t induces a natural isomorphism t from $\sum_{n=0}^{\infty} T^n A$ $\sum_{n=1}^{\infty} T^n A$. But as endofunctors of $\underline{C}_+$ we have $1 \oplus \sum_{n=1}^{\infty} T^n \cong \sum_{n=0}^{\infty} T^n$. ence as self-maps of the H-space $B\underline{C}_+$ we have

$$1 \sim \left(\sum_{n=0}^{\infty} T^n\right) - \left(\sum_{n=1}^{\infty} T^n\right) \overset{t}{\simeq} 0 .$$

is shows that B is contractible. But then $Spt_o(\underline{C}_+)$ is contractible Lemma (2.2).

roof that Theorem B implies Theorem A. Write $\hat{B}_i$ for $Spt_o(\hat{\underline{C}}_i)$. nce we have $\pi_o(\hat{B}_i) = K_{-i}(R)$ by (1.6.1) and since translations are ithful in $\hat{\underline{C}}_i$, it follows that $\hat{B}_i$ is homotopy equivalent to $\times K_{-i}(R)$. Since $\Omega B_i = \Omega\hat{B}_i$, the result is now immediate.

We now begin the proof of theorem B by making a series of eductions. Since

$$\pi_o(B_i) = \pi_o Spt_o(\underline{A}_i) = K_o(\underline{A}_i) ,$$

onnectedness of the B_i for $i \neq 0$ follows from (1.3.1). Now $\underline{C}_i$ is ll and cofinal in $\hat{\underline{C}}_i$ by (1.4.2), so by (2.3) the connected space $= Spt_o(\hat{\underline{C}}_i)$. By construction (or by (2.4.1)), $\hat{B}_o = Spt_o(\hat{\underline{A}})$ is the roup completion of $B\hat{\underline{A}}$. Thus the proof of Theorem B is reduced to howing that $\Omega\hat{B}_{i+1} = \hat{B}_i$ for $i \geq 0$.

Next, observe that $\hat{\mathcal{C}}_{i+1}(\mathcal{A}) = \hat{\mathcal{C}}_1\hat{\mathcal{C}}_i(\mathcal{A})$, so that $_{i+1} = Spt_o(\hat{\underline{C}}_1(\hat{\mathcal{C}}_i(\mathcal{A}))$ and $\hat{B}_i = Spt_o(\hat{\mathcal{C}}_i(\mathcal{A}))$.

Since we can replace $\mathcal{A}$ by $\hat{\mathcal{C}}_i(\mathcal{A})$, it is enough to prove that $\hat{B}_1 = \hat{B}_o = Spt_o(\hat{\underline{A}})$.There is also no loss in generality in assuming hat $\mathcal{A}$ is idempotent complete, since

$$\Omega\,\hat{B}_1 = \Omega\,Spt_o(\hat{\underline{C}}_1(\mathcal{A})) = \Omega\,Spt_o(\hat{\underline{C}}_1(\hat{\mathcal{A}}))$$

y (2.3). In fact, by (2.3) we also have

$$\Omega\,Spt_o(\hat{\underline{C}}_1) = \Omega\,Spt_o(\underline{C}_1) .$$

herefore, Theorem B will follow from:

<u>Theorem (3.2)</u>. Let $\mathcal{A}$ be a small, filtered additive category which is idempotent complete. Then $\Omega\,Spt_o(\underset{=}{C}_1)$ is homotopy equivalent to $Spt_o(\underset{=}{A})$.

<u>Lemma (3.3)</u>. Let $\mathcal{A}$ be a small filtered additive category. Recall that $\underset{=}{C}_+$ and $\underset{=}{C}_-$ are subcategories of $\underset{=}{C}_1$ whose intersection is $\underset{=}{A}$. Let $\underset{=}{P}$ be the simplified double mapping cylinder construction applied to $\underset{=}{A} \to \underset{=}{C}_-$ and $\underset{=}{A} \to \underset{=}{C}_+$. Then $\Omega\,Spt_o(\underset{=}{P})$ is homotopy equivalent to $Spt_o(\underset{=}{A})$.

<u>Proof</u>. This is immediate from Thomason's Theorem (2.5), since by (3.1) the spaces $Spt_o(\underset{=}{C}_+)$ and $Spt_o(\underset{=}{C}_-)$ are contractible.

By the universal mapping property of $\underset{=}{P}$ (see p. 1648 of [T]), there is a strong symmetric monoidal functor $\Sigma : \underset{=}{P} \to \underset{=}{C}_1$. This functor is defined on objects by

$$\Sigma(A^-, A, A^+) = A_- \oplus A \oplus A_+$$

where A^-, A, A^+ are objects of $\underset{=}{C}_+$, $\underset{=}{A}$ and $\underset{=}{C}_-$, respectively. A morphism $(\psi^-, \psi, \psi^+, U^-, U^+)$ in $\underset{=}{P}$ from (A^-, A, A^+) to (B^-, B, B^+) is sent by Σ to the composite

$$A^- \oplus A \oplus A^+ \xrightarrow{1 \oplus \psi \oplus 1} A^- \oplus U^- \oplus A \oplus U^+ \oplus A^+ \xrightarrow{\psi^- \oplus 1 \oplus \psi^+} B^- \oplus B \oplus B^+ .$$

<u>Theorem (3.4)</u>. Let $\mathcal{A}$ be idempotent complete, and let $\underset{=}{P}$ be the double mapping cylinder of Lemma (3.3). Then the functor $\Sigma : \underset{=}{P} \to \underset{=}{C}_1$ induces a homotopy equivalence between the classifying spaces $B\underset{=}{P}$ and $B\underset{=}{C}_1$.

Note that Theorem (3.4) immediately implies Theorem (3.2) by (3.3) and (2.2). Thus we have reduced the proof of Theorem B to the proof of Theorem (3.4).

<u>Proof</u>. We will show that this functor satisfies the conditions of Quillen's Theorem A from [Q]. Fix an object Y of $\underset{=}{C}_1$; we need to show that $Y \downarrow \Sigma$ is a contractible category. To do this, we use the bound d

or $\mathcal{C}_1(\mathcal{A})$ to filter $Y\downarrow\Sigma$ as the increasing union of subcategories il$_d$, and show that each Fil_d has an initial object $*_d$. Therefore il$_d$ is contractible; their union $Y\downarrow\Sigma$ must also be contractible by tandard topology.

The category Fil_d is the full subcategory of all $: Y \to \Sigma(A^-,A,A^+)$ where both α and α^{-1} are bounded by d. Define Y_d, $_d^-$ and Y_d^+ in $\underline{A}$, $\underline{C}_-$ and $\underline{C}_+$ respectively by setting

$$Y_d \quad = Y(-d)\oplus\ldots\oplus Y(d) \text{ in } \underline{A}$$

$$Y_d^- = Y(j) \quad \text{if } j < -d \text{ , and } = 0 \text{ otherwise}$$

$$Y_d^+ = Y(j) \quad \text{if } j > -d \text{ , and } = 0 \text{ otherwise.}$$

he obvious isomorphism $\sigma : Y \cong Y_d^-\oplus Y_d\oplus Y_d^+$ in $\underline{C}_1$ is bounded by d, and orms the object $*_d : Y \to \Sigma(Y_d^-,Y_d,Y_d^+)$ of Fil_d. We will show that $*_d$ s an initial object of Fil_d.

Given the object $\alpha : Y \to \Sigma(A^-,A,A^+)$, we have to show that there s a unique morphism

$$\eta = (\psi,\psi^-,\psi^+,e_-(Y_d),e_+((Y_d)) : (Y_d^-,Y_d,Y_d^+) \to (A^-,A,A^+)$$

in $\underline{P}$ so that $\Sigma(\eta) = \alpha\sigma^{-1}$ in $\underline{C}_1$. Let pr_-, pr, pr_+ be the projections f $\Sigma(A^-,A,A^+)$ onto A^-, A and A^+, respectively. Since α^{-1} is bounded y d, $\alpha^{-1}(A)$ is contained in Y_d, or rather in the image $\sigma^{-1}(Y_d)$ of $_d$ in Y. Hence it makes sense to let e be $\sigma\alpha^{-1}(pr)\alpha\sigma^{-1}$ restricted to $_d$, and it is clear that e is an idempotent of Y_d. Similarly, $\alpha^{-1}(A^-)$ is contained in $Y_d^-\oplus Y_d$, and $\alpha^{-1}(A^+)$ is contained in $Y_d\oplus Y_d^+$. et e_- and e_+ be $\sigma\alpha^{-1}(pr_-)\alpha\sigma^{-1}$ and $\sigma\alpha^{-1}(pr_+)\alpha\sigma^{-1}$ restricted to Y_d. hese maps are also idempotents of Y_d, and it is easy to see that $_- + e + e_+ = 1$. Since $\mathcal{A}$ is idempotent complete, the composition

$$Y_d \cong e_-(Y_d) \oplus e(Y_d) \oplus e_+(Y_d)$$

akes sense in $\mathcal{A}$. Define ψ to be the composite

$$Y_d \cong e_-(Y_d) \oplus e(Y_d) \oplus e_+(Y_d) \xrightarrow{1\oplus\alpha\oplus 1} e_-(Y_d)\oplus A\oplus e_+(Y_d)$$

imilarly, define maps

$$\psi^- : Y_d^- \oplus e_-(Y_d) \xrightarrow{\alpha\sigma^{-1}} A^- \text{ in } \underline{C}_-$$

$$\psi^+ : e_+(Y_d) \oplus Y_d^+ \xrightarrow{\alpha\sigma^{-1}} A^+ \text{ in } \underline{C}_+.$$

This completes the definition of the map $\eta : (Y_d^-, Y_d, Y_d^+) \to (A^-, A, A^+)$ in $\underline{P}$. By definition of Σ we have $\Sigma(\eta) = \alpha\sigma^{-1}$. Because all maps in $\underline{A}$, $\underline{C}_-$ and $\underline{C}_+$ are isomorphisms, it is an easy task to verify that η is the unique map with $\Sigma(\eta) = \alpha\sigma^{-1}$. It follows that $*_d$ is an initial object of Fil_d. Q.E.D.

4. An overview.

To place our construction in perspective, it is appropriate to review a little history. The definition of the functors $K_{-i}(R)$ was given by Bass [B] in 1966 during an attempt to formalize his decomposition of $K_1(R[t_1, t_1^{-1}, \ldots, t_n, t_n^{-1}])$. In 1967, Karoubi [K-1] gave another definition of $K_{-i}(R)$ by defining $K_{-i}(\mathcal{A})$ for any abelian category. A third and fourth definition of $K_{-i}(R)$ were given independantly by Karoubi Villamayor [K-V] using the ring $S(R)$ and by Wagoner [W-1] using the subring $\mu(R)$ of $S(R)$. Happily all these definitions were shown to agree by Karoubi's axiomatic treatment in [K-136].

In 1971, Gersten [Ger] constructed a nonconnective delooping of $K_o(R) \times BGl^+(R)$ using the fact that $\Omega BGl^+(S(R)) = K_o(R) \times BGl^+(R)$. Wagoner [W-2] then constructed the Ω-spectrum $K_o(\mu^i(R)) \times BGl^+(\mu^i(R))$ and showed that the inclusions $\mu(R) \to S(R)$ induced an equivalence of spectra. To our knowledge, nonconnective deloopings of the K-theory of other additive categories besides $\mathcal{F}$ has not been studied until now.

The construction in [P] is very much in the spirit of the early definitions of the $K_{-i}(R)$, but works for any additive category. Needless to say, an open questiom in our work is whether or not the $\Omega BQ\mathcal{C}_n(\mathcal{A})\hat{}$ yield a nonconnective delooping of any (idempotent complete) additive category with exact sequences. A major difference between the categories $\mathcal{C}_i(\mathcal{A})$ and Karoubi's categories $S^i\mathcal{A}$ is that $S\mathcal{A}$ is defined as a quotient of the flasque category $C\mathcal{A}$ (see [K-136]) while $\mathcal{C}_1(\mathcal{A})$ may be viewed as an enlargement of the flasque category

$_+(\mathcal{A})$. It would be interesting to see if the natural inclusion of $C\mathcal{A}$ in $\mathcal{C}_+(\mathcal{A})$ could be made to induce an isomorphism between K-groups.

References.

[A] J.F. Adams, Infinite Loop Spaces, Princeton University Press, Princeton, 1978.

[B] H. Bass, Algebraic K-theory, Benjamin, New York, 1968.

[F] P. Freyd, Abelian Categories, Harper and Row, New York, 1964.

[Ger] S.M. Gersten, On the spectrum of algebraic K-theory, Bull. AMS 78 (1972), 216-219.

[G] D. Grayson, Higher algebraic K-theory: II (after D. Quillen), Lecture Notes in Math. No 551, Springer-Verlag, 1976.

[K-1] M. Karoubi, La periodicite de Bott en K-theorie generale, Ann. Sci. Ec. Norm. Sup. (Paris) 4 (1971), 63-95.

[K-136] M. Karoubi, Foncteur derives et K-theorie, Lecture Notes in Math. No. 136, Springer-Verlag, 1970.

[K-V] M. Karoubi and O. Villamayor, K-Theorie algebrique et K-thorie topologique, C.R. Acad. Sci. (Paris) 269, serie A (1969), 416-419.

[Mac] S. Maclane, Categories for the Working Mathematician, Springer-Verlag, 1971.

[P] E.K. Pedersen, On the K_{-i}functors, J. Algebra 90 (1984), 461-475.

[Q] D. Quillen, Higher algebraic K-theory: I, Lecture Notes in Math. No 341, Springer-Verlag, 1978.

[T] R. Thomason, First quadrant spectral sequences in algebraic K-theory via homotopy colimits, Comm. in Alg. 10 (1982) 1589-1668.

[W-1] J.B. Wagoner, On K_2 of the Laurent polynomial ring, Amer. J. Math. 93 (1971), 123-138.

[W-2] J.B. Wagoner, Delooping classifying spaces in algebnraic K-theory, Topology 11 (1972), 349-370.

[Gr] D. Grayson, Localization for flat modules in Algebraic K-theory, J. of Algebra 61 (1979), 463-496.

Geometric Algebra

Frank Quinn
Department of Mathematics
Virginia Polytechnic Institute and State University
Blacksburg, Virginia USA 24061

Introduction

In this paper we develop a hybrid sort of algebra, whose morphisms involve paths in a space. The primary purpose is to elucidate and extend the algebra with ϵ estimates developed in [3, 4-6, 1]. The setting is also fruitful for investigating relationships between the topology of a space X, and the algebra of $R[\pi_1 X]$ modules.

The first section presents the definitions of geometric R-modules on a space, and their morphisms. We show that by allowing appropriate "homotopies" of morphisms we can recover either ordinary $R[\pi_1 X]$ homomorphisms of free modules, or ϵ homomorphisms of geometric modules. Then we show that if $K \longrightarrow X$ is a map from a CW complex to a space, the cellular chains of K can be seen in a very natural way as a geometric chain complex on X.

Section two gives decompositions of $R[\pi]$ chain complexes, corresponding to amalgamated free product decompositions of π. The approach is to geometrically realize the free product structure as the structure induced on the fundamental group of a space X by a codimension 1 subspace Y. Then we use the equivalence of 1.1 to represent chain complexes by geometric ones on X. Intersections of the geometric structure of the complex with Y then show how to decompose the complex. The main purpose of this is to illustrate the technique, which we anticipate will apply to algebraic K and L theory.

Finally in section three, geometric versions of the Whitehead group are defined. These are shown to be the obstruction groups for the thin h-cobordism theorem, a controlled version of the usual result.

Section 1: Geometric modules and morphisms

Suppose that X is a topological space and R a ring. A *Geometric* R-*module* on X is defined to be a free module R[S] and a map of the basis $f:S \longrightarrow X$. We require that geometric modules be *locally finite* in the sense that every point in X has a neighborhood whose preimage in S is finite. So for example a geometric module on a compact space has a finite basis.

A *geometric morphism* of geometric modules is defined to be a locally finite

---

Partially supported by the National Science Foundation

algebraic sum of paths between generators. More specifically suppose $f_i:S_i \longrightarrow X$ are bases for geometric modules, $i = 0,1$. A morphism $h:R[S_0] \longrightarrow R[S_1]$ is a sum $\Sigma m_j\lambda_j$ where $m_j \in R$ and λ_j is a path. The data for a path consists of elements $x_i \in S_i$ and a map $\lambda:[0,t] \longrightarrow X$ with $\lambda(0)=f_0(x_0)$ and $\lambda(t)=f_1(x_1)$. Here t is a real number $t \geq 0$. Finally we require that for each $y \in S_0$ there are only finitely many paths λ_j starting at y which have nonzero coefficient.

In a morphism we allow deletion of a path with coefficient 0 (or conversely insertion of such a path). We also identify $(m+n)\lambda$ with $(m\lambda)+(n\lambda)$.

We describe how to compose two morphisms. If $f = \Sigma m_j\lambda_j:R[S_1] \longrightarrow R[S_2]$, and $g = \Sigma n_k\alpha_k:R[S_0] \longrightarrow R[S_1]$, then $fg = \Sigma(m_jn_k)(\lambda_j\alpha_k)$. The sum is taken over all pairs (j,k) such that the end of α_j is equal to the starting point of λ_k in S_1. The Moore composition of paths is used: given $\lambda:[0,t] \longrightarrow X$ and $\alpha:[0,u] \longrightarrow X$ then $\lambda\alpha:[0,t+u] \longrightarrow X$ is defined by $\lambda\alpha(s)=\alpha(s)$ for $s \leq u$ and $\lambda\alpha(s)=\lambda(s-t)$ for $s \geq u$. Notice we are writing compositions of paths from right to left, so that it will agree with the notation for composition of functions.

The composition is associative, and there is a unit (the unit in the ring times the constant paths defined on [0,0]). Geometric modules and morphisms therefore form a category. This is not a directly useful category because there are too many paths.

There is a forgetful functor from geometric morphisms to ordinary R-module homomorphisms, defined by forgetting the paths. Explicitly, if $h = \Sigma m_j\lambda_j:R[S_0] \longrightarrow R[S_1]$ then we can define an R-homomorphism $h':R[S_0] \longrightarrow R[S_1]$ by $h'(s) = \Sigma_i(\Sigma_j m_j)t_i$. Here the outer summation is over $t_i \in S_1$, and the inner summation is over j such that the path λ_j goes from s to t_i.

We will define several notions of "homotopy" of geometric morphisms. The goal is to obtain useful intermediate stages between the rigidity of geometric morphisms and the laxity of ordinary algebra over R.

1.1 Unrestricted homotopy of morphisms A homotopy of a morphism is obtained by changing all the paths in the morphism by homotopy holding the endpoints fixed. Since we are using Moore paths, a "homotopy" is allowed to change the interval on which the path is defined. Form the category whose morphisms are homotopy classes of morphisms of geometric R-modules on X. We claim that if X is connected and locally 1-connected then this category is naturally equivalent to the category of free $R[\pi_1X]$ modules, with a restriction on rank. (If X is compact, the modules are finitely generated. If X is noncompact and separable the modules are countably generated, etc.) To simplify the discussion assume that X is compact.

Let $\tilde{X}$ denote the universal cover of X, which exists since X is locally 1-connected. Given a geometric module R[S], with map $S \longrightarrow X$, form the pullback

$$\begin{array}{ccc} \tilde{S} & \longrightarrow & \tilde{X} \\ \downarrow & & \downarrow \\ S & \longrightarrow & X \end{array}$$

then the action of $\pi_1 X$ on $\tilde{S}$ gives $R[\tilde{S}]$ the structure of a finitely generated free $R[\pi_1 X]$ module.

Next suppose that $\Sigma m_j \lambda_j : R[S_0] \longrightarrow R[S_1]$ is a morphism of geometric R-modules on X. The paths lift into the universal cover to give a $\pi_1 X$ equivariant family of paths from $\tilde{S}_0$ to $\tilde{S}_1$. This defines a lift of the morphism itself to an equivariant morphism $R[\tilde{S}_0] \longrightarrow R[\tilde{S}_1]$. Forget the paths to get a $R[\pi_1 X]$ homomorphism.

Notice that there is a unique homotopy class of paths between any two points in $\tilde{X}$, so no information is lost in forgetting the paths in the lifted morphism.

Now we go the other way, from $\pi_1 X$ modules to geometric modules. To a free module $R[\pi_1 X][S]$ we associate the geometric module $R[S]$, with $S \longrightarrow X$ the map to the basepoint. To a $\pi_1 X$ homomorphism $\Sigma m_i p_i$ with $p_i \in \pi_1 X$, choose representative loops α_i for p_i and form the geometric morphism $\Sigma m_i \alpha_i$.

It is straightforward to see that the constructions are inverses, and define an equivalence of categories. The benefits of thinking of $\pi_1 X$ modules this way are explored in section 2.

1.2 ϵ homotopy Suppose X is a metric space, and $\epsilon > 0$. We say a *homotopy* $h: Y \times I \longrightarrow X$ *has radius less than* ϵ if for each $y \in Y$ the arc $h(y \times I)$ lies in the ball of radius ϵ about $h(y,0)$. In particular this gives a notion of ϵ homotopy of morphisms of geometric modules. This notion is most useful when the morphisms themselves are small. We say a *morphism has radius less than* ϵ if each path λ_i in the morphism lies in the ball of radius ϵ about its starting point $\lambda_i(0)$.

Notice that ϵ homotopy is not an equivalence relation: the composition of ϵ homotopies has radius at best 2ϵ. In fact the situation is often worse than this. If X is not compact then it is necessary to use control *functions* $\epsilon : X \longrightarrow (0,\infty)$, (in which case the ball of radius ϵ at x means the ball of radius $\epsilon(x)$). When ϵ is a function the composition of two ϵ homotopies may be much larger than 2ϵ. We describe how to deal with this in section 3.1.

If the paths in an ϵ morphism are discarded, we get a homomorphism $f' : R[S_0] \longrightarrow R[S_1]$ with the property that if we write $f'(s_i)$ as $\Sigma m_{i,j} t_j$ then the coefficient $m_{i,j}$ is zero if t_j is not in the ϵ ball about s_i. This is an ϵ homomorphism in the sense of Connell and Hollingsworth [3], and Quinn [4, 5]. Morphisms which are ϵ homotopic determine the same ϵ homomorphism. As with $\pi_1 X$ homomorphisms if X is locally 1-connected there is a converse to this construction, at least in the appropriate ϵ sense.

Suppose X is locally 1-connected. Then given $\epsilon>0$ (a function if X is not compact) there is $\delta>0$ such that any two points within δ can be joined by a path of radius ϵ. This means paths can be chosen to represent a homomorphism of radius less than δ as an ϵ geometric morphism. Similarly there is γ so that loops of radius less than γ are nullhomotopic by homotopies of radius less than ϵ. This means that any two representations of a homomorphism by geometric morphisms of radius less than γ are ϵ homotopic. Together these observations imply that for sufficiently small γ, γ homomorphisms determine geometric morphisms well defined up to ϵ homotopy.

If X is not locally 1-connected then this correspondence between metric and geometric ϵ theories breaks down. For many purposes it is the geometric theory which is more fundamental. A precursor of geometric morphisms was developed by Chapman [1] to allow non-locally 1-connected control spaces X in certain controlled manifold theorems, as in section 3.

1.3 Controlled homotopy This is a combination of 1.1 and 1.2: suppose $f:E \longrightarrow X$ is a map, X is a metric space, and $\epsilon>0$. Consider homotopies of morphisms in E whose compositions with f have radius less than ϵ (in X).

Suppose f is a projection of a product $X \times Y \longrightarrow X$. If Y is locally 1-connected we can use the universal cover as in 1.1 to obtain $R[\pi_1 Y]$ homomorphisms of geometric modules over X. If X is also locally 1-connected, then we can proceed as in 1.2 to see that the geometric theory of R-modules on $X \times Y$ with ϵ control in X is essentially equivalent to the ϵ metric theory of $R[\pi_1 Y]$-modules on X.

In some more general situations we can generalize from the product situation and think of geometric algebra on E with ϵ control in X as being like $R[\pi_1 f^{-1}(x)]$ metric algebra. In other words let the coefficient ring vary from point to point in X. In some cases (eg. if E is a "stratified system of fibrations over X", Quinn [5]) this can be made precise. In general, however, it seems best to stick with the geometric description.

The controlled version will be applied in section 3.

1.4 Geometric cellular chains Suppose that K is a CW complex, and $f:K \longrightarrow X$ is a map. We interpret the cellular chain complex of K as a geometric $\mathbb{Z}$ complex over X.

The cellular chain group $C_k(K)$ is the free abelian group generated by the k-cells of K. To give this the structure of a geometric module we introduce notation for the cells in K. Let K^k denote the k-skeleton. Let $\theta_s:D^k \longrightarrow K^k$ denote inclusions of k-cells, where s is in an index set S_k. C_k is then by definition $\mathbb{Z}[S_k]$. Define functions $S_k \longrightarrow X$ by mapping s to $0 \in D^k$, applying θ_s to get a point in K, and then applying f to get a point in X. To ensure that this is locally finite we should assume something like: each point in X has a neighborhood U such that $f^{-1}(U)$ is contained in a finite subcomplex. Assuming this the C_k become geometric $\mathbb{Z}$-modules.

The next step is to define geometric boundary homomorphisms $\partial: C_k \longrightarrow C_{k-1}$. In the definition of a CW complex the maps of the k-cells carry the boundary into the k-1 skeleton: $\partial\theta_s: S^{k-1} \longrightarrow K^{k-1}$. These are the attaching maps for the k-cells, so in fact K^k is defined to be K^{k-1} with cells attached by these $\partial\theta_s$. If k>1 then the boundary homomorphism is defined by $\partial s_i = \Sigma d_{i,j} t_j$, where $s_i \epsilon S_k$, $t_j \epsilon S_{k-1}$, and $d_{i,j}$ is the degree of the map $\partial\theta_{si}$ on the cell $\theta_{tj}: D^{k-1} \longrightarrow K^{k-1}$.

Assume that the attaching maps for the k-cells are transverse to the center points of the k-1 cells. The inverse image $(\partial\theta_s)^{-1}(0_t)$ is then a finite set of points, and at each point there is a sign +1 or -1 depending on whether $\partial\theta$ preserves or reverses orientation at that point. The degree of $\partial\theta_s$ on the cell θ_t is the sum of these signs. Define paths in X by taking the radial line in D^k from 0 to the points $(\partial\theta_s)^{-1}(0_t)$ and composing with θ_s and f. The geometric boundary morphism is defined by adding up these paths times the sign of $\partial\theta_s$ on the endpoint. It is clear from the construction that forgetting the paths yields the ordinary boundary homomorphism.

The boundary $\partial: C_1 \longrightarrow C_0$ is defined slightly differently, since degrees are not defined for 0-cells. The 1-cells are arcs, and the ordinary boundary of a 1-cell is defined to be the beginning point minus the endpoint. The geometric boundary is defined to be the arc from the center to the beginning, minus the arc from the center to the end.

It is suggested that the reader draw a picture of a 2-simplex, and draw in the geometric chain groups and boundary morphisms.

Next consider the composition $\partial\partial$. In traditional complexes this is equal to zero. In the geometric context there is a homotopy to 0. To see this note that the center points in the k-2 cells are codimension k-2 in K^{k-2}, in the sense that they have neighborhoods which are products with D^{k-2}. Form a 1-complex in K^{k-1} by adding to these points the rays to the centers of the k-1 cells. Since the attaching maps are transverse to the centers, this 1-complex is also codimension k-2, except at the centers of the k-1 cells.

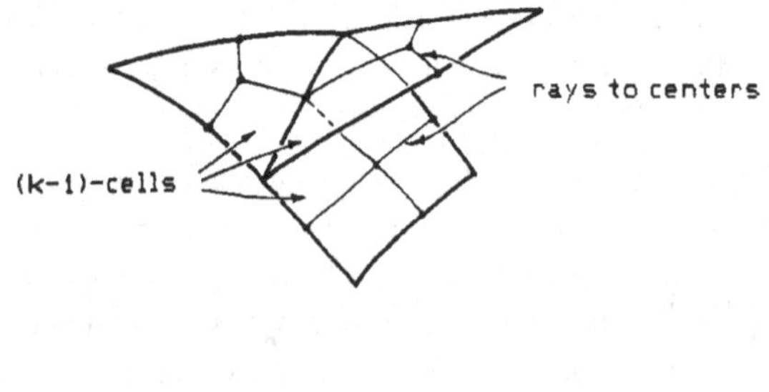

Assume the attaching maps of the k-cells are transverse to this 1-complex. The inverse images in S^{k-1} are then 1-complexes. The vertices are the inverses of the centers of the k-1 cells. There are arcs between these, and disjoint circles. The circles are not useful to us. The cone on the arcs (union of radial lines from the centers of the k-cells) define maps of 2-disks into K.

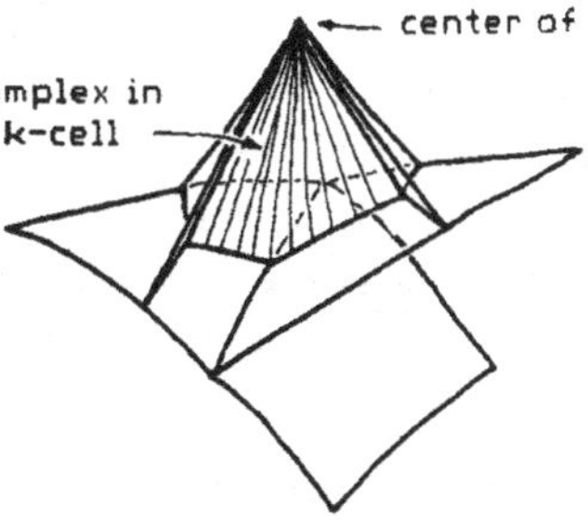

These define a homotopy of $\partial\partial$ to 0. In more detail, each of these 2-disks can be deformed to a map of a square into K. One vertex goes to the center of a k-cell, the adjacent edges to radial lines to centers of (k-1)-cells, and the remaining edges go to radial lines in (k-1)-cells to the center of a (k-2)-cell (see the illustration above). This is a homotopy between two of the paths in the composition $\partial\partial$. Consideration of orientations shows that these paths have opposite sign, so the pair of signed paths are homotopic to a single path with coefficient 0. Therefore up to homotopy they cancel. Finally it is not hard to see that each path in $\partial\partial$ occurs in exactly one of these squares, so the entire composition is homotopic to 0.

This entire collection of data, geometric modules C_k, geometric morphisms ∂, and the homotopy of $\partial\partial$ to 0, forms a geometric chain complex.

Note that if each cell in K has image of sufficiently small diameter in X then the morphisms and homotopies in the geometric complex have radius less than ϵ. Forgetting paths then gives the ϵ chain complexes constructed in Quinn [6, p.271]. Passing to unrestricted homotopy classes gives the classically defined $\mathbb{Z}[\pi_1 X]$ chain complex.

Section 2: Splitting of chain complexes

In this section we suppose that π is a group which is a generalized free product, and construct corresponding splittings of chain complexes over $R[\pi]$. The result itself is not particularly striking. Rather the proof is supposed to suggest benefits of the geometric point of view even in purely algebraic situations. For example it may be that the decomposition theorems for algebraic K-theory (cf. Waldhausen [7, 8]) could be obtained this way. I am told that early preprints of Waldhausen's work have constructions similar to ones used here.

Proposition *Suppose* π *is a homotopy pushout of morphisms* $\alpha,\beta:A \longrightarrow B$ *of groupoids, and the composition* $A \longrightarrow \pi$ *is an injection on each component of* A. *Then any finitely generated free* $R[\pi]$ *chain complex is chain equivalent to a pushout* $a,b:E\otimes_A R[\pi] \longrightarrow F\otimes_B R[\pi]$, *where* E, F *are complexes over* $\mathbb{Z}[A]$, $\mathbb{Z}[B]$, *and* a, b *are chain maps over* α, β.

Proof The homotopy pushout hypothesis on π means that there is a space (CW complex) X with a subspace Y with a neighborhood homeomorphic to $Y\times\mathbb{R}$. The fundamental group of X is π, the fundamental groupoids of Y, X−Y are A, B respectively, and α, β are induced by the inclusions $Y \longrightarrow Y\times(0,\pm\infty) \longrightarrow X-Y$. (Groupoids are disjoint unions of groups. Here they occur as the union of fundamental groups of components of disconnected spaces, see [7].)

Suppose that C_* is a finitely generated free chain complex over $R[\pi_1X]$. Represent C_* as a geometric R-complex on X. Tha data for this are bases $S_i \longrightarrow X$ for the chain groups C_i, geometric morphisms $\partial:R[S_i] \longrightarrow R[S_{i-1}]$, and homotopies $\partial^2\sim 0$. For simplicity we will assume that all paths in the morphisms are defined on the unit interval I. The homotopies then consist of maps of squares $I^2 \longrightarrow X$; the (0,0) corner goes to an element of S_{i+1}, the adjacent sides to paths in ∂_{i+1}, and the remaining sides to paths in ∂_i.

We will say that a geometric complex is "special" if the paths $\lambda:I \longrightarrow X$ have $\lambda^{-1}(Y)$ either I, {1}, or ϕ, and the homotopies $h:I^2 \longrightarrow X$ have $h^{-1}(Y)$ either I^2, or properly contained in $\partial(I^2)$. We claim that such a complex splits in the desired way.

Suppose C_* has this special form. Define E_* to be the submodule of C_* generated by basis elements which map into Y. E_* and the restriction of the boundary morphisms in C_* define a geometric complex on Y; by hypothesis if a path in ∂ starts in Y it stays in Y. The composition ∂^2 is homotopic to 0 in X, but the homotopies are squares with entire boundary mapping to Y. Such squares are required to map to Y, so ∂^2 is nullhomotopic in Y.

Next define F_* by "doubling" E inside C: replace each basis element of C with image in Y by two elements, with images $y\times\{-1\}$ and $y\times\{+1\}$ in $Y\times\mathbb{R} \subset X$. The boundary morphism is that of F in each copy of F, and unchanged in the rest of C except for paths which terminate at a point in Y. Such paths by hypothesis intersect Y in only the final endpoint. Just before the path hits Y, it is either on the + or − side of Y in $Y\times\mathbb{R}$. Push the path off Y, to terminate at the appropriate $y\times\{\pm 1\}$. The homotopies of ∂^2 to 0 also can be pushed off Y. F is therefore a geometric complex over X−Y.

There are chain maps $a,b:E_* \longrightarrow F_2$ defined by inclusions of the copies of E over $Y\times\{\pm 1\}$. C_* is the quotient of F_* by the image $(a-b)E_*$, hence chain equivalent to the pushout. Passing to homotopy classes of morphisms gives complexes over $R[\pi_1Y]$, $R[\pi_1(X-Y)]$. This gives the decomposition required for the proposition.

The proposition therefore will follow if we show that an arbitrary geometric chain complex on X is equivalent to a special one.

The *underlying 2-complex* of a geometric chain complex C_* is formed from the geometric data. The vertices are the union of the generators for the chain groups. The edges are the paths with nonzero coefficient in the boundary homomorphisms. The 2-cells are the squares in the homotopy $\partial^2 \sim 0$. Denote this underlying complex by UC. The maps of the pieces fit together to give a map $UC \longrightarrow X$. The underlying complex is filtered by dimension in the chain complex; define U_iC to be $U(C_*, * \leq i)$. Finally note that the 1-cells are oriented, in that one end has lower filtration than the other.

We say a filtered 2-complex mapping to X is "special" if $f^{-1}(Y)$ is a subcomplex, and if the vertex of highest filtration of a cell is in $f^{-1}(Y)$ then the entire cell is also. In these terms the proof of the proposition is reduced to: show that every geometric chain complex is equivalent to one whose underlying 2-complex is special.

Assume, as an induction hypothesis, that the i-1 filtration $U_{i-1}C$ is special. By small homotopy holding U_{i-1} fixed we may assume that there is a neighborhood N of U_{i-1} in U_i such that $N-U_{i-1} \subset X-Y$. Then we may assume that U_i-U_{i-1} is transverse to Y. The inverse image will therefore be a 1-complex. Squares in U_i-U_{i-1} intersect this 1-complex in arcs with ends on the upper edges, and circles.

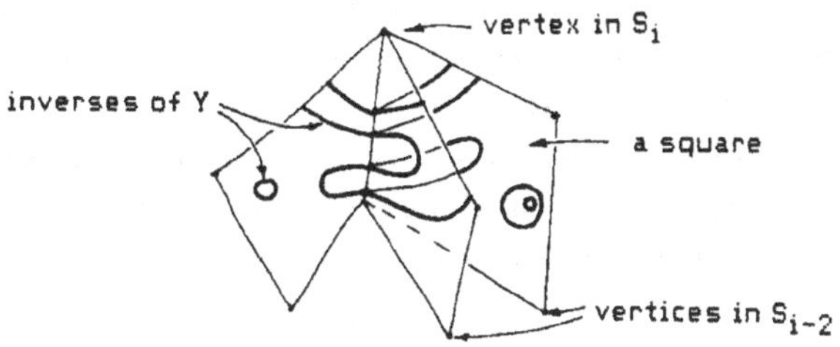

The first step in simplifying the intersection is to note that if there is an arc with both ends on one side of a square, then it encloses a disk in the square. We can push the edge across this disk (draging along any other squares which share that edge). This operation may generate new arcs and circles, but it reduces the number of intersections with the edges.

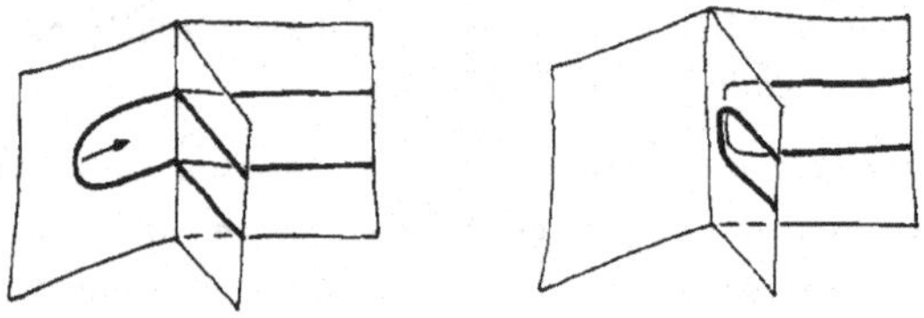

push across

By induction on the intersections with edges, we may assume there are no such arcs.

Next we eliminate the circles interior to each square. These circles map to Y, and the map on the disk the circle bounds in the square gives a nullhomotopy in X. By the hypothesis of injectivity of the fundamental group of Y, these circles are also nullhomotopic in Y. Using the nullhomotopy in Y we can redefine the map to take the disk the circle bounds to Y. This disk can then be pushed off Y.

These changes in the CW complex define changes in the chain complex. Note that the edges, and therefore the boundary morphisms, are changed by homotopy. The homotopies $\partial^2 \sim 0$ are changed by more than homotopy, but that is acceptible; only their existence is part of the data.

Now consider one edge with vertex v in S_i, and consider the point intersection nearest to v of the edge with the inverse of Y. Let $L \subset U_i$ be the component of inverse of Y containing this intersection point. We claim that the region in U_i between v and L is isomorphic with the cone v*L (see the illustration below). For this it is sufficient to show that every intersection of L with an edge is the first intersection of the edge with the inverse of Y. To see this, suppose there is one which is not the first. Choose a path (=sequence of 1-cells) in L from the intersection which is a first, to one which is not. Somewhere in the path there is a single arc so that one end is a first intersection and the other is not. This implies the existence of an arc with both ends on one edge, which contradicts the earlier improvement.

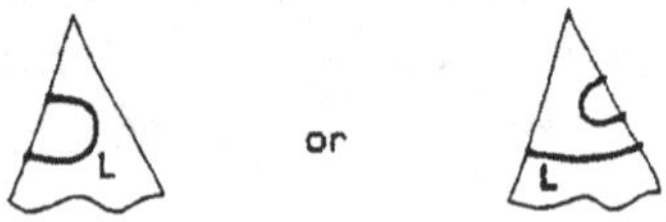

Now construct a new complex U'_i by inserting an arc between v and the cone on L. There is a map $U_i \longrightarrow U'_i$ defined by mapping the cone v*L to the arc. We modify the map $U \longrightarrow X$ so that $U_i \longrightarrow X$ factors through this map.

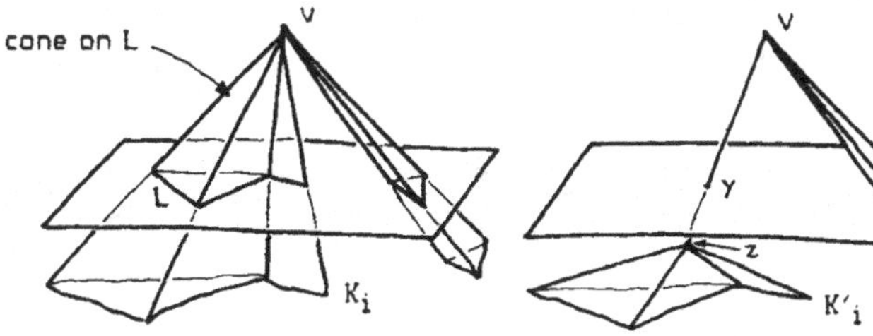

Choose a maximal tree in L, and a collapse of it to a vertex w. Let vw denote the edge between v and w. Then the collapse defines a homotopy of the cone on the tree into L∪(vw). The remaining edges in L define loops in Y which are nullhomotopic in X. The injectivity hypothesis implies that they are also nullhomotopic in Y. A nullhomotopy gives a new map of the cone on this edge, into Y union the image of vw. This factors the map through U′. Push U′ off Y as in the picture above, to leave an intersection point with the arc.

The next step is to define a new complex C′ equivalent to C, which has U'_i as filtration i in the underlying 2-complex. For this we introduce some notation in U′. Let z denote the new cone point, so the arc has been inserted between v and z. Let y denote the intersection point of the arc vz with Y, and let vy, zy denote the paths in vz from the endpoints to y. Let D denote the complex with R[y] in dimension i-1, R[z] in dimension i, and boundary 1(zy). C′ will be defined by modifying the boundary homomorphisms in C⊕D.

Write $\partial: C_i \longrightarrow C_{i-1}$ as $\partial = a+b$, where a consists of the pieces of ∂ whose paths pass through L in U_i, and b is all the rest. Note that the homotopy $\partial(a+b)\sim 0$ breaks into homotopies $\partial a\sim 0$ and $\partial b\sim 0$: since L is an entire component of the intersection with Y, any square in the homotopy with one edge on a path in a must have the other edge on a path in a as well. Note a is defined on the module R[v]. Define $a': R[z] \longrightarrow C_{i-1}$ so that a is homotopic to a′(vz). We use this to construct isomorphisms of the chain groups of C⊕D, and define C′ to have boundary morphisms obtained by conjugation by these isomorphisms. Explicitly C′ is the bottom line in the diagram:

$$\begin{array}{ccccccccc}
\longrightarrow C_{i+1} & \xrightarrow{\left[\begin{smallmatrix}\partial\\ 0\end{smallmatrix}\right]} & C_i\oplus R[z] & \xrightarrow{\left[\begin{smallmatrix}a+b & 0\\ 0 & zy\end{smallmatrix}\right]} & C_{i-1}\oplus R[y] & \xrightarrow{[\partial\ 0]} & C_{i-2}\longrightarrow \\
\downarrow 1 & & \downarrow \left[\begin{smallmatrix}1 & 0\\ vz & 1\end{smallmatrix}\right] & & \downarrow \left[\begin{smallmatrix}1 & a'(yz)\\ 0 & 1\end{smallmatrix}\right] & & \downarrow 1 \\
\longrightarrow C_{i+1} & \xrightarrow[\left[\begin{smallmatrix}\partial\\ (vz)\partial\end{smallmatrix}\right]]{} & C_i\oplus R[z] & \xrightarrow[\left[\begin{smallmatrix}b & a'\\ -vy & zy\end{smallmatrix}\right]]{} & C_{i-1}\oplus R[y] & \xrightarrow[{[\partial\ 0]}]{} & C_{i-2}\longrightarrow
\end{array}$$

The vertex y is special in C′, and there are fewer components of the inverse image of Y in $U_i C'$. Therefore by iterating the construction we can get a complex equivalent to C with filtration i special.

This completes the induction step, and shows that we can find an equivalent complex whose entire underlying 2-complex is special. As indicated above, this implies the proposition.

Section 3: The controlled h-cobordism theorem

The classical h-cobordism theorem states that an h-cobordism with vanishing Whitehead torsion is isomorphic to a product. In the contolled version of this there is a map to a metric space X, $\epsilon>0$ is given, and we want a product structure such that the image in X of each product arc lies in the ball of radius ϵ about its beginning point. Another way to say this is that the product structure has radius less than ϵ as a homotopy. We will see that the obstruction to this lies in a controlled version of the Whitehead group.

The section begins with some generalities on control functions, necessary on noncompact control spaces. Then ϵ Whitehead groups are defined, and the theorem is proved.

3.1 Control functions Suppose X is a metric space, and $\epsilon: X \longrightarrow (0,\infty)$ is a map. If $f,g: Y \longrightarrow X$ are functions then we say g is *within* ϵ *of* f, and write $d(f,g)<\epsilon$, if $d(f(y),g(y)) < \epsilon(f(y))$ for each $y\in Y$. Notice that this usually does not imply that $d(g,f)<\epsilon$, and the triangle inequality does not hold. To deal with this we introduce some notation.

Suppose α,β are maps $X \longrightarrow (0,\infty)$. Define $\alpha\#\beta(x)$ to be $\max\{\alpha(x)+\beta(y),\ d(x,y)<\alpha(x)\}$. If α, β are constant then $\alpha\#\beta = \alpha+\beta$.

It follows easily that if $d(f,g)<\alpha$ and $d(g,h)<\beta$ then $d(f,h) < \alpha\#\beta$. Also for functions α,β,δ, $\alpha\#(\beta\#\delta) \leq (\alpha\#\beta)\#\delta$ (both expressions are maxima of $\alpha(x)+\beta(y)+\delta(z)$, and more values of z are allowed in the second expression). Denote by $n*\beta$ the n-fold iteration of this operation, with parentheses arranged to give the largest value. For example $4*\beta$ means $((\beta\#\beta)\#\beta)\#\beta$, and $m*(n*\beta)) \leq (mn)*\beta$. If β is constant then $n*\beta$ is the ordinary product $n\beta$.

These expressions are usually used as upper bounds. Note that if an expression with any arrangement of parentheses is an upper bound, then the largest arrangement is an upper bound as well. This is why the notation $n*\beta$ is useful.

In the geometric algebra context note that if f,g are geometric morphisms with radius less than α,β respectively, and gf is defined, then gf has radius less than $\alpha*\beta$. Compositions of homotopies behave similarly.

3.2 Whitehead groups Suppose E is a space. The Whitehead group $Wh(R[\pi_1 E])$ is defined to be the set of equivalence classes of isomorphisms of free based modules over $R[\pi_1 E]$. The equivalence relation is generated by direct sums with identity isomorphisms, and by composition with triangular automorphisms. In this context an automorphism is triangular if there is an ordering of the basis of the module so that the matrix expression is zero below the diagonal, and the diagonal entries are units in R times elements of $\pi_1 E$.

Usually a triangular matrix is required to have diagonal entries all equal to 1. The group obtained with this definition is the reduced K-group $\tilde{K}_1(R[\pi_1 E])$. The Whitehead group is obtained from this by dividing by the subgroup generated by the automorphisms of $R[\pi_1 E]$ given by products (unit in R)(element of $\pi_1 E$). But this is equivalent to allowing such products on the diagonal of triangular matrices.

Using the equivalence of section 1.1 we can describe $Wh(R[\pi_1 E])$ as equivalence classes of geometric isomorphisms of finitely generated geometric R-modules on E. We obtain a version with ϵ control by adding ϵ to this description, as in 1.3.

Suppose $p:E \longrightarrow X$ is a map, and X is a metric space. A *geometric ϵ isomorphism* of geometric R-modules on Eis a morphism of radius $<\epsilon$ (measured in X) with an "inverse" also of radius $<\epsilon$, such that the compositions are ϵ homotopic to the identity morphisms. Unfortunately it is possible to have a morphism of radius $<\epsilon$ which is an isomorphism, but whose inverse has very large radius. Therefore the estimate on the radius of the inverse must be included in the definition.

Suppose $M = R[S]$ is a geometric R-module on E. A geometric morphism $A:M \longrightarrow M$ is *(upper) triangular* provided there is an ordering of the basis of M such that A has no paths from t to s unless $t \leq s$, and if $t = s$ there is exactly one path, whose coefficient is a unit in R.

We observe that a triangular morphism is an isomorphism: it can be written as $D(I+B)$ where D is diagonal with entries a unit times a loop, and B has entries strictly above the diagonal. D^{-1} is obtained by inverting the units and reversing the loops, and $(I+B)^{-1} = I+\Sigma_1^n(-B)^i$, where n is large enough so that $B^{n+1} = 0$. Note that the radius of the inverse depends on the radius of the original, and this n.

We define a *deformation* of a geometric module to be a sequence $A_n A_{n-1} \cdots A_1$ of triangular morphisms. We write it as a product, and think of it that way, but actually need to keep track of a little more information than is retained in the product. What is needed is a refined version of the radius.

The *underlying 1-complex* of a morphism is the collection of paths which occur in the morphism. We think of these as trees eminating from the basis of the module, by identifying the beginning points of all paths coming from a given basis element. The underlying 1-complex of a sequence $A_2 A_1$ again consists of a tree for each basis element: start with the tree for A_1 beginning at the element, and at the end of each branch add a copy of the tree of A_2 which begins there. Trees for a sequence with n terms are built up similarly. We now say that *a sequence* $A_n A_{n-1} \dots A_1$ *has radius less than* ϵ if for each basis element s the tree in the underlying 1-complex starting at s lies inside the ball of radius $\epsilon(s)$ about s. As with the radius of isomorphisms, we must require that the trees for the inverse sequence $A^{-1}{}_1 A^{-1}{}_2 \dots A^{-1}{}_n$ lie inside the ϵ balls about their starting points as well.

Notice that the paths which occur in the composition of the sequence are just paths in these trees. The radius of the composition is therefore less than or equal to the radius of the sequence. The difference is that in the composition we allow deletion of paths when the coefficients cancel, whereas no cancellations are allowed in the underlying 1-complex. Therefore the radius of the composition may be strictly smaller.

Finally if A is a geometric morphism, an ϵ deformation of A is a composition $D_0 A D_1$, where D_0, D_1 are ϵ deformations of the range and domain modules of A.

Definition *Suppose* $p: E \longrightarrow X$ *is a map to a metric space, and* $\epsilon: X \longrightarrow (0,\infty)$ *is given. Then* $Wh(X,p,\epsilon)$ *is defined to be equivalence classes of geometric isomorphisms on* E *with radius* $<\epsilon$ *in* X, *with equivalence relation generated by direct sum with identity morphisms, and homotopies and deformations of radius* $< 3*\epsilon$.

The next lemma shows this to be a convenient place to work. However, see the comments following the proof.

Lemma *Direct sum induces an abelian group structure on* $Wh(X,p,\epsilon)$. *Further if* ϵ *isomorphisms are equivalent in this sense, then there is a* $9*\epsilon$ *deformation between isomorphisms* $9*\epsilon$ *homotopic to appropriate stabilizations of the originals.*

Proof of the lemma Since direct sum clearly induces an abelian monoid structure, the point of the first statement is that there are inverses. If A is an isomorphism there is a matrix identity

$$\begin{bmatrix} I & A \\ 0 & I \end{bmatrix} \begin{bmatrix} I & 0 \\ -A^{-1} & I \end{bmatrix} \begin{bmatrix} I & A \\ 0 & I \end{bmatrix} \begin{bmatrix} A & 0 \\ 0 & A^{-1} \end{bmatrix} \begin{bmatrix} I & -I \\ 0 & I \end{bmatrix} \begin{bmatrix} I & 0 \\ I & I \end{bmatrix} \begin{bmatrix} I & -I \\ 0 & I \end{bmatrix} = \begin{bmatrix} I & 0 \\ 0 & I \end{bmatrix}$$

If A is a geometric morphism of radius $<\epsilon$, the left side of the equation is a $3*\epsilon$ deformation of $A \oplus A^{-1}$. (The left three, and right three, terms are triangular.) When the left side is multiplied out there are terms like $A - AA^{-1}A$, so the composition is actually $3*\epsilon$ homotopic to $I \oplus I$ rather than equal to it. This shows that A^{-1} is a $3*\epsilon$ additive inverse for A.

Next suppose that there is a sequence A_i of ϵ isomorphisms, such that there is a $3*\epsilon$ deformation from A_i to A_{i+1}, and i goes from 1 to n+1. Consider the sequence of deformations

$$A_1 \sim A_1 \oplus \Sigma_1^n (I \oplus I) \sim A_1 \oplus \Sigma_1^n (A_i{}^{-1} \oplus A_i) \sim A_1 \oplus \Sigma_1^n (A_i{}^{-1} \oplus A_{i+1}) =$$
$$[\Sigma_1^n (A_i \oplus A_i{}^{-1})] \oplus A_{n+1} \sim [\Sigma_1^n (I \oplus I)] \oplus A_{n+1} \sim A_{n+1}.$$

The first and last are stabilizations, the second and fifth are the $3*\epsilon$ deformations which cancel inverses, and the third is the sum of the deformations $A_i \sim A_{i+1}$. The composition of these give a $9*\epsilon$ deformation from a stabilization of A_1 to a stabilization of A_{n+1}.

Usually there will be stabilizations in the deformations $A_i \sim A_{i+1}$. These are easily incorperated in the above argument.

Remarks The point of the second statement in the lemma is that geometric control is not lost by allowing arbitrarily many deformations. If we simply compose the sequence of deformations $A_i \sim A_{i+1}$ we get a deformation of radius $3n*\epsilon$, which can be arbitrarily large. The method for getting a short deformation comes from Quinn [5, lemma 4.4]; its use in this context is due to Chapman [2, theorem 3.5].

In general the groups $Wh(X,p,\epsilon)$ are quite mysterious. If ϵ is much larger than δ then the image of $Wh(X,p,\delta)$ in $Wh(X,p,\epsilon)$ is sometimes more accessible. Chapman [2] gives criteria for this image to be trivial, in terms of vanishing of ordinary Whitehead groups of $\pi_1(p^{-1}(U))$, for open sets U in X. In a more rigid setting (p a stratified system of fibrations) but no condition on $\pi_1(p^{-1}(U))$, Quinn [5] shows the image to be a generalized homology group of X.

3.3 Controlled h-cobordisms Suppose $\delta: X \longrightarrow (0,\infty)$ is given. Then a manifold triad $(W,\partial_0 W,\partial_1 W)$ with a map $f:W \longrightarrow X$ is a *(δ,h)-cobordism* if f is a proper map, and there are deformation retractions of W to $\partial_i W$ which have radius $<\delta$ in X.

The question we consider is: when does a (δ,h)-cobordism have a product structure $W \simeq (\partial_i W)\times I$ of radius $<\epsilon$ in X? This has been considered at length in the literature; the objective here is just to see the obstructions in geometric algebraic terms.

Some local control of the fundamental group is necessary. For this fix a map $p:E \longrightarrow X$. A map $f:W \longrightarrow E$ is *relatively δ,1 connected* (over X) if for every relative 2-complex (K,L) and commutative diagram

$$\begin{array}{ccc} L & \longrightarrow & W \\ \downarrow & & \downarrow f \\ K & \longrightarrow & E \end{array}$$

there is a map $K \longrightarrow W$ which agrees with the given map on L, and whose composition with f is within ϵ of the given map into E, measured in X.

Theorem *Suppose* $f:W \longrightarrow E$, $p:E \longrightarrow X$, *and* $\delta:X \longrightarrow (0,\infty)$ *are given, so that* pf *is a (δ,h)-cobordism over* X. *Then there is a well-defined invariant* $q_1(W,\partial_0 W) \in Wh(X,p,9*\delta)$ *which vanishes if* W *has a* δ *product structure. Conversely there is a function* $k(n)$ *such that if* $n\geq 6$ *and* $q_1(W,\partial_0 W) = 0$ *then* W *has a product structure of radius* $< k(n)*\delta$.

Proof This is theorem 3.1 of Quinn [5], with some minor refinements in estimates and the use of geometric instead of metric algebra. We outline that proof. A similar statement, with a more geometric definition of the Whitehead group, is given by Chapman [2, 14.2].

Choose a handlebody structure on $(W,\partial_0 W)$ with handles whose images in X have diameter less than δ. By *diameter less than* δ we mean here that if x and y are in the set, then $d(x,y) < \delta(x)$. If W is smooth or PL such a handle structure can be defined from a fine triangulation.

A handlebody structure has a spine, which is a CW complex structure on $(W,\partial_0 W)$. For example the spine of a handlebody structure obtained from a triangulation is just the triangulation itself. We require that the CW structure also have cells of diameter less than δ. This is automatic if the handles are small enough.

We also require the CW structure to be *saturated*. This means that the attaching map $\theta:S^n \longrightarrow K^n$ for an n+1 cell has image a union of cells. In other words, if a cell intersects the image of θ, it is contained in the image. If the CW structure is a triangulation it is automatically saturated. In the topological case it is not hard to arrange saturation.

The saturation condition is used to push things rapidly into skeleta. Suppose that $K \longrightarrow X$ is a saturated complex of dimension n whose cells have diameter less than δ in X. Suppose $h:L \longrightarrow K$ is a map, L a j-complex with $j<n$. L can be pushed off the n-cells of K, to obtain a map h_{n-1} of L into the n-1 skeleton with $d(h,h_{n-1}) < \delta$ (measured in X, as always). Similarly we can push L off the n-1 cells, and repeat until we have h_j mapping L into the j-skeleton of K. Since L has been moved n-j times, in general we only know that $d(h,h_j) < (n-j)*\delta$. However if K is saturated, then a point which is moved out of the interior of a cell in one push stays in the image of the boundary of that cell during later pushes. Therefore $d(h,h_j) < \delta$.

Choose a δ deformation retraction of W to $\partial_0 W$, $h:W\times I \longrightarrow W$. Put the product CW structure on $(W\times I,(\partial_0 W)\times I)$. Use the fact that the structure on W is saturated to get a deformation retraction h' with $d(h,h') < \delta$, $d(h',h) < \delta$, and which preserves skeleta. In particular note that images of cells under h' have diameter $< 3*\delta$.

Apply the cellular chain construction of 1.4 to obtain $C_* = C_*(W,\partial_0 W)$, a geometric chain complex of radius $<\delta$. The deformation retraction h' defines a chain homotopy $s:C_* \longrightarrow C_{*+1}$ of radius $<3*\delta$, such that $s\partial+\partial s$ is $4*\delta$ homotopic to the identity map of C_*.

Consider the morphism $(s\partial s+\partial s\partial):\Sigma_{(j\ \mathrm{even})}C_j \longrightarrow \Sigma_{(j\ \mathrm{odd})}C_j$. This is a $9*\delta$ isomorphism; it has radius $<7*\delta$ and the same formula from odd j back to the even ones is a $9*\delta$ inverse. We define $q_1(W,\partial_0 W)$ to be the equivalence class $[s\partial s+\partial s\partial]$ in $Wh(X,p,9*\delta)$.

The first step in proving the theorem is to show the invariance. If s' is another $4*\delta$ chain contraction then s' is $4*\delta$ homotopic to $s+(s's\partial-\partial s's)$. Using this we can write

$(s'\partial s'+\partial s'\partial)$ as $(I+D)(s\partial s+\partial s\partial)$, where for example $D = \partial ss'\partial ss'\partial s\partial s$. D raises dimension in $\Sigma_{(j\ odd)}C_j$ so I+D is triangular. It has radius less than $3(9)*\delta$ so is an equivalence in $Wh(X,p,9*\delta)$. This shows that q_1 is independent of the deformation retraction.

Now suppose there is another handle decomposition of $(W,\partial_0 W)$ satisfying the conditions above. Then there is a 1-parameter family of handlebody structures, all of diameter less than δ, joining them. In this family the handles can change by isotopy, cancelling pairs can be introduced or be cancelled, and handle additions can occur. Isotopy changes the chain complex only by homotopy. The other changes occur at isolated points in the parameter arc. Choose points between these and apply the construction. This gives a sequence of $9*\delta$ isomorphisms, with adjacent ones related by homotopy and a single modification. Cancelling pairs change the complex by addition of an identity morphism. Handle additions change the boundary morphisms by product with a triangular morphism, so change the isomorphism in the same way. We conclude that adjacent isomorphisms in the sequence are equivalent, hence the ends are equivalent in $Wh(X,p,9*\delta)$.

This shows that q_1 is well defined. If W has a δ product structure then the product structure is a handle structure with no handles. The chain complex is therefore trivial, and the invariant equal to 0.

The converse is essentially proved in section 6 of Quinn [4], which is independent of the rest of that paper. The estimate there is $k(n) = (54+3n)^n$, but this can be improved quite a lot using the "saturation" idea above.

The idea of the proof is to first show that $(W,\partial_0 W)$ has a handlebody structure with handles only in two adjacent dimensions. In this case the boundary morphism between the geometric chain groups is an isomorphism, which when mapped to E represents $\pm q_1$. The hypothesis that $q_1 = 0$ implies, by the lemma in 3.2, that there is an $81*\delta$ deformation of the image of the boundary morphism to the empty morphism. This deformation takes place in E. Since $W \longrightarrow E$ is relatively $(\delta,1)$-connected, the paths and homotopies of the deformation lift back to give a deformation in W. The proof of [4, section 6] then shows how to use such a deformation to cancel the remaining handles.

This completes the sketch of the controlled h-cobordism theorem. We remark that recent work of M. Freedman and the author shows that this theorem also applies to 5-dimensional topological h-cobordisms, provided the local fundamental groups are "poly-(finite or cyclic)".

References

[1] T. A. Chapman, *Controlled boundary and h-cobordism theorems,* Trans. AMS 280 (1983) 73-95.

[2] T. A. Chapman, *Controlled simple homotopy theory and applications,* Lecture Notes in Mathematics 1009 (1983) Springer-Verlag.

[3] E.H. Connell and J. Hollingsworth, *Geometric groups and Whitehead torsion,* Transactions AMS 140 (1969) 161-181.

[4] F. Quinn, *Ends of Maps I,* Annals of Math 110 (1979) 275-331.

[5] F. Quinn, *Ends of Maps II,* Invent. Math 68 (1982) 353-424.

[6] F. Quinn, *Resolutions of homology manifolds, and the topological characterization of manifolds,* Invent. Math. 72 (1983) 267-284.

[7] F. Waldhausen, *Whitehead groups of generalized free products* Springer Lecture Notes 342 (1973) 155-179.

[8] F. Waldhausen, *Algebraic K-theory of generalized free products,* Annals of Math. 108 (1978) 135-256.

THE ALGEBRAIC THEORY OF TORSION I. FOUNDATIONS

by Andrew Ranicki,
Edinburgh University

Introduction

The algebraic theory of torsion developed here takes values in the absolute K_1-group $K_1(A)$ of a ring A, with a torsion invariant $\tau(f) \in K_1(A)$ for a chain equivalence $f:C \longrightarrow D$ of finite chain complexes of based f.g. free A-modules with zero Euler characteristic.

Whitehead [24] defined the torsion $\tau(C) \in K_1(A)$ of a contractible finite chain complex C of based f.g. free A-modules, assuming (as we do here) that A is such that f.g. free A-modules have well-defined rank. The algebraic mapping cone $C(f)$ of a chain equivalence $f:C \longrightarrow D$ of finite chain complexes of based f.g. free A-modules is a contractible chain complex, so that the torsion $\tau(C(f)) \in K_1(A)$ is defined. However, the expected sum formula for the composite $gf:C \longrightarrow D \longrightarrow E$ of chain equivalences $f:C \longrightarrow D$, $g:D \longrightarrow E$

$$\tau(C(gf)) = \tau(C(f)) + \tau(C(g)) \in K_1(A)$$

only holds in general on passing to the reduced K_1-group

$$\tilde{K}_1(A) = \mathrm{coker}(K_1(\mathbb{Z}) \longrightarrow K_1(A)) = K_1(A)/\{\tau(-1:A \longrightarrow A)\} \ .$$

The reduced torsion of the algebraic mapping cone

$$\tau(f) = \tau(C(f)) \in \tilde{K}_1(A)$$

is the torsion invariant usually associated to a chain equivalence f. In particular, the Whitehead torsion $\tau(f) \in Wh(\pi)$ $(\pi = \pi_1(X))$ of a homotopy equivalence $f:X \longrightarrow Y$ of finite CW complexes is the image of $\tau(\tilde{f}:C(\tilde{X}) \longrightarrow C(\tilde{Y})) \in \tilde{K}_1(\mathbb{Z}[\pi])$ in the Whitehead group $Wh(\pi) = K_1(\mathbb{Z}[\pi])/\{\pm\pi\}$. The theory of torsion developed here can be used in certain circumstances to lift the Whitehead torsion to an absolute torsion invariant $\tau(f) \in K_1(\mathbb{Z}[\pi])$, which enters into product formulae for Whitehead torsion.

The Euler characteristic of a finite chain complex C of f.g. free A-modules is defined as usual by

$$\chi(C) = \sum_{r=0}^{\infty} (-)^r \mathrm{rank}_A(C_r) \in \mathbb{Z} \ .$$

The complex C is <u>round</u> if

$$\chi(C) = 0 \in \mathbb{Z} \ .$$

The assumption on A that f.g. free A-modules have well-defined rank ensures that $K_0(\mathbb{Z}) \longrightarrow K_0(A)$ is injective, so that the Euler characteristic may be identified with the absolute projective class

$$\chi(C) = [C] \in \mathbb{Z} = K_0(\mathbb{Z}) \subseteq K_0(A) .$$

The <u>absolute torsion</u> of a chain equivalence $f:C \longrightarrow D$ of round finite chain complexes of based f.g. free A-modules is defined in §4 by a formula of the type

$$\tau(f) = \tau(C(f)) + \beta\tau(-1:A \longrightarrow A) \in K_1(A)$$

with the sign term $\beta = 0$ or 1 depending only on the ranks (mod 2) of the chain modules of C and D. It is quite reasonable that a K_1-valued invariant should only be defined when K_0-valued obstructions vanish! Actually, the absolute torsion is also defined if C,D are such that the Euler characteristic is 0(mod 2). For contractible C,D the torsion of f is just the difference of the torsions of C and D

$$\tau(f) = \tau(D) - \tau(C) \in K_1(A) .$$

The main result of Part I is the logarithmic property of absolute torsion with respect to composition

$$\tau(gf:C \longrightarrow D \longrightarrow E) = \tau(f:C \longrightarrow D) + \tau(g:D \longrightarrow E) \in K_1(A) .$$

As such this is not very prepossessing. The applications of absolute torsion are more interesting, but will be dealt with elsewhere. Parts II and III will deal with products and lower K-theory. Some of the applications to L-theory are contained in a forthcoming joint paper with Ian Hambleton and Larry Taylor on "Round L-theory".

The following preview of the applications of the absolute torsion to topology may help to motivate the paper.

Define a connected finite CW complex X to be <u>round</u> if $\chi(X) = 0 \in \mathbb{Z}$ and the cellular f.g. free $\mathbb{Z}[\pi_1(X)]$-module chain complex $C(\tilde{X})$ of the universal cover $\tilde{X}$ is equipped with a choice of base in the canonical class of bases determined by the cell structure of X up to the multiplication of each base element by $\pm g$ $(g \in \pi_1(X))$. Thus $C(\tilde{X})$ is a round finite chain complex of based f.g. free $\mathbb{Z}[\pi_1(X)]$-modules. The <u>absolute torsion</u> of a homotopy equivalence $f:X \longrightarrow Y$ of round finite CW complexes is defined by

$$\tau(f) = \tau(\tilde{f}:C(\tilde{X}) \longrightarrow C(\tilde{Y})) \in K_1(\mathbb{Z}[\pi_1(X)]) ,$$

and is such that the reduction $\tau(f) \in Wh(\pi_1(X))$ is the usual Whitehead torsion of f. A <u>round finite structure</u> on a topological space X is an equivalence class of pairs

(round finite CW complex K, homotopy equivalence $f: K \longrightarrow X$)

under the equivalence relation

$$(K,f) \sim (K',f') \text{ if } \tau(f'^{-1}f: K \longrightarrow X \longrightarrow K') = 0 \in K_1(\mathbb{Z}[\pi_1(X)]) .$$

For example, the mapping torus of a self map $\zeta: X \longrightarrow X$ of a finitely dominated CW complex X

$$T(\zeta) = X \times [0,1]/\{(x,0) = (\zeta(x),1) \mid x \in X\}$$

has a canonical round finite structure, by a generalization of the trick of Mather [9], with $T(f\zeta g: Y \longrightarrow Y)$ a round finite CW complex in the round finite homotopy type of $T(\zeta)$ for any domination of X

$$(Y, f: X \longrightarrow Y, g: Y \longrightarrow X, h: gf \simeq 1: X \longrightarrow X)$$

by a finite CW complex Y. (Furthermore, if $X = \overline{M}$ is an infinite cyclic cover of a compact manifold M with $\zeta: X \longrightarrow X$ a generating covering translation then the projection $T(\zeta) \longrightarrow M$ is a homotopy equivalence such that the Whitehead torsion $\tau \in Wh(\pi_1(M))$ is the obstruction of Farrell [3] and Siebenmann [20] to fibering M over S^1, giving M the finite homotopy type determined by a handlebody decomposition and assuming $\dim(M) \geqslant 6$). The <u>product structure theorem</u> is that the product $F \times B$ of a finitely dominated CW complex F and a round finite CW complex B has a canonical round finite structure, such that the absolute torsion of a product homotopy equivalence is given by

$$\tau(f \times b: F \times B \longrightarrow F' \times B') = [F] \otimes \tau(b)$$
$$\in K_1(\mathbb{Z}[\pi_1(F \times B)]) = K_1(\mathbb{Z}[\pi_1(F)] \otimes \mathbb{Z}[\pi_1(B)]) ,$$

with $[F] = [F'] \in K_0(\mathbb{Z}[\pi_1(F)])$ the absolute projective class and $\tau(b) \in K_1(\mathbb{Z}[\pi_1(B)])$ the absolute torsion. The circle

$$S^1 = T(\mathrm{id.}: \{\mathrm{pt.}\} \longrightarrow \{\mathrm{pt.}\})$$

has the canonical round finite structure in which the base elements $\tilde{e}^i \in C(\tilde{S}^1)_i = \mathbb{Z}[\pi_1(S^1)] = \mathbb{Z}[z,z^{-1}]$ $(i = 0,1)$ are such that

$$d(\tilde{e}^1) = \tilde{e}^0 - z\tilde{e}^0 .$$

For any finitely dominated CW complex F the product round finite structure on $F \times S^1 = T(1: F \longrightarrow F)$ agrees with the mapping torus round finite structure. Ferry [4] defined a geometric injection

$$\overline{B}' : \tilde{K}_0(\mathbb{Z}[\pi]) \rightarrowtail Wh(\pi \times \mathbb{Z}) ; [F] \longmapsto \tau(1 \times -1: F \times S^1 \longrightarrow F \times S^1)$$

for any finitely presented group π, with $[F] \in \tilde{K}_0(\mathbb{Z}[\pi])$ the Wall finiteness obstruction of a finitely dominated CW complex F with

$\pi_1(F) = \pi$. The image of $\bar{B}'$ consists of the elements $\tau \in Wh(\pi \times \mathbb{Z})$ invariant under the transfer maps associated to the finite covers of S^1. The map $-1:S^1 \longrightarrow S^1$ reflecting the circle in a diameter has absolute torsion

$$\tau(-1:S^1 \longrightarrow S^1) = \tau(-z:\mathbb{Z}[z,z^{-1}] \longrightarrow \mathbb{Z}[z,z^{-1}]) \in K_1(\mathbb{Z}[z,z^{-1}]) ,$$

so that by the product structure theorem $\bar{B}'$ is given algebraically by

$$\begin{aligned} \bar{B}' = -\otimes\tau(-z) \ :\ & \tilde{K}_0(\mathbb{Z}[\pi]) \rightarrowtail Wh(\pi \times \mathbb{Z}) \ ; \\ & [P] \longmapsto \tau(-z:P[z,z^{-1}] \longrightarrow P[z,z^{-1}]) \end{aligned}$$

with $[P]$ the reduced projective class of a f.g. projective $\mathbb{Z}[\pi]$-module P. Thus $\bar{B}'$ does not coincide with the traditional algebraic injection of Bass, Heller and Swan [2]

$$\begin{aligned} \bar{B} = -\otimes\tau(z) \ :\ & \tilde{K}_0(\mathbb{Z}[\pi]) \rightarrowtail Wh(\pi \times \mathbb{Z}) \ ; \\ & [P] \longmapsto \tau(z:P[z,z^{-1}] \longrightarrow P[z,z^{-1}]) \ . \end{aligned}$$

The recent algebraic description due to Lück [8] of the transfer map $p_1^!:K_1(\mathbb{Z}[\pi_1(B)]) \longrightarrow K_1(\mathbb{Z}[\pi_1(E)])$ induced in the K_1-groups by a Hurewicz fibration

$$F \longrightarrow E \xrightarrow{\ p\ } B$$

with finitely dominated fibre F allows the product structure theorem to be extended to the twisted case: the total space E of a fibration with finitely dominated fibre F and round finite base B has a canonical round finite homotopy type, and if

$$\begin{array}{ccc} F & \xrightarrow{\ f\ } & F' \\ \downarrow & & \downarrow \\ E & \xrightarrow{\ e\ } & E' \\ {\scriptstyle p}\downarrow & & \downarrow{\scriptstyle p'} \\ B & \xrightarrow{\ b\ } & B' \end{array}$$

is a fibre homotopy equivalence of such fibrations the homotopy equivalence $e:E \longrightarrow E'$ has absolute torsion

$$\tau(e) = p_1^!(\tau(b)) \in K_1(\mathbb{Z}[\pi_1(E)]) \ .$$

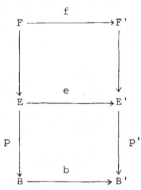

The <u>absolute torsion</u> of a round finite n-dimensional geometric Poincaré complex B is defined by

$$\tau(B) = \tau([B]\cap - : C(\widetilde{B})^{n-*} \longrightarrow C(\widetilde{B})) \in K_1(\mathbb{Z}[\pi_1(B)]) \quad ,$$

satisfying the usual duality $\tau(B)^* = (-)^n\tau(B)$. The Poincaré complex version of the twisted product structure theorem is that the total space of a fibration $F \longrightarrow E \xrightarrow{p} B$ with a round finite n-dimensional Poincaré base B and a finitely dominated m-dimensional Poincaré fibre F is an (m+n)-dimensional Poincaré complex E with a canonical round finite structure, with respect to which the torsion of E is given by

$$\tau(E) = p_1^!(\tau(B)) \in K_1(\mathbb{Z}[\pi_1(E)]) \quad .$$

In particular, for the trivial fibration $E = F\times B$ this is a product formula

$$\tau(F\times B) = [F]\otimes\tau(B) \in K_1(\mathbb{Z}[\pi_1(F\times B)]) \quad .$$

The torsion of the circle S^1 with respect to the canonical round finite structure is

$$\tau(S^1) = \tau(-z:\mathbb{Z}[z,z^{-1}] \longrightarrow \mathbb{Z}[z,z^{-1}]) \in K_1(\mathbb{Z}[\pi_1(S^1)]) = K_1(\mathbb{Z}[z,z^{-1}]) \; ,$$

so that for any finitely dominated m-dimensional Poincaré complex F

$$\begin{aligned}\tau(F\times S^1) &= [F]\otimes\tau(S^1) = [F]\otimes\tau(-z) = \overline{B}'([F]) \\ &\in K_1(\mathbb{Z}[\pi\times\mathbb{Z}]) = K_1(\mathbb{Z}[\pi][z,z^{-1}]) \qquad (\pi = \pi_1(F))\end{aligned}$$

with $\overline{B}':K_0(\mathbb{Z}[\pi]) \rightarrowtail K_1(\mathbb{Z}[\pi][z,z^{-1}])\;;\;[P] \longmapsto \tau(-z:P[z,z^{-1}] \longrightarrow P[z,z^{-1}])$ the absolute version of the injection $\overline{B}':K_0(\mathbb{Z}[\pi]) \rightarrowtail Wh(\pi\times\mathbb{Z})$ described above. More generally, the mapping torus $T(\zeta)$ of a self homotopy equivalence $\zeta:F \longrightarrow F$ is the total space of a fibration over S^1

$$F \longrightarrow T(\zeta) \xrightarrow{p} S^1$$

such that $\pi_1(T(\zeta)) = \pi\times_\alpha\mathbb{Z}$ $(\alpha = \zeta_* : \pi \longrightarrow \pi)$, and $T(\zeta)$ is an (m+1)-dimensional geometric Poincaré complex with a canonical round finite structure with respect to which

$$\begin{aligned}\tau(T(\zeta)) &= p_1^!\tau(S^1) = \tau(-z\widetilde{\zeta}:C(\widetilde{F})_\alpha[z,z^{-1}] \longrightarrow C(\widetilde{F})_\alpha[z,z^{-1}]) \\ &\in K_1(\mathbb{Z}[\pi\times_\alpha\mathbb{Z}]) = K_1(\mathbb{Z}[\pi]_\alpha[z,z^{-1}]) \\ &(gz = z\alpha(g)\;\;(g\in\pi)\,,\;\widetilde{\zeta} : \alpha_!C(\widetilde{F}) \longrightarrow C(\widetilde{F}))\;.\end{aligned}$$

The algebraic theory of surgery of Ranicki [17] has a version for round finite algebraic Poincaré complexes, corresponding to the variant L-groups of Wall [22] in which only based f.g. free modules of even rank are considered (cf. the joint work with Hambleton and Taylor mentioned above). In particular, the round L-theory shows that the algebraic injections of Ranicki [16]

$$\bar{B} : L_n^j(\pi) \rightarrowtail L_{n+1}^k(\pi \times \mathbb{Z}) \qquad ((j,k) = (h,s) \text{ or } (p,h))$$

do not coincide with the geometric injections

$$\bar{B}' : L_n^j(\pi) \rightarrowtail L_{n+1}^k(\pi \times \mathbb{Z}) \; ; \; \sigma_*^j((f,b):M \longrightarrow X) \longmapsto \sigma_*^k((f,b)\times 1:M\times S^1 \longrightarrow X\times S^1)$$

of Shaneson [19] (for (h,s)) and Pedersen and Ranicki [14] (for (p,h)). The algebraic expression for $\bar{B}'$ is given by product with the round finite symmetric Poincaré complex of S^1, defined using the canonical round finite structure on S^1.

This paper is a sequel to the algebraic theory of the Wall finiteness obstruction developed in Ranicki [18]. As there we work with chain complexes in an arbitrary additive category $\mathcal{A}$, although the case $\mathcal{A}$ = {based f.g. free A-modules} for a ring A is the one of main interest

In §1 the isomorphism torsion group $K_1^{iso}(\mathcal{A})$ of an additive category $\mathcal{A}$ is defined by analogy with the automorphism torsion group $K_1^{aut}(\mathcal{A}) = K_1(\mathcal{A})$, using all the isomorphisms in $\mathcal{A}$. §2 is devoted to the isomorphism torsion properties of the permutation isomorphisms $M\oplus N \longrightarrow N\oplus M \; ; \; (x,y) \longmapsto (y,x)$. §3 deals with the torsion of contractible chain complexes. In §4 there is defined the torsion $\tau(f) \in K_1^{iso}(\mathcal{A})$ of a chain equivalence $f:C \longrightarrow D$ of finite chain complexes in $\mathcal{A}$ which are round, that is $[C] = [D] = 0 \in K_0(\mathcal{A})$. In §5 it is shown that if $\mathcal{A}$ is such that stably isomorphic objects are related by canonical stable isomorphisms then $K_1(\mathcal{A})$ is canonically a direct summand of $K_1^{iso}(\mathcal{A})$. In particular, such is the case for $\mathcal{A}$ = {based f.g. free A-modules}, allowing the definition of the absolute torsion $\tau(f) \in K_1(\mathcal{A}) = K_1(A)$ for a chain equivalence $f:C \longrightarrow D$ of round finite chain complexes of based f.g. free A-modules.

I am grateful to Chuck Weibel for a critical reading of an earlier version of the paper, and for several suggestions of a categorical nature (such as the use of permutative categories to avoid potential problems with coherence isomorphisms).

Contents

§1. The isomorphism torsion group $K_1^{iso}(\mathcal{A})$

In order to define the torsion of a chain equivalence it is necessary to first define the torsion of an isomorphism. To this end we shall now define the isomorphism torsion group $K_1^{iso}(\mathcal{A})$ of an additive category, by analogy with the automorphism torsion group $K_1^{aut}(\mathcal{A}) = K_1(\mathcal{A})$.

Let then $\mathcal{A}$ be an additive category, with direct sum $\oplus$. The $\left\{\begin{matrix}\text{isomorphism}\\ \text{automorphism}\end{matrix}\right.$ torsion group $\begin{matrix}K_1^{iso}(\mathcal{A})\\ K_1^{aut}(\mathcal{A})\end{matrix}$ is the abelian group with one generator $\tau(f)$ for each $\left\{\begin{matrix}\text{isomorphism } f:M \longrightarrow N\\ \text{automorphism } f:M \longrightarrow M\end{matrix}\right.$ in $\mathcal{A}$, subject to the relations

$$\text{i)}\left\{\begin{array}{l}\tau(gf:M \longrightarrow N \longrightarrow P) = \tau(f:M \longrightarrow N) + \tau(g:N \longrightarrow P)\\ \tau(gf:M \longrightarrow M \longrightarrow M) = \tau(f) + \tau(g),\ \tau(ifi^{-1}:M' \longrightarrow M \longrightarrow M \longrightarrow M') = \tau(f)\end{array}\right.$$

$$\text{ii)}\left\{\begin{array}{l}\tau(f\oplus f':M\oplus M' \longrightarrow N\oplus N') = \tau(f:M \longrightarrow N) + \tau(f':M' \longrightarrow N')\\ \tau(f\oplus f':M\oplus M' \longrightarrow M\oplus M') = \tau(f:M \longrightarrow M) + \tau(f':M' \longrightarrow M')\ .\end{array}\right.$$

The automorphism torsion group $K_1^{aut}(\mathcal{A})$ is just the Whitehead group of $\mathcal{A}$ in the sense of Bass [1,p.348]. There is defined a forgetful map

$$K_1^{aut}(\mathcal{A}) \longrightarrow K_1^{iso}(\mathcal{A})\ ;\ \tau(f) \longmapsto \tau(f)$$

which in certain circumstances (investigated in §5 below) is a split injection.

Remark: In order to avoid having to keep track of the coherence isomorphisms $(M\oplus N)\oplus P \longrightarrow M\oplus(N\oplus P)$ in $K_1^{iso}(\mathcal{A})$ we shall assume that $\mathcal{A}$ is a permutative category, so that $(M\oplus N)\oplus P = M\oplus(N\oplus P)$. There is a standard procedure for replacing any symmetric monoidal category by an equivalent permutative category (cf. Proposition 4.2 of May [10]).

[]

Let now $\mathcal{E}$ be an exact category. The torsion group $K_1(\mathcal{E})$ was defined by Bass [1,p.390] to be the abelian group with one generator $\tau(f)$ for each automorphism $f:M \longrightarrow M$ in $\mathcal{E}$, subject to the relations

i) $\tau(gf:M \longrightarrow M) = \tau(f:M \longrightarrow M) + \tau(g:M \longrightarrow M)$

ii) $\tau(f'':M'' \longrightarrow M'') = \tau(f:M \longrightarrow M) + \tau(f':M' \longrightarrow M')$ for any automorphism of a short exact sequence in $\mathcal{E}$

$$\begin{array}{ccccccccc} 0 & \longrightarrow & M & \xrightarrow{i} & M'' & \xrightarrow{j} & M' & \longrightarrow & 0 \\ & & \downarrow{\scriptstyle f} & & \downarrow{\scriptstyle f''} & & \downarrow{\scriptstyle f'} & & \\ 0 & \longrightarrow & M & \xrightarrow{i} & M'' & \xrightarrow{j} & M' & \longrightarrow & 0 \end{array}$$

An additive category $\mathcal{A}$ can be given the structure of an exact category by declaring a sequence in $\mathcal{A}$

$$0 \longrightarrow M \xrightarrow{i} M'' \xrightarrow{j} M' \longrightarrow 0$$

to be exact if $ji = 0 : M \longrightarrow M'$ and there exists a morphism $k : M' \longrightarrow M''$ such that

i) $jk = 1_{M'} : M' \longrightarrow M'$

ii) $(i\ k) : M \oplus M' \longrightarrow M''$ is an isomorphism.

We shall always use this exact structure.

Weibel [23] showed that the torsion group $K_1(\mathcal{A})$ of an additive category $\mathcal{A}$ with the above exact structure agrees with the case $i = 1$ of the general definition $K_i(\mathcal{E}) = \pi_{i+1}(B\mathcal{E}^{-1}\mathcal{E})$ $(i \geqslant 0)$ due to Quillen (Grayson [6]) of the algebraic K-groups of an exact category $\mathcal{E}$.

<u>Proposition 1.1</u> (Bass [1,p.397]) There is a natural identification of torsion groups $K_1^{aut}(\mathcal{A}) = K_1(\mathcal{A})$ for an additive category $\mathcal{A}$.

<u>Proof</u>: In order to verify that the natural abelian group morphism

$$K_1^{aut}(\mathcal{A}) \longrightarrow K_1(\mathcal{A}) \ ; \ \tau(f) \longmapsto \tau(f)$$

is an isomorphism it suffices to show that for any morphism $e : M' \longrightarrow M$ in $\mathcal{A}$ the elementary automorphism

$$f = \begin{pmatrix} 1 & e \\ 0 & 1 \end{pmatrix} : M \oplus M' \longrightarrow M \oplus M'$$

is such that $\tau(f) = 0 \in K_1^{aut}(\mathcal{A})$. The automorphisms

$$g = \begin{pmatrix} 1 & 0 & 1 \\ 0 & 1 & 0 \\ 0 & 0 & 1 \end{pmatrix} : M \oplus M' \oplus M \longrightarrow M \oplus M' \oplus M$$

$$h = \begin{pmatrix} 1 & 0 & 0 \\ 0 & 1 & 0 \\ 0 & e & 1 \end{pmatrix} : M \oplus M' \oplus M \longrightarrow M \oplus M' \oplus M$$

are such that

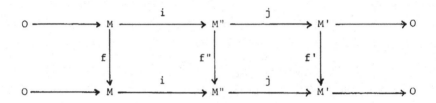

$$f\oplus 1_M = ghg^{-1}h^{-1} : M\oplus M'\oplus M \longrightarrow M\oplus M'\oplus M$$

(a particular example of a Steinberg relation). It follows that

$$\tau(f) = \tau(f\oplus 1_M) = \tau(ghg^{-1}h^{-1}) = 0 \in K_1^{aut}(\mathcal{A}) .$$

[]

Example Let A be an associative ring with 1 such that f.g. free A-modules have well defined rank (e.g. a group ring $\mathbb{Z}[\pi]$). Let $\mathcal{A}$ be the additive category of based f.g. free A-modules and A-module morphisms. The automorphism torsion group of $\mathcal{A}$ is just the usual Whitehead group of A

$$K_1^{aut}(\mathcal{A}) = K_1(\mathcal{A}) = K_1(A) = GL(A)/E(A) .$$

The isomorphism torsion group $K_1^{iso}(\mathcal{A})$ contains $K_1(A)$ as a direct summand, with the natural map $K_1(A) \longrightarrow K_1^{iso}(\mathcal{A})$ split by the surjection

$$K_1^{iso}(\mathcal{A}) \longrightarrow\!\!\!\!\rightarrow K_1(A) \ ; \ \tau(f:M \longrightarrow N) \longmapsto \tau((f_{ij}))$$

sending the isomorphism torsion $\tau(f:M \longrightarrow N) \in K_1^{iso}(\mathcal{A})$ to the torsion $\tau((f_{ij})) \in K_1(A)$ of the invertible $n\times n$ matrix $(f_{ij}) \in GL_n(A)$ $(n = \text{rank}_A M = \text{rank}_A N)$ representing f.

[]

The isomorphism torsion group $K_1^{iso}(\mathcal{A})$ of an additive category $\mathcal{A}$ is considerably larger than the automorphism torsion group $K_1(\mathcal{A})$, and is introduced here for the sole purpose of providing a home for the torsion $\tau(f) \in K_1^{iso}(\mathcal{A})$ of a chain equivalence.

§2. Signs

In dealing with the torsion of chain complexes and chain equivalences we shall be making frequent use of the following elements in $K_1^{iso}(\mathcal{A})$.

The sign of an ordered pair (M,N) of objects of $\mathcal{A}$ is the isomorphism torsion

$$\varepsilon(M,N) = \tau(\begin{pmatrix} 0 & 1_N \\ 1_M & 0 \end{pmatrix} : M\oplus N \longrightarrow N\oplus M) \in K_1^{iso}(\mathcal{A}) .$$

Example Let $\mathcal{A}$ = {based f.g. free A-modules}. The sign of objects M,N in $\mathcal{A}$ is given by

$$\varepsilon(M,N) = \text{rank}_A(M)\,\text{rank}_A(N)\,\tau(-1:A \longrightarrow A) \in K_1(A) \subset K_1^{iso}(\mathcal{A}) ,$$

depending only on the parities of the ranks of M and N.

<u>Proposition 2.1</u> The sign function $(M,N) \longmapsto \varepsilon(M,N)$ has the following properties, for any additive category $\mathcal{A}$:

i) $\varepsilon(M\oplus M',N) = \varepsilon(M,N) + \varepsilon(M',N) \in K_1^{iso}(\mathcal{A})$,

ii) $\varepsilon(M,N) = \varepsilon(M',N) \in K_1^{iso}(\mathcal{A})$ if M is isomorphic to M',

iii) $\varepsilon(M,N) = -\varepsilon(N,M) \in K_1^{iso}(\mathcal{A})$,

iv) $\varepsilon(M,M) = \tau(-1_M : M \longrightarrow M) \in K_1^{iso}(\mathcal{A})$.

<u>Proof</u>: i) For any objects M, M', N of $\mathcal{A}$

$$\varepsilon(M\oplus M',N) = \tau\left(\begin{pmatrix} 0 & 0 & 1_N \\ 1_M & 0 & 0 \\ 0 & 1_{M'} & 0 \end{pmatrix}\right.$$

$$: M\oplus M'\oplus N \xrightarrow{1_M \oplus \begin{pmatrix} 0 & 1_N \\ 1_{M'} & 0 \end{pmatrix}} M\oplus N\oplus M' \xrightarrow{\begin{pmatrix} 0 & 1_N \\ 1_M & 0 \end{pmatrix} \oplus 1_{M'}} N\oplus M\oplus M' \Big)$$

$$= \varepsilon(M,N) + \varepsilon(M',N) \in K_1^{iso}(\mathcal{A}) .$$

ii) Let $f : M \longrightarrow M'$ be an isomorphism in $\mathcal{A}$, and let N be an object. It follows from the commutative diagram of isomorphisms in $\mathcal{A}$

$$\begin{array}{ccc} M\oplus N & \xrightarrow{\begin{pmatrix} 0 & 1_N \\ 1_M & 0 \end{pmatrix}} & N\oplus M \\ {\scriptstyle f\oplus 1_N}\downarrow & & \downarrow{\scriptstyle 1_N\oplus f} \\ M'\oplus N & \xrightarrow{\begin{pmatrix} 0 & 1_N \\ 1_{M'} & 0 \end{pmatrix}} & N\oplus M' \end{array}$$

that

$$\begin{aligned} \varepsilon(M',N) - \varepsilon(M,N) &= \tau(1_N\oplus f) - \tau(f\oplus 1_N) \\ &= \tau(f) - \tau(f) = 0 \in K_1^{iso}(\mathcal{A}) . \end{aligned}$$

iii) For any objects M, N in $\mathcal{A}$

$$\begin{aligned} \varepsilon(M,N) + \varepsilon(N,M) &= \tau\left(\begin{pmatrix} 0 & 1_N \\ 1_M & 0 \end{pmatrix} : M\oplus N \longrightarrow N\oplus M\right) + \tau\left(\begin{pmatrix} 0 & 1_M \\ 1_N & 0 \end{pmatrix} : N\oplus M \longrightarrow M\oplus N\right) \\ &= \tau\left(\begin{pmatrix} 0 & 1_M \\ 1_N & 0 \end{pmatrix}\begin{pmatrix} 0 & 1_N \\ 1_M & 0 \end{pmatrix} = 1_{M\oplus N} : M\oplus N \longrightarrow M\oplus N\right) \\ &= 0 \in K_1^{iso}(\mathcal{A}) . \end{aligned}$$

iv) It is immediate from Proposition 1.1 and the identity

$$\begin{pmatrix} 0 & 1 \\ 1 & 0 \end{pmatrix} = \begin{pmatrix} 1 & 0 \\ -1 & 1 \end{pmatrix}\begin{pmatrix} 1 & 1 \\ 0 & 1 \end{pmatrix}\begin{pmatrix} 1 & 0 \\ -1 & 1 \end{pmatrix}\begin{pmatrix} -1 & 0 \\ 0 & 1 \end{pmatrix} : M\oplus M \longrightarrow M\oplus M$$

that

$$\varepsilon(M,M) = \tau(\begin{pmatrix} -1 & 0 \\ 0 & 1 \end{pmatrix} : M\oplus M \longrightarrow M\oplus M) = \tau(-1:M \longrightarrow M) \in K_1^{iso}(\mathcal{A}) .$$

[]

The isomorphism class group $K_0(\mathcal{A})$ of an additive category $\mathcal{A}$ is defined as usual to be the abelian group with one generator $[M]$ for each isomorphism class of objects M in $\mathcal{A}$, subject to the relations

$$[M\oplus N] = [M] + [N] \in K_0(\mathcal{A}) .$$

Example The projective class group of a ring A is the isomorphism class group of the additive category $\mathcal{P}$ = {f.g. projective A-modules},

$$K_0(A) = K_0(\mathcal{P}) .$$

[]

Example The isomorphism class group $K_0(\mathcal{A})$ of the additive category $\mathcal{A}$ = {based f.g. free A-modules} is such that there is defined an isomorphism

$$K_0(\mathcal{A}) \longrightarrow \mathbb{Z} ; [M] \longmapsto \mathrm{rank}_A(M)$$

(assuming as always that the rank of a f.g. free A-module is well defined).

[]

Proposition 2.2 Sign defines a symplectic form on the isomorphism class group $K_0(A)$ of an additive category $\mathcal{A}$ taking values in the isomorphism torsion group $K_1^{iso}(\mathcal{A})$

$$\varepsilon : K_0(\mathcal{A})\otimes K_0(\mathcal{A}) \longrightarrow K_1^{iso}(\mathcal{A}) ; [M]\otimes[N] \longmapsto \varepsilon(M,N) .$$

Proof: Immediate from Proposition 2.1.

[]

The reduced isomorphism torsion group of $\mathcal{A}$ is the quotient group of $K_1^{iso}(\mathcal{A})$ defined by

$$\tilde{K}_1^{iso}(\mathcal{A}) = \mathrm{coker}(\varepsilon : K_0(\mathcal{A})\otimes K_0(\mathcal{A}) \longrightarrow K_1^{iso}(\mathcal{A})) .$$

Example The reduced isomorphism torsion group $\tilde{K}_1^{iso}(\mathcal{A})$ of $\mathcal{A}$ = {based f.g. free A-modules} contains the reduced torsion group $\tilde{K}_1(A) = \mathrm{coker}(K_1(\mathbb{Z}) \longrightarrow K_1(A)) = K_1(A)/\{\tau(-1:A \longrightarrow A)\}$ as a direct summand, with the natural map $\tilde{K}_1(A) \longrightarrow \tilde{K}_1^{iso}(\mathcal{A}) ; \tilde{\tau}(f) \longmapsto \tilde{\tau}(f)$ split by

$$\tilde{K}_1^{iso}(\mathcal{A}) \longrightarrow\!\!\!\!\rightarrow \tilde{K}_1(A) ; \tilde{\tau}(f:M \longrightarrow N) \longmapsto \tilde{\tau}((f_{ij}))$$

$$(1 \leqslant i,j \leqslant n = \mathrm{rank}_A(M) = \mathrm{rank}_A(N)) .$$

[]

§3. Torsion for chain complexes

Let iso($\mathcal{A}$) denote the set of isomorphisms in an additive category $\mathcal{A}$, and let K be an abelian group. A function $\tau:\text{iso}(\mathcal{A}) \longrightarrow K$ is logarithmic if for all $(f:M \longrightarrow N),(g:N \longrightarrow P) \in \text{iso}(\mathcal{A})$

$$\tau(gf) = \tau(f) + \tau(g) \in K .$$

A function $\tau:\text{iso}(\mathcal{A}) \longrightarrow K$ is additive if for all $(f:M \longrightarrow N)$, $(f':M' \longrightarrow N') \in \text{iso}(\mathcal{A})$

$$\tau(f \oplus f') = \tau(f) + \tau(f') \in K .$$

The isomorphism torsion function

$$\tau : \text{iso}(\mathcal{A}) \longrightarrow K_1^{iso}(\mathcal{A}) \ ; \ f \longmapsto \tau(f)$$

is both logarithmic and additive, by construction, and is universal with respect to functions with these properties.

We shall now define logarithmic torsion functions $\tau:\text{iso}(\mathcal{C}) \longrightarrow K$ for various additive categories $\mathcal{C}$ of chain complexes in an additive category $\mathcal{A}$ (with morphisms either chain maps or chain homotopy classes of chain maps), such that K is one of the K_1-groups of $\mathcal{A}$ considered in §§1,2. In general these torsion functions will not be additive.

We refer to Ranicki [18] for an exposition of the chain homotopy theory of chain complexes in an additive category $\mathcal{A}$, adopting the same terminology and sign conventions.

Let $\mathcal{C}(\mathcal{A})$ be the additive category of finite chain complexes in $\mathcal{A}$

$$C : \ldots \longrightarrow 0 \longrightarrow C_n \xrightarrow{d} C_{n-1} \longrightarrow \ldots \longrightarrow C_1 \xrightarrow{d} C_0$$

and chain maps.

The torsion of an isomorphism $f:C \longrightarrow D$ in $\mathcal{C}(\mathcal{A})$ is defined by

$$\tau(f) = \sum_{r=0}^{\infty} (-)^r \tau(f:C_r \longrightarrow D_r) \in K_1^{iso}(\mathcal{A}) .$$

Proposition 3.1 The torsion function

$$\tau : \text{iso}(\mathcal{C}(\mathcal{A})) \longrightarrow K_1^{iso}(\mathcal{A}) \ ; \ f \longmapsto \tau(f)$$

is logarithmic and additive.

Proof: Immediate from the logarithmic and additive properties of $\tau:\text{iso}(\mathcal{A}) \longrightarrow K_1^{iso}(\mathcal{A})$.

The torsion of a contractible finite chain complex C in $\mathcal{A}$ is defined by

$$\tau(C) = \tau(d+\Gamma = \begin{pmatrix} d & 0 & 0 & \dots \\ \Gamma & d & 0 & \dots \\ 0 & \Gamma & d & \dots \\ \vdots & \vdots & \vdots & \end{pmatrix}$$

$$: C_{odd} = C_1 \oplus C_3 \oplus C_5 \oplus \dots \longrightarrow C_{even} = C_0 \oplus C_2 \oplus C_4 \oplus \dots)$$

$$\in K_1^{iso}(\mathcal{A}) \ ,$$

using any chain contraction $\Gamma : 0 \simeq 1 : C \longrightarrow C$ of C. The morphism $d+\Gamma : C_{odd} \longrightarrow C_{even}$ is an isomorphism since there is defined an inverse

$$(d+\Gamma)^{-1} = \begin{pmatrix} 1 & 0 & 0 & \dots \\ \Gamma^2 & 1 & 0 & \dots \\ 0 & \Gamma^2 & 1 & \dots \\ \vdots & \vdots & \vdots & \end{pmatrix}^{-1} \begin{pmatrix} \Gamma & d & 0 & \dots \\ 0 & \Gamma & d & \dots \\ 0 & 0 & \Gamma & \dots \\ \vdots & \vdots & \vdots & \end{pmatrix}$$

$$: C_{even} = C_0 \oplus C_2 \oplus C_4 \oplus \dots \longrightarrow C_{odd} = C_1 \oplus C_3 \oplus C_5 \oplus \dots \ .$$

If $\Gamma' : 0 \simeq 1 : C \longrightarrow C$ is another chain contraction of C the morphisms defined by

$$\Delta = (\Gamma' - \Gamma)\Gamma \ : \ C_r \longrightarrow C_{r+2} \quad (r \geqslant 0)$$

are such that

$$\Delta d - d\Delta = \Gamma' - \Gamma \ : \ C_r \longrightarrow C_{r+1} \quad (r \geqslant 0)$$

(defining a homotopy of chain homotopies $\Delta : \Gamma \simeq \Gamma' : 0 \simeq 1 : C \longrightarrow C$). The simple automorphisms

$$h_{even} = \begin{pmatrix} 1 & 0 & 0 & \dots \\ \Delta & 1 & 0 & \dots \\ 0 & \Delta & 1 & \dots \\ \vdots & \vdots & \vdots & \end{pmatrix}$$

$$: C_{even} = C_0 \oplus C_2 \oplus C_4 \oplus \dots \longrightarrow C_{even} = C_0 \oplus C_2 \oplus C_4 \oplus \dots \ ,$$

$$h_{odd} = \begin{pmatrix} 1 & 0 & 0 & \dots \\ \Delta & 1 & 0 & \dots \\ 0 & \Delta & 1 & \dots \\ \vdots & \vdots & \vdots & \end{pmatrix}$$

$$: C_{odd} = C_1 \oplus C_3 \oplus C_5 \oplus \dots \longrightarrow C_{odd} = C_1 \oplus C_3 \oplus C_5 \oplus \dots$$

are such that the diagram of isomorphisms

$$\begin{array}{ccc} C_{odd} & \xrightarrow{d+\Gamma} & C_{even} \\ h_{odd}\downarrow & & \downarrow h_{even} \\ C_{odd} & \xrightarrow{d+\Gamma'} & C_{even} \end{array}$$

commutes up to a simple automorphism of the type

$$(d+\Gamma)^{-1}h_{even}^{-1}(d+\Gamma')h_{odd} = \begin{pmatrix} 1 & 0 & 0 & \cdots \\ ? & 1 & 0 & \cdots \\ ? & ? & 1 & \cdots \\ \vdots & \vdots & \vdots & \end{pmatrix}$$

$$: C_{odd} = C_1 \oplus C_3 \oplus C_5 \oplus \ldots \longrightarrow C_{odd} = C_1 \oplus C_3 \oplus C_5 \oplus \ldots \ .$$

As usual, simple means $\tau = 0$. It follows that the torsion of C is independent of the choice of chain contraction Γ, with

$$\tau(C) = \tau(d+\Gamma : C_{odd} \longrightarrow C_{even}) = \tau(d+\Gamma' : C_{odd} \longrightarrow C_{even}) \in K_1^{iso}(\mathcal{A}) \ .$$

Example For $\mathcal{A}$ = {based f.g. free A-modules} the component of the isomorphism torsion $\tau(C) \in K_1^{iso}(\mathcal{A})$ in the automorphism torsion group is the torsion $\tau(C) \in K_1^{aut}(\mathcal{A}) = K_1(A)$ originally defined by Whitehead [24], with C a contractible finite based f.g. free A-module chain complex.

[]

Proposition 3.2 The torsion of an isomorphism $f:C \longrightarrow D$ of contractible finite chain complexes in an additive category $\mathcal{A}$ is given by

$$\tau(f) = \tau(D) - \tau(C) \in K_1^{iso}(\mathcal{A}) \ .$$

Proof: Given a chain contraction $\Gamma_C : 0 \simeq 1 : C \longrightarrow C$ of C define a chain contraction of D by

$$\Gamma_D = f\Gamma_C f^{-1} : 0 \simeq 1 : D \longrightarrow D \ .$$

There is then defined a commutative diagram of isomorphisms in $\mathcal{A}$

$$\begin{array}{ccc} C_{odd} = C_1 \oplus C_3 \oplus C_5 \oplus \ldots & \xrightarrow{d_C+\Gamma_C} & C_{even} = C_0 \oplus C_2 \oplus C_4 \oplus \ldots \\ f_{odd} = f_1 \oplus f_3 \oplus f_5 \oplus \ldots \Big\downarrow & & \Big\downarrow f_{even} = f_0 \oplus f_2 \oplus f_4 \oplus \ldots \\ D_{odd} = D_1 \oplus D_3 \oplus D_5 \oplus \ldots & \xrightarrow{d_D+\Gamma_D} & D_{even} = D_0 \oplus D_2 \oplus D_4 \oplus \ldots \end{array} ,$$

so that

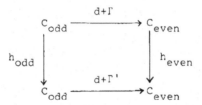

$$\begin{aligned}\tau(D)-\tau(C) &= \tau(d_D+\Gamma_D:D_{odd}\longrightarrow D_{even}) - \tau(d_C+\Gamma_C:C_{odd}\longrightarrow C_{even})\\ &= \tau(f_{even}:C_{even}\longrightarrow D_{even}) - \tau(f_{odd}:C_{odd}\longrightarrow D_{odd})\\ &= \tau(f)\in K_1^{iso}(\mathcal{A})\ .\end{aligned}$$

[]

The <u>intertwining</u> of finite chain complexes C,D in $\mathcal{A}$ is the linear combination of signs defined by

$$\beta(C,D) = \sum_{i>j}(\varepsilon(C_{2i},D_{2j}) - \varepsilon(C_{2i+1},D_{2j+1})) \in K_1^{iso}(\mathcal{A})\ .$$

This invariant plays an important role in quantifying the failure of the torsion of chain complexes to be additive. Note that $\beta(C,D)$ is the difference of the torsions of the permutation isomorphisms $(C\oplus D)_{even}\longrightarrow C_{even}\oplus D_{even}$ and $(C\oplus D)_{odd}\longrightarrow C_{odd}\oplus D_{odd}$.

<u>Proposition 3.3</u> The torsions of contractible finite chain complexes in an additive category $\mathcal{A}$ appearing in a short exact sequence

$$0\longrightarrow C\xrightarrow{\ \ i\ \ }C''\xrightarrow{\ \ j\ \ }C'\longrightarrow 0$$

are related by the <u>sum formula</u>

$$\tau(C'') = \tau(C)+\tau(C')+\sum_{r=0}^{\infty}(-)^r\tau((i\ k):C_r\oplus C'_r\longrightarrow C''_r)+\beta(C,C')$$

$$\in K_1^{iso}(\mathcal{A})\ ,$$

with $\{k:C'_r\longrightarrow C''_r\,|\,r\geqslant 0\}$ any sequence of splitting morphisms such that $jk=1:C'_r\longrightarrow C'_r$ $(r\geqslant 0)$ and each $(i\ k):C_r\oplus C'_r\longrightarrow C''_r$ $(r\geqslant 0)$ is an isomorphism.

<u>Proof</u>: Consider first the special case

$$i = \begin{pmatrix}1\\0\end{pmatrix} : C_r\longrightarrow C''_r = C_r\oplus C'_r\ ,$$

$$j = (0\ 1) : C''_r = C_r\oplus C'_r\longrightarrow C'_r\ ,$$

$$k = \begin{pmatrix}0\\1\end{pmatrix} : C'_r\longrightarrow C''_r = C_r\oplus C'_r\ ,$$

so that

$$d'' = \begin{pmatrix}d & e\\ 0 & d'\end{pmatrix} : C''_r = C_r\oplus C'_r\longrightarrow C''_{r-1} = C_{r-1}\oplus C'_{r-1}$$

for some morphisms $e:C'_r\longrightarrow C_{r-1}$ $(r\geqslant 1)$ such that $de+ed'=0$.

Given chain contractions of C and C'

$$\Gamma : 0\simeq 1 : C\longrightarrow C\ ,\quad \Gamma' : 0\simeq 1 : C'\longrightarrow C'$$

define a chain contraction of C''

$$\Gamma'' : 0\simeq 1 : C''\longrightarrow C''$$

by

$$\Gamma'' = \begin{pmatrix} \Gamma & -\Gamma(e\Gamma'+\Gamma e) \\ 0 & \Gamma' \end{pmatrix} : C''_r = C_r \oplus C'_r \longrightarrow C''_{r+1} = C_{r+1} \oplus C'_{r+1} \quad .$$

There is then defined an isomorphism of short exact sequences in $\mathcal{A}$

$$\begin{array}{ccccccccc} 0 & \longrightarrow & C_{odd} & \xrightarrow{i_{odd}} & C''_{odd} & \xrightarrow{j_{odd}} & C'_{odd} & \longrightarrow & 0 \\ & & \downarrow{\scriptstyle d+\Gamma} & & \downarrow{\scriptstyle d''+\Gamma''} & & \downarrow{\scriptstyle d'+\Gamma'} & & \\ 0 & \longrightarrow & C_{even} & \xrightarrow{i_{even}} & C''_{even} & \xrightarrow{j_{even}} & C'_{even} & \longrightarrow & 0 \end{array} ,$$

so that

$$\begin{aligned} \tau(C'') &= \tau(d''+\Gamma'':C''_{odd} \longrightarrow C''_{even}) \\ &= \tau(d+\Gamma:C_{odd} \longrightarrow C_{even}) + \tau(d'+\Gamma':C'_{odd} \longrightarrow C'_{even}) \\ &\quad + \tau((i_{even}\ k_{even}) : C_{even} \oplus C'_{even} \longrightarrow C''_{even}) \\ &\quad - \tau((i_{odd}\ k_{odd}) : C_{odd} \oplus C'_{odd} \longrightarrow C''_{odd}) \\ &= \tau(C) + \tau(C') + \beta(C,C') \in K_1^{iso}(\mathcal{A}) \ , \end{aligned}$$

verifying the sum formula in the special case.

In the general case let $\bar{C}''$ be the finite chain complex defined by

$$\bar{d}'' : \bar{C}''_r = C_r \oplus C'_r \xrightarrow{(i\ k)} C''_r \xrightarrow{d''} C''_{r-1} \xrightarrow{(i\ k)^{-1}} C_{r-1} \oplus C'_{r-1} = \bar{C}''_{r-1}$$

so that there are defined an isomorphism of chain complexes

$$(i\ k) : \bar{C}'' \longrightarrow C''$$

and a short exact sequence of contractible finite chain complexes

$$0 \longrightarrow C \xrightarrow{\bar{i}} \bar{C}'' \xrightarrow{\bar{j}} C' \longrightarrow 0$$

with

$$\bar{i} = \begin{pmatrix} 1 \\ 0 \end{pmatrix} : C_r \longrightarrow \bar{C}''_r = C_r \oplus C'_r \ ,$$

$$\bar{j} = (0\ 1) : \bar{C}''_r = C_r \oplus C'_r \longrightarrow C'_r \ .$$

By the special case

$$\tau(\bar{C}'') = \tau(C) + \tau(C') + \beta(C,C') \in K_1^{iso}(\mathcal{A})$$

and by Proposition 3.2

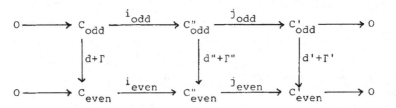

$$\tau(C'') - \tau(\bar{C}'') = \sum_{r=0}^{\infty} (-)^r \tau((i\ k):C_r \oplus C'_r \longrightarrow C''_r) \in K_1^{iso}(\mathcal{A}) .$$

The sum formula in the general case follows.

[]

The <u>reduced torsion</u> $\tilde{\tau}(C) \in \tilde{K}_1^{iso}(\mathcal{A})$ of a contractible finite chain complex C in $\mathcal{A}$ is the reduction of the absolute torsion $\tau(C) \in K_1^{iso}(\mathcal{A})$. The intertwining term $\beta(C,C')$ in the sum formula of Proposition 3.3 vanishes in the reduced torsion group, so that

$$\tilde{\tau}(C'') = \tilde{\tau}(C) + \tilde{\tau}(C') + \sum_{r=0}^{\infty} (-)^r \tilde{\tau}((i\ k):C_r \oplus C'_r \longrightarrow C''_r) \in \tilde{K}_1^{iso}(\mathcal{A}) .$$

<u>Remark</u> For $\mathcal{A}$ = {based f.g. free A-modules} the sum formula for reduced torsions in $\tilde{K}_1(A)$ was first obtained by Milnor [11], and the sum formula for absolute torsions in $K_1(A)$ was first obtained by Fossum, Foxby and Iversen [5].

[]

Let $\mathcal{C}^f(\mathcal{A})$ be the additive category of finite chain complexes in $\mathcal{A}$ and chain homotopy classes of chain maps, i.e. the derived category. The isomorphism set $\mathrm{iso}(\mathcal{C}^f(\mathcal{A}))$ consists of the chain homotopy classes of chain equivalences. The appearance of the intertwining term $\beta(C,C')$ in the sum formula of Proposition 3.3 implies that it is not in general possible to extend the universal isomorphism torsion function

$$\tau : \mathrm{iso}(\mathcal{A}) \longrightarrow K_1^{iso}(\mathcal{A}) \ ; \ f \longmapsto \tau(f)$$

to an additive function

$$\tau : \mathrm{iso}(\mathcal{C}^f(\mathcal{A})) \longrightarrow K_1^{iso}(\mathcal{A})$$

such that for every contractible finite chain complex C in $\mathcal{A}$

$$\tau(0 \longrightarrow C) = \tau(C) \in K_1^{iso}(\mathcal{A}) .$$

If there were such an extension, and if C,C' are contractible finite chain complexes in $\mathcal{A}$ such that $\beta(C,C') \neq 0 \in K_1^{iso}(\mathcal{A})$, then

$$\begin{aligned} \tau(0 \longrightarrow C\oplus C') &= \tau(C\oplus C') \\ &= \tau(C) + \tau(C') + \beta(C,C') \\ &\neq \tau(0 \longrightarrow C) + \tau(0 \longrightarrow C') \in K_1^{iso}(\mathcal{A}) , \end{aligned}$$

a contradiction.

Example Let $\mathcal{A}$ = {based f.g. free A-modules} for some ring A (such as a group ring $\mathbb{Z}[\pi]$) for which $\mathbb{Z} \longrightarrow A ; 1 \longmapsto 1$ induces an injection

$$K_1(\mathbb{Z}) = \mathbb{Z}_2 \rightarrowtail K_1(A) \; ; \; \tau(-1:\mathbb{Z} \longrightarrow \mathbb{Z}) \longmapsto \tau(-1:A \longrightarrow A) \; .$$

The contractible finite chain complexes in $\mathcal{A}$ defined by

$$C : \ldots \longrightarrow 0 \longrightarrow A \xrightarrow{1} A \longrightarrow 0$$

$$C' : \ldots \longrightarrow 0 \longrightarrow 0 \longrightarrow A \xrightarrow{1} A$$

are such that $\beta(C,C') \neq 0 \in K_1^{iso}(\mathcal{A})$, with automorphism torsion component

$$\beta(C,C') = \tau(-1:A \longrightarrow A) \neq 0 \in K_1^{aut}(\mathcal{A}) = K_1(A) .$$

(On the other hand $\beta(C',C) = 0 \in K_1^{iso}(\mathcal{A})$).

[]

In §4 below we shall define a logarithmic torsion function $\tau:\mathrm{iso}(\mathcal{E}^r(\mathcal{A})) \longrightarrow K_1^{iso}(\mathcal{A})$ on a certain full subcategory $\mathcal{E}^r(\mathcal{A}) \subset \mathcal{C}^f(\mathcal{A})$. We shall be making frequent use of the following properties of β.

Proposition 3.4 The intertwining function $(C,D) \longmapsto \beta(C,D) \in K_1^{iso}(\mathcal{A})$ is such that

i) $\beta(C \oplus C',D) = \beta(C,D) + \beta(C',D)$,

ii) $\beta(C,D \oplus D') = \beta(C,D) + \beta(C,D')$,

iii) $\beta(C,D) - \beta(D,C) + \sum_{r=0}^{\infty} (-)^r \varepsilon(C_r,D_r)$

$$= \varepsilon(C_{even},D_{even}) - \varepsilon(C_{odd},D_{odd}) ,$$

iv) $\beta(C,SC) + \sum_{r=0}^{\infty} (-)^r \varepsilon(C_r,C_{r-1}) = \varepsilon(C_{even},C_{odd})$ where $SC_r = C_{r-1}$,

v) $\beta(SC,C) = \varepsilon(C_{odd},C_{even})$,

vi) $\beta(SC,SD) = -\beta(C,D)$,

vii) $\beta(C,D) = \beta(C',D')$ if C is isomorphic to C' and D is isomorphic to D'.

Proof: These properties of β follow from the properties of the sign function $(M,N) \longmapsto \varepsilon(M,N)$ obtained in Proposition 2.1.

[]

§4. Torsion for chain equivalences

The algebraic mapping cone of a chain equivalence $f:C \longrightarrow D$ is a contractible chain complex $C(f)$. The torsion $\tau(f) \in K_1^{iso}(\mathcal{A})$ will now be defined in the case when C and D are finite complexes such that $[C] = [D] = 0 \in K_0(\mathcal{A})$, as the sum of the torsion $\tau(C(f))$ and a sign term.

The algebraic mapping cone $C(f)$ of a chain map $f:C \longrightarrow D$ in $\mathcal{A}$ is the chain complex in $\mathcal{A}$ defined as usual by

$$d_{C(f)} = \begin{pmatrix} d_D & (-)^{r-1}f \\ 0 & d_C \end{pmatrix}$$

$$: C(f)_r = D_r \oplus C_{r-1} \longrightarrow C(f)_{r-1} = D_{r-1} \oplus C_{r-2} \quad .$$

A chain map f is a chain equivalence if and only if $C(f)$ is chain contractible.

A chain homotopy in $\mathcal{A}$

$$g : f \simeq f' : C \longrightarrow D$$

determines an isomorphism of the algebraic mapping cones

$$h : C(f) \longrightarrow C(f')$$

with

$$h = \begin{pmatrix} 1 & (-)^r g \\ 0 & 1 \end{pmatrix} : C(f)_r = D_r \oplus C_{r-1} \longrightarrow C(f')_r = D_r \oplus C_{r-1} \; .$$

(The sign convention is that $d_D g + g d_C = f' - f : C_r \longrightarrow D_r$).

<u>Proposition 4.1</u> The algebraic mapping cone $C(f)$ of a chain equivalence $f:C \longrightarrow D$ of finite chain complexes in $\mathcal{A}$ is a contractible finite chain complex $C(f)$ in $\mathcal{A}$ such that the torsion $\tau(C(f)) \in K_1^{iso}(\mathcal{A})$ is a chain homotopy invariant of f, with $\tau(C(f)) = \tau(C(f'))$ for chain homotopic $f,f':C \longrightarrow D$.

<u>Proof</u>: Given a chain homotopy $g:f \simeq f':C \longrightarrow D$ apply Proposition 3.2 to the isomorphism $h:C(f) \longrightarrow C(f')$ defined above, to obtain

$$\begin{aligned} \tau(C(f')) - \tau(C(f)) &= \tau(h) \\ &= \sum_{r=0}^{\infty} (-)^r \tau(h:C(f)_r \longrightarrow C(f')_r) \\ &= 0 \in K_1^{iso}(\mathcal{A}) \; . \end{aligned}$$

[]

The following results determine the behaviour of the torsion $\tau(C(f)) \in K_1^{iso}(\mathcal{A})$ under the composition and addition of chain equivalences.

Proposition 4.2 i) The torsion of the algebraic mapping cone $C(gf)$ of the composite $gf:C \longrightarrow D \longrightarrow E$ of chain equivalences $f;C \longrightarrow D$, $g:D \longrightarrow E$ of finite chain complexes in $\mathcal{A}$ is given by

$$\tau(C(gf)) = \tau(C(f)) + \tau(C(g)) + \gamma(C,D,E) \in K_1^{iso}(\mathcal{A}) ,$$

with the sign term γ defined by

$$\begin{aligned}\gamma(C,D,E) = \beta(E,SC) - \beta(D,SC) - \beta(E,SD)& \\ + (\varepsilon(D_{even},C_{odd}) - \varepsilon(D_{odd},C_{even}))& \\ + (\varepsilon(D_{even},E_{even}) - \varepsilon(D_{odd},E_{odd}))& \\ + (\varepsilon(C_{odd},E_{even}) - \varepsilon(C_{even},E_{odd}))& \\ + (\varepsilon(D_{even},D_{odd}) - \varepsilon(D_{even},D_{even}))& \\ \in \mathrm{im}(\varepsilon : K_0(\mathcal{A})\otimes K_0(\mathcal{A}) \longrightarrow K_1^{iso}(\mathcal{A}))& .\end{aligned}$$

ii) The torsion of the algebraic mapping cone $C(f\oplus f')$ of the sum $f\oplus f':C\oplus C' \longrightarrow D\oplus D'$ of chain equivalences $f:C \longrightarrow D$, $f':C' \longrightarrow D'$ of finite chain complexes in $\mathcal{A}$ is given by

$$\begin{aligned}\tau(C(f\oplus f')) = \tau(C(f)) + \tau(C(f')) + \beta(D\oplus SC,D'\oplus SC')& \\ + \sum_{r=0}^{\infty}(-)^r\varepsilon(C_{r-1},D'_r) \in K_1^{iso}(\mathcal{A})& .\end{aligned}$$

iii) For a chain equivalence $f:C \longrightarrow D$ of contractible finite chain complexes in $\mathcal{A}$

$$\tau(C(f)) = \tau(D) - \tau(C) + \beta(D,SC) \in K_1^{iso}(\mathcal{A}) .$$

iv) The torsion of the algebraic mapping cone $C(1)$ of the identity chain map $1:C \longrightarrow C$ on a finite chain complex C in $\mathcal{A}$ is given by

$$\tau(C(1)) = \beta(C,SC) + \varepsilon(C_{odd},C_{odd}) - \varepsilon(C_{even},C_{odd}) \in K_1^{iso}(\mathcal{A}) .$$

Proof: i) Given a chain complex C let ΩC be the chain complex defined by

$$d_{\Omega C} = d_C : \Omega C_r = C_{r+1} \longrightarrow \Omega C_{r-1} = C_r .$$

Given chain equivalences $f:C \longrightarrow D$, $g:D \longrightarrow E$ of finite chain complexes in $\mathcal{A}$ define a chain map

$$h : \Omega C(g) \longrightarrow C(f)$$

by

$$h = \begin{pmatrix} 0 & -1 \\ 0 & 0 \end{pmatrix} : \Omega C(g)_r = E_{r+1} \oplus D_r \longrightarrow C(f)_r = D_r \oplus C_{r-1} \quad .$$

The algebraic mapping cone $C(h)$ is a contractible finite chain complex which fits into two short exact sequences of such complexes

$$0 \longrightarrow C(f) \xrightarrow{\quad i \quad} C(h) \xrightarrow{\quad j \quad} C(g) \longrightarrow 0$$

$$0 \longrightarrow C(gf) \xrightarrow{\quad i' \quad} C(h) \xrightarrow{\quad j' \quad} C(-1_D : D \longrightarrow D) \longrightarrow 0$$

with

$$i = \begin{pmatrix} 1 \\ 0 \end{pmatrix} : C(f)_r \longrightarrow C(h)_r = C(f)_r \oplus C(g)_r \ ,$$

$$j = (0 \ \ 1) : C(h)_r = C(f)_r \oplus C(g)_r \longrightarrow C(g)_r \ ,$$

$$i' = \begin{pmatrix} 0 & 0 \\ 0 & 1 \\ 1 & 0 \\ 0 & f \end{pmatrix} : C(gf)_r = E_r \oplus C_{r-1} \longrightarrow C(h)_r = D_r \oplus C_{r-1} \oplus E_r \oplus D_{r-1} \quad ,$$

$$j' = \begin{pmatrix} 1 & 0 & 0 & 0 \\ 0 & -f & 0 & 1 \end{pmatrix}$$

$$: C(h)_r = D_r \oplus C_{r-1} \oplus E_r \oplus D_{r-1} \longrightarrow C(-1_D)_r = D_r \oplus D_{r-1} \quad .$$

The morphisms j, j' are split by the morphisms

$$k = \begin{pmatrix} 0 \\ 1 \end{pmatrix} : C(g)_r \longrightarrow C(h)_r = C(f)_r \oplus C(g)_r \ ,$$

$$k' = \begin{pmatrix} 1 & 0 \\ 0 & 0 \\ 0 & 0 \\ 0 & 1 \end{pmatrix}$$

$$: C(-1_D)_r = D_r \oplus D_{r-1} \longrightarrow C(h)_r = D_r \oplus C_{r-1} \oplus E_r \oplus D_{r-1}$$

and

$$\tau((i\ k) = \begin{pmatrix} 1 & 0 \\ 0 & 1 \end{pmatrix} : C(f)_r \oplus C(g)_r \longrightarrow C(h)_r = C(f)_r \oplus C(g)_r) = 0 ,$$

$$\tau((i'\ k') = \begin{pmatrix} 0 & 0 & 1 & 0 \\ 0 & 1 & 0 & 0 \\ 1 & 0 & 0 & 0 \\ 0 & f & 0 & 1 \end{pmatrix}$$

$$: C(gf)_r \oplus C(-1_D)_r = E_r \oplus C_{r-1} \oplus D_r \oplus D_{r-1}$$

$$\longrightarrow C(h)_r = D_r \oplus C_{r-1} \oplus E_r \oplus D_{r-1})$$

$$= \varepsilon(E_r \oplus C_{r-1}, D_r) + \varepsilon(E_r, C_{r-1}) \in K_1^{iso}(\mathcal{A}) .$$

Applying the sum formula of Proposition 3.3 twice

$$\tau(C(h)) = \tau(C(f)) + \tau(C(g)) + \sum_{r=0}^{\infty} (-)^r \tau((i\ k) : C(f)_r \oplus C(g)_r \longrightarrow C(h)_r)$$

$$+ \beta(C(f), C(g))$$

$$= \tau(C(gf)) + \tau(C(-1_D)) + \beta(C(gf), C(-1_D))$$

$$+ \sum_{r=0}^{\infty} (-)^r \tau((i'\ k') : C(gf)_r \oplus C(-1_D)_r \longrightarrow C(h)_r)$$

$$\in K_1^{iso}(\mathcal{A}) .$$

Eliminating $\tau(C(h))$, substituting the values obtained above for $\tau((i\ k))$, $\tau((i'\ k'))$ and also

$$\tau(C(-1_D)) = \varepsilon(D_{even}, D_{even}) - \sum_{r=0}^{\infty} (-)^r \varepsilon(D_r, D_{r-1}) ,$$

$$\tau(C(f), C(g)) = \beta(D \oplus SC, E \oplus SD) ,$$

$$\tau(C(gf), C(-1_D)) = \beta(E \oplus SC, D \oplus SD) \in K_1^{iso}(\mathcal{A})$$

leads to the required expression for $\tau(C(gf)) \in K_1^{iso}(\mathcal{A})$.

ii) The algebraic mapping cone $C(f \oplus f')$ of the sum $f \oplus f' : C \oplus C' \longrightarrow D \oplus D'$ of chain equivalences fits into a short exact sequence of contractible finite chain complexes

$$0 \longrightarrow C(f) \xrightarrow{\ i\ } C(f \oplus f') \xrightarrow{\ j\ } C(f') \longrightarrow 0$$

with

$$i = \begin{pmatrix} 1 & 0 \\ 0 & 0 \\ 0 & 1 \\ 0 & 0 \end{pmatrix}$$

$$: C(f)_r = D_r \oplus C_{r-1} \longrightarrow C(f\oplus f')_r = D_r \oplus D'_r \oplus C_{r-1} \oplus C'_{r-1} \ ,$$

$$j = \begin{pmatrix} 0 & 1 & 0 & 0 \\ 0 & 0 & 0 & 1 \end{pmatrix}$$

$$: C(f\oplus f')_r = D_r \oplus D'_r \oplus C_{r-1} \oplus C'_{r-1} \longrightarrow C(f')_r = D'_r \oplus C'_{r-1} \ .$$

Define a splitting morphism for j by

$$k = \begin{pmatrix} 0 & 0 \\ 1 & 0 \\ 0 & 0 \\ 0 & 1 \end{pmatrix}$$

$$: C(f')_r = D'_r \oplus C'_{r-1} \longrightarrow C(f\oplus f')_r = D_r \oplus D'_r \oplus C_{r-1} \oplus C'_{r-1} \ ,$$

with

$$\tau((i \ k) = \begin{pmatrix} 1 & 0 & 0 & 0 \\ 0 & 0 & 1 & 0 \\ 0 & 1 & 0 & 0 \\ 0 & 0 & 0 & 1 \end{pmatrix}$$

$$: C(f)_r \oplus C(f')_r = D_r \oplus C_{r-1} \oplus D'_r \oplus C'_{r-1}$$

$$\longrightarrow C(f\oplus f')_r = D_r \oplus D'_r \oplus C_{r-1} \oplus C'_{r-1})$$

$$= \varepsilon(C_{r-1}, D'_r) \in K_1^{iso}(\mathcal{A}) \ .$$

It is now immediate from the sum formula of Proposition 3.3 that

$$\tau(C(f\oplus f')) = \tau(C(f)) + \tau(C(f')) + \beta(D\oplus SC, D'\oplus SC') + \sum_{r=0}^{\infty} (-)^r \varepsilon(C_{r-1}, D'_r) \in K_1^{iso}(\mathcal{A}) \ .$$

iii) Set E = 0 in the composition formula i).

iv) Set $f = 1 : C \longrightarrow D = C$, $g = 1 : D = C \longrightarrow E = C$ in the composition formula i).

[]

The reduced torsion of a chain equivalence $f:C \longrightarrow D$ of finite chain complexes in $\mathcal{A}$ is defined by

$$\tilde{\tau}(f) = \tilde{\tau}(C(f)) \in \tilde{K}_1^{iso}(\mathcal{A}) ,$$

that is the reduction of the absolute torsion $\tau(C(f)) \in K_1^{iso}(\mathcal{A})$ of the algebraic mapping cone $C(f)$.

Example For $\mathcal{A}$= {based f.g. free A-modules} the automorphism component $\tilde{\tau}(f) \in \tilde{K}_1(A)$ of the reduced torsion is just the torsion of a chain equivalence $f:C \longrightarrow D$ in the sense of Whitehead [24] and Milnor [11].

[]

Proposition 4.3 i) The reduced torsion function

$$\tilde{\tau} : \mathrm{iso}(\mathcal{C}^f(\mathcal{A})) \longrightarrow \tilde{K}_1^{iso}(\mathcal{A}) \ ; \ f \longmapsto \tilde{\tau}(f)$$

is logarithmic and additive.

ii) The reduced torsion of an isomorphism $f:C \longrightarrow D$ is the reduction of the absolute torsion $\tau(f) = \sum_{r=0}^{\infty} (-)^r \tau(f:C_r \longrightarrow D_r) \in K_1^{iso}(\mathcal{A})$, that is

$$\tilde{\tau}(f) = \sum_{r=0}^{\infty} (-)^r \tilde{\tau}(f:C_r \longrightarrow D_r) \in \tilde{K}_1^{iso}(\mathcal{A}) .$$

iii) The reduced torsion of a chain equivalence $f:C \longrightarrow D$ of contractible finite chain complexes is the difference of the reduced torsions of C and D

$$\tilde{\tau}(f) = \tilde{\tau}(D) - \tilde{\tau}(C) \in \tilde{K}_1^{iso}(\mathcal{A}) .$$

Proof: i) Immediate from the formulae of Proposition 4.2, since all the sign terms vanish on passing to the reduced torsion group $\tilde{K}_1^{iso}(\mathcal{A})$.

ii) Define an isomorphism of contractible finite chain complexes

$$1 \oplus f : C(f) \longrightarrow C(1:D \longrightarrow D)$$

and apply Proposition 3.2.

iii) Apply the logarithmic property of $\tilde{\tau}$ given by i) to the composite

$$f : C \longrightarrow 0 \longrightarrow D$$

(up to chain homotopy).

[]

The class of a finite chain complex C in $\mathcal{A}$ is the element of the isomorphism class group of $\mathcal{A}$ defined by

$$[C] = \sum_{r=0}^{\infty} (-)^r [C_r] = [C_{even}] - [C_{odd}] \in K_0(\mathcal{A}) \quad ,$$

a chain homotopy invariant of C.

Example For $\mathcal{A}$ = {based f.g. free A-modules} the class of a finite chain complex C is just the Euler characteristic of C

$$[C] = \chi(C) = \sum_{r=0}^{\infty} (-)^r \mathrm{rank}_A(C_r) \in K_0(\mathcal{A}) = \mathbb{Z} \quad .$$

[]

A finite chain complex C in $\mathcal{A}$ is round if

$$[C] = 0 \in K_0(\mathcal{A}) \quad .$$

In particular, a contractible finite chain complex is round.

The torsion of a chain equivalence $f:C \longrightarrow D$ of round finite chain complexes in $\mathcal{A}$ is defined by

$$\tau(f) = \tau(C(f)) - \beta(D,SC) \in K_1^{iso}(\mathcal{A}) \quad .$$

Remark This formula can be used to define the torsion $\tau(f) \in K_1^{iso}(\mathcal{A})$ of a chain equivalence $f:C \longrightarrow D$ of any finite chain complexes in $\mathcal{A}$, but the resulting function $\tau:\mathrm{iso}(\mathcal{C}^f(\mathcal{A})) \longrightarrow K_1^{iso}(\mathcal{A})$ is neither logarithmic nor additive (cf. Proposition 4.2, and the Example just before Proposition 3.4). There does not appear to be a reasonable way to define either a logarithmic or an additive torsion function $\tau:\mathrm{iso}(\mathcal{C}^f(\mathcal{A})) \longrightarrow K_1^{iso}(\mathcal{A})$ in general.

[]

Let $\mathcal{C}^r(\mathcal{A})$ be the additive category of round finite chain complexes in $\mathcal{A}$ and chain homotopy classes of chain maps, a full subcategory of the derived category $\mathcal{C}^f(\mathcal{A})$.

Proposition 4.4 i) The torsion function

$$\tau \ : \ \mathrm{iso}(\mathcal{C}^r(\mathcal{A})) \longrightarrow K_1^{iso}(\mathcal{A}) \ ; \ f \longmapsto \tau(f)$$

is logarithmic, that is $\tau(gf) = \tau(f) + \tau(g)$.

ii) The torsion function $\tau:\mathrm{iso}(\mathcal{C}^r(\mathcal{A})) \longrightarrow K_1^{iso}(\mathcal{A})$ is not additive in general, with the torsion of a sum $f\oplus f':C\oplus C' \longrightarrow D\oplus D'$ given by

$$\tau(f\oplus f') = \tau(f) + \tau(f') - \beta(C,C') + \beta(D,D') \in K_1^{iso}(\mathcal{A}) \quad .$$

iii) The torsion of an isomorphism $f:C \longrightarrow D$ of round finite chain complexes agrees with the previous definition

$$\tau(f) = \sum_{r=0}^{\infty} (-)^r \tau(f:C_r \longrightarrow D_r) \in K_1^{iso}(\mathcal{A}) \quad .$$

iv) The torsion of a chain equivalence $f:C \longrightarrow D$ of contractible finite chain complexes is the difference of the torsions of C and D

$$\tau(f) = \tau(D) - \tau(C) \in K_1^{iso}(\mathcal{A}) .$$

v) The torsion of a chain equivalence $f:C \longrightarrow D$ of round finite chai complexes which fits into a short exact sequence

$$0 \longrightarrow C \xrightarrow{f} D \xrightarrow{g} E \longrightarrow 0$$

is related to the torsion of the contractible finite chain complex E by the formula

$$\tau(f) = \tau(E) + \sum_{r=0}^{\infty} (-)^r \tau((f\ h):C_r \oplus E_r \longrightarrow D_r) + \beta(C,E) \in K_1^{iso}(\mathcal{A}) \quad ,$$

with $\{h:E_r \longrightarrow D_r \mid r \geqslant 0\}$ splitting morphisms for $\{g:D_r \longrightarrow E_r \mid r \geqslant 0\}$.

Proof: i) For round C,D,E the sign term $\gamma(C,D,E)$ in the composition formula of Proposition 4.2 i) is given by

$$\gamma(C,D,E) = \beta(E,SC) - \beta(D,SC) - \beta(E,SD) \in K_1^{iso}(\mathcal{A}) .$$

ii) By the sum formula of Proposition 4.2 ii)

$$\begin{aligned}
\tau(f \oplus f') &= \tau(C(f \oplus f')) - \beta(D \oplus D', SC \oplus SC') \\
&= \tau(C(f)) + \tau(C(f')) - \beta(D \oplus D', SC \oplus SC') \\
&\qquad + \beta(D \oplus SC, D' \oplus SC') + \sum_{r=0}^{\infty} (-)^r \varepsilon(C_{r-1}, D'_r) \\
&= \tau(C(f)) + \tau(C(f')) - \beta(D,SC) - \beta(D',SC') \\
&\qquad - \beta(C,C') + \beta(D,D') \\
&\qquad \text{(by Proposition 3.4)} \\
&= \tau(f) + \tau(f') - \beta(C,C') + \beta(D,D') \in K_1^{iso}(\mathcal{A}) .
\end{aligned}$$

iii) Given an isomorphism $f:C \longrightarrow D$ of round finite chain complexes in $\mathcal{A}$ define an isomorphism of contractible finite chain complexes

$$f' = 1 \oplus f : C' = C(f) \longrightarrow D' = C(1_D : D \longrightarrow D) .$$

By Proposition 3.2

$$\begin{aligned}
\tau(D') - \tau(C') &= \sum_{r=0}^{\infty} (-)^r \tau(f':C'_r \longrightarrow D'_r) \\
&= \sum_{r=0}^{\infty} (-)^r \tau(1 \oplus f : D_r \oplus C_{r-1} \longrightarrow D_r \oplus D_{r-1}) \\
&= \sum_{r=0}^{\infty} (-)^r \tau(f:C_{r-1} \longrightarrow D_{r-1}) \\
&= -(\sum_{r=0}^{\infty} (-)^r \tau(f:C_r \longrightarrow D_r)) \in K_1^{iso}(\mathcal{A}) .
\end{aligned}$$

By the logarithmic property of torsion proved in i)

$$\begin{aligned}\tau(f) &= \tau(f) - \tau(1_D)\\ &= (\tau(C') - \beta(D,SC)) - (\tau(D') - \beta(D,SD))\\ &= \tau(C') - \tau(D')\\ &= \sum_{r=0}^{\infty}(-)^r\tau(f:C_r \longrightarrow D_r) \in K_1^{iso}(\mathcal{A}) .\end{aligned}$$

iv) Immediate from the logarithmic property of τ applied to the composite $0:C \xrightarrow{f} D \longrightarrow 0$, noting that $\tau(C \longrightarrow 0) = -\tau(C) \in K_1^{iso}(\mathcal{A})$.

v) Apply the sum formula of Proposition 3.3 to the short exact sequence of contractible finite chain complexes

$$0 \longrightarrow C(1_C) \xrightarrow{\ i\ } C(f) \xrightarrow{\ j\ } E \longrightarrow 0$$

with

$$i = \begin{pmatrix} f & 0 \\ 0 & 1 \end{pmatrix} : C(1_C)_r = C_r \oplus C_{r-1} \longrightarrow C(f)_r = D_r \oplus C_{r-1} ,$$

$$j = (g\ \ 0) : C(f)_r = D_r \oplus C_{r-1} \longrightarrow E_r$$

to obtain

$$\begin{aligned}\tau(C(f)) &= \tau(C(1_C)) + \tau(E) + \sum_{r=0}^{\infty}(-)^r\tau(\begin{pmatrix} f & 0 & h \\ 0 & 1 & 0 \end{pmatrix}:C_r\oplus C_{r-1}\oplus E_r \longrightarrow D_r\oplus C_{r-1})\\ &\qquad + \beta(C(1_C),E)\\ &= \beta(C,SC) + \tau(E) + \sum_{r=0}^{\infty}(-)^r(\tau((f\ \ h):C_r\oplus E_r \longrightarrow D_r) + \varepsilon(C_{r-1},E_r))\\ &\qquad + \beta(C\oplus SC,E) \in K_1^{iso}(\mathcal{A}) .\end{aligned}$$

It follows that

$$\begin{aligned}\tau(f) &= \tau(C(f)) - \beta(D,SC)\\ &= \tau(E) + \sum_{r=0}^{\infty}(-)^r\tau((f\ \ h):C_r\oplus E_r \longrightarrow D_r) + \beta(C,E)\\ &\qquad + (\beta(SC,E) - \beta(E,SC) + \sum_{r=0}^{\infty}(-)^r\varepsilon(C_{r-1},E_r)) \in K_1^{iso}(\mathcal{A}) .\end{aligned}$$

By Proposition 3.4 iii)

$$\begin{aligned}\beta(SC,E) - \beta(E,SC) + \sum_{r=0}^{\infty}(-)^r\varepsilon(C_{r-1},E_r)\\ &= \varepsilon(C_{odd},E_{even}) - \varepsilon(C_{even},E_{odd})\\ &= 0 \in K_1^{iso}(\mathcal{A}) \text{ (since } C,E \text{ are round)}.\end{aligned}$$

[]

An element $x \in K_0(\mathcal{A})$ is <u>even</u> if

$$\varepsilon(x,y) = 0 \in K_0(\mathcal{A}) \quad ,$$

for every $y \in K_0(\mathcal{A})$. The even elements of $K_0(\mathcal{A})$ define a subgroup, the kernel of the adjoint map of the sign form of Proposition 2.2

$$K_0(\mathcal{A}) \longrightarrow \mathrm{Hom}_{\mathbb{Z}}(K_0(\mathcal{A}), K_1^{iso}(\mathcal{A})) \ ;$$
$$[M] \longmapsto ([N] \longmapsto \varepsilon(M,N)) \ .$$

<u>Example</u> For $\mathcal{A}$ = {based f.g. free A-modules} the isomorphism

$$K_0(\mathcal{A}) \longrightarrow \mathbb{Z} \ ; \ [M] - [N] \longmapsto \mathrm{rank}_A(M) - \mathrm{rank}_A(N)$$

sends the subgroup of even elements in $K_0(\mathcal{A})$ to the subgroup $2\mathbb{Z} \subset \mathbb{Z}$ of even integers.

[]

A finite chain complex C in $\mathcal{A}$ is <u>even</u> if the class $[C] \in K_0(\mathcal{A})$ is even. In particular, a round finite complex is even, since $0 \in K_0(\mathcal{A})$ is an even element.

Let $\mathcal{C}^e(\mathcal{A})$ be the additive category of even finite chain complexes in $\mathcal{A}$ and chain homotopy classes of chain maps. Thus $\mathcal{C}^e(\mathcal{A})$ is a full subcategory of $\mathcal{C}^f(\mathcal{A})$, and $\mathcal{C}^r(\mathcal{A})$ is a full subcategory of $\mathcal{C}^e(\mathcal{A})$.

The <u>torsion</u> of a chain equivalence $f:C \longrightarrow D$ of even finite chain complexes in $\mathcal{A}$ is defined in exactly the same way as for round complexes, by the formula

$$\tau(f) = \tau(C(f)) - \beta(D,SC) \in K_1^{iso}(\mathcal{A}) \ .$$

<u>Proposition 4.5</u> The torsion function $\tau:\mathrm{iso}(\mathcal{C}^e(\mathcal{A})) \longrightarrow K_1^{iso}(\mathcal{A})$ has all the properties stated for $\tau:\mathrm{iso}(\mathcal{C}^r(\mathcal{A})) \longrightarrow K_1^{iso}(\mathcal{A})$ in Proposition 4.4, in particular the logarithmic property.

<u>Proof</u>: The proof of Proposition 4.4 depended on the sign properties of round complexes which are the same for even complexes.

[]

Given an object A of $\mathcal{A}$ and an integer $n \geqslant 0$ define the elementary contractible finite chain complex in $\mathcal{A}$

$$A(n,n+1) \; : \; \ldots \longrightarrow 0 \longrightarrow A \xrightarrow{\;1\;} A \longrightarrow 0 \longrightarrow \ldots$$

concentrated in degrees n,n+1. For any finite chain complex C in $\mathcal{A}$ the inclusion

$$i = \begin{pmatrix} 1 \\ 0 \end{pmatrix} : C \longrightarrow C \oplus A(n,n+1)$$

is a chain equivalence such that

$$\begin{aligned} \tau(C(i)) &= \tau(C(1_C)) + (-)^n(\varepsilon(C_{n-1},A) - \varepsilon(C_n,A) + \varepsilon(C_{n+1},A)) \\ &\in \operatorname{im}(\varepsilon : K_0(\mathcal{A}) \otimes K_0(\mathcal{A}) \longrightarrow K_1^{iso}(\mathcal{A})) \ , \end{aligned}$$

and such that for round finite C

$$\tau(i) = \sum_{r > n+1} (-)^r \varepsilon(C_r,A) \in K_1^{iso}(\mathcal{A}) \ .$$

Working exactly as in Whitehead [24] (the special case $\mathcal{A}$ = {based f.g. free A-modules}) it can be shown that the reduced torsion $\tilde{\tau}(f) = \tilde{\tau}(C(f)) \in \tilde{K}_1^{iso}(\mathcal{A})$ of a chain equivalence $f:C \longrightarrow D$ of finite chain complexes in $\mathcal{A}$ is such that $\tilde{\tau}(f) = 0$ if and only if there exist elementary complexes $A_i(m_i,m_i+1)$ $(1 \leqslant i \leqslant p)$, $B_j(n_j,n_j+1)$ $(1 \leqslant j \leqslant q)$ such that the chain equivalence

$$f \oplus 0 \; : \; C' = C \oplus \sum_{i=1}^{p} A_i(m_i,m_i+1) \longrightarrow D' = D \oplus \sum_{j=1}^{q} B_j(n_j,n_j+1)$$

is chain homotopic to an isomorphism $f':C' \longrightarrow D'$ such that

$$\tilde{\tau}(f':C'_r \longrightarrow D'_r) = 0 \in \tilde{K}_1^{iso}(\mathcal{A}) \quad (r \geqslant 0) \ .$$

There does not appear to be a corresponding interpretation of the vanishing $\tau(f) = 0$ of the absolute torsion $\tau(f) \in K_1^{iso}(\mathcal{A})$ of a chain equivalence $f:C \longrightarrow D$ of round finite chain complexes, except in the trivial case when the classes $[C_r],[D_r] \in K_0(\mathcal{A})$ $(r \geqslant 0)$ are all even and the sign terms vanish.

§5. Canonical structures

The isomorphism torsion group $K_1^{iso}(\mathcal{A})$ is too large (and insufficiently functorial) for practical applications, as compared to the automorphism torsion group $K_1^{aut}(\mathcal{A}) = K_1(\mathcal{A})$. We shall now investigate structures on an additive category $\mathcal{A}$ which ensure that the natural map $K_1(\mathcal{A}) \longrightarrow K_1^{iso}(\mathcal{A})$ is a canonically split injection, with a splitting map $K_1^{iso}(\mathcal{A}) \longrightarrow K_1(\mathcal{A})$ allowing an automorphism torsion component $\tau^{aut} \in K_1(\mathcal{A})$ to be split off from any isomorphism torsion $\tau \in K_1^{iso}(\mathcal{A})$.

A canonical structure ϕ on an additive category $\mathcal{A}$ is a collection of isomorphisms $\{\phi_{M,N}: M \longrightarrow N\}$, one for each ordered pair (M,N) of isomorphic objects in $\mathcal{A}$, such that

i) $\phi_{M,M} = 1 : M \longrightarrow M$,

ii) $\phi_{M,P} = \phi_{N,P}\phi_{M,N} : M \longrightarrow N \longrightarrow P$,

iii) $\phi_{M\oplus M',N\oplus N'} = \phi_{M,N}\oplus\phi_{M',N'} : M\oplus M' \longrightarrow N\oplus N'$.

Example Let $\mathcal{A}$ = {based f.g. free A-modules}, assuming (as always) that A is such that f.g. free A-modules have well defined rank. Based f.g. free A-modules M,N are isomorphic if and only if they have the same rank, n say, in which case there is defined a canonical isomorphism

$$\phi_{M,N} : M \longrightarrow N \ ; \ \sum_{r=1}^{n} a_r x_r \longmapsto \sum_{r=1}^{n} a_r y_r \quad (a_r \in A)$$

with $(x_1,x_2,\ldots,x_n)$, $(y_1,y_2,\ldots,y_n)$ the given bases of M,N. The collection $\phi = \{\phi_{M,N}\}$ defines a canonical structure on $\mathcal{A}$.

[]

Proposition 5.1 A canonical structure ϕ on an additive category $\mathcal{A}$ determines a splitting of the natural map $K_1(\mathcal{A}) \longrightarrow K_1^{iso}(\mathcal{A})$

$$K_1^{iso}(\mathcal{A}) \longrightarrow\!\!\!\!\rightarrow K_1(\mathcal{A}) \ ; \ \tau(f:M \longrightarrow N) \longmapsto \tau(\phi_{N,M}f:M \longrightarrow N \longrightarrow M) ,$$

so that $K_1^{iso}(\mathcal{A}) = K_1(\mathcal{A})\oplus ?$.

Proof: Trivial.

[]

In fact, canonical stable isomorphisms are sufficient to split $K_1(\mathcal{A}) \longrightarrow K_1^{iso}(\mathcal{A})$, as follows.

A <u>stable isomorphism</u> between objects M,N in an additive category $\mathcal{A}$

$$[f] : M \longrightarrow N$$

is an equivalence class of isomorphisms $f: M\oplus X \longrightarrow N\oplus X$ under the equivalence relation

$$(f: M\oplus X \longrightarrow N\oplus X) \sim (g: M\oplus Y \longrightarrow N\oplus Y) \text{ if the automorphism}$$

$$h : M\oplus X\oplus Y \xrightarrow{f\oplus 1_Y} N\oplus X\oplus Y \xrightarrow{1_N\oplus\begin{pmatrix} 0 & 1_Y \\ 1_X & 0 \end{pmatrix}} N\oplus Y\oplus X \xrightarrow{g^{-1}\oplus 1_X} M\oplus Y\oplus X \xrightarrow{1_M\oplus\begin{pmatrix} 0 & 1_X \\ 1_Y & 0 \end{pmatrix}} M\oplus X\oplus Y$$

is simple, that is $\tau(h) = 0 \in K_1(\mathcal{A})$.

<u>Proposition 5.2</u> Stable isomorphisms are the morphisms of a category $\mathcal{A}^s$, with the same objects as $\mathcal{A}$.

<u>Proof</u>: The composite of the stable isomorphisms

$$[f] : M \longrightarrow N \quad , \quad [g] : N \longrightarrow P$$

is the stable isomorphism

$$[g][f] = [e] : M \longrightarrow P$$

represented by the isomorphism

$$e : M\oplus X\oplus Y \xrightarrow{f\oplus 1_Y} N\oplus X\oplus Y \xrightarrow{1_N\oplus\begin{pmatrix} 0 & 1_Y \\ 1_X & 0 \end{pmatrix}} N\oplus Y\oplus X \xrightarrow{g\oplus 1_X} P\oplus Y\oplus X \xrightarrow{1_P\oplus\begin{pmatrix} 0 & 1_X \\ 1_Y & 0 \end{pmatrix}} P\oplus X\oplus Y \ .$$

[]

Although the stable category $\mathcal{A}^s$ is not additive it is possible to define the <u>sum</u> of stable isomorphisms $[f]: M \longrightarrow N$, $[f']: M' \longrightarrow N'$ to be the stable isomorphism

$$[f]\oplus[f'] = [f''] : M\oplus M' \longrightarrow N\oplus N'$$

represented by the isomorphism

$$f'' : M\oplus M'\oplus X\oplus X' \xrightarrow{1_M\oplus\begin{pmatrix} 0 & 1_X \\ 1_{M'} & 0 \end{pmatrix}\oplus 1_{X'}} M\oplus X\oplus M'\oplus X' \xrightarrow{f\oplus f'} N\oplus X\oplus N'\oplus X' \xrightarrow{1_N\oplus\begin{pmatrix} 0 & 1_{N'} \\ 1_X & 0 \end{pmatrix}\oplus 1_{X'}} N\oplus N'\oplus X\oplus X' \ .$$

The torsion of a stable $\begin{cases} \text{isomorphism} & [f]:M \longrightarrow N \\ \text{automorphism} & [f]:M \longrightarrow M \end{cases}$

is defined by

$$\begin{cases} \tau([f]) = \tau(f:M\oplus X \longrightarrow N\oplus X) \in K_1^{iso}(\mathcal{A}) \\ \tau([f]) = \tau(f:M\oplus X \longrightarrow M\oplus X) \in K_1^{aut}(\mathcal{A}) = K_1(\mathcal{A}) \ , \end{cases}$$

using any representative isomorphism f. In both cases

$$\tau([g][f]) = \tau([f]) + \tau([g]) \quad , \quad \tau([f]\oplus[f']) = \tau([f]) + \tau([f']) \ .$$

A canonical stable structure $[\phi]$ on an additive category $\mathcal{A}$ is a collection of stable isomorphisms $\{[\phi_{M,N}]:M \longrightarrow N\}$, one for each ordered pair (M,N) of stably isomorphic objects in $\mathcal{A}$, such that

i) $[\phi_{M,M}] = [1_M] : M \longrightarrow M$,

ii) $[\phi_{M,P}] = [\phi_{N,P}][\phi_{M,N}] : M \longrightarrow N \longrightarrow P$,

iii) $[\phi_{M\oplus M',N\oplus N'}] = [\phi_{M,N}]\oplus[\phi_{M',N'}] : M\oplus M' \longrightarrow N\oplus N'$.

Thus $[\phi]$ is a canonical structure on the stable category $\mathcal{A}^s$. An actual canonical structure ϕ on $\mathcal{A}$ determines a canonical stable structure $[\phi]$ on $\mathcal{A}$ with

$$[\phi_{M,N}] = [\phi_{M\oplus X,N\oplus X}] : M \longrightarrow N$$

for any objects M,N,X in $\mathcal{A}$ such that $M\oplus X$ is isomorphic to $N\oplus X$.

Proposition 5.3 A canonical stable structure $[\phi]$ on an additive category $\mathcal{A}$ determines a splitting of the natural map $K_1(\mathcal{A}) \longrightarrow K_1^{iso}(\mathcal{A})$

$$K_1^{iso}(\mathcal{A}) \longrightarrow K_1(\mathcal{A}) \ ; \ \tau(f:M \longrightarrow N) \longmapsto \tau([\phi_{N,M}][f]:M \longrightarrow N \longrightarrow M) \ ,$$

so that $K_1^{iso}(\mathcal{A}) = K_1(\mathcal{A})\oplus ?$.

Proof: Trivial.

[]

An additive category $\mathcal{A}$ which is equipped with a sufficiently additive "Eilenberg swindle" has a canonical stable structure, as follows.

A flasque structure $\{\Sigma,\sigma,\rho\}$ on an additive category $\mathcal{A}$ consists of

i) an object ΣM for each object M of $\mathcal{A}$,

ii) an isomorphism $\sigma_M:M\oplus\Sigma M \longrightarrow \Sigma M$ for each object M of $\mathcal{A}$,

iii) an isomorphism $\rho_{M,N}: \Sigma(M\oplus N) \longrightarrow \Sigma M\oplus\Sigma N$ for each pair of objects M,N in $\mathcal{A}$, such that

$$\sigma_{M\oplus N} : M\oplus N\oplus\Sigma(M\oplus N) \xrightarrow{1_{M\oplus N}\oplus\rho_{M,N}} M\oplus N\oplus\Sigma M\oplus\Sigma N$$

$$\xrightarrow{1_M\oplus\begin{pmatrix} 0 & 1_{\Sigma M} \\ 1_N & 0\end{pmatrix}\oplus 1_{\Sigma N}} M\oplus\Sigma M\oplus N\oplus\Sigma N \xrightarrow{\sigma_M\oplus\sigma_N} \Sigma M\oplus\Sigma N \xrightarrow{\rho^{-1}_{M,N}} \Sigma(M\oplus N) .$$

The terminology derives from Karoubi [7,p.147].

An additive category $\mathcal{A}$ admits a structure $\{\Sigma,\sigma\}$ satisfying i) and ii) (but not necessarily iii)) if and only if $K_0(\mathcal{A}) = 0$, or equivalently if each object M is stably isomorphic to 0. The isomorphisms $\sigma_M : M\oplus\Sigma M \longrightarrow \Sigma M$ represent stable isomorphisms $[\sigma_M] : M \longrightarrow 0$.

Example If $\mathcal{A}$ is an additive category with countable direct sums then $K_0(\mathcal{A}) = 0$ by the original Eilenberg swindle (cf. Swan [21,p.66]), which is incorporated in the flasque structure $\{\Sigma,\sigma,\rho\}$ defined on $\mathcal{A}$ by

i) $\Sigma P = \sum_{1}^{\infty} P = P\oplus P\oplus P\oplus\ldots$,

ii) $\sigma_P : P\oplus\Sigma P \longrightarrow \Sigma P \; ; \; (x,(y_1,y_2,\ldots)) \longmapsto (x,y_1,y_2,\ldots)$

iii) $\rho_{P,Q} : \Sigma(P\oplus Q) \longrightarrow \Sigma P\oplus\Sigma Q \; ;$
$((x_1,y_1),(x_2,y_2),\ldots) \longmapsto ((x_1,x_2,\ldots),(y_1,y_2,\ldots))$.

In particular, $\mathcal{A}$ = {projective A-modules} is an additive category with countable direct sums, for any ring A.

[]

Remark In the above example Σ can be extended to an exact endofunctor $\Sigma : \mathcal{A} \longrightarrow \mathcal{A}$ such that σ defines a natural equivalence of functors

$$\sigma : 1_{\mathcal{A}}\oplus\Sigma \simeq \Sigma : \mathcal{A} \longrightarrow \mathcal{A} ,$$

by defining $\Sigma(f:P \longrightarrow Q)$ to be

$$\Sigma f : \Sigma P \longrightarrow \Sigma Q \; ; \; (x_1,x_2,\ldots) \longmapsto (f(x_1),f(x_2),\ldots) \quad .$$

It follows that $K_*(\mathcal{A}) = 0$. A flasque category in the sense of Karoubi [7] is in particular an additive category $\mathcal{A}$ for which there exists an exact endofunctor $\Sigma : \mathcal{A} \longrightarrow \mathcal{A}$ such that $1_{\mathcal{A}}\oplus\Sigma$ is naturally equivalent to Σ. Such structures were considered in connection with formal delooping procedures abstracting the Bott periodicity theorem. In the lower algebraic K-theory examples below the flasque structures $\{\Sigma,\sigma,\rho\}$ are such that Σ does not in general extend to morphisms, and the flasque structure only guarantees that $K_0(\mathcal{A}) = 0$ for the additive categories $\mathcal{A}$ in question.

[]

Proposition 5.4 A flasque structure $\{\Sigma,\sigma,\rho\}$ on an additive category $\mathcal{A}$ determines a canonical stable structure $[\phi]$ on $\mathcal{A}$ by

$$[\phi_{M,N}] = [\sigma_N]^{-1}[\sigma_M] : M \longrightarrow O \longrightarrow N ,$$

so that the natural map $K_1(\mathcal{A}) \longrightarrow K_1^{iso}(\mathcal{A})$ splits and $K_1^{iso}(\mathcal{A}) = K_1(\mathcal{A})\oplus ?$.

Proof: The stable isomorphism $[\phi_{M,N}]:M \longrightarrow N$ is represented by the isomorphism

$$\phi_{M,N} : M\oplus\Sigma M\oplus\Sigma N \xrightarrow{\sigma_M\oplus 1_{\Sigma N}} \Sigma M\oplus\Sigma N \xrightarrow{1_{\Sigma M}\oplus\sigma_N^{-1}} \Sigma M\oplus N\oplus\Sigma N \xrightarrow{\begin{pmatrix} O & 1_N \\ 1_{\Sigma M} & O\end{pmatrix}\oplus 1_{\Sigma N}} N\oplus\Sigma M\oplus\Sigma N \ .$$

The conditions i) $[\phi_{M,M}] = [1_M]$, ii) $[\phi_{M,P}] = [\phi_{N,P}][\phi_{M,N}]$ for a canonical stable structure $[\phi]$ are clear from the definition of the stable category $\mathcal{A}^s$ (Proposition 5.1). As for the additivity condition iii) $[\phi_{M\oplus M',N\oplus N'}] = [\phi_{M,N}]\oplus[\phi_{M',N'}]$ this follows on observing that the isomorphism

$$f : \Sigma(M\oplus M')\oplus\Sigma(N\oplus N') \xrightarrow{\rho_{M,M'}\oplus\rho_{N,N'}} \Sigma M\oplus\Sigma M'\oplus\Sigma N\oplus\Sigma N' \xrightarrow{1_{\Sigma M}\oplus\begin{pmatrix} O & 1_{\Sigma N} \\ 1_{\Sigma M'} & O\end{pmatrix}\oplus 1_{\Sigma N'}} \Sigma M\oplus\Sigma N\oplus\Sigma M'\oplus\Sigma N'$$

is such that there is defined a commutative diagram of isomorphisms in $\mathcal{A}$

$$\begin{array}{ccc}
M\oplus M'\oplus\Sigma(M\oplus M')\oplus\Sigma(N\oplus N') & \xrightarrow{1_{M\oplus M'}\oplus f} & M\oplus M'\oplus\Sigma M\oplus\Sigma N\oplus\Sigma M'\oplus\Sigma N' \\
\downarrow{\scriptstyle \phi_{M\oplus M',N\oplus N'}} & & \downarrow{\scriptstyle \phi_{M,N}\oplus\phi_{M',N'}} \\
N\oplus N'\oplus\Sigma(M\oplus M')\oplus\Sigma(N\oplus N') & \xrightarrow{1_{N\oplus N'}\oplus f} & N\oplus N'\oplus\Sigma M\oplus\Sigma N\oplus\Sigma M'\oplus\Sigma N'
\end{array}$$

Thus $[\phi]$ is a canonical stable structure on $\mathcal{A}$, and Proposition 5.3 applies.

[]

Flasque structures arise naturally in lower algebraic K-theory, as follows.

Given a ring A let $\mathcal{C}_i(A)$, $\mathcal{P}_i(A)$ $(i \geq 1)$ be the additive categories defined by Pedersen [12]. The objects of $\mathcal{C}_i(A)$ are $\mathbb{Z}^i$-graded A-modules

$$M = \sum_{J \in \mathbb{Z}^i} M(J)$$

with each $M(J)$ a f.g. free A-module. The morphisms of $\mathcal{C}_i(A)$ are the A-module morphisms

$$f = \sum_{J,K \in \mathbb{Z}^i} f(J,K) : M = \sum_{J \in \mathbb{Z}^i} M(J) \longrightarrow N = \sum_{K \in \mathbb{Z}^i} N(K)$$

which are bounded in the sense that there exists an integer $s \geq 0$ such that

$$f(J,K) = 0 : M(J) \longrightarrow N(K) \text{ if } J = (j_1, j_2, \ldots, j_i),\ K = (k_1, k_2, \ldots, k_i)$$
$$\text{are such that } \max\{|j_r - k_r| \,|\, 1 \leq r \leq i\} > s\ .$$

$\mathcal{P}_i(A)$ is the idempotent completion of $\mathcal{C}_i(A)$, with objects (M,p) the projections $p = p^2 : M \longrightarrow M$ in $\mathcal{C}_i(A)$, and morphisms

$$f : (M,p) \longrightarrow (N,q)$$

defined by morphisms $f : M \longrightarrow N$ in $\mathcal{C}_i(A)$ such that $qfp = f : M \longrightarrow N$. Also, let $\mathcal{C}_0(A) = \{\text{f.g. free A-modules}\}$, and let $\mathcal{P}_0(A)$ be the idempotent completion of $\mathcal{C}_0(A)$, so that up to natural equivalence

$$\mathcal{P}_0(A) = \{\text{f.g. projective A-modules}\}\ .$$

The main result of [12] is that there are natural identifications

$$K_1(\mathcal{C}_{i+1}(A)) = K_0(\mathcal{P}_i(A)) = K_{-i}(A) \qquad (i \geq 0)$$

with $K_{-i}(A)$ $(i \geq 1)$ the lower algebraic K-groups of Bass [1].

<u>Example</u> The bounded $\mathbb{Z}^i$-graded A-module category $\mathcal{C}_i(A)$ $(i \geq 1)$ admits a flasque structure $\{\Sigma,\sigma,\rho\}$, with

$$\Sigma M(j_1,j_2,\ldots,j_i) = \begin{cases} 0 & \text{if } j_1 = -1,0 \\ \sum\limits_{k=0}^{j_1-1} M(k,j_2,\ldots,j_i) & \text{if } j_1 \geq 1 \\ \sum\limits_{k=j_1+1}^{-1} M(k,j_2,\ldots,j_i) & \text{if } j_1 \leq -2 \end{cases}$$

$$\sigma_M : M(j_1,j_2,\ldots,j_i)\oplus\Sigma M(j_1,j_2,\ldots,j_i) \longrightarrow \Sigma M(j_1+1,j_2,\ldots,j_i) \ ;$$

$$(x_{j_1},(x_0,x_1,\ldots,x_{j_1-1})) \longmapsto (x_0,x_1,\ldots,x_{j_1}) \quad \text{if } j_1 \geq 0$$

$$\sigma_M : M(j_1,j_2,\ldots,j_i)\oplus\Sigma M(j_1,j_2,\ldots,j_i) \longrightarrow \Sigma M(j_1+1,j_2,\ldots,j_i) \ ;$$

$$(x_{j_1},(x_{j_1+1},x_{j_1+2},\ldots,x_{-1})) \longmapsto (x_{j_1},x_{j_1+1},\ldots,x_{-1}) \quad \text{if } j_1 \leq -1,$$

$$\rho_{M,N} : \Sigma(M\oplus N) \longrightarrow \Sigma M\oplus\Sigma N \ ; \ \sum_k (x_k,y_k) \longmapsto (\sum_k x_k,\sum_k y_k) \ .$$

This flasque structure (for which I am indebted to Chuck Weibel) determines by Proposition 5.4 a canonical stable structure $[\phi]$ on $\mathcal{C}_i(A)$, and hence a direct sum decomposition

$$K_1^{iso}(\mathcal{C}_i(A)) = K_1^{aut}(\mathcal{C}_i(A))\oplus ? \ .$$

The automorphism torsion component $\tau(C) \in K_1^{aut}(\mathcal{C}_i(A)) = K_{1-i}(A)$ of the isomorphism torsion $\tau(C) \in K_1^{iso}(\mathcal{C}_i(A))$ of a contractible finite chain complex C in $\mathcal{C}_i(A)$ is an absolute version of the reduced torsion invariant $\tilde{\tau}(C) \in \tilde{K}_{1-i}(A)$ $(= K_{1-i}(A)$ for $i > 1)$ obtained by Pedersen [13]. In particular, for $i = 1$ the splitting map is given explicitly by

$$K_1^{iso}(\mathcal{C}_1(A)) \longrightarrow\!\!\!\!\!\rightarrow K_1^{aut}(\mathcal{C}_1(A)) = K_0(A) \ ;$$

$$\tau(f:M \longrightarrow N) \longmapsto [(\sum_{j=-\infty}^{s-1} M(j)) \cap f^{-1}(\sum_{j=0}^{\infty} N(j))] - [\sum_{j=0}^{s-1} M(j)]$$

with $s \geq 0$ a bound for $f^{-1}:N \longrightarrow M$, such that

$$f^{-1}(N(j)) \subsetneq \sum_{k=-s}^{s} M(j+k) \qquad (j \in \mathbb{Z}) \ .$$

The flasque structure isomorphisms $\sigma_M : M\oplus\Sigma M \longrightarrow \Sigma M$ are such that $\sigma_M(\sum_{j=0}^{\infty} M(j)\oplus\Sigma M(j)) = \sum_{j=0}^{\infty} \Sigma M(j)$, and σ_M^{-1} has bound $s=1$, so that the isomorphism torsion $\tau(\sigma_M)\in K_1^{iso}(\mathcal{C}_1(A))$ has image 0 in $K_1^{aut}(\mathcal{C}_1(A)) = K_0(A)$.

[]

Given a filtered additive category $\mathcal{A}$ let $\mathcal{C}_i(\mathcal{A})$ $(i\geqslant 0)$ be the filtered additive category of $\mathbb{Z}^i$-graded objects in $\mathcal{A}$ defined by Pedersen and Weibel [15], with $\mathcal{C}_0(\mathcal{A}) = \mathcal{A}$, and let $\mathcal{P}_i(\mathcal{A})$ $(i\geqslant 0)$ be the idempotent completion of $\mathcal{C}_i(\mathcal{A})$. By the main result of [15] there are natural identifications of algebraic K-groups

$$\begin{aligned} K_{n+1}(\mathcal{C}_{i+1}(\mathcal{A})) &= K_n(\mathcal{P}_i(\mathcal{A})) = K_{n-i}(\mathcal{P}_0(\mathcal{A})) \quad \text{for } n,i\geqslant 0 \\ &= K_n(\mathcal{C}_i(\mathcal{A})) \quad \text{for } n\geqslant 1 \\ &= K_{n-i}(\mathcal{A}) \quad \text{for } n-i\geqslant 1 \end{aligned}$$

with the higher K-groups defined using the split exact structure, and the lower K-groups $K_{-j}(\mathcal{P}_0(\mathcal{A}))$ $(j\geqslant 1)$ as defined by Karoubi [7].

<u>Example</u> The bounded $\mathbb{Z}^i$-graded category $\mathcal{C}_i(\mathcal{A})$ $(i\geqslant 1)$ admits a flasque structure $\{\Sigma,\sigma,\rho\}$, defined exactly as in the previous Example, which is the special case $\mathcal{A}=$ {f.g. free A-modules}. The splitting map for $K_1^{aut} \rightarrowtail K_1^{iso}$ in the case $i = 1$ is given by

$$K_1^{iso}(\mathcal{C}_1(\mathcal{A})) \twoheadrightarrow K_1^{aut}(\mathcal{C}_1(\mathcal{A})) = K_0(\mathcal{P}_0(\mathcal{A})) \ ;$$

$$\tau(f:M\longrightarrow N) \longmapsto [\sum_{j=-s}^{s-1} M(j), f^{-1}p_{N^+}f] - [\sum_{j=0}^{s-1} M(j), 1]$$

with p_{N^+} the projection

$$p_{N^+} : N = \sum_{j=-\infty}^{\infty} N(j) \longrightarrow N \ ; \quad \sum_{j=-\infty}^{\infty} x(j) \longmapsto \sum_{j=0}^{\infty} x(j)$$

and $s\geqslant 0$ a bound for $f^{-1}: N\longrightarrow M$,

$$f^{-1}(N(j)) \subseteq \sum_{k=-s}^{s} M(j+k) \qquad (j\in\mathbb{Z}) \ .$$

Again, $\tau(\sigma_M)\in K_1^{iso}(\mathcal{C}_1(\mathcal{A}))$ has image $0\in K_1^{aut}(\mathcal{C}_1(\mathcal{A}))$. The case $i=1$ is the most significant one, since $\mathcal{C}_i(\mathcal{A}) = \mathcal{C}_1(\mathcal{C}_{i-1}(\mathcal{A}))$ for $i\geqslant 1$.

[]

A more detailed account of the applications of the algebraic theory of torsion to lower K-theory will appear elsewhere.

References

[1] H.Bass Algebraic K-theory Benjamin (1968)

[2] , A.Heller and R.G.Swan
The Whitehead group of a polynomial extension
Publ. Math. I.H.E.S. 22, 61 - 79 (1964)

[3] F.T.Farrell
The obstruction to fibering a manifold over a circle
Indiana Univ. J. 21, 315 - 346 (1971)

[4] S.Ferry A simple-homotopy approach to the finiteness obstruction
Springer Lecture Notes 870, 73 - 81 (1981)

[5] R.Fossum, H.-B.Foxby and B.Iversen
The Whitehead torsion of a bounded complex and higher order Mennicke symbols preprint

[6] D.Grayson (after D.Quillen)
Higher algebraic K-theory II
Springer Lecture Notes 551, 217 - 240 (1976)

[7] M.Karoubi Foncteurs dérivés et K-théorie
ibid. 136, 107 - 186 (1970)

[8] W.Lück Eine allgemeine Beschreibung des Transfers für Faserungen auf projektiven Klassengruppen und Whiteheadgruppen
Göttingen Ph.D. thesis (1984)

[9] M.Mather Counting homotopy types of manifolds
Topology 4, 93 - 94 (1965)

[10] J.P.May E_∞-spaces, group completions and permutative categories
L.M.S. Lecture Notes 11, 61 - 94 (1974)

[11] J.Milnor Whitehead torsion Bull. A.M.S. 72, 358 - 426 (1966)

[12] E.K.Pedersen
On the $K_{-i}(-)$ functors J. of Algebra 90, 461 - 475 (1984)

[13] K_{-i}-invariants of chain complexes
Springer Lecture Notes 1060, 174 - 186 (1984)

[14] and A.A.Ranicki
Projective surgery theory Topology 19, 239 - 254 (1980)

[15] and C.Weibel
A nonconnective delooping of algebraic K-theory
these proceedings

[16] A.A.Ranicki
Algebraic L-theory II: Laurent extensions
Proc. L.M.S. (3) 27, 126 - 158 (1973)

[17] The algebraic theory of surgery ibid. 40, 87 - 283 (1980)

[18] The algebraic theory of finiteness obstruction
Math. Scand. (to appear)

[19] J.Shaneson Wall's surgery obstruction groups for $G \times \mathbb{Z}$
Ann. of Maths. 90, 296 - 334 (1969)

[20] L.Siebenmann
A total Whitehead torsion obstruction to fibering over the circle Comm. Math. Helv. 45, 1 - 48 (1970)

[21] R.G.Swan Algebraic K-theory Springer Lecture Notes 76 (1968)

[22] C.T.C.Wall Foundations of algebraic L-theory
ibid. 343, 266 - 300 (1973)

[23] C.Weibel K-theory of Azumaya algebras
Proc. A.M.S. 81, 1 - 7 (1981)

[24] J.H.C.Whitehead
Simple homotopy types
Am. J. Math. 72, 1 - 57 (1950)

Department of Mathematics
Edinburgh University
Edinburgh EH9 3JZ
Scotland, U.K.

Equivariant Moore Spaces

by
Justin R. Smith*

Introduction.

This paper studies the following problem, originally proposed by Steenrod in 1960:

Given a group π, a right $\mathbf{Z}\pi$-module M and an integer $n>1$, does there exist a topological space X with the properties:

1. $\pi_1(X)=\pi$;
2. $H_i(\tilde{X}) = 0$, $i\neq 0, n$;
3. $H_0(\tilde{X}) = \mathbf{Z}$;
4. $H_n(\tilde{X}) = M$?

where $\tilde{X}$ is the universal covering space of X, equipped with the usual π-action. The space X, if it exists, is called an *equivariant Moore space of type (M, n; π)* or just a *space* of type $(M, n; \pi)$. A triple $(M, n; \pi)$ for which such a space exists will be said to be *topologically realizable*.

Section 1 of the present paper develops an obstruction theory for the existence of equivariant Moore spaces and proves that:

Theorem: Let $(M, n; \pi)$ *be a triple as described above and suppose that the* $n+2$-*dimensional homological k-invariant of the chain complex* $K^+(M, n)\otimes Z_*$ *is nonzero. Then there doesn't exist a topological space,* X, *with the property that* $\pi_1(X) = \pi$, $H_i(X; \mathbf{Z}\pi) = 0$, $0 < i < n$, *or* $i=n+1, n+2$, $H_n(X;\mathbf{Z}\pi) = M$. *In particular, no equivariant Moore space of type* $(M, n; \pi)$ *exists.* □

Remarks: 1. $K^+(M,n)$ is the quotient of the chain complex of a $K(M,n)$ by the 0-dimensional chain module and Z_* is a $\mathbf{Z}\pi$-projective resolution of $\mathbf{Z}$. The tensor product in the hypothesis is over $\mathbf{Z}$ and equipped with the *diagonal* π-action.

2. The statement about the homological k-invariant of $Y=K^+(M,n)\otimes Z_*$ is equivalent to the statement that the evaluation map $H^{n+2}(Y; H_{n+2}(Y)) \to \mathrm{Hom}_{\mathbf{Z}\pi}(H_{n+2}(Y), H_{n+2}(Y))$ is *not surjective*.

* The author was partially supported by NSF Grant #MCS81-16614.
AMS (MOS) Subject Classification (1980): Primary 55S45; Secondary 18G55.

Sections 2 and 3 of the present paper develop an algorithm for the computation of this homological k-invariant and show that:

Theorem: The hypotheses of the theorem above are satisfied if $\pi = \mathbb{Z}/2\mathbb{Z} \oplus \mathbb{Z}/2\mathbb{Z}$ *(on generators* s *and* t) *and* M *is the* $\mathbb{Z}\pi$*-module whose underlying abelian group is* $\mathbb{Z}\oplus\mathbb{Z}\oplus\mathbb{Z}$ *with* s *and* t *acting via right multiplication by the matrices.*

$$\begin{bmatrix} 0 & 1 & 1 \\ 1 & 0 & 1 \\ 0 & 0 & -1 \end{bmatrix} \quad \textit{and} \quad \begin{bmatrix} -1 & 0 & 0 \\ -1 & 0 & -1 \\ 1 & -1 & 0 \end{bmatrix} \textit{ respectively.}$$

The first counterexample to the Steenrod conjecture was due to Gunnar Carlsson in [2]. The present counterexample has the advantage that the $\mathbb{Z}$-rank of the module is *minimal* and the fundamental group is the smallest possible - Peter Kahn (in unpublished work) proved that any module whose underlying abelian group is $\mathbb{Z}\oplus\mathbb{Z}$ is topologically realizable. On the positive side of the Steenrod conjecture we have:

Theorem: Let M *be a* $\mathbb{Z}\pi$*-module of homological dimension* k *and suppose that* $M/p\bullet M = M_p = 0$ *for all primes* $p < 1 + k/2$. *Then there exist equivariant Moore spaces of type* $(M, n; \pi)$, *where* n *is any integer* $>k$. □

Here M_p denotes the p-torsion submodule of M.

The Steenrod problem has been studied before by Frank Quinn, James Arnold, Peter Kahn, and Gunnar Carlsson.

Frank Quinn developed an obstruction theory to putting a suitable group action on a pre-existing (non equivariant) Moore space. The main drawback to his theory is that the obstructions (and even the obstruction groups) do not seem to be readily computable -- see [14].

James Arnold, in [19], developed a form of homological algebra based upon *permutation modules* rather than projective modules and used it to prove that all modules over a cyclic group are topologically realizable. Peter Kahn developed an obstruction theory to the existence of equivariant Moore spaces for $\mathbb{Z}$-torsion free modules. When coupled with the results of Kiyoshi Igusa (see [10] and [11]) it implies that the $\mathbb{Z}GL_4(\mathbb{Z})$-module $\mathbb{Z}^4$ (where the group acts by matrix multiplication) is not topologically realizable in any dimension. Unfortunately the results of Igusa don't provide for an easy computation of the obstruction.

The work of Gunnar Carlsson (which resulted in the first counterexample) hinged upon an argument involving cohomology operations that doesn't appear to generalize beyond the examples given in his paper (see [2]). Carlsson approached the problem from the point of view of group actions on CW complexes and the induced actions on homology. The present paper was originally written in 1980 after Carlsson's result.

It was felt that Carlsson's result laid the Steenrod problem to rest but in recent years there has been renewed interest in the approach of the present paper. This is due to connections between the Steenrod problem and the theory of *transformation groups*. The present paper develops an obstruction theory for equivariant Moore spaces that is completely different from all of the theories discussed above and which appears to be much more tractable from a computational point of view. It also turns out that the obstruction theory presented here generalizes to an obstruction theory to topologically realizing *chain complexes* that have nonvanishing homology in more than one dimension. Such an obstruction theory provides a first obstruction to imposing a group-action upon a space (e.g., a manifold). Certainly, if no group-action exists on a CW-complex homotopy-equivalent to the desired space then it can't exist on the desired space either. Also, if one can demonstrate the existence of some desired group-action on a CW-complex homotopy equivalent to a manifold one can explore the (surgery-theoretic) problem of smoothing the action (for instance) to get a similar action on the manifold.

Section 3 of this paper gives an *explicit formula* for the obstruction for all modules whose underlying abelian group is $\mathbb{Z}^3$. This formula can be readily generalized to *all* $\mathbb{Z}$-free modules.

I am indebted to Sylvain Cappell and Andrew Ranicki for their encouragement.

§1 The Obstruction Theory

In this section we will describe the obstructions to the existence of equivariant Moore spaces. Essentially they will turn out to be obstructions to adjoining terms to a partial Postnikov tower in such a way that the first non-vanishing homology module above the one we want to realize, is *annihilated*.

Definition 1.1: Let M be a right $\mathbb{Z}\pi$-module and n be an integer > 1. The equivariant Eilenberg-MacLane space $K_\pi(M, n)$ is defined to be a space homotopy-equivalent to $(K(M,n) \times \tilde{K}(\pi,1))/\pi$ where:

a. the second factor is the universal cover of a $K(\pi,1)$;

b. the cartesian product above is equipped with the *diagonal* π-action. □

Remarks: 1. Note that $K_\pi(M, n)$ has the following properties:

i. $\pi_1(K_\pi(M,n)) = \pi$;

ii. $\pi_i(K_\pi(M,n)) = 0$, $i \neq 1, n$;

iii. $\pi_n(K_\pi(M,n)) = M$, as $\mathbf{Z}\pi$-modules, i.e. the action of the fundamental group on π_n coincides with the action of π on M.

2. In the definition of $K_\pi(M,n)$ above we could have used the cellular bar construction of Milgram for $K(M,n)$ instead of the semi-simplicial complex of Eilenberg and MacLane -- see [13].

Let X be a topological space with fundamental group π and consider a map inducing an isomorphism of fundamental groups:

1.2: $X \xrightarrow{f} K_\pi(M,n)$

The homotopy class of this map defines a cohomology class [f] in $H^n(X;M)$ and it is well-known that the map $f_*:H_n(X;\mathbf{Z}\pi) \to M$ is the image of [f] under the evaluation map:

$$H^n(X;M) \xrightarrow{e} \mathrm{Hom}_{\mathbf{Z}\pi}(H_n(X;\mathbf{Z}\pi), M)$$

-- i.e. [f] is just the element of $H^n(X;M)$ given by $f^*(\iota)$, where $\iota \in H^n(K(M,n);M)$ is an element whose image under the evaluation map is the *identity map* of M -- see [18, chapter 8]. Furthermore the map f is the classifying map for a fibration over X with fiber a $K(M,n-1)$. If E is the total space of this fibration its homology fits into the Serre exact sequence of a fibration:

$$\ldots \to H_n(K(M,n-1)) \to H_n(E;\mathbf{Z}\pi) \to H_n(X;\mathbf{Z}\pi) \xrightarrow{\alpha} H_{n-1}(K(M,n-1))$$
$$\to H_{n-1}(E;\mathbf{Z}\pi) \to H_{n-1}(X;\mathbf{Z}\pi) \to 0$$

where the map α can be regarded as coinciding with f_* or e[f] since it is essentially the pullback of the transgression homomorphism for the universal fibration over $K(M,n)$ -- which can be regarded as the identity map of M.

Lemma 1.3: Let $H_{n-1}(X;\mathbf{Z}\pi) = 0$. *Then there exists a* $K(M,n-1)$*-fibration over* X *with total space* E *such that* $H_n(E;\mathbf{Z}\pi) = H_{n-1}(E;\mathbf{Z}\pi) = 0$ *if and only if* $M = H_n(X;\mathbf{Z}\pi)$ *and there exists a map* $f:X \to K_\pi(M,n)$ *inducing an isomorphism of* π_1, *and an isomorphism in homology in dimension* n. *A map* f *with those properties exists if and only if the evaluation map with local coefficients in* M --

$e: H^n(X;M) \to \mathrm{Hom}_{\mathbb{Z}\pi}(M,M)$, *where* $M = H_n(X;\mathbb{Z}\pi)$, *is surjective.*

Proof: Most of this follows immediately from the Serre exact sequence above. We need only prove the last statement about the evaluation map. Let Λ denote $\mathrm{Hom}_{\mathbb{Z}\pi}(M,M)$, regarded as a *ring.* From the remarks following 1.1 it is clear that a map $f: X \to K_\pi(M,n)$ inducing an isomorphism in homology in dimension n represents a cohomology class $[f] \in H_n(X;M)$ whose image under the evaluation map is an *automorphism* of M, and conversely, the existence of such a cohomology class implies the existence of the map. Since the evaluation map $H^n(X;M) \to \mathrm{Hom}_{\mathbb{Z}\pi}(M,M)$ is *natural* with respect to changes of coefficients it follows that it is a homomorphism of Λ-modules (where Λ acts on the right by *changes of coefficients* in the cohomology, and by *composition* in the Hom-group). Since Λ is generated, as a module over itself, by any automorphism of M it follows that an automorphism is in the image of the evaluation map if and only if that map is *surjective.* □

Lemma 1.4: A topological realization for the triple $(M,n;\pi)$ *exists if and only if there exists a sequence of spaces* $\{X_i\}$ *such that:*

1. X_1 *is a* $K(M,n)$*-fibration over a* $K(\pi,1)$;

2. X_{i+1} *is a* $K(N_i,n+i)$*-fibration over* X_i *with* $N_i = H_{n+i}(X_i;\mathbb{Z}\pi)$ *and whose characteristic class is a cohomology class of* $H^{n+i}(X_i;N_i)$ *whose image under the evaluation map is an automorphism of* N_i.

Remarks: 1. It is clear from lemma 1.3 that X_{i+1} can't exist unless the evaluation map $H^{n+i}(X_i;N_i) \to \mathrm{Hom}_{\mathbb{Z}\pi}(N_i,N_i)$ is surjective. Consequently the ith obstruction to the existence of a topological realization of the triple $(M,n;\pi)$ is *defined* if and only if the previous i-1 obstructions *vanish.*

2. Since the evaluation map for *integral* cohomology is well-known to be surjective it is easy to see why all of the obstructions in the theory presented here vanish if π is the *trivial group.*

Proof: The *if* part of the statement follows from 1.3 which implies that

$H_0(X_i;\mathbb{Z}\pi) = \mathbb{Z}$;

$H_j(X_i;\mathbb{Z}\pi) = 0$ if $0 < j < n$;

$H_n(X_i;\mathbb{Z}\pi) = M$;

$H_j(X_i; \mathbb{Z}\pi) = 0$ if $n < j < n+i$.

The X_i form a convergent sequence of fibrations whose limit will be a suitable equivariant Moore space.

The *only if* part of the argument is a consequence of the existence and uniqueness of equivariant Postnikov towers -- see [1]. □

The remainder of this section will be spent developing algebraic criteria for the *surjectivity of the evaluation map* -- since the results above show that the equivariant Moore space can be constructed if this map is surjective. Essentially we will show that it is possible to define *obstructions* to the surjectivity of the evaluation map -- these will turn out to be closely related to the *homological k-invariants* of chain complexes defined by Heller in [8].

Definition 1.5: Let C_* be a projective $\mathbb{Z}\pi$-chain complex with the following properties:

i. $H_i(C_*) = 0$, $1 < n$;

ii. $H_n(C_*) = M$;

iii. $H_i(C_*) = 0$, $n < i < n+k$;

iv. $H_{n+k}(C_*) = N$.

Let P_* be a projective resolution for M. Then there exists a unique chain-homotopy class of chain maps $c: C_* \to \Sigma^n P_*$ inducing an isomorphism in homology in dimension n. Let $\mathfrak{A}(c)$ denote the algebraic mapping cone of c. We have the exact sequence:

$$0 \to \Sigma^n P_* \to \mathfrak{A}(c) \to \Sigma C_* \to 0$$

and $H_i(\mathfrak{A}(c)) = 0$ for $i < n+k+1$ and $H_{n+k+1}(\mathfrak{A}(c)) = N$ so that there exists a map $\mathfrak{A}(c) \to \Sigma^{n+k+1} Q_*$, where Q_* is a projective resolution of N. By composition there is a chain map $\Sigma^n P_* \to \Sigma^{n+k+1} Q_*$ defining a class $x \in \mathrm{Ext}_{\mathbb{Z}\pi}^{k+1}(M,N)$. This will be called the *homological k-invariant of* C_* *in dimension* n+k. □

Remarks: 1. It is not hard to see that this definition agrees with that of Heller in [8]: the pair $(\Sigma^{-n} C_*, \Sigma^{-n}\mathfrak{A}(c))$ can be regarded as a *0-truncated segregated pair* as in §6 of [8] -- see §3 of that paper also; C_* is regarded as a triangular complex such that $T_{*,i} = 0$ if $i > 0$.

2. It is also clear that if C_* is the (cellular or singular) chain complex of a connected topological space and n=0, the homological k-invariant defined above agrees with the *topological k-invariant* of the topological space.

Definition 1.6: Let $f:C_* \to D_*$ be a chain map of chain complexes. Then $\mathfrak{S}(f)$ is defined to be $\sum^{-1}\mathfrak{A}(f)$ -- the *desuspension* of the algebraic mapping cone. □

Remark: We have the usual short exact sequence of chain complexes:

$0 \to \sum^{-1}D_* \to \mathfrak{S}(f) \to C_* \to 0$. Let C_* be a chain complex such that

$H_0(C_*) = \mathbf{Z}$;
$H_n(C_*) = M$;
$H_i(C_*) = 0,\ n < i < n+k$;
$H_{n+k}(C_*) = H$

where M and H are $\mathbf{Z}\pi$-modules, n and k are positive integers and $n > 1$.

There exists a unique chain-homotopy class of chain maps $f_0:C_* \to Z_*$ inducing an isomorphism of H_0, where Z_* is some projective resolution of $\mathbf{Z}$ over $\mathbf{Z}\pi$. Then $H_0(\mathfrak{S}(f)) = 0$ and the canonical projection $\mathfrak{S}(f_0) \to C_*$ induces homology isomorphisms in all higher dimensions. If P_* is a projective resolution of M then there exists a unique chain-homotopy class of chain maps $f_n:\mathfrak{S}(f_0) \to \sum^n P_*$ inducing an isomorphism of homology in dimension n.

Proposition 1.7: Under the conditions discussed above the following diagram commutes, with all horizontal and vertical sequences exact:

$$\begin{array}{ccccc}
\mathrm{Ext}^{k+1}_{\mathbf{Z}\pi}(M,H) & \xleftarrow{\lambda} & H^{n+k+1}(V_*;H) & \longleftarrow & H^{n+k+1}(\pi,H) \\
\Vert & & \uparrow{\scriptstyle \xi} & & \uparrow \\
\mathrm{Ext}^{k+1}_{\mathbf{Z}\pi}(M,H) & \xleftarrow{\zeta} & \mathrm{Hom}_{\mathbf{Z}\pi}(H,H) & \xleftarrow{\delta} & H^{n+k}(\mathfrak{S}(f_0);H) \\
 & & \uparrow{\scriptstyle e} & & \uparrow{\scriptstyle \rho} \\
 & & H^{n+k}(C_*;H) & & H^{n+k}(C_*;H)
\end{array}$$

where e *and* δ *are the evaluation maps, respectively, of* C_* *and* $\mathfrak{S}(f_0)$. If $c = \xi(1_H)$, *then the evaluation map of* C_* *is surjective if and only if* $c = 0$. *Furthermore,* $\lambda(c)$ *is the homological k-invariant of* $\mathfrak{S}(f_0)$ *in dimension* n+k. *The complex* V_* *is the algebraic mapping cone of the composite* $\sum^{-1}Z_* \subset \mathfrak{S}(f_0) \xrightarrow{\alpha} \sum^n P_*$, *where* $\alpha = f_n$ *and* λ *is induced by the canonical inclusion* $\sum^n P_* \subset V_*$.

Remarks: 1. The element c defined above is just the homological k-invariant of C_* in dimension n+k -- the simple definition given in 1.5 doesn't apply here because C_* has homology in dimension 0. See [8] for the general definition of homological k-invariants.

2. When C_* is the chain complex of X_k in 1.4 the element c will be the obstruction to the existence of X_{k+1} and the kth obstruction to the existence of the corresponding equivariant Moore space. In order to prove the nonexistence of a given equivariant Moore space it clearly suffices to show that $\lambda(c) \neq 0$ at some stage of the construction. This will form the basis of the proof of the main results stated in the introduction since $\lambda(c)$ turns out to be more readily computable than c itself.

Proof: The diagram in the statement is the result of applying $H^{n+k}(*;H)$ to the *following* commutative exact diagram of chain complexes (where, as usual, Z_* is a projective resolution of $\mathbb{Z}$):

$$\begin{array}{ccccccccc}
 & & 0 & & 0 & & 0 & & \\
 & & \downarrow & & \downarrow & & \downarrow & & \\
0 & \longrightarrow & \Sigma^{n-1}P_* & \longrightarrow & V' & \longrightarrow & \Sigma^{-1}Z_* & \longrightarrow & 0 \\
 & & \downarrow & & \downarrow & & \downarrow & & \\
0 & \longrightarrow & \Sigma^{n-1}P_* & \longrightarrow & \mathfrak{C}(f_n) & \longrightarrow & \mathfrak{C}(f_0) & \longrightarrow & 0 \\
 & & \downarrow & & \downarrow & & \downarrow & & \\
 & & 0 & \longrightarrow & C_* & \longrightarrow & C_* & \longrightarrow & 0 \\
 & & & & \downarrow & & \downarrow & & \\
 & & & & 0 & & 0 & &
\end{array}$$

where:

1. The middle row and the right column are the canonical exact sequences for $\mathfrak{C}(f_n)$ and $\mathfrak{C}(f_0)$, respectively;
2. $\mathfrak{C}(f_n) \to C_*$ is the composite of the canonical projections $\mathfrak{C}(f_n) \to \mathfrak{C}(f_0)$ and $\mathfrak{C}(f_0) \to C_*$ and V' is defined to be its kernel.

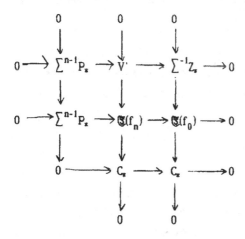

The existence and exactness of the remaining maps in the diagram now follows by a straightforward diagram chase. The fact that $H^{n+k}(\mathfrak{C}(f_n);H) = \mathrm{Hom}_{\mathbb{Z}\pi}(H,H)$ follows from the fact that the homology of $\mathfrak{C}(f_n)$ vanishes below dimension n+k. That e and δ are the evaluation maps followsfrom the naturality of evaluation maps with respect to maps of chain complexes. That $\lambda(c)$ is the n+k-dimensional homological k-invariant of $\mathfrak{C}(f_0)$ folllows from the definition. It is also not hard to see that $V' = \mathfrak{C}(g)$, where g is the composite $\sum^{-1}Z_* \subset \mathfrak{C}(f_0) \to \sum^n P_*$ so that $V_* = \sum V'$. □

It was noted above that the elements $\lambda(c)$ are easier to calculate than the obstructions themselves. The $\lambda(c)$ do, however, occur in a natural geometric setting that will be described now.

Definition 1.8: Let $(M, n; \pi)$ be a triple as in 1.1. A *split topological realization of* $(M, n; \pi)$ is a space X of type $(M, n; \pi)$ such that the characteristic map $c:X \to K(\pi,1)$ (inducing an isomorphism of fundamental groups) is *split* by a map $s:K(\pi,1) \to X$ (i.e. $c \circ s$ is homotopic to the identity). □

Remarks: 1. Note that in the split case the first k-invariant of the space X must vanish. The condition that X be split is stronger than the vanishing of the first k-invariant, however. It essentially implies that there is *no* interaction between the fundamental group and the homotopy groups above π_2.

2. Since the splitting map $s:K(\pi,1) \to X_k$ must lift to X_{k+1} in each stage of the construction it follows that the classifying element of the fibration used to construct X_{k+1} comes from the *relative* cohomology group $H^{n+k}(X_k, K(\pi,1); H)$ -- since the relative chain complex is essentially $\mathfrak{C}(f_0)$ (in the setting of 1.7) -- it follows that the obstructions of the existence of a *split* equivariant Moore space are the images of the *nonsplit* obstructions under λ.

The geometric significance of the split case is connected with Steenrod's original definition of an equivariant Moore space as a CW-complex acted upon by a group, π, such that its equivariant homology had prescribed properties (so, for instance, the equivariant Moore space was generally *simply connected* and corresponded to the *universal cover* of an equivariant Moore space in *our* sense).

Proposition: A triple $(M, n; \pi)$ *has a split topological realization (in the sense of the present paper) if and only if it is realizable (in Steenrod's sense) by a* π*-complex that has a fixed point.*

Proof: Suppose $(M, n; \pi)$ has a split realization X, in our sense. Then there exists a π-equivariant map $z:\tilde{K}(\pi,1) \to \tilde{X}$ whose mapping cone is a *pointed* π-complex realizing $(M, n; \pi)$ in Steenrod's sense.

The converse follows by taking a Steenrod realization and taking the topological product with $\tilde{K}(\pi,1)$ and defining the group action diagonally. The group π acts on the resulting space freely so we can take the quotient to get a realization in *our* sense. The existence of the *fixed point* in the original space implies that the final space will be split (it will contain a copy of $K(\pi,1) = \tilde{K}(\pi,1) \times \text{fixed point}/\pi$). □

Suppose we are at the beginning of the process of constructing an equivariant Moore space of type $(M, n; \pi)$, in the non split case. Then X_1 will be the total space of a $K(M,n)$-fibration over $K(\pi,1)$. The results of V. K. A. M. Gugenheim in [7] imply that the (equivariant) chain-complex of X_1 will be a *twisted tensor product* of the chain complex for a $K(M,n)$ by that of $\tilde{K}(\pi,1)$ (we can use the description of these chain-complexes that appears in [4]). Since in the stable range (all dimensions $<2n$, in this case) a twisted tensor product is the same as an *ordinary* tensor product followed by a *twisted direct sum* it follows that:

Proposition 1.9: Under the hypotheses of 1.4 and 1.7 (with $n>2$, $k = 2$, *and* C_* *the chain complex of* X_1*) the obstruction,* c_3*, to the existence of* X_3 *is an element of* $H^3(V_*; M/2M)$ *that maps under* λ *to the n+k-dimensional homological k-invariant of* $K^+(M,n)\otimes Z_\pi$.

Remarks: 1. This result has the interesting consequence that, if the first obstruction c_2 is nonvanishing in the *split* case it doesn't vanish in the *general* case, i.e., one can't "cancel out" the first obstruction by introducing a non-trivial topological k-invariant in the lowest dimension.

Phrased in the terms of Steenrod's original formulation of the problem this says that if the first obstruction to introducing an appropriate π-action to a pointed complex is nonvanishing, then letting the basepoint move freely won't simplify matters.

2. In the non-split case there will turn out to be *three* essentially different sources of obstruction:

A. *The "homological" obstructions* -- coming from the homological k-invariants of the equivariant Eilenberg-MacLane spaces;

B. *The "topological" obstructions* -- defined when the homological obstructions vanish and coming from the rightmost colume of 1.7. They derive from the effects of cohomology operations on the first topological k-invariant and *vanish identically* in the split case.

C. *The "multiplicative" obstructions* -- derived from the fact that fibrations correspond to *twisted tensor products* rather than twisted direct sums. This obstruction is explored in [18] and is shown to be nonvanishing in general.

3. $K^+(M,n)$ denotes the quotient of $K(M,n)$ by the subcomplex of 0-dimensional elements. See 1.7 for definitions of V_* and λ.

Proof: This follows from the fact that, in the stable range (i.e. dimensions n through 2n-1), $\tilde{K}(\pi,1)\otimes_{\xi}K(M,n)=\tilde{K}(\pi,1)\oplus_{\xi'}\tilde{K}(\pi,1)\otimes K^{+}(M,n)$, (where ξ' is the restriction of ξ to the stable range), and by direct computation of $\mathfrak{S}(f_0)$ in this setting (see the discussion preceding 1.7). □

At this point we are in a position to state sufficient conditions for the existence of equivariant Moore spaces. We will make use of the results involving the homology of Eilenberg-MacLane spaces in [5] and [3].

We begin with the following well-known result (which can be proved directly using the Hurewicz homomorphism):

Corollary 1.10: If M *is a* $\mathbf{Z}\pi$*-module of homological dimension* ≤ 2, *there exists a space of type* (M, n; π) *for any* $n > 1$. □

Remark: Using the obstruction theory described above, this follows from the fact that $H_{n+1}(K(M,n);\mathbf{Z}) = 0$ for all $n>1$ and all abelian groups M -- see [5, §20].

It follows that the first *nontrivial obstruction* is $c_2 \in \mathrm{Ext}_{\mathbf{Z}\pi}^{3}(M,H_{n+2}(K(M,n);\mathbf{Z})$. Since it is proved in [5, §§21, 22] that:

$H_4(K(M,2);\mathbf{Z}) = \Gamma(M);$

$H_{n+2}(K(M,n);\mathbf{Z}) = M/2M$, if $n > 2$.

(where $\Gamma(M)$ is the Whitehead functor).

Corollary 1.11: Let M *be a* $\mathbf{Z}\pi$*-module of homological dimension* ≤ 3 *and suppose that* $\mathrm{Ext}_{\mathbf{Z}\pi}^{3}(M,\Gamma(M)) = 0$. *Then there exists an equivariant Moore space of type* (M, 2; π). □

Theorem 1.12: Let M *be a* $\mathbf{Z}\pi$*-module of homological dimension* k *and suppose* M_p *(the p-torsion submodule)* $= M/p{\bullet}M = 0$ *for all primes* $p < 1 + k/2$. *Then there exist equivariant Moore spaces of type* (M, n; π), *where* n *is any integer* $\geq k$.

Proof: This follows immediately from the results on the homology of Eilenberg-MacLane spaces in the stable range in [3]. Those results imply that the homology of a K(M,n) in dimension n+k is a sum of copies of M_p and $M/p{\bullet}M$ for primes p such that $2(p-1) \leq k$ where $k < n$. □

§ 2 The First Homological k-Invariant of a Chain-Complex.

In this section we will develope methods for computing the homological k-invariants of a chain complex. We will make extensive use of the perturbation theory of DGA-algebras. This theory was developed by H. Cartan in unpublished work and later elaborated by V.K.A.M. Gugenheim (see [7]).

Definition 2.1: Let $f:C \to D$, $g:D \to C$ be maps of chain-complexes. Then:

1. if f maps each C_i to D_{i+k} then f will be called a *map of degree* k;

2. if f is a map of degree k then df is defined to be $d_D \cdot f + (-1)^{k+1} f \cdot d_C$. The map f is defined to be a *chain map* if it is of degree 0 and $df = 0$.

3. if f and g, above, are *both chain maps* and:

   a. $f \cdot g = 1_D$, and $g \cdot f = d\varphi$, where φ is some map of degree +1; and

   b. $f \cdot \varphi = 0$, $\varphi \cdot g = 0$, and $\varphi^2 = 0$;

then *the triple* (f, g, φ) is called a *contraction of* C *onto* D. The map f is called the *projection* of the contraction, and g is called the *injection*. □

Remarks: 1. Since df has the special meaning given above, we will follow Gugenheim in [7] in using $d \cdot f$ to denote the composite.

2. We will also use the convention that if $f:C_1 \to D_1$, $g:C_2 \to D_2$ are maps, and $a \otimes b \in C_1 \otimes C_2$ (where a is a homogenous element), then $(f \otimes g)(a \otimes b) = (-1)^{\deg(g) \cdot \deg(a)} f(a) \otimes g(b)$. This convention simplifies some of the common expressions in homological algebra. For instance the differential, $d_\otimes$, of the tensor product $C \otimes D$ is just $d_C \otimes 1 + 1 \otimes d_D$.

3. It is not difficult to see that the definition of a chain-map given above coincides with the usual definition.

4. The definition of a contraction of chain complexes given here is slightly stronger than the original definition due to Eilenberg and MacLane in [4], since they don't require the chain-homotopy to be *self-annihilating*. The *present* definition is due to Weishu Shih in [16]. Its use in the present paper is justified by the fact that it enables us to use the following lemma, which is central to perturbation theory in differential algebra:

Lemma 2.2 **(Perturbation Lemma)**: *Let* $(f, g, \varphi):C \to D$ *be a contraction of chain complexes with differentials* d_C *and* d_D, *respectively. Suppose* C *is equipped* *with second differential* d' *and an increasing filtration* (F_iC) *such that:*

1. $t=d'-d_C$ *lowers filtration degree by at least* 1;
2. φ *and* d_C *preserve the filtration;*
3. $t(F_0C)=0$.

Then there exists a second differential d" *on* D *and a contraction* $(f',g',\varphi'):(C,d')\to(D,d")$. *The contraction* (f',g',φ') *is defined by:*

1. $f'=f\cdot(1+t\cdot T_{\infty}\cdot\varphi)$;
2. $g'=T_{\infty}\cdot g$;
3. $\varphi'=T_{\infty}\cdot\varphi$;

where $T_{\infty}=1+\sum_{i=1}^{\infty}(\varphi\cdot t)^i$ *and the differential* d", *on* D, *is given by* $d"=d+f\cdot t\cdot T_{\infty}\cdot g$. □

Remarks: 1. The summation above has i going from 1 to infinity. Note that this "infinite series" actually reduces to a *finite sum* when evaluated on elements of C because of the conditions on the filtration of C. Throughout the remainder of this section we will use the notation $T_{\infty}=(1-\varphi\cdot t)^{-1}$. This is *more than just a notational convention* -- the condition of the filtration of C implies that $T_{\infty}\cdot(1-\varphi\cdot t)=1_C$.

2. This lemma first appeared in [7], although it was used *implicitly* in [16].

Definition 2.3: If M and N are $\mathbb{Z}\pi$-modules, F is a free $\mathbb{Z}\pi$-module with preferred basis (y_i) and $f:M\to N$ is a homomorphism of abelian groups that doesn't necessarily preserve the action of π then the *F-extension of* f, denoted $\tilde{f}_F:M\otimes_{\mathbb{Z}}F\to N\otimes_{\mathbb{Z}}F$ (with *diagonal* π-action) is defined to be the $\mathbb{Z}$-linear map for which $\tilde{f}_F(m\otimes(y_i\bullet v))=f(m\bullet v^{-1})\bullet v\otimes(y_i\bullet v)$ for all $m\in M$ and $v\in\pi$. □

Remarks: 1. This construction will be used as a way to convert *maps* into *module homomorphisms* -- it is not difficult to see that $\tilde{f}_F$ is a $\mathbb{Z}\pi$-module homomorphism. The construction was motivated by the *Borel Construction* for making a group action free (i.e. take the product with a space upon which the group acts freely and give the product the *diagonal* action).

2. The F-extension of f clearly depends upon the preferred basis for F that was used in its construction. If f is already a module homomorphism $\tilde{f}_F=f\otimes 1$.

3. The definition above can clearly be generalized to the case where M, N, and F are *chain complexes*. In this case bases for the chain modules of F must be defined in each dimension. If f was *originally a chain map* its F-extension will *also* be a chain map if the *differential* on F is *identically zero*.

Lemma 2.4 **(The Module Lemma):** *Let* C *and* D *be chain complexes and let* $(f, g, \varphi): C \to D$ *be a* $\mathbf{Z}$*-contraction (i.e. the maps involved aren't necessarily module homomorphisms). Let* Z_* *be a free resolution of* $\mathbf{Z}$ *over* $\mathbf{Z}\pi$ *and suppose some preferred basis has been chosen in each dimension. Define:*

1. $\hat{f} = \tilde{f}_Z \cdot (1 - (1 \otimes d_Z) \cdot \hat{\varphi}_Z)^{-1}$;
2. $\hat{g} = (1 - \hat{\varphi}_Z \cdot (1 \otimes d_Z))^{-1}$;
3. $\tilde{\varphi} = (1 - \hat{\varphi}_Z \cdot (1 \otimes d_Z))^{-1}$;
4. $d' = (\tilde{d}_D)_Z + \tilde{f}_Z \cdot (1 \otimes d_Z) \cdot (1 - \hat{\varphi}_Z \cdot (1 \otimes d_Z))^{-1} \cdot \tilde{g}_Z$;
5. $c' = (\tilde{d}_C)_Z + 1 \otimes d_Z$;

Remark: Notice that when the composite $\varphi_Z \cdot (1 \otimes d_Z)$ is evaluated on $a \otimes b \in C_* \otimes Z_*$ the dimension of the first factor is *lowered* by 1 and that of the second factor is *raised* by 1.

Proof: This is a straightforward application of the Perturbation Lemma to the contraction $(\tilde{f}, \tilde{g}, \hat{\varphi}): C_* \otimes Z_* \to D_* \otimes Z_*$, where the differentials of $C_* \otimes Z_*$ and $D_* \otimes Z_*$ are taken to be $(\tilde{d}_C)_Z$ and $(\tilde{d}_D)_Z$, respectively. The "perturbation", t, is $1 \otimes d_Z$, which evaluates to $(-1)^{\dim(a)} a \otimes d_Z(b)$ on $a \otimes b$, by the convention regarding evaluation of maps on tensor products. The filtration degree of such an element is defined to be the dimension of b. □

Let C_* be a chain complex over $\mathbf{Z}\pi$ and suppose its lowest dimensional non-vanishing homology module is H_n (in dimension n). Furthermore suppose the next non-vanishing homology module is H_{n+k} (in dimension n+k) with $k \geq 1$. Let D_* be a $\mathbf{Z}\pi$-chain complex with:

1. $D_i = 0, i < n$;
2. $D_n = H_n$ (as a $\mathbf{Z}\pi$-module);
3. $D_i = 0, n < i < n+k$;
4. D_* is $\mathbf{Z}$-chain homotopy equivalent to C_*.

(to simplify the discussion somewhat we'll assume that the boundary homomorphisms of D_* commute with the action of π, although this isn't necessary).

The theory of chain-complexes over a PID ($\mathbf{Z}$ in this case) guarantees the existence of such a D_* and a contraction (see theorem 5.1.15 on p. 164 of [9]):

2.5: $(f, g, \varphi): C_* \to D_*$

over $\mathbf{Z}$. If Z_* is a free $\mathbf{Z}\pi$-resolution of $\mathbf{Z}$ with preferred bases for its chain modules chosen (so $\tilde{f}_Z$, $\tilde{g}_Z$, and $\hat{\varphi}_Z$ can be defined as in 2.3) then 2.4 implies the existence of a contraction over $\mathbf{Z}\pi$ -- $(\hat{f}, \hat{g}, \tilde{\varphi}): C_* \otimes Z_* \to (D_* \otimes Z_*, d')$ where $\hat{f}$, $\hat{g}$, and $\tilde{\varphi}$ are defined as in 2.4.

Corollary 2.6: Under the conditions in the discussion above, the first non-trivial homological k-invariant of $C_* \otimes Z_*$ *is given by the cocycle:*

$$(-1)^{n+k} p \cdot \tilde{f}_Z \cdot (1 \otimes d_Z) \cdot (\hat{\varphi}_Z \cdot (1 \otimes d_Z))^k \cdot \tilde{g}_Z : H_n \otimes Z_{k+1} \to H_{n+k}$$

where $p: D_{n+k} \otimes Z_0 \to H_{n+k}$ *is the projection of the cycle module to the homology module.*

Remarks: 1. We may consider this cocycle as being defined in either $\mathrm{Hom}_{\mathbf{Z}\pi}(H_n \otimes Z_*, H_{n+k})$ which defines an element of $\mathrm{Ext}_{\mathbf{Z}\pi}^{k+1}(H_n, H_{n+k})$ or $\mathrm{Hom}_{\mathbf{Z}\pi}(Z_*, \mathrm{Hom}_{\mathbf{Z}}(H_n, H_{n+k}))$, which gives rise to the isomorphic group $H^{k+1}(\pi, \mathrm{Hom}_{\mathbf{Z}}(H_n, H_{n+k}))$ -- see [17].

2. By the remarks following 1.4 and 1.7 it follows that, if $C_* = K(M,n)$ with M a $\mathbf{Z}$-free $\mathbf{Z}\pi$-module then $C_* \otimes Z_*$ is the (equivariant) chain complex of $K_\pi(M,n)$ and the homological k-invariant in question is the first obstruction to the existence of an equivariant Moore space of type $(M,n;\pi)$.

3. Note that this cocycle *vanishes* if any of the $\mathbf{Z}$-homomorphisms f, g, or φ is also *$\mathbf{Z}\pi$-linear*, in the dimension range of the formula.

4. It should be kept in mind that the boundary map in the resolution, $H_n \otimes Z_*$, of H_n *is not* $1 \otimes d_Z$ -- it is $(\tilde{f}_Z)_n \cdot (1 \otimes d_Z) \cdot (\tilde{g}_Z)_n$ (it is not difficult to see that $H_n \otimes Z_*$, with this differential, is *still* a resolution of H_n -- at least up to dimension n+k). This twisted differential coincides with $1 \otimes d_Z$ if and only if f is $\mathbf{Z}\pi$-linear, i.e. $\tilde{f}_Z = f \otimes 1$ in dimension n.

Proof: Throughout this argument the term characteristic map of a chain complex will refer to the canonical map (up to a chain homotopy) from a chain complex to a projective resolution of its lowest dimensional homology module, i.e. if the lowest dimensional nonvanishing homology module of C_* is in dimension n and has a projective resolution P_* then the characteristic map of C_* is a chain map $C_*\to\Sigma^n P_*$.

We will prove that the cocycle in the statement of the theorem represents the first homological k-invariant of $D''=(D_*\otimes_{\mathbf{Z}} Z_*, d')$, which is chain homotopy equivalent to $C_*\otimes Z_*$. First note that D'' can be regarded as the direct sum $D''=H_n\otimes Z_*\oplus D'$, where D' has no nonvanishing chain modules below dimension n+k. This is not necessarily a direct sum decomposition of *chain complexes*. In fact, by the description of d' in 2.4 it follows that:

1. $D'\otimes Z_*$ is a chain subcomplex of D'';
2. The boundary of $H_n\otimes Z_*$ may contain components in $D'\otimes Z_*$.

This follows from the existence of a corresponding direct sum decomposition of *chain complexes* when the unperturbed differential is used, and the fact the perturbation terms in d' *lower* the dimension of the Z_*-factor and *raise* that of the D_*-factor. Let d_B denote "the portion of the boundary of $H_n\otimes Z_*$ that lies in D', i.e. the composite $H_n\otimes Z_*\to D''\to_{d'} D''\to D'$, where the leftmost map is the inclusion and the rightmost map is the projection.

It is not hard to see that there is a chain homotopy equivalence $h:\mathfrak{A}(c)\to\Sigma D'$ (recall that $\mathfrak{A}(c)$ is the algebraic mapping cone of c), where $c:D''\to H_n\otimes Z_*$ is the projection, which is also the *characteristic map*. The map h can be described on the various direct summands of $\mathfrak{A}(c)$ as follows (where D'' is regarded as the direct sum $H_n\otimes Z_*\oplus D'$):

1. $h|\Sigma D' = 1: \Sigma D'\to \Sigma D'$;
2. $h|\Sigma H_n\otimes Z_* = 0: \Sigma H_n\otimes Z_*\subset\mathfrak{A}(c)\to \Sigma D'$;
3. $h|H_n\otimes Z_i=(-1)^{i+1}d_B:H_n\otimes Z_*\subset\mathfrak{A}(c)\to\Sigma D'$.

where in the last statement d_B is regarded as a map from $H_n\otimes Z_i$ to $D'_{i-1} = (\Sigma D')_i$.

The conclusion now follows from the fact that the component of d' that maps $H_n\otimes Z_{k+1}$ to $D_{n+k}\otimes Z_0$ is the cocycle given in the statement of this theorem. □

Definition 2.7: Let n and k be integers ≥ 1, let $E=(f,g,\varphi):C_* \to D_*$ be a $\mathbb{Z}$-contraction of $\mathbb{Z}\pi$-chain complexes and let $\mu_1,\ldots,\mu_{k+1} \in \pi$. The *c-symbol of the elements* $\mu_1,\ldots,\mu_{k+1}$, *for the contraction* E, *and in dimension* n, is *denoted* $\mathfrak{C}_n(E;\mu_1,\ldots,\mu_{k+1})$ and is defined to be the element of $Hom_{\mathbb{Z}}(D_n,D_{n+k})$ given by:

$$\mathfrak{C}_n(E;\mu_1,\ldots,\mu_{k+1}) = (-1)^{n+k} f\cdot(\alpha_k \cdot \mu_1^{-1})\cdot \mu_1 \cdot \ldots \mu_{k+1}$$

where α_k is defined inductively by $\alpha_i = \varphi\cdot(\alpha_{i-1}\cdot\mu_{k+2-i}^{-1})$ and $\alpha_0 = g$.□

Remarks: 1. By abuse of notation we have used the *symbol* for μ_i to denote the *homomorphism* of C_* or D_* *induced* by μ_i.

2. Note that, due to the self- and mutual annihilating properties of f, g, and φ, the c-symbol will *vanish* if any of the μ_i is equal to the *identity element*.

3. The definition above will be extended to the case where the μ_i are arbitrary elements of the *group-ring* $\mathbb{Z}\pi$. This is done by simply defining it to be *Z-linear* in each argument, μ_i.

Definition 2.8: Let Z_* be a free resolution of $\mathbb{Z}$ over $\mathbb{Z}\pi$ and suppose preferred basis elements have been chosen in each dimension. If $a \in Z_t$ is a preferred basis element its *boundary tree with respect to* Z_* is defined to be a tree whose nodes are labelled with triples (k,λ,b) as follows:

1. k is an integer, called the *dimension* of the node;
2. $\lambda \in \mathbb{Z}\pi$ is called its *multiplier;*
3. b is a preferred basis element of Z_t, called the *base* of the node.

The boundary tree of a is constructed inductively as follows:

1. There is one node of dimension t labelled with $(t,0,a)$;
2. Given a node, n, of dimension i with label (i,λ,b), suppose the boundary of b is equal to $\sum\lambda_j c_j$, $\lambda_j \neq 0$, where the c_j are *preferred basis elements.* Then there is one *descendant* of n for each *term* of that linear combination and these descendant nodes are labelled $(i-1,\lambda_j,c_j)$, respectively. Each node of dimension $i<t$ is joined to a *unique* node of dimension $i+1$.
3. The process described above terminates in nodes of dimension 0. □

Remarks: This is simply a way of keeping track of all the terms that arise from taking boundaries of elements and then taking boundaries of the individual terms of the linear combinations that arise. This notational device will turn out to be indispensible in performing explicit computations.

Definition 2.9: 1. A *track* through a boundary tree is defined to be a path that starts at the node of highest dimension and proceeds to a node of dimension 0 without ever covering a given edge more than *once*.

2. A track will be called *essential* if the number $1 \in \mathbf{Z}\pi$ never occurs as a multiplier. Otherwise it will be called *inessential*.

3. A boundary tree that has no inessential tracks will be called *reduced*. □

Remarks: 1. Since a track isn't allowed to double back on itself, dimension is clearly a monotone decreasing function of distance along a track.

2. Given an arbitrary boundary tree it is clearly possible to *reduce* it, i.e. to find a subtree containing all of the essential tracks of the original tree and not containing any inessential tracks. Simply delete from the original tree any subtree whose root has a multiplier of 1.

3. Given a track, T, its *multiplier sequence* is defined to be the sequence of multipliers encountered in traversing the track from the root to the end. This sequence is assumed to begin at one dimension below the top dimension. If T is a track in the boundary tree of $a \in Z_t$, as in 2.8, then the multiplier sequence is denoted $(T_{t-1},\dots,T_0)$.

Our main result is the following:

Theorem 2.10: Let $\mathrm{E}=(f,g,\varphi):C_*\to D_*$ *be a* $\mathbf{Z}$*-contraction of* $\mathbf{Z}\pi$*-chain complexes such that:*

1. The lowest-dimensional nonvanishing homology module of C_* *is* H_n *in dimension* n*, and it is* $\mathbf{Z}$*-torsion free;*

2. The next nonvanishing homology module of C_* *is* H_{n+k} *in dimension* $n+k$;

3. $D_i=0,\ i<n,\ D_n=H_n,\ D_i=0,\ n<i<n+k.$

Let Z_* *be a free resolution of* $\mathbf{Z}$ *over* $\mathbf{Z}\pi$ *with preferred basis elements chosen in each dimension and with the property that the augmentation* $Z_0\to\mathbf{Z}$ *maps preferred basis elements to* 1. *Then the first homological k-invariant of* $C_*\otimes Z_*$ *is an element of* $H^{k+1}(\pi,\mathrm{Hom}_{\mathbf{Z}}(H_n,H_{n+k}))=\mathrm{Ext}_{\mathbf{Z}\pi}^{k+1}(H_n,H_{n+k})$

represented by a cocycle that maps a preferred basis element $b \in Z_{k+1}$ *to*

$$\sum_T \mathfrak{C}_n(E;T_0,\ldots,T_k)$$

where the sum is taken over all essential tracks in a boundary tree of b *with respect to* Z_*.

Remarks: 1. The condition involving the augmentation homomorphism of Z_* won't prove to be very restrictive.

2. Let Z_* be the *right bar resolution* of $\mathbf{Z}$ with the symbols $[\mu_1|\ldots|\mu_i]$ as preferred basis elements (where i is any positive integer and the μ's run over all elements of π). Then it isn't hard to see that a *reduced boundary tree* for $[\mu_1|\ldots|\mu_{k+1}]$ is:

$$(k+1,0,[\mu_1|\ldots|\mu_{k+1}]) \rightarrow (k,\mu_{k+1},[\mu_1|\ldots|\mu_k]) \rightarrow \ldots \rightarrow (0,\mu_1,[])$$

so that the first homological k-invariant is a cocycle whose value on $[\mu_1|\ldots|\mu_{k+1}]$ is $p\circ\mathfrak{C}_n(E;\mu_1,\ldots,\mu_{k+1}) \in \mathrm{Hom}_{\mathbf{Z}}(D_n,H_{n+k})$ where $D_n = H_n$ and $p:D_{n+k} \rightarrow D_{n+k}/\partial D_{n+k+1} = H_{n+k}$, is the projection.

Proof: First note that the defintions of the two quantities equated in the statement of the theorem *never* make use of the self-annihilating property of the boundary homomorphism d_Z. Thus, in principle, it is possible to define the terms in the theorem with d_Z an *arbitrary sequence of homomorphisms*, $d_i:Z_i \rightarrow Z_{i-1}$. We will, consequently, separate the proof of the theorem into two cases:

Case I: We assume that the boundary homomorphisms d_Z have the property that $d_Z(b) = \sum m_j b_j \mu_j$ if b_j is a preferred basis element, where $\mu_j \in \pi$, $m_j \in \mathbf{Z}$. (In case II the coefficients of the b_j will be arbitrary elements of $\mathbf{Z}\pi$).

Define:

1. B_i to be the subtree of the boundary tree of b spanned by all the nodes of dimension $\geq i$;

2. The *end* of a track, T, to be the base of its lowest-dimensional node, denoted e(T). See 2.8 for a definition of the base of a node in a boundary tree.

3. $A_i(T_k,\ldots,T_1)$ (where each $T_j = m_j v_j$ for some $m_j \in \mathbf{Z}$, $v_j \in \pi$) to be $\alpha_{k+1-i}\cdot T_i \cdot\ldots\cdot T_k$, where α_i is defined as in 2.7 using $\mu_j = v_{j-1}$ and T is some track in B_i;

4. V_i to be $(\hat{\varphi}_Z \cdot (1 \otimes d_Z))^{k+1-i} \cdot \tilde{g}_Z$, as in 2.6;

Remarks: 1. Note that $\tilde{g}_Z(x \otimes b)$, where $x \in D_n$, is equal to $g(x) \otimes b$ if b is a preferred basis element -- see 2.3.

2. We will *actually* give an inductive proof of the *following* statement:

Claim: Under the hypotheses of the theorem

$$\sum A_i(T_k,\ldots,T_i)(x) \otimes e(T) \bullet \hat{T}_i \ldots \hat{T}_k = V_i(x \otimes b)$$

(where b is a preferred basis element and the sum is over all tracks in B_i)

for all $x \in D_n$ *and all* $1 \le i < k+1$, *where* $\hat{T}_j$ *denotes the element* $v \in \pi$ *whenever* T_j *is of the form* mv *and* $m \in \mathbb{Z}$.

Remarks: 1. Since the c-symbol vanishes *identically* on tracks that *aren't essential* (and this doesn't depend upon d_Z being self-annihilating) the sum above can be regarded as a sum over *essential* tracks.

2. Proving the claim above proves the theorem in Case I because, when $i=1$ we simply take the boundary d_Z one more time and take $\tilde{f}_Z$ and apply the *augmentation*, which maps all preferred basis elements to 1 (by hypothesis).

Proof of claim: First we will verify the claim in the case where $i=k$. Suppose $d_Z(b)=\sum m_j b_j \mu_j$, where $m_j \in \mathbb{Z}$, b_j are preferred basis elements of Z_* and $\mu_j \in \pi$. Then $V_k(x \otimes b) = \hat{\varphi}_Z \cdot (1 \otimes d_Z)(g(x) \otimes b)$ (see remark 1 preceding the claim) $= \hat{\varphi}_Z(\sum m_j g(x) \otimes b_j \mu_j) = \sum m_j (\varphi(g(x)\mu_j^{-1})\mu_j \otimes b_j \mu_j)$ (see 2.3) $= \sum A_k(T_k) \otimes e(T)\hat{T}_k$ (where the sum is over all $T \in B_k$). The last equality is a consequence of the definition of a boundary tree (2.8) which implies that the ends of tracks in B_k are in a 1-1 correspondence with the b_j.

Now we will assume the inductive hypothesis and assume the claim is true in dimensions $\ge i$. Note that $V_{i-1}(x \otimes b) = \hat{\varphi}_Z \cdot (1 \otimes d_Z) V_i(x \otimes b)$. By hypothesis $V_i(x \otimes b) = \sum A_i(T_k,\ldots,T_i) \otimes e(T)\hat{T}_i \ldots \hat{T}_k$. The inductive definition of a boundary tree implies that $(1 \otimes d_Z)\sum A_i(T_k,\ldots,T_i) \otimes e(T)\hat{T}_i \ldots \hat{T}_k$ (summed over all $T \in B_i$) $= \sum A_i(T_k,\ldots,T_i) \otimes e(T) T_{i-1} \hat{T}_i \ldots \hat{T}_k$ (summed over all $T \in B_{i-1}$--see 2.8) and evaluating $\hat{\varphi}_Z$ on *this* gives

$$(1) \quad \sum \varphi(A_i(T_k,\ldots,T_i)\hat{T}_k^{-1} \ldots \hat{T}_{i-1}^{-1})\, \hat{T}_{i-1}\, \hat{T}_i \ldots \hat{T}_k \otimes e(T) T_{i-1}\, \hat{T}_i \ldots \hat{T}_k$$

(summed over all $T \in B_{i-1}$)

Let $T_{i-1} = m\hat{T}_{i-1}$. Then $\varphi(A_i(T_k,\ldots,T_i)\hat{T}_k^{-1}\ldots\hat{T}_{i-1}^{-1})\hat{T}_{i-1}\hat{T}_i\ldots\hat{T}_k = \varphi(\alpha_{k+1-i}T_i\ldots T_k\hat{T}_k^{-1}\ldots\hat{T}_{i-1}^{-1})\hat{T}_{i-1}\hat{T}_i\ldots\hat{T}_k = \varphi(\alpha_{k+1-i}\hat{T}_{i-1}^{-1})\hat{T}_{i-1}T_i\ldots T_k = \alpha_{k+2-i}\hat{T}_{i-1}T_i\ldots T_k = A_{i-1}(T_k,\ldots,T_{i-1})/m$. Substituting this into formula (1) gives

$$(1)\quad \sum A_{i-1}(T_k,\ldots,T_{i-1})\otimes e(T)\hat{T}_{i-1}\hat{T}_i\ldots\hat{T}_k$$

(summed over all $T\in B_{i-1}$)

which proves the induction step and, by the remarks following the claims, also proves the theorem in case I. Case II follows from case I by noting that each of the terms in the formula

$$(2)\quad \sum_T \mathfrak{C}_n(E;T_0,\ldots,T_k)(x) = (-1)^{n+k}p\cdot\tilde{f}_Z\cdot(1\otimes d_Z)\cdot(\hat{\varphi}_Z\cdot(1\otimes d_Z))^k\cdot\tilde{g}_Z(x\otimes b)$$

(summed over all $T\in B$)

is *linear* in the boundary maps in the following sense:

1. Suppose d, d', d" are sequences of homomorphisms $Z_i\to Z_{i-1}$ that are identical except that, in dimension j $d_j = d'_j + d''_j$. Then the value of the right-hand side of equation (2) calculated using $d_Z = d$ (in all dimensions) will be the *sum* of the values obatined using d' and d".

2. The same is true of the *left* hand side if we define the boundary trees of d, d', d" to have the same underlying tree structure (with the possibility of many of the multipliers begin 0).

Since, in each dimension, we can decompose the boundary homomorphism $(d_Z)_i$ into linear combinations of homomorphisms that satisfy the conditions of case I, it follows that the theorem is true in all cases. □

We will now give a second example of how to calculate and use boundary trees (recall that the first example used the bar resolution of $\mathbb{Z}$ and appeared immediately after the *statement* of the theorem). This second example will be more important than the first since it will be used extensively in the next section.

We will consider the case where Z_* is the *Gruenberg Resolution* of $\mathbb{Z}$ over $\mathbb{Z}\pi$. Suppose the group π has the presentation $\langle x_1,\ldots,x_s;\, r_1,\ldots,r_t\rangle$ (we are only assuming that the presentation is finite to simplify the discussion). Then theorem 10.9 on p.271 of [15] implies the existence of a free resolution of $\mathbb{Z}$ over $\mathbb{Z}\pi$ with chain modules generated by the symbols:

$(R_{i1}\ldots R_{ij})$, in dimension 2j, and $(R_{i1}\ldots R_{ij}X_{ij+1})$, in dimension 2j+1 where the R-symbols are in a 1-1 correspondence with a set of *free generators for the relation subgroup* -- this is the *normal closure* of the relations $\{r_i\}$ in the free group generated by the x_i. They may be obtained by computing a Reidemeister-Schreier system of generators. The X-symbols are in a 1-1 correspondence with the generators $\{x_i\}$.

In order to define the boundary of an element it is necessary to recall the notion of a *Fox Derivative* or *free derivative:*

These are symbols $\{\partial/\partial x_i\}$ that operate on the words in the $\{x_i\}$ via the following rules:

1. $\partial x_i/\partial x_j = \delta_{ij}$;
2. $\partial(w_1 w_2)/\partial x_i = \partial w_1/\partial x_i + w_1 \bullet \partial w_2/\partial x_i$

where w_1 and w_2 are arbitrary words in the x_i -- see [6,§2] as a general reference. The formula

$$w-1 = \sum \partial w/\partial x_i \bullet (x_i - 1)$$

was proved in [6, p.551]. We will need another version of this formula though. Let $^-$ be the anti-involution on the group ring of the free group that maps all group elements to their inverses. If w is a word in the free group then

$w^- - 1 = \sum \partial w^-/\partial x_i \bullet (x_i - 1)$, and taking $^-$ of both sides gives

$w - 1 = \sum (x_i^- - 1) \bullet (\partial w^-/\partial x_i)^-$, which implies

2.11: $w - 1 = \sum (x_i - 1) \bullet \bar{\partial}_i w$, *where we have written* $\bar{\partial}_i w = -x_i^{-1}(\partial w^-/\partial x_i)^-$.

The $\bar{\partial}_i$ are similar to the Fox derivatives -- they satisfy the following relations (which are sufficient to *define* them):

2.12: $\bar{\partial}_j x_i = \delta_{ij}$;

$\bar{\partial}_j(w_1 w_2) = (\bar{\partial}_j w_1) w_2 + \bar{\partial}_j w_2$.

Statement 2.11 above and the definition of the boundary maps for the Gruenberg resolution on p.271 of [15] imply that the boundary maps are given by:

1. $d(R_{i1} \dots R_{ij} X_{ij+1}) = R_{i1} \dots R_{ij}[x_{ij+1} - 1]_\pi$;

2. $d(R_{i1}...R_{ij}) = \sum R_{i1}...R_{ij-1}X_k[\bar{\partial}_k(r_{ij})]_\pi$;

where $[*]_\pi$ denotes the image in $\mathbb{Z}\pi$ under the homomorphism $\mathbb{Z}F_s \to \mathbb{Z}\pi$ defined by the presentation for π given above, where F_s is the free group on the symbols $\{x_i\}$. Theorem 2.10, coupled with the descriptions of the boundary maps in the Gruenberg Resolution immediately implies:

Corollary 2.13: Under the hypotheses of theorem 2.10, if π *has a presentation* $\langle x_1,...x_s;\ r_1,...,r_t\rangle$ *then the first homological k-invariant of* $C_*\otimes\mathbb{Z}_*$ *is an element of* $H^{k+1}(\pi,\mathrm{Hom}_{\mathbb{Z}}(H_n,H_{n+k}))$ *represented by a cochain on the Gruenberg Resolution of* $\mathbb{Z}$ *corresponding to the presentation above, as follows:*

1. *if* $k=2m$ *then the value of the cocycle on* $R_{i1}...R_{im}X_{im+1}$ *is* $p\cdot\sum \mathfrak{C}_n(E;[x_{im+1}]_\pi,[\bar{\partial}_{j1}(r_{im})]_\pi,..., [\bar{\partial}_{jm}(r_{i1})]_\pi, [x_{im+1}]_\pi)$;
2. *if* $k=2m-1$ *then the value of the cocycle on* $R_{i1}...R_{im}$ *is* $p\cdot\sum \mathfrak{C}_n(E;[\bar{\partial}_{j1}(r_{im})]_\pi,..., [\bar{\partial}_{jm}(r_{i1})]_\pi, [x_{jm}]_\pi)$;

where $p{:}D_{n+k} \to D_{n+k}/d(D_{n+k+1}) = H_{n+k}$ *is the projection.* □

Remarks: 1. In the summations above all of the j_i vary *independantly* so that they are *m-fold* summations.

2. Note that we have replaced $[x_{ji}-1]_\pi$ by $[x_{ji}]_\pi$. This is permissible because the c-symbols in question are multilinear and they vanish if any of their arguments is equal to 1.

We will conclude this section with an example. Suppose π is the group $\mathbb{Z}/2\mathbb{Z}\oplus\mathbb{Z}/2\mathbb{Z}$ presented by $\langle s,t;\ s^2,\ t^2,\ (ts)^2\rangle$. Then the Reidemeister-Schreier theory implies that the relation subgroup of the free group on s and t is generated by the following words:

$$r_1 = s^2,\ r_2 = t^2,\ r_3 = tsts,\ r_4 = tsts^{-1},\ r_5 = ts^2t^{-1},\ r_6 = sts^{-1}t^{-1}$$

Corollary 2.13 implies that the first homological k-invariant of $C_*\otimes\mathbb{Z}_*$ in the case where $k=2$ is a cochain whose value on R_1T is $p\cdot\mathfrak{C}_n(E;t,[\partial_s s^2]_\pi,s)$ (since $\partial_t s^2$ is clearly 0). The ∂-symbol, is easliy calculated using 2.12 -- for instance $\partial_s s^2 = 1 + s$, and we may drop the 1-term.

§3 The first homological k-invariant of an Equivariant Eilenberg-MacLane Space.

In this section we will use the results of the preceding section to compute the first homological k-invariant of an equivariant Eilenberg-MacLane space. This computation, coupled with the results of section 1 will imply the existence of a triple (M, n; π) where $M=\mathbf{Z}^3$ and $\pi=\mathbf{Z}/2\mathbf{Z}\oplus\mathbf{Z}/2\mathbf{Z}$, for which the corresponding equivariant Moore space doesn't exist.

The methods of this section will be generally applicable to any triple ($\mathbf{Z}^3$, n; π), where π is any group acting on $\mathbf{Z}^3$, although the final calculation at the end of the section will be performed with $\pi=\mathbf{Z}/2\mathbf{Z}\oplus\mathbf{Z}/2\mathbf{Z}$ using the presentation given at the end of section 2 with s and t acting via right multiplication by:

$$\begin{bmatrix} 0 & 1 & 1 \\ 1 & 0 & 1 \\ 0 & 0 & -1 \end{bmatrix} \quad \textit{and} \quad \begin{bmatrix} -1 & 0 & 0 \\ -1 & 0 & -1 \\ 1 & -1 & 0 \end{bmatrix} \textit{respectively.}$$

We begin by constructing a contraction of DGA-algebras $(a,b,G):A(\mathbf{Z}^3,2)\rightarrow P(x,y,z)$, where $P(x,y,z)$ is the *divided polynomial algebra* -- the $\mathbf{Z}$-subalgebra of $\mathbf{Q}[x,y,z]$ generated by the elements $(\gamma_i(x)=x^i/i!,\ \gamma_j(y)=y^j/j!,\ \gamma_k(z)=z^k/k!)$ for all values of i, j, and k, and $A(\mathbf{Z}^3,n)$ is the n-fold bar construction $\bar{B}^n(\mathbf{Z}[\mathbf{Z}^3])$ -- see [4, §14].

The DGA-algebra $P(x,y,z)$ is the chain-complex that will be used for D_* in the application of 2.11. The π-action on $P(x,y,z)$ can be regarded as being induced by that on $\mathbf{Q}[x,y,z]$ (where x, y, and z are regarded as generating the module $\mathbf{Z}^3$ and the action on the powers of these elements is defined so that the group π acts via *algebra homomorphisms*).

We begin with the following application of the Perturbation Lemma in the preceding section:

Corollary 3.1: Let $(f,g,\varphi):C\rightarrow D$ *be a contraction of DGA-algebras. Then* $(\bar{B}(f),\ \bar{B}(g),\tilde{\varphi}):\ \bar{B}(C)\rightarrow\ \bar{B}(D)$ *is a contraction of DGA-Hopf algebras, where* $\tilde{\varphi}$ =

$(1-\hat{\varphi}\cdot d_s)^{-1}\cdot\hat{\varphi}$ *and* $\hat{\varphi}$ *is defined by* $\hat{\varphi}([a|u]) = -[\varphi(a)|u] + (-1)^{\dim(a)}[g{\circ}f(a)|\hat{\varphi}(u)]$. □

Remarks: This is a straightforward consequence of the Perturbation Lemma, where $\hat{\varphi}$ is the homotopy for $\bar{B}(C)$ defined using only the *tensor boundary* and the *simplicial boundary* is regarded as a *perturbation* -- see [4].

The construction of the contraction (a,b,G) will involve several steps. We will initially construct a contraction from $A(\mathbb{Z}^3,1)$ onto $\Lambda(x,y,z)$ (the *exterior algebra*).

We begin with the contraction

(p,q,θ): $A(\mathbb{Z},1)\rightarrow\Lambda(x)$

where $\Lambda(x)$ is the exterior algebra over $\mathbb{Z}$ on one generator x. The maps are defined by:

1. $p([n_1|\ldots|n_k]) = 0$ is $k>1$;
   $p([n]) = nx$;

2. $q(x) = [1]$;

3. $\theta([n_1|\ldots|n_k]) = 0$ if $n_1 = 1$;
   $\theta([n_1|\ldots|n_k]) = \sum[1|j|n_2|\ldots|n_k]$ if $n_1>1$, where the summation has j going from 1 to $n_1 - 1$.
   $\theta([n_1|\ldots|n_k]) = -\sum[1|-j|n_2|\ldots|n_k]$ if $n_1 < 0$, where the summation has j going from 1 to $|n_1|$.
   (See [5, p.95])

We now take the bar construction of this and use 3.1 to get the contraction:

3.3: (p,q,Θ):$A(\mathbb{Z},2)\rightarrow P(x)=\bar{B}(\Lambda(x))$,

where, by abuse of notation we are denoting $\bar{B}(p)$ and $\bar{B}(q)$ by p and q, respectively (we won't be using the original definitions of p and q any longer). The chain-homotopy Θ is defined by $\Theta=(1-\Theta'\cdot d_s)^{-1}\cdot\Theta'$ (by 2.3) where Θ' is defined by $\Theta'[a|_2u]=-[\theta(a)|_2u] + (-1)^{\dim(a)}[q\cdot p(a)|_2\Theta'(u)]$.

Remark: The perturbation term $(1-\Theta'\cdot d_s)^{-1}$ will not be significant here because we will only apply Θ' to elements of dimension ≤ 3. In fact we can just assume that $\Theta([a]) =$

$-[\theta(a)]$ because the elements we will work with won't even have a $|_2$.

In the case of $\mathbb{Z}^3$ we have

$$(\Theta'', \hat{p}, \hat{q}): \otimes_1{}^3 A(\mathbb{Z}_i, 2) \to P(x,y,z)$$

(i runs from 1 to 3 in the tensor product), where $P(x,y,z) = P(x) \otimes P(y) \otimes P(z)$ and we have numbered the copies of $\mathbb{Z}$ for the sake of definiteness.

The maps are defined by: $\hat{p} = p_1 \otimes p_2 \otimes p_3$ and $\hat{q} = q_1 \otimes q_2 \otimes q_3$, where $(p_i, q_i, \Theta_i): A(\mathbb{Z}_i, 2) \to P(*)$, with $* = x$ if $i = 1$, y if $i = 2$, and z if $i = 3$, and the contractions are as defined in the statements following 3.3.

3.4: $\Theta'': \otimes_1{}^3 A(\mathbb{Z}_i, 2) \to \otimes_1{}^3 A(\mathbb{Z}_i, 2)$ is defined by

$$\Theta''(U \otimes V \otimes W) = \Theta_1(U) \otimes V \otimes W - (-1)^{\dim(U)} q_1 \cdot p_1(U) \otimes \Theta_2(V) \otimes W - (-1)^{\dim(U)+\dim(V)} q_1 \cdot p_1(U) \otimes q_2 \cdot p_2(V) \otimes \Theta_3(W).$$

Now we will develop a contraction

$$(\hat{f}, \hat{g}, \hat{\Psi}): \bar{B}(\otimes_1{}^3 A(\mathbb{Z}_i, 1)) \to \otimes_1{}^3 A(\mathbb{Z}_i, 2)$$

This will be done in two stages using the results of chapter I of [5]. First we will construct a contraction

$$(f_1, g_1, \Psi_1): \bar{B}(\otimes_1{}^3 A(\mathbb{Z}_i, 1)) \to A(\mathbb{Z}_1, 2) \otimes \bar{B}(A(\mathbb{Z}_2, 1) \otimes A(\mathbb{Z}_3, 1))$$

The maps involved will only be discussed in the dimension range of interest (i.e. dimensions ≤ 3):

$f_1([A \otimes B]) = 0$ unless A or B is 1;

$f_1([1 \otimes B]) = 1 \otimes [B]$, $f_1([A \otimes 1]) = [A] \otimes 1$;

$g_1(1 \otimes [B]) = [1 \otimes B]$, $g_1([A] \otimes 1) = [A \otimes 1]$;

$\Psi_1([A \otimes B]) = 0$ if either A or B are 1;

$\Psi_1([A \otimes B]) = [1 \otimes B | A \otimes 1]$, otherwise;

where $A \in A(\mathbb{Z}_1, 1)$, $B \in A(\mathbb{Z}_2, 1) \otimes A(\mathbb{Z}_3, 1)$.

Remarks: The statement about Ψ_1 follows directly from the formula given at the bottom of p. 53 of [5] for φ and the definitions of the face and degeneracy operators on the bar construction of [4]. Recall that the formula for φ in [5] is only sensitive to the *simplicial dimension* of an element of the bar construction -- and $A\otimes B$ has simplicial dimension 1 (*whatever* the dimensions of A and B might be).

Now we define

$$(f_2,g_2,\Psi_2): \bar{B}(\otimes_2{}^3A(\mathbf{Z}_i,1)) \to A(\mathbf{Z}_2,2)\otimes A(\mathbf{Z}_3,2)$$

in *exactly the same way* (let $A\in A(\mathbf{Z}_2,1)$, $B\in A(\mathbf{Z}_3,1)$ in the formula above). The two contractions are combined to give $(\hat{f},\hat{g},\hat{\Psi})$ where $\hat{f}=(1\otimes f_2)\cdot f_1$, $\hat{g}=g_1\cdot(1\otimes g_2)$, and $\hat{\Psi}=\Psi_1+g_1\cdot(1\otimes\Psi_2)\circ f_1$:

3.5: 1. $\hat{f}([A_1\otimes A_2\otimes A_3]) = 0$ *unless two out of the three terms are* 1 *in which case* $\hat{f}([\ldots\otimes A_i\otimes\ldots])=[A_i]$, i= 1, 2, or 3;

2. $\hat{g}([A_i]) = [A_i]$, i= 1, 2, *or* 3;

3. $\hat{\Psi}([A_1\otimes A_2\otimes A_3])= 0$ *if two out of the three terms are 1; otherwise* $\hat{\Psi}([A_1\otimes A_2\otimes A_3])= [1\otimes A_2\otimes A_3| A_1\otimes 1\otimes 1]$ *if all three terms are* $\neq 1$;

$\hat{\Psi}([1\otimes A_2\otimes A_3])= [1\otimes 1\otimes A_3| 1\otimes A_2\otimes 1]$, *where* $A_i\in A(\mathbf{Z}_i,1)$, i= 1, 2, *or* 3. □

In the last step we will define a contraction $(\hat{R},\hat{S},\hat{\Xi}):A(\mathbf{Z}^3,2)\to\bar{B}(\otimes_2{}^3A(\mathbf{Z}_i,1))$, and we will compose the three contractions to get (a,b,G). The contraction $(\hat{R},\hat{S},\hat{\Xi})$ will be defined by applying 3.1 to the contraction $(-,\hat{s},\hat{\xi})$: $A(\mathbf{Z}^3,1)\to\otimes_1{}^3A(\mathbf{Z}_i,1)$. As with the contractions above, *this* contraction will be built in two steps:

$$(r_1,s_1,\xi_1): A(\mathbf{Z}_1\oplus\mathbf{Z}_2\oplus\mathbf{Z}_3,1)\to A(\mathbf{Z}_1,1)\otimes A(\mathbf{Z}_2\oplus\mathbf{Z}_3,1)$$
$$(r_2,s_2,\xi_2): A(\mathbf{Z}_2\oplus\mathbf{Z}_3,1)\to A(\mathbf{Z}_2,1)\otimes A(\mathbf{Z}_3,1)$$

We will use triples (u,v,w) to denote elements of $\mathbf{Z}^3$ and, abusing the notation a little, triples with the first term equal to 0 will denote elements of $\mathbf{Z}^2$.

Definition 3.6: A $\leq$ 4-dimensional element of $A(\mathbf{Z}^3,2)$ that is a linear combination of basis elements, each of which contains at least *two adjacent* I_1-symbols, will be called *special.*

Remarks: For instance, $[(1,0,0)|_1(1,2,3)|_1(0,1,1)]$ is special and $[(1,0,0)|_2(1,2,3)]$ isn't. In fact it isn't hard to see that the only non-special 4-dimensional canonical basis elements of $A(\mathbb{Z}^3,2)$ are of the form $[u|_2v]$, with $u, v\in\mathbb{Z}^3$.

The following result will enable us to eliminate some terms in the final formula:

Proposition 3.7: ***The map,*** a, ***in the contraction*** $(a,b,G):A(\mathbb{Z}^3,2)\to P(x,y,z)$ ***maps all special elements to*** 0.

Proof: The map a is the composite of:

$$\hat{R}:A(\mathbb{Z}^3,2)\to\overline{B}(\otimes_1{}^3A(\mathbb{Z}_i,1))$$

$$\hat{f}:\overline{B}(\otimes_1{}^3A(\mathbb{Z}_i,1))\to\otimes_1{}^3A(\mathbb{Z}_i,2)$$

$$\hat{p}:\otimes_1{}^3A(\mathbb{Z}_i,2)\to P(x,y,z)$$

The composite $\hat{f}\cdot\hat{R}$ is already described in theorem 6.1 of [5]. Using the formula presented in the statement of that theorem we get:

$$\hat{f}\cdot\hat{R}([(u_1,v_1,w_1)|_1(u_2,v_2,w_2)]) = [(u_1,0,0)|_1(u_2,0,0)]\otimes1\otimes1 + 1\otimes[(0,v_1,0)|_1(0,v_2,0)]\otimes1 + 1\otimes1\otimes[(0,0,w_1)|_1(0,0,w_2)]$$

This is mapped to 0 by $\hat{p}$ since p_i maps all terms of $A(\mathbb{Z}_i,2)$ of the form $[x|_1y]$ to zero, with $x,y\in\mathbb{Z}_i$ (the point is that such elements are suspensions of elements of $A(\mathbb{Z}_i,1)$ of dimension >1). A similar argument is used in the higher dimensional cases. □

Remarks: Special elements may be ignored in the formula for a chain homotopy, G, since they will have at ***least one*** $|_1$-term in them even ***after*** the boundary is taken(in a bar construction).

Recall that triples (u,v,w) with u=0 denote elements of $\mathbb{Z}_2\oplus\mathbb{Z}_3$. The results of the first chapter of [5] imply that:

3.8: 1. $\dashv([(u,v,w)]) = [(u,0,0)]\otimes1\otimes1 + 1\otimes[(0,v,0)]\otimes1 + 1\otimes1\otimes[(0,0,w)]$;

$$\dashv([(u_1,v_1,w_1)|(u_2,v_2,w_2)]) = [(u_1,0,0)|(u_2,0,0)]\otimes1\otimes1 + [(u_1,0,0)]\otimes[(0,v_2,0)]\otimes1 + [(u_1,0,0)]\otimes1\otimes[(0,0,w_2)] + 1\otimes[(0,v_1,0)|(0,v_2,0)]\otimes1 + 1\otimes[(0,v_1,0)]\otimes[(0,0,w_2)] + 1\otimes1\otimes[(0,0,w_1)|(0,0,w_2)]$$

(see theorem 6.1 in [5]);

2. $\hat{s}([(u,0,0)]\otimes1\otimes1) = [(u,0,0)]$;

$\hat{s}(1\otimes[(0,v,0)]\otimes1) = [(0,v,0)]$;

$\hat{s}(1\otimes 1\otimes[(0,0,w)]) = [(0,0,w)]$;

$\hat{s}([(u,0,0)]\otimes[(0,v,0)]\otimes[(0,0,w)])=$

$\hat{s}([(u,0,0)]\otimes 1\otimes 1)*\hat{s}(1\otimes[(0,v,0)]\otimes 1)*\hat{s}(1\otimes 1\otimes[(0,0,w)])$, *where the * denotes the shuffel product in* $A(\mathbb{Z}^3,1)$ *(see [4, p.74]);*

3. $\hat{\xi} = \xi_1 + s_1\cdot(1\otimes\xi_2)\cdot r_1$. *Note that since* s_1 *involves shuffel products it will map special elements to special elements.* □

Since ξ_1 always increases the number of bars in a canonical basis element of $A(\mathbb{Z}^3,1)$ it follows that ξ_1 can be *disregarded* in dimension 2 (since this gives rise to $\hat{\Xi}$ in dimension 3 and that is going to be plugged into a, which annihilates special elements).

In dimension 1

$\xi_1([(u,v,w)])= [(0,v,w)|_1(u,0,0)]$

$\xi_2([(0,v,w)])= [(0,0,w)|_1(0,v,0)]$

and, using the expression $r_1([(u,v,w)])= [(u,0,0)]\otimes 1+1\otimes[(0,v,w)]$, we get $\hat{\xi}([(u,v,w)])=$ $[(0,v,w)|_1(u,0,0)] + [(0,0,w)|_1(0,v,0)]$ in dimension 1. All of this implies that $(\hat{R},\hat{S},\hat{\Xi})$ is defined by:

3.9: 1. $\hat{R}$ *is given by 3.8 (except that the terms in the right-hand side are enclosed in brackets);*

2. $\hat{S}$ *is as given in 3.8;*

3. $\hat{\Xi}$ *in dimension 2 maps* $[(u,v,w)]$ *to* $[(0,v,w)|_1(u,0,0)]$ - $[(0,0,w)|_1(0,v,0)]$;

4. *In dimension 3,* $\hat{\Xi}$ *is special.* □

Now we are in a position to combine 3.9, 3.5, and 3.3 to get a formula for (a,b,G), where $a=\hat{p}\cdot\hat{f}\cdot\hat{R}$, $b=\hat{S}\cdot\hat{g}\cdot\hat{q}$, and $G=\hat{\Xi}+\hat{S}\cdot\hat{\psi}\cdot\hat{R}+\hat{S}\cdot\hat{g}\cdot\Theta''\cdot\hat{f}\cdot\hat{R}$:

3.10: 1. $a([(u,v,w)]) = u\bullet x + v\bullet y + w\bullet z \in P(x,y,z)$;

$a([(u_1,v_1,w_1)|_1(u_2,v_2,w_2)]) = 0$;

$a([(u_1,v_1,w_1)|_1(u_2,v_2,w_2)|_1(u_3,v_3,w_3)]) = 0$;

$a([(u_1,v_1,w_1)|_2(u_2,v_2,w_2)]) = u_1u_2\bullet\gamma_2(x) + v_1v_2\bullet\gamma_2(y) + w_1w_2\bullet\gamma_2(z) +$

$u_1v_2 \bullet xy + u_1w_2 \bullet xz + v_1w_2 \bullet yz$;

2. $b(u \bullet x + v \bullet y + w \bullet z) = u \bullet [(1,0,0)] + v \bullet [(0,1,0)] + w \bullet [(0,0,1)]$;
3. $G([(u,v,w)]) = -[(0,v,w)|_1(u,0,0)] - [(0,0,w)|_1(0,v,0)] + \Theta_1([(u,0,0)]) + \Theta_2([(0,v,0)]) + \Theta_3([(0,0,w)])$;
4. $G([(u_1,v_1,w_1)|_1(u_2,v_2,w_2)]) = [(0,v_2,0)|_2(u_1,0,0)] + [(0,0,w_2)|_2(u_1,0,0)] + [(0,0,w_2)|_2(0,v_1,0)]$ + *special terms.* □

Remarks: Note that a is $\mathbb{Z}\pi$-linear in dimension 2 so that the differential on the resolution of $M \otimes Z_*$ is *untwisted*-- see remark 4 folowing 2.6. Since a is *not* $\mathbb{Z}\pi$-linear *in dimension* 4, the obstruction isn't *trivially* 0.

We will conclude this section by performing a concrete calculation in the case where $\pi = \mathbb{Z}/2\mathbb{Z} \oplus \mathbb{Z}/2\mathbb{Z}$ using the presentation $\langle s, t;\, s^2, t^2, (ts)^2 \rangle$ given at the end of section 2 with s and t identified with:

$$\begin{bmatrix} 0 & 1 & 1 \\ 1 & 0 & 1 \\ 0 & 0 & -1 \end{bmatrix} \quad \text{and} \quad \begin{bmatrix} -1 & 0 & 0 \\ -1 & 0 & -1 \\ 1 & -1 & 0 \end{bmatrix} \text{ respectively.}$$

The example at the end of section 2 implies that the first homological k-invariant of $A(\mathbb{Z}^3,3) \otimes Z_*$, where Z_* is the Gruenberg resolution of $\mathbb{Z}$ over $\mathbb{Z}\pi$ corresponding to the presentation given above, is a cocycle whose value on the preferred basis element R_1T (where $r_1 = s^2$) is $\mathfrak{C}_3(E;t,s,s) \in \mathrm{Hom}_{\mathbb{Z}}(\mathbb{Z}^3, \mathbb{Z}^3/2\mathbb{Z}^3)$, where $E = \overline{B}(a,b,G): A(\mathbb{Z}^3,3) \to \overline{B}(P(x,y,z))$. Since we will be in the stable range (i.e. the top dimension is 5, which is < 2×the bottom dimension of 3), we can assume $\overline{B}(G) = -G$ and we get:

3.11: $c(R_1T) = p^\circ(G(G(b(x) \bullet s) \bullet s) \bullet t) \bullet t$. □

This lends itself to a straightforward computation:

$$\begin{bmatrix} 1 & 0 & 0 \\ 0 & 1 & 0 \\ 0 & 0 & 1 \end{bmatrix} \xrightarrow{\bullet s} \begin{bmatrix} -1 & 0 & 0 \\ -1 & 0 & -1 \\ 1 & -1 & 0 \end{bmatrix} \xrightarrow{G}$$

$$\begin{bmatrix} \Theta_1([(-1,0,0)]) - [(1,0,0)|(-1,0,0)], \\ -[(0,0,-1)|(-1,0,0)] + \Theta_1([(-1,0,0)]) + \Theta_3([(0,0,-1)]), \\ -[(0,-1,0)|(1,0,0)] + \Theta_2([(0,-1,0)]) \end{bmatrix}$$

Remark: We have deleted all terms containing (0,0,0) and written the main terms in a form suggestive of matrix notation. The ith row of the formula is derived from the ith row of the identity matrix and will give rise to the ith row of the result (i=1,2,3). Continuing, we get:

$$\xrightarrow{-s} \begin{bmatrix} [(-1,0,0)(1,0,0)], \\ -[(-1,1,0)|(1,0,0)]+[(-1,0,0)|(1,0,0)]+[(1,-1,0)|(-1,1,0)], \\ -[(1,0,1)|(-1,0,0)]+[(-1,0,-1)|(1,0,1)] \end{bmatrix}$$

$$\xrightarrow{-G} \begin{bmatrix} 0, & & 0, \\ [(0,1,0)|_2(1,0,0)] & \xrightarrow{-t} & [(1,0,1)|_2(0,1,1)] \\ [(0,0,1)|_2(-1,0,0)] & & [(0,0,-1)|_2(0,-1,-1)] \end{bmatrix}$$

$$\xrightarrow{-a} \begin{bmatrix} 0 & 0 & 0 \\ 0 & 0 & 1 \\ 0 & 0 & 1 \end{bmatrix} \xrightarrow{-t} \begin{bmatrix} 0 & 0 & 0 \\ 0 & 0 & -1 \\ 0 & 0 & -1 \end{bmatrix} \xrightarrow{-p} \begin{bmatrix} 0 & 0 & 0 \\ 0 & 0 & 1 \\ 0 & 0 & 1 \end{bmatrix}$$

Remarks: 1. Note that the first application of G makes use of the formula in line 3 of 3.10 and the second makes use of the formula in line 4.

2. The map p is just reduction mod 2.

Our computations show that:

$$c(R_1,T)= \begin{bmatrix} 0 & 0 & 0 \\ 0 & 0 & 1 \\ 0 & 0 & 1 \end{bmatrix} \in \mathrm{Hom}_{\mathbb{Z}}(\mathbb{Z}^3, \mathbb{Z}^3/2\mathbb{Z}^3)$$

A straightforward calculation shows that this is *not* a coboundary, i.e.:

1. the boundary of R_1T is $R_1(t-1)$, so the value of any coboundary on R_1T is $(t-1)\bullet$some cochain on $R_1 \in Z_2$.

2. If a cochain takes the value

$$M = \begin{bmatrix} A & B & C \\ D & E & F \\ G & H & J \end{bmatrix}$$

on R_1 (here all the letters are 0 or 1 and we are working mod 2), then the coboundary is $t\bullet M\bullet t-M$ (recall that t acts upon the Hom-group by *conjugation* and $t^{-1}=t$). This is

$$\begin{vmatrix} A+E+H & A+B+D & A+B+D+E+F \\ B+D+H & A+E+G & A+B+C+G+H+J+F \\ G+H & G+H & G+H \end{vmatrix}$$

so that all entries on the third row must be the *same*.

References

1. H. Baues, *Obstruction Theory*, Springer-Verlag Lecture Notes in Mathematics 628 (1977).

2. G. Carlsson, "A counterexample to a conjecture of Steenrod," *Invent. Math.*, vol. 64 (1981), 171-174.

3. H. Cartan, "Algebras D'Eilenberg-MacLane et Homotopie," *Seminaire Henri Cartan 1954/55*, ENS, Paris.

4. S. Eilenberg and S. MacLane, "On the groups H(π,n). I," *Ann. of Math.*, vol.58 (1954), 55-106.

5. S. Eilenberg and S. MacLane, "On the groups H(π,n). II," *Ann. of Math.*, vol.60 (1954), 49-139.

6. R. Fox, "Free differential calculus," *Ann. of Math.*, vol.57 (1954), 547-560.

7. V.K.A.M. Gugenheim, "On the chain-complex of a fibration," *Illinois J. of Math.*, vol. 16 (1972), 398-414.

8. A. Heller, "Homological resolutions of complexes with operators," *Ann. of Math.*, vol.60 (1954), 283-303.

9. P. Hilton and S. Wylie, *Homology theory*, Cambridge University Press, 1965.

10. K. Igusa, "The generalized Grassman invariant $K_3(\mathbb{Z}[\pi]) \rightarrow H\bullet(\pi;\mathbb{Z}_2[\pi])$," to appear in *Springer-Verlag Lecture Notes in Mathematics.*

11. ______, "On the algebraic theory of A_∞-ring spaces," Springer-Verlag Lecture Notes in Mathematics, no. 967, 146-194.

12. P. Kahn, "Steenrod's problem and k-invariants of certain classifying spaces," Springer-Verlag Lecture Notes in Mathematics, no. 967, 195-214

13. J. Milgram, "The bar construction and abelian H-spaces," *Illinois J. of Math.*, vol. 11 (1967), 234-241.

14. F. Quinn, "Finite abelian group-actions on finite complexes," *Springer-Verlag Lecture Notes in Mathematics 658*.

15 J. Rotman, *An Introduction to Homological Algebra*, Academic Press, 1979.

16. Shih Weishu, "Homologie des Espaces Fibrès," *I. H. E. S Publ. Math.*, vol. 13 (1962), 93-176.

17. J. Smith, "Group cohomology and equivariant Moore spaces," *J. of Pure and Appl. Alg.*, vol. 24 (1982), 73-77.

18______, "Equivariant Moore spaces. II -- The low-dimensional case," *to appear in the J. of Pure and Appl. Alg.*

19. J. Arnold, "Homological algebra based upon permutation modules," *J. of Algebra*, vol. 70 (1981), 250-260.

Department of Mathematics and Computer Science
Drexel University
Philadelphia, PA 19104

Triviality of the Involution on SK_1 for Periodic Groups

by

Jonathan D. Sondow

Courant Institute of Mathematical Sciences
New York University
New York, N. Y. 10012

Let π be a periodic group or, more generally, a finite group with every p-Sylow subgroup either cyclic, or, for $p = 2$, a dihedral, quaternionic, or semidihedral group. In §1 of this note we show that the involution on $SK_1(\mathbb{Z}\pi)$ is trivial. The proof uses the results of Oliver (3), (4), where he calculates $SK_1(\mathbb{Z}\pi) \cong \mathbb{Z}_2^k$. In §2 we derive some consequences for the groups $H^n(\mathbb{Z}_2;Wh(\pi))$, which have applications in surgery theory (cf. (7, Prop. 4.1)) and the theory of semifree group actions (5, Prop. 3).

By Wall (9) the involution is also trivial on $Wh(\pi)/SK_1(\mathbb{Z}\pi)$. The question remains whether it is trivial on $Wh(\pi)$ for π periodic (see the Remark in §2).

For examples of finite groups with non-trivial involution on SK_1 see (11, proof of Theorem 4.8) and (10, Props. 24 and 25).

Involution on SK_1 for Periodic Groups

§1. The involution on $SK_1(\mathbb{Z}\pi)$.

For any group π the involution on the group ring $\mathbb{Z}\pi$ defined by $x \longmapsto x^{-1}$, $x \in \pi$, induces by conjugate transpose on $GL(\mathbb{Z}\pi)$ an involution on $Wh(\pi)$ and, for finite π, on $SK_1(\mathbb{Z}\pi)$ and $SK_1(\mathbb{Z}\pi)_{(p)}$.

Lemma. Fix a prime p and let π be a finite group such that the involution on $SK_1(\mathbb{Z}\pi')_{(p)}$ is trivial for every p-hyperelementary subgroup $\pi' \subset \pi$. then the involution is trivial on $SK_1(\mathbb{Z}\pi)_{(p)}$.

Proof. This follows immediately from the Dress induction isomorphism

$$\varinjlim_{\pi'} SK_1(\mathbb{Z}\pi')_{(p)} \xrightarrow{\cong} SK_1(\mathbb{Z}\pi)_{(p)}$$

of Oliver [4, p. 302], where the limit is taken with respect to inclusion and conjugation (which commute with the involution) among p-$\mathbb{Q}$-elementary (hence p-hyperelementary) subgroups $\pi' \subset \pi$.

Theorem 1. Let π be a finite group whose 2-Sylow subgroup is cyclic, dihedral, quaternionic, or semidihedral. Then the involution on $SK_1(\mathbb{Z}\pi)_{(2)}$ is trivial.

Proof. First reduction. It is not difficult to show that the hypothesis on π is satisfied by every subgroup of π. Hence by the Lemma it suffices to prove the result when π is 2-hyperelementary.

Second reduction. It is also not difficult to show that the 2-Sylow subgroup π_2 has a normal abelian subgroup with cyclic quotient. Hence by [4, Prop. 9 (ii)] we have $SK_1(\hat{\mathbb{Z}}_2\pi) = 0$ in the exact sequence

$$0 \to C\ell_1(\mathbb{Z}\pi)_{(2)} \to SK_1(\mathbb{Z}\pi)_{(2)} \to SK_1(\hat{\mathbb{Z}}_2\pi) \to 0,$$

where $C\ell_1(\mathbb{Z}\pi)$ denotes the kernel of the natural surjection (see [3, p. 184])

$$SK_1(\mathbb{Z}\pi) \to \sum_p SK_1(\hat{\mathbb{Z}}_p\pi).$$

Thus we need only show that the involution is trivial on $C\ell_1(\mathbb{Z}\pi)_{(2)}$.

Third reduction. $SK_1(\mathbb{Z}\pi) = 0$ if π_2 is cyclic by [3, Theorem 2], so finally we are reduced to proving that the involution is trivial on $C\ell_1(\mathbb{Z}\pi)_{(2)}$ if π is 2-hyperelementary and π_2 is dihedral, quaternionic, or semidihedral.

To begin the proof we have $\pi \cong \mathbb{Z}_n \rtimes \pi_2$ with n odd. Write $\mathbb{Z}_n = \{1, x, \ldots, x^{n-1}\}$ and $\mathrm{Aut}(\mathbb{Z}_n) = \{a \mid 1 \le a < n, (a,n) = 1\}$. The action of π_2 on $\mathbb{Z}_n$ is given by a homomorphism $t: \pi_2 \to \mathrm{Aut}(\mathbb{Z}_n)$ with $gxg^{-1} = x^{t(g)}$ for $g \in \pi_2$. Thus $\mathbb{Z}(\mathbb{Z}_n \rtimes \pi_2) = \mathbb{Z}(\mathbb{Z}_n)(\pi_2)^t$ is a twisted group ring with involution defined by

$$xg \mapsto \overline{xg} = x^{t(g^{-1})}g^{-1}.$$

Now fix $d|n$. Let $\zeta_d = e^{2\pi i/d}$ and let $\mathbb{Z}\zeta_d$ denote the ring of integers in the extension of $\mathbb{Q}$ by the dth roots of unity. Since n is odd and π_2 is a 2-group, we have $(d, t(g)) = 1$ for $g \in \pi_2$. Hence $\zeta_d \mapsto \zeta_d^{t(g)}$ defines an automorphism t_g of $\mathbb{Z}\zeta_d$, and $g \mapsto t_g$ defines an action of π_2 on $\mathbb{Z}\zeta_d$.

Let $\mathbb{Z}\zeta_d(\pi_2)^t$ denote the corresponding twisted group ring, with multiplication given by $\alpha g * \alpha_1 g_1 = \alpha t_g(\alpha_1) g g_1$ for $\alpha, \alpha_1 \in \mathbb{Z}\zeta_d$ and $g, g_1 \in \pi_2$. The involution on $\mathbb{Z}\zeta_d(\pi_2)^t$ is defined by

$$\alpha g \mapsto \overline{\alpha g} = t_{g^{-1}}(\bar{\alpha}) g^{-1} = \overline{t_{g^{-1}}(\alpha)}\, g^{-1},$$

where $\bar{\alpha}$ is the ordinary complex conjugate of $\alpha \in \mathbb{Z}\zeta_d \subset \mathbb{C}$.

Setting $pr_d(x^m g) = \zeta_d^m g$ for $x^m \in \mathbb{Z}_n$ defines the natural projection

$$pr_d : \mathbb{Z}(\mathbb{Z}_n \rtimes \pi_2) \to \mathbb{Z}\zeta_d(\pi_2)^t.$$

One checks easily that pr_d commutes with the involution. Hence the induced homomorphism

$$pr_{d*(2)} : C\ell_1(\mathbb{Z}(\mathbb{Z}_n \rtimes \pi_2))_{(2)} \to C\ell_1(\mathbb{Z}\zeta_d(\pi_2)^t)_{(2)}$$

commutes with the involution on these groups. Therefore

$$\sum_{d|n} pr_{d*(2)} : C\ell_1(\mathbb{Z}(\mathbb{Z}_n \rtimes \pi_2))_{(2)} \to \sum_{d|n} C\ell_1(\mathbb{Z}\zeta_d(\pi_2)^t)_{(2)}$$

commutes with the involution, which leaves each summand on the right invariant. But according to (4, Prop. 11) if π_2 is dihedral, quaternionic, or semidihedral, then $C\ell_1(\mathbb{Z}\zeta_d(\pi_2)^t)_{(2)}$ is either 0 or $\mathbb{Z}_2$, which have only the trivial involution. Hence the involution on the direct sum is trivial. Since $\sum_{d|n} pr_{d*(2)}$ is injective (indeed, bijective) by (4, p.328), it follows that the involution on $C\ell_1(\mathbb{Z}(\mathbb{Z}_n \rtimes \pi_2))_{(2)} \cong C\ell_1(\mathbb{Z}\pi)_{(2)}$ is trivial. This completes the proof of Theorem 1.

Remark. A. Bak has an unpublished proof that the involution is trivial on $C\ell_1(\mathbb{Z}\pi)$ for any finite group π, a general result he conjectured and proved a special case of in (0).

Theorem 2. *Let π be a finite group whose p-Sylow subgroups are cyclic or, for p = 2, dihedral, quaternionic, or semidihedral (e.g. π periodic). Then the involution on $SK_1(\mathbb{Z}\pi)$ is trivial.*

Proof. $SK_1(\mathbb{Z}\pi)_{(p)} = 0$ for $p > 2$ by (3, Theorem 2). So $SK_1(\mathbb{Z}\pi) = SK_1(\mathbb{Z}\pi)_{(2)}$ and the result follows by Theorem 1.

§2. $H^n(Wh(\pi))$

Wall has shown [9, p. 617] (see also [2, Corol. 6.10]) that for a finite group π the involution $\tau \longmapsto \bar{\tau}$ is trivial on $Wh'(\pi) \equiv Wh(\pi)/SK_1(\mathbb{Z}\pi)$. Hence setting $h(\tau) = \tau - \bar{\tau}$ for $\tau \in Wh(\pi)$ defines a homomorphism

$$h: Wh(\pi) \longrightarrow SK_1(\mathbb{Z}\pi).$$

From now on assume π satisfies the hypothesis of Theorem 2.

Remark. Although the involution will then be trivial on both $Wh(\pi)/SK_1(\mathbb{Z}\pi)$ and $SK_1(\mathbb{Z}\pi)$, it does not follow that it is trivial on $Wh(\pi)$. (For example, it might take (1,0) to (1,1) in $\mathbb{Z} \times \mathbb{Z}_2$.) Question: Is it, if π is periodic?

By Theorem 2 we have $h\, SK_1(\mathbb{Z}\pi) = 0$, so there is a unique homomorphism $h': Wh'(\pi) \to SK_1(\mathbb{Z}\pi)$ factoring $h = h' \circ \nu$, where $\nu: Wh(\pi) \to Wh'(\pi)$ is the natural projection.

It follows from Oliver [3, Theorem 2] and [4, Theorem 6] that $SK_1(\mathbb{Z}\pi) \cong \mathbb{Z}_2^k$, where k is the number of conjugacy classes of odd cyclic subgroups $C \subset \pi$ such that (i) the 2-Sylow subgroup of the centralizer of C is nonabelian and (ii) there is no $g \in \pi$ with $gxg^{-1} = x^{-1}$ for all $x \in C$. Combining this with Wall's result [9, Theorems 1.2 and 6.1] (see also [8, §§6 and 7]) that $SK_1(\mathbb{Z}\pi) = tor(Wh(\pi))$, and Bass' theorem [1] on the rank of $Wh(\pi)$, yields $Wh(\pi) \cong \mathbb{Z}_2^k \times \mathbb{Z}^{r-q}$, where (using [6, Chap. 13]) q is the number of conjugacy classes of cyclic subgroups of π, and r is the number of conjugacy classes of unordered pairs $\{x, x^{-1}\}$, $x \in \pi$.

Let m be the $\mathbb{Z}_2$ rank of im(h) = image of h in $SK_1(\mathbb{Z}\pi) \cong \mathbb{Z}_2^k$. Note $m \leq \min(k, r-q)$ since $im(h) = im(h')$ and we may interpret $h': \mathbb{Z}^{r-q} \to \mathbb{Z}_2^k$.

For an abelian group G with involution $g \longmapsto \bar{g}$ set

$$H^n(G) = \frac{\{g \in G \mid g = (-1)^n \bar{g}\}}{\{g + (-1)^n \bar{g} \mid g \in G\}} = H^n(\mathbb{Z}_2; G).$$

These are elementary 2-groups.

Theorem 3. *If π is a finite group whose p-Sylow subgroups are cyclic or, for p = 2, dihedral, quaternionic, or semidihedral, then*

(i) $H^n(SK_1(\mathbb{Z}\pi)) \cong \mathbb{Z}_2^k$ <u>for all</u> n

(ii) $$H^n(Wh(\pi)) \cong \begin{cases} \mathbb{Z}_2^{k-m} & \underline{\text{for}}\ n\ \underline{\text{odd}} \\ \mathbb{Z}_2^{k-m+r-q} & \underline{\text{for}}\ n\ \underline{\text{even}}. \end{cases}$$

<u>Proof</u>. (i) This is immediate since $\sigma = \bar{\sigma} = -\bar{\sigma}$ for $\sigma \in SK_1(\mathbb{Z}\pi) \cong \mathbb{Z}_2^k$.

(ii) <u>n odd</u>. Recall that $\tau = \bar{\tau} + h(\tau)$ with $h(\tau) \in SK_1(\mathbb{Z}\pi)$ for $\tau \in Wh(\pi)$. Hence $\tau = -\bar{\tau}$ implies $2\tau = h(\tau) \in SK_1(\mathbb{Z}\pi) = tor(wh(\pi))$, whence $\tau \in SK_1(\mathbb{Z}\pi)$. Conversely, $\tau \in SK_1(\mathbb{Z}\pi)$ implies $\tau = -\bar{\tau}$ as above, so $H^n(Wh(\pi)) = \{\tau = -\bar{\tau}\} / \{\tau - \bar{\tau}\} = SK_1(\mathbb{Z}\pi)/ im(h) \cong \mathbb{Z}_2^k/\mathbb{Z}_2^m$.

(ii) <u>n even</u>. Choose a basis $\sigma_1,\dots,\sigma_m,\dots,\sigma_k$ for $SK_1(\mathbb{Z}\pi)$ as a vector space over $\mathbb{Z}_2$ such that $\sigma_1,\dots\sigma_m$ is a basis for the subspace $im(h')$. Then by induction on $r-q$ find a basis $\tau'_1,\dots,\tau'_m,\dots,\tau'_{r-q}$ for $Wh'(\pi)$ as a free $\mathbb{Z}$ module such that $h'(\tau'_i) = \sigma_i$ for $i = 1,\dots,m$ and $h'(\tau'_{m+1}) = \dots = h'(\tau_{r-q}) = 0$. Finally, pick $\tau_i \in \nu^{-1}(\tau'_i) \subset Wh(\pi)$, so that $h(\tau_i) = h'(\tau'_i)$, $i = 1,\dots,r-q$. Let $<x_1,\dots,x_s>$ denote the subgroup generated by $x_1,\dots,x_s$. Then, since $\tau = \bar{\tau} + h(\tau)$ for $\tau \in Wh(\pi)$ and $h(\tau) = 0$ if $\tau \in SK_1(\mathbb{Z}\pi)$, we get

$$H^n(wh(\pi)) = \frac{\{\tau = \bar{\tau}\}}{\{\tau + \bar{\tau}\}}$$

$$= \frac{<\sigma_1,\dots,\sigma_k, 2\ _1,\dots 2\ _m,\ _{m+1},\dots,\ _{r-q}}{<2\tau_1+\sigma_1,\dots,2\tau_m+\sigma_m,2\tau_{m+1},\dots,2\tau_{r-q}>}$$

$$\cong \frac{<\sigma_1,\dots,\sigma_k,2\tau_1+\sigma_1,\dots,2\tau_m+\sigma_m>}{<2\tau_1+\sigma_1,\dots,2\tau_m+\sigma_m>} \times \frac{<\tau_{m+1},\dots,\tau_{r-q}>}{2<\tau_{m+1},\dots,\tau_{r-q}>}$$

$$\cong \mathbb{Z}_2^k \times \mathbb{Z}_2^{r-q-m}.$$

From Theorem 3(ii) we derive some consequences which do not depend on the unknown m, but only on the reasily computed r, q, and k.

<u>Corollary</u>. <u>If</u> π <u>is a periodic group, then</u>

(i) the Herbrand quotient $\frac{|H^0(Wh(\pi))|}{|H^1(Wh(\pi))|} = 2^{r-q}$

(ii) for n odd, $H^n(Wh(\pi)) = 0$ only if $k \leq r-q$

(iii) for n even, $H^n(Wh(\pi)) = 0$ if and only if $Wh(\pi) = 0$

Proof: (i) follows trivially, (ii) follows from $m \leq r-q$, and (iii) because $m \leq k$ also, so $k-m+r-q = 0 \iff k = r-q = 0 \iff Wh(\pi) = 0$.

References

0. A. Bak, The involution on Whitehead torsion, General Top. and Appl. 7 (1977), 201-206.

1. H. Bass, The Dirichlet Unit Theorem, induced characters, and Whitehead groups of finite groups, Topology 4, (1966), 391-410.

2. J. Milnor, Whitehead torsion, Bull. AMS 72 (1966), 358-426

3. R. Oliver, SK_1 for finite group rings: I, Invent. Math. 57 (1980), 183-204.

4. ________, SK_1 for finite group rings: III, Proc. Conf. on Algebraic K-Theory (Evanston, 1980), Lecture Notes in Math., vol.854, Springer-Verlag (1981), 299-337.

5. M. Rothenberg and J. Sondow, Nonlinear smooth representations of compact Lie groups, Pacific J. Math. 84 (1979), 427-444.

6. J.-P. Serre, Linear representations of finite groups, Springer, New York (1977).

7. J. Shaneson, Wall's obstruction groups for G x Z, Ann. of Math. 90 (1969), 296-334.

8. M. Stein, Whitehead groups of finite groups, Bull. AMS 84 (1978), 201-212.

9. C.T.C. Wall, Norms of units in group rings, Proc. London Math. Soc. (3) 29 (1974), 593-632.

10. R. Oliver, SK_1 for finite group rings: II, Math. Scand. 47 (1980), 195-231.

11. ________, SK_1 for finite group rings: IV, Proc. London Math. Soc. (3) 46 (1983), 1-37.

THE INVOLUTION IN THE ALGEBRAIC K-THEORY OF SPACES

Wolrad Vogell

The primary purpose of this paper is to study the canonical involution on the algebraic K-theory of spaces functor A(X). From a technical point of view the main result is that several ways of defining such an involution lead in fact to the same result.

The secondary purpose is to establish that the involution on A(X) relates nicely to involutions on related functors, specifically the algebraic K-theory of rings, and concordance spaces. Technically this follows simply by comparing the latter involutions to suitable models of the involution on A(X). In particular there results the expected fact that the involution on concordance spaces corresponds to the involution on the algebraic K-theory of rings.

This is certainly a desirable result, and indeed several applications of it have already been published, cf. [2], [3], [5]. For example, it is possible to obtain some numerical information on the homotopy type of the diffeomorphism groups of some manifolds.

Aside from this technical result, the study of the involution on A(X) also has some interest of its own. We obtain another proof of the theorem that stable homotopy splits off the algebraic K-theory of spaces.

Here is a summary of the contents of the paper.

In § 1 a concept of equivariant Spanier-Whitehead duality is discussed. Here the word 'equivariant' refers to the homotopy theory of spaces over a fixed space.

Using this concept of duality a model for A(X) is developed which lends itself to a natural definition of an involution. Namely, on the level of the categories of spaces used to define A(X) the involution corresponds to the transition from a space over X to its equivariant Spanier-Whitehead dual.

It will be convenient later on to have a different description of duality available. Namely, instead of considering spaces over a fixed space one can equivalently use simplicial sets with an action of a simplicial group. The translation into this framework is given in the second part of § 1.

The concept of Spanier-Whitehead duality that we need is a version of Ranicki duality for simplicial groups. Its relation to the usual concept of equivariant Spanier-Whitehead duality is briefly discussed in the appendix to § 1, cf. also [20].

In § 2 it is described how an involution on A(X) may be defined in various ways. It is shown that these definitions lead to the same involution up to homotopy (cf. cor. 2.10., and the remark after prop. 2.5.).

Using the 'manifold model' [18], the relation of A(X) with the concordance space functor is described, and it is shown that the involutions on both functors correspond to each other.

Using the model of A(X) developed in § 1, the involution is compared with that on the algebraic K-theory of rings (cf. prop. 2.11.).

In § 3 the description of A(X) obtained in § 1 is adapted to give another proof of the splitting theorem: The canonical map $\Omega^\infty S^\infty(X_+) \longrightarrow A(X)$ from the stable homotopy of X to the algebraic K-theory splits up to homotopy. In fact, a very direct description of a splitting map $A(X) \longrightarrow \Omega^\infty S^\infty(X_+)$ is given.

A summary of the contents of paragraphs 1 and 2 has been published in [15].

I wish to thank F. Waldhausen and M. Bökstedt for numerous helpful discussions.

§ 1. Duality in equivariant homotopy theory

In this paragraph we discuss a concept of Spanier-Whitehead duality appropriate in equivariant homotopy theory, meaning the homotopy theory of spaces (= simplicial sets) parametrized by a simplicial set X. The motivation for this is that A(X), the *algebraic K-theory* of X may be defined in terms of certain categories of simplicial sets over X, cf. [16]. Actually we consider two equivalent formulations of this duality, one involving simplicial sets over X, the other one employing simplicial sets with an action of the loop group of X. Both versions are used in § 2.

We introduce some language and notations. If X is a connected simplicial set let $R(X)$ denote the category of *retractive simplicial sets over* X, i.e. an object is a triple (Y,r,s) consisting of a simplicial set Y, a retraction $r:Y \to X$, and a section s of r. A morphism from (Y,r,s) to (Y',r',s') is a map $f:Y \to Y'$ such that $r'f = r$ and $fs = s'$. An h-*equivalence* is a morphism in $R(X)$ which is a weak homotopy equivalence. Let $hR(X)$ denote the category of h-equivalences; $R_f(X)$ is the subcategory of $R(X)$ of those objects (Y,r,s) satisfying that Y-s(X) contains only finitely many non-degenerate simplices. The category $R_{hf}(X)$ is the category of *homotopy finite* objects, i.e. it is the full subcategory of $R(X)$ of those objects which can be related to an object of $R_f(X)$ by a finite chain of h-equivalences. Let $hR_{hf}(X)$ be the intersection of $hR(X)$ and $R_{hf}(X)$, and similarly with $hR_f(X)$. We will be interested in certain subcategories of $hR(X)$: let $hR^\ell_k(X)$ denote the connected component of $hR(X)$ containing the objects

$$X \cup_{\partial D^\ell} \underbrace{\cup \ldots \cup}_{k} {}_{\partial D^\ell} D^\ell \underbrace{\cup \ldots \cup}_{k} D^\ell \longrightarrow X$$

Indeed, all such objects are in the same connected component of $hR(X)$, regardless of the attaching maps.

There is an external pairing

$$_X\wedge_{X'} : R(X) \times R(X') \longrightarrow R(X\times X')$$

$$(Y , Y') \longmapsto Y\times Y' \cup_{X\times Y' \cup Y\times X'} X\times X' .$$

This pairing is natural in X and X', and associative up to canonical isomorphism. The properties of this pairing may be conveniently summarized by saying that it defines a *bi-exact functor* in the sense of [19]. It also preserves finiteness (resp. finiteness up to homotopy), where these terms are defined with regard to the categories $R_f(X)$ (resp. $R_{hf}(X)$).

Notation: For typographical reasons we shall simply write $Y \wedge Y'$ instead of $Y_X\wedge_{X'}Y'$ if there is no risk of confusion. A special case of this pairing is given by the *fibrewise suspension over* X, defined as

$$\Sigma^n_X(Y) = S^n {}_*\wedge_X Y ,$$

where S^n denotes a pointed simplicial set representing the n-sphere.

We can now give the fundamental

Definition:
$$A(X) = \mathbb{Z} \times |\lim_{\substack{\to \\ k,\ell}} hR^\ell_k(X)|^+ .$$

The maps in the direct system are given by Σ_X in the ℓ-variable, and by wedge with an ℓ-sphere in the k-variable; "+" denotes Quillen's construction to abelianize the fundamental group.

To define the concept of Spanier-Whitehead duality in the context of retractive spaces, we fix a d-spherical fibration ξ over X, (i.e. $\text{fibre}(|\xi| \to |X|) \simeq S^d$) with a given section. Let $Th(\xi)$ denote an object of $R(X\times X)$ satisfying

(i) $Th(\xi) \rightleftarrows X\times X$ is in the same component of $hR(X\times X)$ as the object

$$X \times X \cup_X \xi \rightleftarrows X \times X$$

where the maps in the push-out are given by the diagonal map, and the section of ξ;

(ii) $Th(\xi) \to X\times X$ is a (Kan) fibration;

(iii) There is a map $\iota:Th(\xi) \to Th(\xi)$ covering the flip map $X\times X \to X\times X$ and such that ι^2 = identity.

Such a space will be called a *Thom space* of ξ. Note that $Th(\xi)/X^2$ is essentially the Thom space of ξ in the usual sense.

We will also need the suspensions of $Th(\xi)$. Since these are not automatically Kan fibrations again, we use the following modification. There is a functorial way of turning a map $A \to B$ of simplicial sets into a Kan fibration. This can be done by using a relative version of Kan's functor Ex^∞, cf. [6]. We continue to denote this functor Ex^∞. Define

$$Th_n(\xi) = Ex^\infty(\Sigma^n_{X^2}(Th(\xi))), \quad \iota_n = Ex^\infty(\Sigma^n_{X^2}(\iota)).$$

Of course, there always exists a Thom space in this sense: e.g. choose $Th(\xi) = Ex^{\infty}(X^2 \cup_X \xi)$. It will however be convenient later on not to be restricted to one choice of a Thom space.

Spanier-Whitehead duality is defined with respect to a chosen space $Th(\xi)$. Let (Y,r,s) (resp. Y',r',s') be an object of $R(X)$. An n-*duality map* is a map in $R(X\times X)$

$$u:\ Y \wedge Y' \longrightarrow Th_{n-d}(\xi)$$

satisfying that the induced map

$$\begin{array}{rcl}\alpha_u : H_q(Y',X;\mathbb{Z}[\pi_1 X]) & \longrightarrow & H^{n-q}(Y,X;\mathbb{Z}[\pi_1 X]) \\ z & \longmapsto & u^*(t)/z\end{array}$$

is an isomorphism for all q. Here $t \in H^n(Th_{n-d}(\xi),X^2;\mathbb{Z}[\pi_1 X]')$ is a class mapping to a generator of $H^n(p^*Th_{n-d}(\xi),X\times\tilde{X};\mathbb{Z}[\pi_1 X]) \approx \mathbb{Z}[\pi_1 X]$ under the canonical map, where '$\sim$' denotes the universal covering, $p: X\times\tilde{X} \longrightarrow X\times X$ is the canonical projection, and $\mathbb{Z}[\pi_1 X]'$ denotes the right $\mathbb{Z}[\pi_1 X^2]$-module $\mathbb{Z}[\pi_1 X]$ with $\pi_1 X^2$-action given by $x.(g,h) = h^{-1}xg$, $(g,h) \in (\pi_1 X)^2$, $x \in \mathbb{Z}[\pi_1 X]$.
Similarly we require that $\alpha_{u'}$ gives an isomorphism, where u' is the composition of u with the flip map $Y\wedge Y' \to Y'\wedge Y$.
It turns out that this definition gives the correct notion of duality in the context of manifolds and for the purpose of K-theory.

Before stating some elementary facts about duality, we have to mention a technical point. Since in our definition of Thom spaces we insisted on the Kan condition we cannot just identify $Th_{n+1}(\xi)$ with the suspension of $Th_n(\xi)$. These spaces are of course homotopy equivalent, but it is desirable to have a specific homotopy equivalences such that the following diagrams commute.

$$\begin{array}{ccc} S^1{}_*\wedge_*(S^n{}_*\wedge_{X^2}Th(\xi)) & \xrightarrow{\approx} & S^{n+1}{}_*\wedge_{X^2}Th(\xi) \\ \downarrow & & \downarrow \\ S^1{}_*\wedge_{X^2}Th_n(\xi) & \xrightarrow{f_n} & Th_{n+1}(\xi) \end{array}$$

$$\begin{array}{ccccc} S^1{}_*\wedge_*(S^n{}_*\wedge_{X^2}Th(\xi)) & \xrightarrow{\approx} & (S^n{}_*\wedge_*S^1)_*\wedge_{X^2}Th(\xi) & \xrightarrow{\approx} & S^{n+1}{}_*\wedge_{X^2}Th(\xi) \\ \downarrow & & & & \downarrow \\ S^1{}_*\wedge_{X^2}Th_n(\xi) & & \xrightarrow{g_n} & & Th_{n+1}(\xi)\ . \end{array}$$

The vertical arrows are the natural inclusions induced from the map $Y \to Ex^{\infty}(Y)$.

The reason for considering the maps g_n which permute the suspension coordinates is to ensure compatibility with certain constructions to be performed later on. The existence of the maps f_n (resp. g_n) follows from the Kan condition on $Th_{n+1}(\xi)$.

Lemma 1.1. Let $u\colon Y \wedge Y' \to Th_{n-d}(\xi)$ be an n-duality map. Then

(i) $u'\colon Y' \wedge Y \to Th_{n-d}(\xi)$ is also an n-duality map, where

$u'(y' \wedge y) = \iota_{n-d}(u(y \wedge y'))$,

(ii) $\Sigma_\ell u\colon \Sigma_X(Y) \wedge Y' = (S^1{}_* \wedge_X Y) \wedge Y' \to S^1{}_* \wedge_{X^2} Th_{n-d}(\xi) \xrightarrow{f_n}$ $Th_{n-d+1}(\xi)$ and

$\Sigma_r u\colon Y \wedge \Sigma_X(Y') \to S^1{}_* \wedge_{X^2} Th_{n-d}(\xi) \xrightarrow{g_n} Th_{n-d+1}(\xi)$

are (n+1)-duality maps. □

We now want to investigate the dependence of duality on the spherical fibration ξ. Let $R_{fib}(X)$ denote the subcategory of those objects of $R(X)$ satisfying that the structural retraction is a fibration. There is an operation

$$\cdot\colon R_{fib}(X) \times R(X) \longrightarrow R(X)$$

$$(\xi\,,\,Y) \longmapsto \xi \cdot Y := \xi \times_X Y \cup_{\xi \underset{X}{\cup} Y} X\,.$$

We list some of its properties:

(i) If $Y \in R_{fib}(X)$, then $\xi \cdot Y$ is (up to a dimension shift) the fibrewise join over X of ξ and Y.

(ii) If $Y \in R_f(X)$, and fibre($\xi \to X$) is finite, then $\xi \cdot Y \in R_f(X)$.

(iii) $\xi \cdot -\colon hR(X) \longrightarrow hR(X)$ is an exact functor in the sense of [16].

(iv) The operation is compatible with the external pairing:

$$(\xi \cdot Y) \wedge (\eta \cdot Y') \approx (\xi \wedge \eta) \cdot (Y \wedge Y'), \text{ where}$$

$\xi, \eta \in R_{fib}(X)$, $Y, Y' \in R(X)$.

(v) If $\xi = X \times S^r$, then $\xi \cdot Y = \Sigma^r_X(Y)$, the r-fold fibrewise suspension over X.

There is a kind of Thom isomorphism in this setting.
Let ξ be a d-spherical fibration ($d \geq 2$) over X as before, Y an object of $R_{hf}(X)$ satisfying that $\pi_i X \xrightarrow{\approx} \pi_i Y$, $i = 0,1$. Let $t \in H^d(\xi, X)$ be a Thom class of ξ.

Lemma 1.2. There are isomorphisms for all $q \geq 0$

$$\varphi_t\colon H^q(Y,X;\mathbf{Z}[\pi_1 X]) \longrightarrow H^{q+d}(\xi\cdot Y,Y;\mathbf{Z}[\pi_1 X]),\ \varphi_t(a) = t \cup a$$

$$\psi_t\colon H_{q+d}(\xi\cdot Y,X;\mathbf{Z}[\pi_1 X]) \longrightarrow H_q(Y,X;\mathbf{Z}[\pi_1 X]),\ \psi_t(b) = t \cap b\,.$$

Proof: Without loss of generality assume that $Y \in R_f(X)$. Then one can find a filtration $Y_0 = X \subset Y_1 \subset \ldots \subset Y_k = Y$, such that $Y_i/Y_{i-1} \simeq X \vee V S^i$. Now by property (iii) above, $\xi \cdot -$ is an exact functor. So by a five-lemma argument one immediately reduces to the case that $Y = X \vee S^0 = X_+$. But in this case the assertion of the lemma is obvious. □

If $Th(\xi)$ is a Thom space of ξ, and η is another spherical fibration (with a section), then by properties (iii) and (iv) above, $(\eta \wedge \varepsilon) \cdot Th(\xi)$ is a Thom space of $\eta \cdot \xi$, $(\varepsilon = X \times S^0)$; in shorthand notation $(\eta \wedge \varepsilon) \cdot Th(\xi) = Th(\eta \cdot \xi)$.

Corollary 1.3. Let $u: Y \wedge Y' \to Th_{n-d}(\xi)$ be an n-duality map with respect to ξ, and let η be a d'-spherical fibration (orientable and with a section). Then the map

$$(\eta \wedge \varepsilon) \cdot u: \eta \cdot Y \wedge Y' \to (\eta \wedge \varepsilon) \cdot Th_{n-d}(\xi) = Th_{n-d-d'}(\eta \cdot \xi)$$

is an (n+d')-duality map with respect to $\eta \cdot \xi$.

Proof: We have a commutative diagram

$$\begin{array}{ccc} H_q(\eta \cdot Y, X; \mathbf{Z}[\pi_1 X]) & \longrightarrow & H^{n-q+d'}(Y', X; \mathbf{Z}[\pi_1 X]) \\ \downarrow \approx & & \downarrow = \\ H_{q-d'}(Y, X; \mathbf{Z}[\pi_1 X]) & \longrightarrow & H^{n-q+d'}(Y', X; \mathbf{Z}[\pi_1 X]) \end{array}$$

where the vertical map on the left is the isomorphism of lemma 1.2., and the horizontal maps are given by slant product with a Thom class of ξ (resp. $\eta \cdot \xi$). By assumption the lower of these maps is an isomorphism, hence so is the upper one. Interchanging the roles of homology and cohomology gives another commutative diagram for which the same argument applies. □

Let ξ be as before. Define a category $\mathcal{DR}^n_{Th(\xi)}(X)$, in which an object is given by a triple (Y, Y', u), where Y (resp. Y') is an object of $R_{hf}(X)$, subject to the technical condition that the inclusion of X in Y (resp. Y') induces an isomorphism on π_0 and π_1, and $u: Y \wedge Y' \to Th_{n-d}(\xi)$ is an n-duality map.

A morphism $(Y, Y', u) \to (Z, Z', v)$ is a pair of morphisms in $R_{hf}(X)$

$$f: Y \to Z, \quad f': Z' \to Y'$$

such that the diagram

$$\begin{array}{ccc} Y \wedge Z' & \xrightarrow{f \wedge id} & Z \wedge Z' \\ {\scriptstyle id \wedge f'} \downarrow & & \downarrow {\scriptstyle v} \\ Y \wedge Y' & \xrightarrow[u]{} & Th_{n-d}(\xi) \end{array}$$

commutes.

A morphism (f,f') in $\mathcal{DR}^n_{Th(\xi)}(X)$ is called an h-equivalence if f and f' are h-equivalences. The subcategory of h-equivalences will be denoted $h\mathcal{DR}^n_{Th(\xi)}(X)$.

The category $h\mathcal{DR}^n_{Th(\xi)}(X)$ does not essentially depend on the particular choice of the space $Th(\xi)$. Namely, suppose $Th'(\xi)$ is another model for the Thom space of ξ in the sense defined above. The conditions on such a space imply that there is a fibre homotopy equivalence $Th(\xi) \xrightarrow{\simeq} Th'(\xi)$.

Lemma 1.4. A fibre homotopy equivalence $\alpha: Th(\xi) \longrightarrow Th'(\xi)$ induces a functor $\bar{\alpha}: h\mathcal{DR}^n_{Th(\xi)}(X) \to h\mathcal{DR}^n_{Th'(\xi)}(X)$ which is a homotopy equivalence.

Proof: It is clear that α induces such a functor. To see that this functor is a homotopy equivalence choose an inverse of α, $\alpha': Th'(\xi) \longrightarrow Th(\xi)$. This defines a functor $\bar{\alpha}'$ in the other direction. Let $g: Th(\xi) \wedge I_+ \to Th(\xi)$ be a homotopy over X from $\alpha'\alpha$ to id. Define an endofunctor f of $h\mathcal{DR}^n_{Th(\xi)}(X)$ by

$$(Y,Y',u) \longmapsto (Y\ {}_X\wedge_* I_+, Y', g\circ(u\ {}_X\wedge_*\ id)) .$$

There are two natural transformations

$$id \longrightarrow f \longleftarrow \bar{\alpha}'\alpha .$$

These provide the required homotopy $\bar{\alpha}'\bar{\alpha} \simeq id$, cf. [11]. Similarly, $\bar{\alpha}\bar{\alpha}' \simeq id$. □

In view of this lemma the choice of the Thom space $Th(\xi)$ does not really matter. In the following we will always assume that a definite choice of $Th(\xi)$ has been made. To simplify the notation we will usually write $h\mathcal{DR}^n_\xi(X)$ instead of the more precise $h\mathcal{DR}^n_{Th(\xi)}(X)$ whenever there is no danger of confusion.

By lemma 1.1. there are two suspension functors

$$\Sigma_\ell: \mathcal{DR}^n_\xi(X) \longrightarrow \mathcal{DR}^{n+1}_\xi(X),\ (Y,Y',u) \mapsto (\Sigma_X Y, Y', \Sigma_\ell u), \text{ resp.}$$

$$\Sigma_r: \mathcal{DR}^n_\xi(X) \longrightarrow \mathcal{DR}^{n+1}_\xi(X),\ (Y,Y',u) \mapsto (Y, \Sigma_X Y', \Sigma_r u) .$$

Stably the category $\mathcal{DR}^n_\xi(X)$ does not depend on the spherical fibration ξ. Namely, in view of cor. 1.3.

$$(Y,Y',u) \longmapsto (\xi\cdot Y, Y', (\xi\wedge\varepsilon)\cdot u)$$

defines a functor

$$\varphi_\xi : h\mathcal{DR}^n_\varepsilon(X) \longrightarrow h\mathcal{DR}^{n+d}_\xi(X),$$

where $\varepsilon = X \times S^o$ is the trivial spherical fibration.

Lemma 1.5. The functor φ_ξ induces a weak homotopy equivalence

$$\varinjlim_{\Sigma_\ell} h\mathcal{DR}^n_\varepsilon(X) \longrightarrow \varinjlim_{\Sigma_\ell} h\mathcal{DR}^n_\xi(X) .$$

Proof: First assume that X is finite up to homotopy. Then we can find an inverse η of ξ such that $\xi\cdot\eta \simeq X \times S^r$. Multiplication with η defines a functor

$$\varphi_\eta : hDR^n_\xi(X) \longrightarrow hDR^{n+r-d}_\varepsilon(X) .$$

The composite $\varphi_\xi\varphi_\eta$ (resp. $\varphi_\eta\varphi_\xi$) is the same up to homotopy as the r-fold suspension

$$hDR^n_\varepsilon(X) \longrightarrow hDR^{n+r}_\varepsilon(X) \quad (\text{resp. } hDR^n_\xi(X) \longrightarrow hDR^{n+r}_\xi(X))$$

(cf. property (v) of the operation •). The general case of the lemma follows by a direct limit argument. □

Remark: There is an analogous assertion with Σ_ℓ replaced by Σ_r throughout. □

Since $\Sigma_\ell\Sigma_r = \Sigma_r\Sigma_\ell$ it makes sense to talk about the limit

$$\lim_{\substack{\to \\ \Sigma_\ell,\Sigma_r}} hDR^n_\varepsilon(X).$$

We have

Proposition 1.6. The forgetful functor

$$\delta : \lim_{\substack{\to \\ \Sigma_\ell,\Sigma_r}} hDR^n_\varepsilon(X) \longrightarrow \lim_{\substack{\to \\ \Sigma_X}} hR_{hf}(X)$$

$$(Y,Y',u) \longmapsto Y$$

is a weak homotopy equivalence.

Proof: This will be proved below after prop. 1.15. □

Remark: By imposing a condition on the homotopy type as in the definition of the categories $hR^\ell_k(X)$ one defines categories $hDR^{\ell,m}_k(X)$, i.e. an object is a triple (Y,Y',u), where Y (resp. Y') has the homotopy type of a wedge of X and k spheres of dimension ℓ (resp. m). (The spherical fibration ξ is suppressed in the notation of these categories).

The map δ restricts to a weak homotopy equivalence

$$\delta: \lim_{\substack{\to \\ \ell,m}} hDR^{\ell,m}_k(X) \longrightarrow \lim_{\substack{\to \\ \ell}} hR^\ell_k(X) .$$ □

A different setting for the duality just described is provided by using simplicial sets with a group action instead of retractive simplicial sets. The group in question is the *loop group* of X, cf. [7]. This setting is sometimes more convenient to work in. We have to give a few definitions first.

Let G be a simplicial group. $U(G)$ is the category of pointed simplicial sets with right (simplicial) G-action. $U_f(G)$ is the subcategory of those G-sets which are free (in the pointed sense, i.e. $xg = x$ implies $g = 1$ or $x = *$) and finitely generated over G, i.e. they are generated as a G-set by finitely many simplices. An h-equivalence is a G-map which is a weak homotopy equivalence of the underlying simplicial sets, $hU(G)$ is the subcategory of h-equivalences, $U_{hf}(G)$ is the subcategory of $U(G)$ of those G-sets which are related to objects of $U_f(G)$ by a finite chain of homotopy equivalences; $h\mathcal{U}_{hf}(G) := \mathcal{U}_{hf}(G) \cap h\mathcal{U}(G)$.

Let M and M' denote objects of $\mathcal{U}(G)$. An n-*duality map* is a pointed right (G×G)-map

$$u\colon M \wedge M' \longrightarrow Ex^{\infty}(S^n \wedge G_+)$$

satisfying that it induces an isomorphism of $\mathbb{Z}[\pi_o G]$-modules

$$\alpha_u\colon H_q^G(M') \longrightarrow H_G^{n-q}(M)$$

$$z \longmapsto u^*(t)/z, \quad 0\leq q\leq n,$$

where $t \in H^n_{G\times G}(Ex^{\infty}(S^n \wedge G_+);\mathbb{Z}[\pi_o G]')$ is a class mapping to a generator of $H^n_{G\times G_o}(Ex^{\infty}(S^n\wedge G_+);\mathbb{Z}[\pi_o G]) \approx \mathbb{Z}[\pi_o G]$ under the canonical map, where G_o denotes the identity component of G, and $\mathbb{Z}[\pi_o G]'$ denotes the $\mathbb{Z}[\pi_o G^2]$-module $\mathbb{Z}[\pi_o G]$ with $\pi_o G^2$-action given by $x.(a,b) = b^{-1}xa$, $(a,b) \in (\pi_o G)^2$, $x \in \mathbb{Z}[\pi_o G]$.

Here we consider $S^n\wedge G_+$ as a right simplicial (G×G)-set via $(x\wedge g)\cdot(h,k)=x\wedge k^{-1}gh$, $x \in S^n$, $g,h,k \in G$. This induces a G×G-action on $Ex^{\infty}(S^n\wedge G_+)$.

Similarly we ask that $\alpha_{u'}$ is an isomorphism, where u' is defined as the composite

$$M'\wedge M \xrightarrow{\approx} M \wedge M' \xrightarrow{u} Ex^{\infty}(S^n\wedge G_+) \xrightarrow{\iota_n} Ex^{\infty}(S^n\wedge G_+)$$

and the map ι_n is induced by $g \longmapsto g^{-1}$.

By definition $H_*^G(M;A) = H_*(M\times^G E, *\times^G E;A)$, $H_G^*(M;A) = H^*(M\times^G E, *\times^G E;A)$, where E is a universal G-bundle, and A is a $\pi_o G$-module.

Ex^{∞} denotes the functor of [6] which turns a simplicial set into a Kan simplicial set.

Note that in case that M and M' are finite (up to homotopy) the second condition on a duality map is implied by the first, and vice versa.

We call M' an n-*dual* of M if there exists an n-duality map $u\colon M \wedge M' \longrightarrow Ex^{\infty}(S^n \wedge G_+)$.

Example 1.7. The map $\mu\colon (S^k \wedge G_+) \wedge (S^{n-k} \wedge G_+) \longrightarrow S^n \wedge G_+ \longrightarrow Ex^{\infty}(S^n \wedge G_+)$ induced from the map $G \times G \longrightarrow G$, $(g,h) \longmapsto h^{-1}g$ is an n-duality map. □

Just as in the case of retractive spaces, duality is compatible with suspension. The rigorous statement is as follows. Choose sequences a_n and b_n of homotopy equivalences such that the following diagrams commute.

$$\begin{array}{ccc} S^1 \wedge Ex^{\infty}(S^n \wedge G_+) & \xrightarrow{a_n} & Ex^{\infty}(S^{n+1} \wedge G_+) \\ \uparrow & & \uparrow \\ S^1 \wedge S^n \wedge G_+ & \xrightarrow{\approx} & S^{n+1} \wedge G_+ \end{array} \qquad \begin{array}{ccc} Ex^{\infty}(S^n \wedge G_+) \wedge S^1 & \xrightarrow{b_n} & Ex^{\infty}(S^{n+1} \wedge G_+) \\ \uparrow & & \uparrow \\ S^n \wedge G_+ \wedge S^1 \xrightarrow{\approx} S^n \wedge S^1 \wedge G_+ & \xrightarrow{\approx} & S^{n+1} \wedge G_+ \end{array}.$$

Here the vertical arrows denote the canonical inclusions. The maps a_n, b_n exist by the Kan condition on $Ex^{\infty}(S^{n+1} \wedge G_+)$.

We now can state the analogue of lemma 1.1.

Lemma 1.8. Let $u: M \wedge M' \longrightarrow Ex^{\infty}(S^n \wedge G_+)$ be an n-duality map; $\iota_n: Ex^{\infty}(S^n \wedge G_+) \longrightarrow Ex^{\infty}(S^n \wedge G_+)$ denotes the inversion map induced by $g \mapsto g^{-1}$. Then

(i) $u': M \wedge M \longrightarrow Ex^{\infty}(S^n \wedge G_+)$, $m' \wedge m \mapsto \iota_n u(m \wedge m')$ is an n-duality map;

(ii) $\Sigma_{\ell}(u): (S^1 \wedge M) \wedge M' \xrightarrow{id \wedge u} S^1 \wedge Ex^{\infty}(S^n \wedge G_+) \xrightarrow{a_n} Ex^{\infty}(S^{n+1} \wedge G_+)$

and

$\Sigma_r(u): M \wedge (S^1 \wedge M') \xrightarrow{\approx} (M \wedge M') \wedge S^1 \xrightarrow{u \wedge id} Ex^{\infty}(S^n \wedge G_+) \wedge S^1 \xrightarrow{b_n} Ex^{\infty}(S^{n+1} \wedge G_+)$

are (n+1)-duality maps. □

We want to give a function space description of duality now, analogous to that of [14]. Let $F_G^n(M)$ denote the simplicial set of pointed right G-equivariant maps from M to $Ex^{\infty}(S^n \wedge G_+)$. G acts freely (pointed) from the left on this function space. Convert this to a right action using $g \mapsto g^{-1}$. $F_G^n(M)$ is a Kan simplicial set, since $Ex^{\infty}(S^n \wedge G_+)$ satisfies the Kan condition.

The evaluation map

$$e: F_G^n(M) \wedge M \longrightarrow Ex^{\infty}(S^n \wedge G_+)$$

induces a map

$$\alpha_e: H_q^G(F_G^n(M)) \longrightarrow H_G^{n-q}(M).$$

Let $M \in \mathcal{U}_f(G)$ be of G-dimension k.

Lemma 1.9. The map α_e is an isomorphism in the range $0 \leqq q \leqq 2(n-k) - 1$.

Proof: By induction. The assertion is trivially true in the case $M = *$. For $M = G_+$ we have $F_G^n(M) = Map_*(S^0, S^n \wedge G_+) = S^n \wedge G_+$, and the evaluation map $G_+ \wedge S^n \wedge G_+ \xrightarrow{\approx} S^n \wedge G_+ \wedge G_+ \longrightarrow S^n \wedge G_+$ is a special case of the map of example 1.7.. So α_e is an isomorphism in that case. Since M was supposed to be finite, it has a G-skeleton filtration $* = M_0 \subset M_1 \subset \ldots \subset M_k = M$, such that we have cofibration sequences

$$M_{i-1} \rightarrowtail M_i \twoheadrightarrow \bigvee_{\alpha} S^i \wedge G_+,$$

$\alpha \in$ some finite index set. The general case then follows by a five lemma argument and the fact that the canonical map $S^{n-i} \wedge G_+ \longrightarrow \Omega^i(S^n \wedge G_+)$ is $(2(n-i) - 1)$-connected. □

Corollary 1.10. Let M be an object of $\mathcal{U}_f(G)$. Suppose that $u: M \wedge M' \longrightarrow Ex^{\infty}(S^n \wedge G_+)$ is an n-duality map. Let $n > \dim M$. Then the map

$$\hat{u}: M' \longrightarrow F_G^n(M)$$

adjoint to u is $(2(n-\dim M) - 1)$-connected.

Proof: There is a commutative diagram

$$\begin{array}{ccc} M \wedge M' & \xrightarrow{\;u\;} & \\ {\scriptstyle id \wedge \hat{u}}\downarrow & & Ex^{\infty}(S^n \wedge G_+) \\ M \wedge F^n_G(M) & \xrightarrow{\;e\;} & \end{array}$$

which implies another one

$$\begin{array}{ccc} H^G_q(M') & \xrightarrow{\;\alpha_u\;} & \\ \downarrow & & H^{n-q}_G(M) \\ H^G_q(F^n_G(M)) & \xrightarrow{\;\alpha_e\;} & \end{array}$$

By assumption α_u is an isomorphism for all q; α_e is an isomorphism in a certain range by lemma 1.9. Hence so is $\hat{u}$. □

Remark: If M is finite up to homotopy only, define dim M to be the least of all dimensions of finite G-sets which can be related to M by finite chains of homotopy equivalences. Then the assertion of the corollary also holds in this case.

Suppose that we are given two n-duality maps $u: M \wedge M' \longrightarrow Ex^{\infty}(S^n \wedge G_+)$ (resp. $v: N \wedge N' \longrightarrow Ex^{\infty}(S^n \wedge G_+)$). Let $f: M \longrightarrow N$ denote a morphism in $\mathcal{U}(G)$. If there exists a morphism $f': N' \longrightarrow M'$ such that the following diagram commutes up to homotopy

$$\begin{array}{ccc} M \wedge N' & \xrightarrow{\;f \wedge id\;} & N \wedge N' \\ {\scriptstyle id \wedge f'}\downarrow & & \downarrow{\scriptstyle v} \\ M \wedge M' & \xrightarrow{\;u\;} & Ex^{\infty}(S^n \wedge G_+) \end{array}$$

then f' will be called an n-*dual of* f.

Suppose that M (resp. N') is homotopy equivalent to a G-set of G-dimension at most k (resp. n). Let $n > 2k+1$. Further suppose that M' satisfies the Kan condition.

Lemma 1.11. In this situation there exists an n-dual of f.

Proof: The condition on a dual map $f': N' \longrightarrow M'$ is equivalent to asking if there exists an arrow $N' \longrightarrow M'$ such that the following diagram commutes up to homotopy.

$$\begin{array}{ccc} M' & \xrightarrow{\;\hat{u}\;} & F^n_G(M) \\ \uparrow & & \uparrow{\scriptstyle f^*} \\ N' & \xrightarrow{\;\hat{v}\;} & F^n_G(N) \end{array}$$

where $\hat{u}$ (resp. $\hat{v}$) denotes the adjoint of u (resp. v).
First assume that N' is n-dimensional (not just up to homotopy). By cor. 1.10. the map $\hat{u}$ is $(2(n-k)-1)$-connected. Hence, since M' is Kan, one can construct a lifting

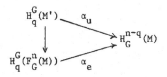

up to homotopy as indicated by the broken arrow. Furthermore, this lifting is unique up to homotopy. In the general case, where N' is only homotopy equivalent to an n-dimensional G-set, one can still find a Kan simplicial set $\bar{N}$, and an n-dimensional G-set $\tilde{N}$ such that $N' \xrightarrow{\simeq} \bar{N} \xleftarrow{\simeq} \tilde{N}$.
Now one first finds a map $\tilde{N} \longrightarrow M'$; this extends to $\bar{N} \longrightarrow M'$ by the Kan condition on M'. Restricting this map to N' gives the desired map. □

Our next goal is to show that for every object of $\mathcal{U}_{hf}(G)$ there exists an n-dual if n is sufficiently large. To show this we need the following technical lemma.

Lemma 1.12. Let $M \xrightarrow{f} N \xrightarrow{g} N \cup_f CM$ be a cofibration sequence of objects of $\mathcal{U}_{hf}(G)$. (CM denotes the cone on M.) Let M' (resp. N') denote an n-dual of M (resp. N), and let $f'\colon N' \longrightarrow M'$ denote an n-dual of f. Then there exists an (n+1)-duality map

$$w\colon N \cup_f CM \wedge M' \cup_{f'} CN' \longrightarrow Ex^\infty(S^{n+1} \wedge G_+).$$

Proof: Let $u\colon M \wedge M' \longrightarrow Ex^\infty(S^n \wedge G_+)$ (resp. $v\colon N \wedge N' \longrightarrow Ex^\infty(S^n \wedge G_+)$) denote an n-duality map. The map w is constructed as follows.

Let α denote the composite

$$\Sigma(M \wedge N') \xrightarrow{\mu} (\Sigma M \wedge N') \vee (M \wedge \Sigma N') \xrightarrow{(id \wedge f', f \wedge -id)} (\Sigma M \wedge M') \vee (N \wedge \Sigma N'),$$

where μ denotes the comultiplication, and $-id$ is a homotopy inverse. By definition of f' the following composite map is nullhomotopic:

$$\Sigma(M \wedge N') \xrightarrow{\alpha} (\Sigma M \wedge M') \vee (N \wedge \Sigma N') \xrightarrow{(\Sigma_\ell u, \Sigma_r v)} Ex^\infty(S^{n+1} \wedge G_+).$$

Hence the map $(\Sigma_\ell u, \Sigma_r v)$ may be extended to a map

$$\bar{w}\colon (\Sigma M \wedge M') \wedge (N \wedge \Sigma N') \cup_\alpha C\Sigma(M \wedge N') \longrightarrow Ex^\infty(S^{n+1} \wedge G_+).$$

The left-hand side is isomorphic to $((N \cup_f CM) \wedge (M' \cup_{f'} CN'))/N \wedge M'$.

Let w denote the composite

$$N \cup_f CM \wedge M' \cup_{f'} CN' \longrightarrow\!\!\!\!\!\!\!\rightarrow ((N \cup_f CM) \wedge (M' \cup_{f'} CN'))/N \wedge M' \xrightarrow{\bar{w}} Ex^\infty(S^{n+1} \wedge G_+).$$

By construction of w the following diagrams commute up to homotopy.

$$\begin{array}{ccc}
N \cup_f CM \wedge M' & \xrightarrow{h \wedge id} & \Sigma M \wedge M' \\
\downarrow{\scriptstyle id \wedge g'} & & \downarrow{\scriptstyle \Sigma_\ell u} \\
N \cup_f CM \wedge M' \cup_{f'} CN' & \xrightarrow{w} & Ex^\infty(S^{n+1} \wedge G_+) \\
 & & \\
N \wedge M' \cup_{f'} CN' & \xrightarrow{id \wedge h'} & N \wedge \Sigma N' \\
\downarrow{\scriptstyle g \wedge id} & & \downarrow{\scriptstyle \Sigma_r v} \\
N \cup_f CM \wedge M' \cup_{f'} CN' & \xrightarrow{w} & Ex^\infty(S^{n+1} \wedge G_+) .
\end{array}$$

Here h: $N \cup_f CM \longrightarrow M$ (resp. h': $M' \cup_{f'} CN' \longrightarrow N'$) denotes the canonical map in the cofibre sequence of f (resp. f').

These diagrams imply the following commutative diagram of (co-)homology groups.

$$\begin{array}{ccccccccc} H_q^G(N') & \longrightarrow & H_q^G(M') & \longrightarrow & H_q^G(M' \cup_{f'} CN') & \longrightarrow & H_q^G(\Sigma N') & \longrightarrow & H_q^G(\Sigma M') \\ \downarrow \alpha_{\Sigma_\ell v} & & \downarrow \alpha_{\Sigma_\ell u} & & \downarrow \alpha_w & & \downarrow \alpha_{\Sigma_r v} & & \downarrow \alpha_{\Sigma_r u} \\ H_G^p(\Sigma N) & \longrightarrow & H_G^p(\Sigma M) & \longrightarrow & H_G^p(N \cup_f CM) & \longrightarrow & H_G^p(N) & \longrightarrow & H_G^p(M) \end{array}$$

$(p = n+1-q)$.

Since by assumption u and v are duality maps, all the vertical arrows except possibly α_w are isomorphisms. Hence by the five lemma so is α_w, as was to be shown. □

Proposition 1.13. Let M be an object of $\mathcal{U}_{hf}(G)$. There exists an n-dual of M, if n is sufficiently large.

Proof: We say that an object N of $\mathcal{U}_{hf}(G)$ is obtained from M by *attaching of a* k-*cell* if N is isomorphic to the pushout of the following diagram of pointed G-maps

$$M \longleftarrow \partial\Delta^k \quad G_+ \rightarrowtail \Delta^k \wedge G_+$$

where $\partial\Delta^k$ denotes the boundary of the k-simplex, and the map on the right is the natural inclusion.

Since M is finite up to homotopy one can find a Kan simplicial set $\bar{M}$, and a finite G-set $\tilde{M}$ such that $M \overset{\simeq}{\rightarrowtail} \bar{M} \overset{\simeq}{\leftarrowtail} \tilde{M}$.

Suppose we have found an n-duality map $\tilde{M} \wedge M' \longrightarrow Ex^\infty(S^n \wedge G_+)$. Then we may extend this to a map $\bar{M} \wedge M' \longrightarrow Ex^\infty(S^n \wedge G_+)$, since the inclusion $\tilde{M} \wedge M' \longrightarrow \bar{M} \wedge M'$ is a homotopy equivalence, and because $Ex^\infty(S^n \wedge G_+)$ satisfies the Kan condition. This extended map is clearly also an n-duality map. Restricting this map to the subspace $M \wedge M'$ finally gives a duality map for M. This argument shows that there is no loss of generality in assuming M to be finite.

Now any finite object of $\mathcal{U}(G)$ may be obtained from the base point by attaching of a finite number of cells. Hence an n-dual of M can be constructed inductively, the inductive step being provided by lemma 1.12. □

Define a category $\mathcal{DU}^n(G)$ in which an object is a triple (M,M',u), where M and M' are objects of $\mathcal{U}_{hf}(G)$, and u: $M \wedge M' \longrightarrow Ex^\infty(S^n \wedge G_+)$ is an n-duality map. We add the technical condition that $\pi_i M = \pi_i M' = 0$, $i = 0,1$. A morphism from (M,M',u) to (N,N',v) is a pair of morphisms in $\mathcal{U}(G)$, f: $M \longrightarrow N$, and f': $N' \longrightarrow M'$ such that the following diagram commutes

$$\begin{array}{ccc} N' \wedge M & \xrightarrow{\ id \wedge f\ } & N' \wedge N \\ {\scriptstyle f' \wedge id}\downarrow & & \downarrow{\scriptstyle v} \\ M' \wedge M & \xrightarrow[\ u\ ]{} & Ex^\infty(S^n \wedge G_+) \end{array}$$.

A morphism (f,f') is called an h-equivalence if both f and f' are h-equivalences. By lemma 1.8. there are two suspension functors Σ_ℓ (resp. Σ_r): $\mathcal{DU}^n(G) \longrightarrow \mathcal{DU}^{n+1}(G)$ given by suspending M (resp. M').

We are now going to compare the two settings for duality. Let X denote a connected simplicial set, and let G be its loop group in the sense of [7]. Let E be a universal G-bundle. There is an adjoint functor pair (cf. [16]):

$$\Phi_X: h\mathcal{R}(X) \longrightarrow h\mathcal{U}(G)$$
$$(Y,r,s) \longmapsto Y \times_X E/E$$
$$\Psi_G: h\mathcal{U}(G) \longrightarrow h\mathcal{R}(X)$$
$$M \longmapsto (M\times^G E \rightleftarrows * \times^G E).$$

Let ε again denote the trivial fibration $X \times S^0$. The spaces $\Psi_{G^2}(Ex^\infty(S^n \wedge G_+))$ can be used as Thom spaces $Th_n(\varepsilon)$ in the sense defined above.
Here G_+ is considered as an object of $\mathcal{U}(G\times G)$.
Define a functor

$$D\Phi : h\mathcal{DR}^n_\varepsilon(X) \longrightarrow h\mathcal{DU}^n(G)$$
$$(Y,Y',u) \longmapsto (\Phi_X(Y),\Phi_X(Y'),u'),$$

where u' is the composite

$$\Phi_X(Y') \wedge \Phi_X(Y) \longrightarrow \Phi_{X^2}(Th_n(\varepsilon)) = \Phi_{X^2}\Psi_{G^2}(Ex^\infty(S^n \wedge G_+)) \longrightarrow Ex^\infty(S^n \wedge G_+) .$$

Similarly,

$$D\Psi : h\mathcal{DU}^n(G) \longrightarrow h\mathcal{DR}^n_\varepsilon(X)$$

$(u: M \wedge M' \longrightarrow Ex^\infty(S^n \wedge G_+) \longrightarrow (\Psi_G(M) \wedge \Psi_G(M') \longrightarrow \Psi_{G^2}(Ex^\infty(S^n \wedge G_+) = Th_n(\varepsilon)).$

Proposition 1.14. $D\Phi$ and $D\Psi$ are mutually inverse homotopy equivalences.

Proof: We first remark that $D\Phi$ and $D\Psi$ are not adjoint. Let f: $h\mathcal{DR}^n_\varepsilon(X) \longrightarrow h\mathcal{DR}^n_\varepsilon(X)$ be given by $(Y,Y',u) \longrightarrow (\Psi\Phi(Y),Y,\bar{u})$, where $\bar{u}$: $\Psi\Phi(Y) \wedge Y' \longrightarrow \Psi\Phi(Y \wedge Y') \xrightarrow{\Psi\Phi(u)} \Psi\Phi\Psi(Ex^\infty(S^n \wedge G_+)) \longrightarrow (Ex^\infty(S^n \wedge G_+))$ and Φ (resp. Ψ) is short for Φ_X (resp. Ψ_G). Similarly, f' is the corresponding endofunctor of $h\mathcal{DR}^n_\varepsilon(X)$ defined by a condition on Y'. There is a natural transformation from the identity to f, and another one from f' to the identity. Since $D\Psi \cdot D\Phi = f'f$, the composite $D\Psi \cdot D\Phi$ is therefore homotopic to the identity; similarly with the other composition. □

This proposition shows that both settings for duality are actually equivalent. In the next proposition it is shown that the choice of duality data is in fact a "contractible choice".

Proposition 1.15. The forgetful functor

$$\varepsilon : \varinjlim_{\Sigma_\ell,\Sigma_r} h\mathcal{DU}^n(G) \longrightarrow \varinjlim_{\Sigma} h\mathcal{U}_{hf}(G)$$

$$(M,M',u) \longmapsto M$$

is a weak homotopy equivalence.

Proof: To prove the assertion we need a stronger finiteness condition on the objects of $h\mathcal{DU}^n(G)$. We adapt an argument of [19] to show that this condition may be assumed without loss of generality.

Let $h\mathcal{DU}^n(G)'$ denote the full subcategory of $h\mathcal{DU}^n(G)$ of those objects (M,M',u) which satisfy that M' is actually finite, not just finite up to homotopy.
The inclusion $h\mathcal{DU}^n(G)' \subset h\mathcal{DU}^n(G)$ is a homotopy equivalence. To see this we introduce two further subcategories. Namely, let $h\mathcal{DU}^n(G)_{Kan}$ denote the full subcategory consisting of those objects which satisfy that (the underlying simplicial set of) M' is a Kan simplicial set; let $h\mathcal{DU}^n(G)''$ be the full subcategory of those objects of $h\mathcal{DU}^n(G)$ which lie in either $h\mathcal{DU}^n(G)'$ or $h\mathcal{DU}^n(G)_{Kan}$. The inclusion $h\mathcal{DU}^n(G)'' \subset h\mathcal{DU}^n(G)$ is a homotopy equivalence. This may be seen from the existence of the functor Ex^∞. Ex^∞ extends to a functor $h\mathcal{DU}^n(G) \longrightarrow h\mathcal{DU}^n(G)_{Kan}$. There is a natural transformation $Ex^\infty \longrightarrow Id$, given by the canonical map $M' \longrightarrow Ex^\infty M'$. This shows that $h\mathcal{DU}^n(G)_{Kan} \subset h\mathcal{DU}^n(G)$ is a homotopy equivalence. By the same argument $h\mathcal{DU}^n(G)'' \subset h\mathcal{DU}^n(G)$ is a homotopy equivalence.

The next step is to show that the inclusion $i: h\mathcal{DU}^n(G)' \longrightarrow h\mathcal{DU}^n(G)''$ is a homotopy equivalence.
We use Quillen's theorem A, cf. [10]. So we have to show that the right fibre $(M,M',u)/i$ over any object (M,M',u) of $h\mathcal{DU}^n(G)''$ is contractible.
It suffices to prove that any finite diagram $\mathcal{D} \longrightarrow (M,M',u)/i$ in the fibre is contractible.
Let $(f_i,f_i^!)$: $(M,M',u) \longrightarrow (M_i,M_i^!,u_i)$ represent such a diagram. If M' is already finite there is nothing to prove since the diagram then has an obvious initial object. So assume M' is Kan. Since M' is also finite up to homotopy, one can find a diagram $M \overset{\simeq}{\rightarrowtail} \bar{M} \overset{\simeq}{\leftarrowtail} M'$, where $\tilde{M}$ is finite and $\bar{M}$ is obtained from $\tilde{M}$ by filling horns. Since the $M_i^!$ are all finite, one can find another subset $\tilde{\tilde{M}}$ of $\bar{M}$ containing the images $f_i^!(M_i^!)$ of all the simplicial sets $M_i^!$ and which can be obtained from $\tilde{\tilde{M}}$ by filling finitely many (G)-horns. Hence $\tilde{\tilde{M}}$ is itself finite (as a G-set).
Since M' is a Kan set one can find a retraction $\bar{M} \longrightarrow M'$. This gives a map $q: \tilde{\tilde{M}} \twoheadrightarrow M'$.
Define a duality map $\tilde{\tilde{u}}$ as the composite $M \wedge \tilde{\tilde{M}} \longrightarrow M \wedge M' \xrightarrow{u} Ex^\infty(S^n \wedge G_+)$. Then the object $((M,\tilde{\tilde{M}},u);\ (id,q)\colon (M,M',u) \longrightarrow (M,\tilde{\tilde{M}},\tilde{\tilde{u}}))$ of the fibre is an initial object for the diagram $\mathcal{D}$. Hence the diagram is contractibe, whence i is a homotopy equivalence. This in turn implies that the inclusion $h\mathcal{DU}^n(G)' \subset h\mathcal{DU}^n(G)$ is a homotopy equivalence.

We are now reduced to proving that the restriction of the map ε of the proposition

to the subcategory $\varinjlim h\mathcal{D}\mathcal{U}^n(G)'$ is a homotopy equivalence.
Again we use theorem A of Quillen. Let N be an object of $\varinjlim h\mathcal{U}_{hf}(G)$. We have to show that the right fibre N/ε is contractible. Let $\mathcal{D}: I \longrightarrow N/\varepsilon$ be a finite diagram in the fibre. $\mathcal{D}$ is represented by $(M_i, M_i^!, u_i: M_i \wedge M_i^! \longrightarrow Ex^\infty(S^n \wedge G_+);\ a_i: N \xrightarrow{\simeq} M_i)_{i\in I}$.
By proposition 1.13. there exists an m-dual of N for some large m. We may assume that n is large, and in particular that there exists an n-dual of N. Let $v: N \wedge N' \longrightarrow Ex^\infty(S^n \wedge G_+)$ denote an n-duality map with $N' \in \mathcal{U}_{hf}(G)$. Assume that N' is a Kan set.
Consider the following diagram

$$\begin{array}{ccc} & & N' \\ & \nearrow & \downarrow \hat{v} \\ M_i^! & \xrightarrow{b_i} & F_G^n(N) \end{array}$$

where $\hat{v}$ denotes the adjoint of v, and b_i is the composite

$$M_i^! \xrightarrow{\hat{u}_i} F_G^n(M_i) \xrightarrow{a_i^*} F_G^n(N) .$$

Here $\hat{u}_i$ is the adjoint of u_i and a_i^* is induced from a_i.
Let $n > 2\max\{\dim M_i, \dim N\}$. Further assume that $\dim M_i^! \leqq n$. By coro. 1.10. the map $\hat{v}$ is $(2(n-\dim N)-1)$-connected, hence by assumption on n it is at least n-connected.
Since $\dim M_i^! < n$, and N' was assumed to be Kan, by obstruction theory therefore there exist liftings up to homotopy $c_i: M_i^! \longrightarrow N'$ of b_i as indicated by the broken arrow in the diagram. For each i choose a specific homotopy $h_i: M_i \times \Delta^1 \longrightarrow F_G^n(N)$ such that $h_i|M_i^! \times 0 = \hat{v}c_i$, and $h_i|M_i^! \times 1 = b_i$.
The map b_i is at least n-connected since a_i^* is a homotopy equivalence and $\hat{u}_i$ is n-connected. Since $M_i^!$ and N' have no G-homology in dimensions $>n$, and by the Hurewicz theorem, c_i is therefore a homotopy equivalence.
Let $N^\S$ denote the pushout of the following diagram

$$N' \xleftarrow{\amalg c_i} \amalg M_i^! \times 0 \longrightarrow \amalg M_i^! \times \Delta^1 .$$

There is a map $\hat{v}^\S: N^\S \longrightarrow F_G^n(N)$ given by $\hat{v}$ on N', and h_i on $M_i^! \times \Delta^1$. Further there is a commutative diagram

$$\begin{array}{ccccc} & & N^\S & \longrightarrow & Ex^\infty(N^\S) \\ & \overset{d_i}{\nearrow} & \downarrow \hat{v}^\S & \swarrow \hat{v}^\dagger & \\ M_i^! & \xrightarrow{b_i} & F_G^n(N) & & \end{array}$$

where $d_i: M_i^! = M_i^! \times 1 \longrightarrow M_i^! \times \Delta^1 \longrightarrow N^\S$, and $\hat{v}^\dagger$ is some extension of $\hat{v}^\S$ (which exists by the Kan condition on $F_G^n(N)$).

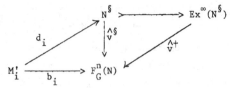

Since $Ex^{\infty}(N^{\S})$ is Kan, the inclusion $\bigcup_i d_i(M_i^!) \subset Ex^{\infty}(N^{\S})$ factors as

$$\bigcup_i d_i(M_i^!) \longrightarrow \bar{N} \xrightarrow{\simeq} Ex^{\infty}(N^{\S}),$$

where the second arrow is a homotopy equivalence and $\bar{N}$ is obtained from $\bigcup_i d_i(M_i^!)$ by attaching of finitely many G-cells. Now this union is a finite G-set. Therefore $\bar{N}$ is also finite. The map

$$f_i: M_i^! \xrightarrow{d_i} \bigcup_i d_i(M_i^!) \longrightarrow \bar{N}$$

is a homotopy equivalence, since its composition with $\bar{N} \xrightarrow{\simeq} Ex^{\infty}(N^{\S})$ is one.
Let $\hat{\bar{v}}$ denote the composite $\bar{N} \longrightarrow Ex^{\infty}(N^{\S}) \xrightarrow{\hat{v}^{\dagger}} F_G^n(N)$ and let $\bar{v}: N \wedge \bar{N} \longrightarrow Ex^{\infty}(S^n \wedge G_+)$ denote the adjoint of this map. By construction $\bar{v}$ is an n-duality map. The object $(N,\bar{N},\bar{v})$ of $h\mathcal{DU}^n(G)'$ maps to the diagram $(M_i,M_i^!,u_i,a_i)$ by (a_i,f_i). Hence $((N,\bar{N},\bar{v});\ id: N \longrightarrow N)$ is a cone point for the diagram $\mathcal{D}$.
This proves that any finite diagram in N/ε is nullhomotopic, as was to be shown.
□

Remark: Just as in the case of the categories $\mathcal{R}(X)$ (resp. $\mathcal{DR}_{\varepsilon}^n(X)$) certain subcategories $\mathcal{U}_k^{\ell}(G)$ (resp. $\mathcal{DU}_k^{\ell,m}(G)$) of $\mathcal{U}(G)$ (resp. $\mathcal{DU}^n(G)$) may be defined. By restriction to a connected component one obtains from proposition 1.15. another homotopy equivalence

$$\lim_{\substack{\rightarrow \\ \ell,m}} h\mathcal{DU}_k^{\ell,m}(G) \longrightarrow \lim_{\substack{\rightarrow \\ \ell}} h\mathcal{U}_k(G).$$ □

Proof of prop. 1.6.: There is a commutative diagram

$$\begin{array}{ccc} h\mathcal{DR}_{\varepsilon}^n(X) & \xrightarrow{D\Phi} & h\mathcal{DU}^n(G) \\ \downarrow \delta & & \downarrow \varepsilon \\ h\mathcal{R}_{hf}(X) & \xrightarrow{\Phi} & h\mathcal{U}_{hf}(G) \end{array}.$$

Φ is a homotopy equivalence because it has an adjoint, $D\Phi$ is a homotopy equivalence by prop. 1.14.; ε becomes a homotopy equivalence after passing to the limit by prop. 1.15., therefore so does δ, as was to be shown.
□

As a corollary to proposition 1.6. and 1.15. one obtains the following descriptions of A(X).

Corollary 1.16.

$$A(X) \simeq \mathbf{Z} \times |\lim_{\substack{\rightarrow \\ k,\ell,m}} h\mathcal{DR}_k^{\ell,m}(X)|^+$$

$$\simeq \mathbf{Z} \times |\lim_{\substack{\rightarrow \\ k,\ell,m}} h\mathcal{DU}_k^{\ell,m}(G)|^+$$

$(G = G(X))$. □

Appendix

In this appendix we first prove that the categories $\mathcal{DR}^n_\xi(X)$ are covariantly functorial in X. Then we use this result to compare the concept of Spanier-Whitehead duality developed in this paragraph with other concepts of duality.

Let ξ_i denote a d-spherical fibration (with a section), $i = 1,2$. Let $\bar{f}: \xi_1 \longrightarrow \xi_2$ be a map covering $f: X_1 \longrightarrow X_2$; $\bar{f}$ induces a map $X_1^2 \cup_{X_1} \xi_1 \longrightarrow X_2^2 \cup_{X_2} \xi_2$, and hence a map $\bar{\bar{f}}: Th(\xi_1) \longrightarrow Th(\xi_2)$.

Proposition 1.17. In this situation there is a functor

$$f_*: \mathcal{DR}^n_{\xi_1}(X_1) \longrightarrow \mathcal{DR}^n_{\xi_2}(X_2)$$

given by

$$(Y_1, Y_1', u_1) \longmapsto (Y_1 \underset{X_1}{\cup} X_2,\ Y_1' \underset{X_1}{\cup} X_2, \bar{u}_1)$$

where $\bar{u}_1$ denotes the composite

$$(Y_1 \underset{X_1}{\cup} X_2) \wedge (Y_1' \underset{X_1}{\cup} X_2) = (Y_1 \wedge Y_1') \underset{X_1^2}{\cup} X_2^2 \longrightarrow Th_{n-d}(\xi_1) \underset{X_1^2}{\cup} X_2^2 \longrightarrow Th_{n-d}(\xi_2)$$

the last map being induced by $\bar{\bar{f}}$.

Proof: The fact which requires proof is that $\bar{u}_1$ indeed defines an n-duality map. Let $G_i = \pi_1 X_i$, $i = 1,2$. We have to show that the map

$$\bar{u}_1: H_q(Y_1' \underset{X_1}{\cup} X_2, X_2; \mathbf{Z}[G_2]) \longrightarrow H^{n-q}(Y_1 \underset{X_1}{\cup} X_2, X_2; \mathbf{Z}[G_2])$$

given by slant product with a certain class t in $H^d(Th(\xi_1), X_1^2; \mathbf{Z}[G_1])$ is an isomorphism for all $q \leqq n$.

Let $C_* = C_*(\tilde{Y}_1, \tilde{X}_1)$ (resp. $C_*' = C_*(\tilde{Y}_1', \tilde{X}_1)$) denote the chain complex of the universal cover of the pair (Y_1, X_1) (resp. $Y_1', X_1)$). Define $D_* = \mathrm{Hom}_{\mathbf{Z}[G_1]}(C_{n-*}, \mathbf{Z}[G_1])$. The chain complexes C_*, C_*', and D_* consist of free $\mathbf{Z}[G_1]$-modules of finite rank. There is a (degree 0) map $C_*' \longrightarrow D_*$, given by $c' \longrightarrow z/c'$, where $c' \in C_*'$, and z is a cocycle in $\mathrm{Hom}_{\mathbf{Z}[G_1 \times G_1]}(C_n((Y_1, X_1) \times (Y_1', X_1)); \mathbf{Z}[G_1])$ representing the image under u_1^* of the class t in $H^n(Y_1 \wedge Y_1', X_1 x X_1; \mathbf{Z}[G_1])$. Using this map define a map of double complexes

$$\varphi_{u_1}: F_* \underset{G_1}{\otimes} C' \longrightarrow F_* \underset{G_1}{\otimes} D_*$$

where F_* denotes a free $\mathbf{Z}[G_1]$-resolution of the G_1-module $\mathbf{Z}[G_2]$. There a spectral sequences associated to these double complexes which are given by

$$E^1_{p,q} = F_p \underset{G_1}{\otimes} H_q(C_*') = F_p \underset{G_1}{\otimes} H_q(Y_1', X_1; \mathbf{Z}[G_1])$$

$$\Longrightarrow H_{p+q}(C_* \underset{G_1}{\otimes} \mathbf{Z}[G_2]) = H_{p+q}(Y_1 \underset{X_1}{\cup} X_2, X_2; \mathbf{Z}[G_2]),$$

and

$$E'^1_{p,q} = F_p \underset{G_1}{\otimes} H_q(D_*) = F_p \underset{G_1}{\otimes} H^{n-q}(\mathrm{Hom}_{\mathbf{Z}[G_1]}(C_*, \mathbf{Z}[G_1]))$$

$$= F_p \underset{G_1}{\otimes} H^{n-q}(Y_1, X_1; \mathbf{Z}[G_1])$$

$$\Longrightarrow H_{p+q}(D_* \underset{G_1}{\otimes} \mathbf{Z}[G_2]) = H^{n-(p+q)}(Y_1 \underset{X_1}{\cup} X_2, X_2; \mathbf{Z}[G_2]).$$

By assumption u_1 is an n-duality map, therefore φ_{u_1} induces an isomorphism $E^1_{p,q} \xrightarrow{\approx} E'^1_{p,q}$. Hence we obtain an isomorphism of the abutments as was to be shown.
□

The proposition has the following immediate consequence. Given an n-duality map $u: Y \wedge Y' \longrightarrow Th_{n-d}(\varepsilon)$ in $\mathcal{DR}^n_\varepsilon(X)$ the induced map

$$\bar{u}: Y/X \wedge Y'/X \longrightarrow Th_{n-d}(\varepsilon)/X^2 \simeq S^n \wedge X_+ \longrightarrow S^n$$

is an ordinary n-duality map.

Indeed, this is the special case $X_1 = X$, $X_2 = *$ of the proposition.

The concept of duality is closely linked with some sort of *finiteness condition*. The definition of duality used in this paper requires that the cofibre of the inclusion $X \to Y$ is finite (at least up to homotopy). In view of this the concept of duality defined here might be called 'cofibre-wise' duality.
A different finiteness condition would be to ask that the fibre of the retraction $Y \to X$ be finite. The ensuing concept of duality would accordingly have to be called 'fibre-wise' duality. In the theory of 'fibre-wise' Spanier-Whitehead duality one starts with a space Y over X, satisfying that the structural map $Y \to X$ is a fibration with fibre homotopy equivalent to a finite complex. The Spanier-Whitehead dual is then defined by taking the ordinary dual of each fibre, cf. [8], [9], [20].
In contrast with prop. 1.17. it turns out that this kind of duality depends contravariantly on the base space. Moreover, one has an 'operation' of 'fibre-wise' duality on 'cofibre-wise' duality. This is induced from the action of the category $\mathcal{R}_{fib}(X)$ on $\mathcal{R}(X)$, cf. the definition after lemma 1.1.

The concepts of cofibre-wise duality and of fibre-wise duality are different in general: Even if both kinds of duals are defined, they may not coincide. As an example consider the case where X is a compact n-dimensional manifold with boundary ∂. Let $Y = X \times S^0$. Y is a space over X via the projection. Under these circumstances both concepts of duality are defined. A fibre-wise n-dual is given by $X \times S^n$, whereas a cofibre-wise n-dual is given by $X \underset{\partial}{\cup} X$, the double of X, as one may see from the geometric description of duality given below in lemma 2.8.

§ 2. The canonical involution in the algebraic K-theory of spaces

One of the reasons for studying algebraic K-theory is that it gives information about the *concordance spaces* (or synonymously: pseudo-isotopy spaces) of manifolds.

Reacall the definition. Let X be a compact manifold with boundary ∂X. Let $C(X) = \mathrm{Aut}(X \times [0,1] \text{ rel } X \times 0 \cup \partial X \times [0,1])$, where Aut(...) denotes CAT-automorphisms of X. There is a canonical involution $\iota\colon C(X) \longrightarrow C(X)$ given by

$$\iota(f) = (\mathrm{id} \times r) \circ f \circ (\mathrm{id} \times r) \circ ((f|X \times 1)^{-1} \times \mathrm{id}),$$

where $r\colon [0,1] \to [0,1]$ denotes the reflection at the midpoint. This involution on $C(X)$ gives (after localizing away from 2) a splitting up to homotopy into the two eigenspaces of ι:

$$C(X) \simeq C(X)^s \times C(X)^a,$$

where $C(X)^s$ (resp. $C(X)^a$) denotes the symmetric (resp. anti-symmetric) part of $C(X)$. The interest in this splitting comes from the fact that the factors have a meaning of their own, and can be treated with different methods, cf. [4].

There is a stabilization map

$$\Sigma\colon C(X) \longrightarrow C(X \times I)$$

given by product with the interval (or rather a technical modification of this, because of the condition of standard behaviour at the boundary). The *stable concordance space* of X is defined as $\underline{C}(X) = \lim\limits_{\vec{k}} C(X \times I^k)$. By [4], $\underline{C}(X)$ is a homotopy functor.

It is this space that can be related to the algebraic K-theory of X. Following [18], we shall describe this relationship. Let X denote a compact manifold of dimension d with boundary ∂X; I denotes the interval [a,b]. A *partition* is a triple (M,F,N), where M is a compact codimension zero submanifold of $X \times I$, N is the closure of the complement of M, and $F = M \cap N$. F is to be standard in a neighborhood of $\partial X \times I$, i.e. there exists a number $t \in I$ such that F equals $X \times t$ in this neighborhood.

Let $P(X)$ denote the simplicial set in which a p-simplex is a (CAT-) locally trivial family of partitions parametrized by the p-simplex Δ^p. Let $H(X)$ denote the simplicial subset of $P(X)$ defined by the condition that M is an h-cobordism rel. boundary between $X \times 0$ and F. $H(X)$ is called the h-*cobordism-space* of X.

Proposition 2.1. $H(X)$ is a classifying space for $C(X)$.

Proof: To prove this, we construct a free action of the simplicial group $C(X)$ on some contractible space $E(X)$ such that the orbit space is the component of $H(X)$ containing the trivial cobordism. Let $E(X)$ denote the space (=simplicial set) of embeddings $X \times [0,1] \longrightarrow X \times [a,b]$ restricting to the identity on $X \times a$. This is a space of collars and hence is contractible. $C(X)$ acts on $E(X)$ by composition of maps. $E(X)$ may also be viewed as the space of trivial h-cobordisms together with a given trivialization. The action of $C(X)$ then just changes the trivialization. Hence the orbit space is indeed the 0-component of $H(X)$, as was to be shown. □

There is an obvious involution on the space $H(X)$, given ty turning a partition upside down. One may ask what is the relation between the involutions on $C(X)$ and $H(X)$.

Proposition 2.2. The involutions on $C(X)$ and on $H(X) = BC(X)$ agree up to homotopy.

Proof: Define a map $\varphi: C(X) \longrightarrow E(X)$ by $\varphi(f)(x,t) = \alpha(f(x,t/2))$, where $\alpha: X\times[0,1] \to X\times[a,b]$ is the canonical linear isomorphism. Similarly, let $\psi: C(X) \longrightarrow E(X)$ take f to $\psi(f)$: $(x,t) \longmapsto \alpha(\iota(f)(x,t/2)$. There is a natural map $p: E(X) \longrightarrow H(X)$, defined by forgetting the product structure of a collar:

$$f: X\times[0,1] \longrightarrow X\times[a,b] \longrightarrow (\mathrm{im}(f),\ldots) .$$

Let $\iota': H(X) \longrightarrow H(X)$ denote the involution on $H(X)$. We obtain a pull-back diagram

$$\begin{array}{ccc} C(X) & \xrightarrow{\psi} & E(X) \\ {\scriptstyle\varphi}\downarrow & & \downarrow{\scriptstyle\iota' p} \\ E(X) & \xrightarrow[p]{} & H(X) \end{array} .$$

Clearly the involution on $C(X)$ can be described by interchanging the corners of the diagram and applying the involution ι' on $H(X)$. In view of prop. 2.1. the diagram is also homotopy cartesian. This proves the proposition because from the diagram one obtains homotopy equivalences

$$E(X) \times_{H(X)} E(X) \xrightarrow{\simeq} E(X) \times_{H(X)} H(X)^I \times_{H(X)} E(X) \xleftarrow{\simeq} \Omega H(X)$$

which are compatible with the involutions if the middle term is given the involution defined by

$(f,w: I \to H(X),g) \longmapsto (g, \iota' \circ w \circ r, f)$, $r: I \to I$ the reflection map.

□

On the simplicial set of partitions $P(X)$ define a partial ordering by letting $(M,F,N) < (M',F',N')$ if firstly M is contained in M', and secondly the maps

$$F' \longrightarrow M' - (M - F) \longleftarrow F$$

are homotopy equivalences. This defines a simplicial partially ordered set, and hence a simplicial category which will be denoted $hP(X)$. We have a particular partition given by attaching k trivial m-handles to $X\times[a,a']$ in such a way that the complementary (d-m)-handles are trivially attached to $X\times[b,b']$, $a<a'<b'<b$. Let $hP_k^{\ell,m}(X)$ be the connected component of $hP(X)$ containing this particular partition, $(\ell = d-m)$.

An (anti-)involution on $hP(X)$ is defined by the contravariant functor

$$T': hP(X) \longrightarrow hP(X), \quad (M,F,N) \longmapsto (N^*,F^*,M^*),$$

where M^* (resp. N^*) is the image of M (resp. N) under the map $\mathrm{id}\times r: X\times I \longrightarrow X\times I$. It restricts to a contravariant functor

$$T': hP_k^{\ell,m}(X) \longrightarrow hP_k^{m,\ell}(X) .$$

In [18] it is proved that the categories $hP_k^{\ell,m}(X)$ approximate $A(X)$. To fit these approximations together one needs a stabilization process. There are two ways to stabilize a partition (M,F,N), namely taking the lower (resp. upper) part to its product with an interval. These disagree because of the condition of standard behaviour near the boundary. We have to consider a technical modification of the various spaces of partitions. We fix some standard choices. Let $X' \subset \text{Int}\, X$ be a submanifold of X such that $\text{Cl}(X - X')$ is a collar on ∂X. Similarly, let J denote an interval containing two subintervals J', J'' such that $J' \subset \text{Int}\, J$, $J'' \subset \text{Int}\, J'$, further let $[a',b']$ be a symmetric subinterval of I.

Let $\underline{P}(X)$ be the simplicial subset of $P(X)$ of those partitions satisfying that

$$F \subset X \times [a',b'] \;;\; F \cap (X - X') \times I = (X - X') \times a'.$$

The inclusion $\underline{P}(X) \subset P(X)$ (resp. $h\underline{P}(X) \subset hP(X)$) is a homotopy equivalence. Define the *lower stabilization* as the map

$$\sigma_\ell : h\underline{P}(X) \longrightarrow h\underline{P}(X \times J)$$

which takes the lower part of a partition (M,F,N) to

$$M \times J' \cup X \times [a,a'] \times J \subset X \times I \times J.$$

The upper part of a partition is mapped by σ_ℓ to the fibrewise suspension of N considered as a space over X.

The *upper stabilization* is the map

$$\sigma_u : h\underline{P}(X) \longrightarrow h\underline{P}(X \times J)$$

defined by

$$M \longmapsto M \times J' \cup X' \times [a,b'] \times \text{Cl}(J' - J'') \cup X \times [a,a'] \times J \subset X \times I \times J.$$

The involution T' does not restrict to a map $\underline{P}(X) \longrightarrow \underline{P}(X)$ (because of the standard behaviour near the boundary). So a slight modification of T' is necessary. Choose a map $j: P(X) \longrightarrow \underline{P}(X)$ homotopy inverse to the natural inclusion $i: \underline{P}(X) \longrightarrow P(X)$. Letting $T = jT'i$ define a map $T: \underline{P}(X) \longrightarrow \underline{P}(X)$ (resp. a contravariant functor $T: h\underline{P}(X) \longrightarrow h\underline{P}(X)$), and one verifies

Lemma 2.3. (i) T is an involution up to homotopy, i.e. $T^2 \simeq \text{id}$;
(ii) $\sigma_u T \simeq T\sigma_\ell$; (iii) $T\sigma_u \simeq \sigma_\ell T$. □

Now consider the limit $\lim\limits_{\substack{\rightarrow \\ \ell,m}} h\underline{P}(X \times J^{\ell+m-d})$ where the maps in the direct system are given by σ_ℓ (resp. σ_u). Using a mapping cylinder argument, T can be defined as a map

$$\lim_{\substack{\rightarrow \\ \ell,m}} h\underline{P}(X \times J^{\ell+m-d}) \longrightarrow \lim_{\substack{\rightarrow \\ \ell,m}} h\underline{P}(X \times J^{\ell+m-d})$$

and from lemma 2.3. we have

Lemma 2.4. T is a weak involution in the sense that the restriction of T^2 to any compactum is homotopic to the restriction of the identity; in particular, T induces an involution on homotopy groups. □

In [18] it is proved that a connected component of A(X) can be obtained by performing the + construction on the space

$$\left| \lim_{\substack{\rightarrow \\ k,\ell,m}} h\underline{P}_k^{\ell,m}(X \times J^{\ell+m-d}) \right| .$$

Hence lemma 2.4. provides a weak involution on A(X). We continue to denote this involution with the letter T.

The relation between algebraic K-theory and concordance spaces may be described by a certain commutative diagram

$$\begin{array}{ccc} \lim\limits_{\substack{\rightarrow \\ \ell,m}} \underline{H}(X \times J^{\ell+m-d}) & \longrightarrow & \lim\limits_{\substack{\rightarrow \\ k,\ell,m}} \underline{P}_k^{\ell,m}(X \times J^{\ell+m-d}) \\ \downarrow & & \downarrow \\ \lim\limits_{\substack{\rightarrow \\ \ell,m}} h\underline{H}(X \times J^{\ell+m-d}) & \longrightarrow & \lim\limits_{\substack{\rightarrow \\ k,\ell,m}} h\underline{P}_k^{\ell,m}(X \times J^{\ell+m-d}) \end{array}$$

where $\underline{H}(\ldots)$ denotes the intersection of $H(\ldots)$ with $\underline{P}(\ldots)$, $h\underline{H}(\ldots)$ is the simplicial subcategory of $h\underline{P}(\ldots)$ with $\underline{H}(\ldots)$ as simplicial set of objects. The vertical maps of the diagram are the natural inclusions, and the horizontal maps are given by the identification of $\underline{H}(\ldots)$ (resp. $h\underline{H}(\ldots)$) with $\underline{P}_o^{\ell,m}(\ldots)$ (resp. $h\underline{P}_o^{\ell,m}(\ldots)$). In [18] it is shown that after performing the + construction the diagram is homotopy cartesian in a range of dimensions. Further the term in the lower left corner is contractible.

Each of the terms in the diagram has a description in terms of spaces of partitions. Further the operation of turning a partition upside down gives an involution on each of these spaces. The vertical maps in the diagram are compatible with the involution by definition of the involution on the categories $h\underline{P}(X)$, resp. $h\underline{H}(X)$. The horizontal maps are given by the canonical map from a member of a direct system to its limit. They are compatible with the involution because the involution is defined on each of the spaces in the direct system.

Our next goal will be to relate the involution T to another one defined in a quite different manner. We return to the setting of § 1. The description of A(X) given there (in particular cor. 1.14.) provides a natural involution on A(X) in a straightforward way. The details are as follows.

Let X be a simplicial set, $\xi \longrightarrow X$ an (orientable) d-spherical fibration (with a section); Th(ξ) is a Thom space of ξ in the sense of § 1. Let ρ_n denote the self map of S^n given by the following permutation of factors:

$$S^n = S_1^1 \wedge S_2^1 \wedge \ldots \wedge S_n^1 \xrightarrow{\approx} S_n^1 \wedge S_{n-1}^1 \wedge \ldots \wedge S_1^1 = S^n .$$

Define a contravariant functor

$$\tau_{\xi,n}: h\mathcal{DR}^n_\xi(X) \longrightarrow h\mathcal{DR}^n_\xi(X)$$
$$(Y,Y',u) \longmapsto (Y',Y,\tau_{\xi,n}(u)),$$

where $\tau_{\xi,n}(u): Y' \wedge Y \xrightarrow{\approx} Y \wedge Y' \xrightarrow{u} S^{n-d} \wedge Th(\xi) \xrightarrow{\rho_{n-d} \wedge \iota} S^{n-d} \wedge Th(\xi) = Th_{n-d}(\xi)$.

By lemma 1.1. this is a duality map again. Clearly, $\tau^2_{\xi,n} = id$. The map ι in this definition ensures that $\tau_{\xi,n}$ is a map over $X \times X$, and the map ρ_{n-d} is introduced to guarantee compatibility with the suspension functors. Indeed, one easily verifies that

(i) $\Sigma_\ell \tau_{\xi,n} = \tau_{\xi,n+1}\Sigma_r$, (ii) $\Sigma_r \tau_{\xi,n} = \tau_{\xi,n+1}\Sigma_\ell$.

Therefore one has a well-defined functor

$$\tau_\xi: \varinjlim_{\Sigma_\ell,\Sigma_r} h\mathcal{DR}^n_\xi(X) \longrightarrow \varinjlim_{\Sigma_\ell,\Sigma_r} h\mathcal{DR}^n_\xi(X)^{op} ,$$

and in particular τ restricts to a functor

$$\tau_\xi: \varinjlim_{\ell,m} h\mathcal{DR}^{\ell,m}_k(X) \longrightarrow \varinjlim_{\ell,m} h\mathcal{DR}^{\ell,m}_k(X)^{op} .$$

By cor. 1.16. the categories $h\mathcal{DR}^{\ell,m}_k(X)$ approximate $A(X)$. So one finally obtains an involution τ_ξ on $A(X)$ depending on the spherical fibration ξ. By prop. 1.17. this involution is natural for maps of pairs $(X,\xi) \longrightarrow (X',\xi')$, where $\xi \longrightarrow \xi'$ covers $X \longrightarrow X'$.

We now want to investigate the dependence of the involution τ_ξ on the spherical fibration ξ.

Recall from § 1 that there is an operation of (spherical) fibrations on spaces over X, given by $(Y,\xi) \longmapsto \xi \cdot Y$. Let $\xi\cdot: A(X) \longrightarrow A(X)$ denote the map induced from this operation. Let ξ^{-1} denote an inverse of ξ; ε is the trivial spherical fibration $X \times S^0 \longrightarrow X$.

Proposition 2.5. The following diagram commutes up to homotopy:

$$\begin{array}{ccc} A(X) & \xrightarrow{\ \tau_\varepsilon\ } & A(X) \\ {\scriptstyle \xi^{-1}\cdot}\downarrow & \searrow^{\tau_\xi} & \downarrow{\scriptstyle \xi\cdot} \\ A(X) & \xrightarrow[\ \tau_\varepsilon\ ]{} & A(X) \end{array}$$

Proof: We prove that the right triangle commutes up to homotopy. The proof for the other triangle is entirely analogous. By cor. 1.3. there are maps φ_ξ (resp. ψ_ξ): $h\mathcal{DR}^n_\varepsilon(X) \longrightarrow h\mathcal{DR}^{n+d}_\xi(X)$ given by

$$(Y,Y',u) \longmapsto (\xi \cdot Y, Y', (\xi \wedge \varepsilon) \cdot u)$$

$$(\text{resp.} \quad (Y,Y',u) \longmapsto (Y, \xi \cdot Y', (\varepsilon \wedge \xi) \cdot u)).$$

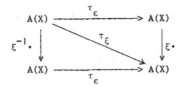

Let $\delta_\xi\colon h\mathcal{DR}^n_\xi(X) \longrightarrow h\mathcal{R}_{hf}(X)$ denote the forgetful functor $(Y,Y',u) \longmapsto Y$. There is a commutative diagram

$$\begin{array}{ccccccc}
h\mathcal{R}_{hf}(X) & \xleftarrow{\delta_\varepsilon} & h\mathcal{DR}^n_\varepsilon(X) & \xrightarrow{\tau_\varepsilon} & h\mathcal{DR}^n_\varepsilon(X) & \xrightarrow{\delta_\varepsilon} & h\mathcal{R}_{hf}(X) \\
\| & & \downarrow{\scriptstyle \psi_\xi} & & \downarrow{\scriptstyle \varphi_\xi} & & \downarrow{\scriptstyle \xi\cdot} \\
h\mathcal{R}_{hf}(X) & \xleftarrow[\delta_\xi]{} & h\mathcal{DR}^{n+d}_\xi(X) & \xrightarrow[\tau_\xi]{} & h\mathcal{DR}^{n+d}_\xi(X) & \xrightarrow[\delta_\xi]{} & h\mathcal{R}_{hf}(X)
\end{array} .$$

Restricting to the connected components of 'spherical objects', and passing to the limit gives homotopy equivalences δ_ε (resp. δ_ξ) by prop. 1.6. together with lemma 1.5., and also φ_ξ (resp. ψ_ξ) by lemma 1.5. The upper row of the diagram represents the involution τ_ε on A(X), the lower row represents τ_ξ. This proves the proposition. □

Of course, an involution on A(X) can also be defined using simplicial sets with group action. It is induced by the contravariant functor, also denoted τ_n,

$$h\mathcal{DU}^n(G) \longrightarrow h\mathcal{DU}^n(G), \qquad (M,M',u) \longmapsto (M',M,\tau_n(u)),$$

where

$$\tau_n(u)\colon M' \wedge M \xrightarrow{\approx} M \wedge M' \xrightarrow{u} Ex^\infty(S^n \wedge G_+) \xrightarrow{Ex^\infty(\rho_n \wedge \iota)} Ex^\infty(S^n \wedge G_+),$$

$$(\iota\colon G_+ \longrightarrow G_+,\ g \longmapsto g^{-1}).$$

One easily verifies that the functors $D\Phi$ and $D\Psi$ are equivariant with respect to this involution. By prop. 1.14. this involution is therefore the same, up to homotopy, as the involution τ_ε defined just before.

In the following it will be convenient to have a slightly different description of the categories $h\mathcal{DR}^n_\xi(X)$ available. Namely, instead of working with simplicial sets one could as well use spaces having the homotopy type of CW complexes and continuous maps throughout to define the categories $h\mathcal{DR}^n_\xi(X)$. Geometric realization induces a functor $h\mathcal{DR}^n_\xi(X) \longrightarrow h\mathcal{DR}^n_\xi(|X|)$ which is a weak homotopy equivalence.
In the following the symbol $h\mathcal{DR}^n_\xi(X)$ will have either of these two meanings depending on whether X is a simplicial set or a topological space.

To compare the involution defined on $h\underline{P}(X)$ with that defined on $h\mathcal{DR}^n_\xi(X)$ one has to relate both categories. Now $h\mathcal{DR}^n_\xi(X)$ is a category while $h\underline{P}(X)$ is a simplicial category. So to compare both one has to make $h\mathcal{DR}^n_\xi(X)$ a simplicial category as well.
Let $h\mathcal{DR}^n_\xi(X)_p$ denote the category with objects locally trivial p-parameter families Y,Y' of objects of $h\mathcal{R}_{hf}(X)$ together with a p-parameter family of n-duality maps

$$\begin{array}{ccc}
Y \wedge Y' & \longrightarrow & Th_{n-d}(\xi) \times \Delta^p \\
& \searrow \quad \swarrow & \\
& \Delta^p &
\end{array}$$

Similarly, morphisms are given by p-parameter families of morphisms of $h\mathcal{DR}^n_\xi(X)$.
(Here $Y \wedge Y'$ denotes a p-parameter version of the fibrewise smash product over X,

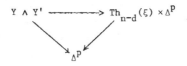

namely

$$Y \wedge Y' = Y \times_{\Delta^p} Y' \cup_{(X\times\Delta^p)\times_{\Delta^p} Y' \cup Y \times_{\Delta^p} (X\times\Delta^p)} X \times \Delta^p \times X \quad .)$$

The categories $h\mathcal{DR}^n_\xi(X)_p$ assemble to a simplicial category which is denoted $h\mathcal{DR}^n_\xi(X)_\bullet$. Forgetting part of the structure we also have a simplicial category $h\mathcal{R}_{hf}(X)_\bullet$. Identifying $h\mathcal{DR}^n_\xi(X)$ with $h\mathcal{DR}^n_\xi(X)_o$, the total degeneracy map gives an inclusion $h\mathcal{DR}^n_\xi(X) \longrightarrow h\mathcal{DR}^n_\xi(X)_\bullet$, and we have

Lemma 2.6. The inclusion $h\mathcal{DR}^n_\xi(X) \longrightarrow h\mathcal{DR}^n_\xi(X)_\bullet$ is a weak homotopy equivalence.

Proof: Indeed, this follows from the fact that the maps $h\mathcal{DR}^n_\xi(X) \longrightarrow h\mathcal{DR}^n_\xi(X)_k$ are weak homotopy equivalences for all k. □

Now let X denote a compact (orientable) manifold of dimension d. There is a map $h\underline{P}(X) \longrightarrow h\mathcal{R}(X)_\bullet$ given by $(M,F,N) \longmapsto M$, where M is considered as a space over $X \times a$. We want to lift this to a map $h\underline{P}(X) \longrightarrow h\mathcal{DR}^n_\xi(X)_\bullet$ (for suitable ξ and n) in such a way that it is compatible with the involution on both terms. To do so one associates to a partition (M,F,N) a duality map as follows. Let $M' = M-F$, $N' = N-F$. The inclusion

$$i\colon M' \times N' \longrightarrow (X\times[a,b])^2 - \text{diagonal}$$

induces a map over $X\times X$

$$j\colon M' \wedge N' \longrightarrow ((X\times[a,b])^2 - \Delta) \cup_{X^2\times[a,b]\times b \,\cup\, X^2\times a\times[a,b]} X^2$$

(Δ = diagonal).

If (M,F,N) is a p-parameter family of partitions, one replaces the product $M' \times N'$ by the fibre product $M' \times_{\Delta^p} N'$, and $M' \wedge N'$ by the p-parameter version of the smash-product defined above.

Let Z denote the target of the map j. Z is an object of $\mathcal{R}(X^2)$ by the obvious projection map, and the inclusion given by

$$X \times X \longrightarrow X\times a\times X\times b \longrightarrow Z \;.$$

Let ξ denote the tangent microbundle of X (resp. an $\mathbf{R}^n$-bundle to which it corresponds by the Kister-Mazur theorem). Let ξ^+ denote the fibrewise one-point-compactification of ξ. ξ^+ is an orientable d-spherical fibration with a section. Convert the map $Z \longrightarrow X^2$ into a fibration $Z' \longrightarrow X^2$.

Lemma 2.7. $Z' \longrightarrow X^2$ is a Thom space of ξ^+ in the sense of § 1.

Proof: We first show that $Z \longrightarrow X^2$ is in the same component of $h\mathcal{R}(X\times X)$ as the object $X\times X \cup_X \xi^+$. We represent ξ^+ as follows. Let η denote a neighborhood of the diagonal in $X\times X$ which is an $\mathbf{R}^n$-bundle; η' is a smaller neighborhood satisfying the same requirement. Then $\xi^+ = \eta \cup_{\eta-\eta'} X$. We have the following chain of homotopy equivalences

$$Z' \xleftrightarrow{\simeq} Z \xleftarrow{\simeq} (X\times[a,b])^2 - \Delta \xleftarrow{\simeq} (X\times[a,b])^2 - U$$

$U = \eta' \times \Delta'$, Δ' = diagonal of $[a,b] \times [a,b]$)

$$= (X^2 - \eta') \times [a,b]^2 \cup_{(X^2 - \eta') \times ([a,b]^2 - \Delta')} X^2 \times ([a,b]^2 - \Delta')$$

$$\xrightarrow{\cong} (X^2 - \eta') \cup_{(X^2 - \eta') \times S^o} X^2 \times S^o = X^2 \cup_{X^2 - \eta'} X^2$$

$$= X^2 \cup_{(X^2 - \eta') - (X^2 - \eta)} X^2 - (X^2 - \eta) = X^2 \cup_{\eta - \eta'} \eta \ .$$

Now the inclusion $\eta - \eta' \longrightarrow X^2$ is homotopic to the composite $\eta - \eta' \xrightarrow{pr} X \xrightarrow{\Delta} X^2$, as one sees from the diagram

$$\begin{array}{ccccc} \eta-\eta' & \longrightarrow & \eta & \longrightarrow & X^2 \\ & {\scriptstyle pr}\searrow & \downarrow{\scriptstyle pr} & \nearrow{\scriptstyle \Delta} & \\ & & X & & \end{array}$$

where the left triangle is commutative, and the right triangle commutes up to homotopy since $\eta \xrightarrow{pr} X$ is a homotopy equivalence. Therefore $X^2 \cup_{\eta-\eta'} \eta$ is in the same connected component of $h\mathcal{R}(X^2)$ as the object

$$\text{pushout } (X^2 \longleftarrow X \longleftarrow \eta - \eta' \longrightarrow \eta) =$$

$$X^2 \cup_X (X \cup_{\eta-\eta'} \eta) = X^2 \cup_X \xi^+ \ .$$

Define a map $\iota: Z \longrightarrow Z$ by $(x,x',s,t) \longmapsto (x',x,r(t),r(s))$, where $s,t \in [a,b]$ and $r: [a,b] \to [a,b]$ is the reflection map. Clearly, $\iota^2 = \mathrm{id}$. It induces a map $\iota': Z' \longrightarrow Z'$ with the same property. Therefore Z' has all properties required of a Thom space of ξ^+ .

□

Lemma 2.8. The map j is a d-duality map.

Proof: Let $t \in H^{d+1}((X \times [a,b])^2, (X \times [a,b])^2 - \Delta)$ be a Thom class of the tangent microbundle of $X \times [a,b]$. The exact sequence of the triple $((X \times [a,b])^2, Z, X^2)$ identifies t with a generator t' of $H^d(Z, X^2)$. There is a commutative diagram ($q \leqq d$)

$$\begin{array}{ccc} H_q(N', X \times b) & \xrightarrow{\alpha_j} & H^{d-q}(M', X \times a) \approx H^{d-q}(M, X \times a) \\ \downarrow \approx & & \downarrow \approx \\ H_q(X \times [a,b] - M, X \times [a,b] - X \times [a,b)) & \xrightarrow[\gamma_t]{} & H^{d+1-q}(X \times [a,b), M) \end{array}$$

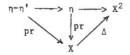

where $\alpha_j(z) = j^*(t')/z$. (All homology groups have $\mathbf{Z}[\pi_1 X]$-coefficients.) The vertical isomorphism on the right comes from the exact sequence of the triple $(X \times [a,b], M, X \times a)$. The bottom map γ_t is the usual Alexander duality isomorphism. (One has to be a little careful, since the assumptions of the duality theorem are not quite satisfied here, e.g. $X \times [a,b)$ is not compact, and, more seriously, M is not contained in the interior of $X \times [a,b]$. But in the special situation at hand this does not affect the result because the intersection of M with the boundary of $X \times [a,b]$ is homotopy equivalent to X.) Hence α_j is an isomorphism as asserted.

The last two lemmas provide a map

$$f\colon h\mathcal{P}(X) \longrightarrow h\mathcal{DR}^d_{\xi^+}(X).$$

$$(M,F,N) \longmapsto (M',N',j)$$

which is compatible with the involutions T' (resp. τ). Of course, f restricts to a map

$$f\colon h\mathcal{P}^{\ell,m}_k(X) \longrightarrow h\mathcal{DR}^{\ell,m}_k(X). \quad (\ell+m=d).$$

We would like to stabilize this map with respect to dimension. In order to do so one first has to replace the categories $h\mathcal{P}^{\ell,m}_k(X)$ by $h\underline{\mathcal{P}}^{\ell,m}_k(X)$. Secondly, one modifies the suspension maps on the catefories $h\mathcal{DR}^n_\xi(X)$. Namely, let

$$\Sigma'_\ell\colon \mathcal{DR}^n_\xi(X) \longrightarrow \mathcal{DR}^n_\xi(X\times J')$$

$$(Y,Y',u' \longmapsto (Y\times J' \cup_{Y\times \partial J'} X\times\partial J', Y'\times J',\ldots),$$

and similarly with Σ_r. (J' denotes some interval). We obtain a diagram

$$\begin{array}{ccc} h\underline{\mathcal{P}}^{\ell,m}_k(X) & \xrightarrow{\quad f \quad} & h\mathcal{DR}^{\ell,m}_k(X). \\ \sigma_\ell\downarrow & & \downarrow\Sigma_r \\ h\underline{\mathcal{P}}^{\ell,m}_k(X\times J) & \xrightarrow[\quad f' \quad]{} & h\mathcal{DR}^{\ell,m}_k(X\times J'). \end{array}$$

where f' takes a partition (M,F,N) in $h\underline{\mathcal{P}}(X\times J)$ to

$$(M'-F\cup_{X\times J} X\times J',\ N'-F\cup_{X\times J} X\times J',\ldots), \text{ and}$$

and $X\times J \xrightarrow{\approx} X\times J'$ is given by some fixed isomorphism $J \xrightarrow{\approx} J'$. This diagram commutes up to homotopy since there is a natural transformation $\Sigma_r f \longrightarrow f'\sigma_\ell$ which is given by

$$M\times J' \longmapsto M\times J' \cup X\times[a,a']\times J$$

$$N\times J' \cup_{N\times\partial J'} X\times\partial J' \longmapsto N\times J \cup_{N\times Cl(J-J')} X\times\partial J'$$

There is a similar diagram with σ_ℓ (resp. Σ_r) replaced by σ_u (resp. Σ_ℓ).

Hence one obtains a map in the limit

$$f\colon \varinjlim_{k,\ell,m} h\underline{\mathcal{P}}^{\ell,m}_k(X\times J^{\ell+m-d}) \longrightarrow \varinjlim_{k,\ell,m} h\mathcal{DR}^{\ell,m}_k(X\times J'^{\ell+m-d}),$$

which is well-defined up to homotopy. Standard mapping cylinder arguments now show that f is compatible with the involutions up to weak homotopy, i.e. the restrictions of τf (resp. fT) to any compactum are homotopic.

Lemma 2.9. The map f is a weak homotopy equivalence.

Proof: Composing f with the forgetful map

$$g: \lim_{\substack{\rightarrow \\ k,\ell,m}} h\mathcal{DR}_k^{\ell,m}(X \times J'^{\,\ell+m-d})_{\circ} \longrightarrow \lim_{\substack{\rightarrow \\ k,\ell,m}} h\mathcal{R}_k^{\ell}(X \times J'^{\,\ell+m-d})_{\circ}$$

gives (up to a minor modification) the map proved to be a homotopy equivalence in [18, prop. 5.4.]. The map g is a homotopy equivalence by prop. 1.6. and lemma 2.6. Hence f is also a homotopy equivalence, as was to be shown. □

Corollary 2.10. The involutions defined by T and τ on A(X) agree up to weak homotopy. □

Our next goal is to show that the involution on A(X) gives upon 'linearization' the usual involution on the K-theory of (group) rings. We first have to explain the meaning of this statement. Let R be a ring. We define the K-theory of R to be

$$K(R) = \mathbb{Z} \times BGl(R)^+ ,$$

that is, we replace the class group by $\mathbb{Z}$ in order to make the analogy with A(X) more transparent. Let R be equipped with an anti-involution $\bar{\ }: R \to R$. For a typical example let $R = \mathbb{Z}[G]$, the group ring of a group G, and the anti-involution being defined by $g \mapsto g^{-1}$, $g \in G$. There is an induced involution on $Gl_k(R)$ given by $A \mapsto (\bar{A}^t)^{-1}$, the conjugate transpose inverse of A. This defines the usual involution on K(R).

There is a canonical map $A(X) \longrightarrow K(R)$, called 'linearization', where $R = \mathbb{Z}[\pi_1 X]$, cf. [16]. In order to define this map we use a slightly different description of K(R). Namely, let $iso\mathcal{F}_k(R)$ denote the category of free (right) R-modules of rank k and their isomorphisms. The canonical inclusion $Gl_k(R) \longrightarrow iso\mathcal{F}_k(R)$ is an equivalent of categories. This allows one to define

$$K(R) = \mathbb{Z} \times |\lim_{\substack{\rightarrow \\ k}} iso\mathcal{F}_k(R)|^+ .$$

The linearization map is induced by the functors

$$h\mathcal{U}_k^{\ell}(G) \longrightarrow iso\mathcal{F}_k(R)$$
$$M \longmapsto H_{\ell}^{G}(M),$$

$k,\ell \geqq 0$, $G = G(X)$, the loop group of X.

Proposition 2.11. The linearization map $A(X) \longrightarrow K(\mathbb{Z}[\pi_1 X])$ is equivariant with respect to the involution on both terms.

Proof: Let $R = \mathbb{Z}[\pi_1 X] = \mathbb{Z}[\pi_0 G]$. Let $iso\mathcal{DF}_k(R)$ denote the category of triples (A,A',u), where A and A' are free (right) R-modules of rank k, and $u: A \otimes A' \longrightarrow R$ is an $R \otimes R$-map defining a non-singular pairing. (R is a right $R \otimes R$-module by letting $r.(s \otimes t) = \bar{t}\, r\, s$.) A morphism $(A,A',u) \longrightarrow (B,B',v)$ is a pair of isomorphisms $f: A \longrightarrow B$, $f': B' \longrightarrow A'$, such that $u(f' \otimes id) = v(id \otimes f)$. The category $iso\mathcal{DF}_k(R)$ has an involution defined by $(A,A',u) \mapsto (A',A,\bar{u})$. The canonical inclusion $Gl_k(R) \longrightarrow iso\mathcal{DF}_k(R)$ is compatible with the involutions on both terms. Further there is a functor

$$h\mathcal{DU}_k^{\ell,m}(G) \longrightarrow iso\mathcal{DF}_k(R)$$
$$(M,M',u) \longmapsto (H_\ell^G(M),\ H_m^G(M'),\ H_{\ell+m}^G(u)).$$

The category $iso\mathcal{DF}_k(R)$ was designed in exactly such a way as to make this map equivariant with respect to the involutions. Altogether we obtain a commutative diagram

$$\begin{array}{ccccc} h\mathcal{DU}_k^{\ell,m}(G) & \longrightarrow & iso\mathcal{DF}_k(R) & \longleftarrow & Gl_k(R) \\ \downarrow & & \downarrow & & \| \\ h\mathcal{U}_k^{\ell}(G) & \longrightarrow & iso\mathcal{F}_k(R) & \longleftarrow & Gl_k(R). \end{array}$$

The middle vertical map is an equivalence of categories; the vertical map on the left becomes a homotopy equivalence after passing to the limit with respect to ℓ and m by prop. 1.13. Therefore the upper left arrow is an approximation to the linearization map. We have seen above that the arrows in the upper row of the diagram preserve the involution. This proves the proposition. □

Remark: The linearization map considered above is actually a special case of a more general natural transformation from the K-theory of spaces to the K-theory of simplicial rings in the sense of [16]. The K-theory of a simplicial ring R can be defined from the category of free simplicial R-modules in a way formally quite similar to the construction of A(X) from the category of free pointed simplicial G-sets. The natural transformation is then given by the map $A(X) \longrightarrow K(\mathbb{Z}[G])$ (G=simplicial loop group of X), which associates to a free pointed simplicial G-set M the simplicial $\mathbb{Z}[G]$-module $\tilde{\mathbb{Z}}[M]$, the underlying simplicial abelian group of which is freely generated by the non-basepoint elements of M. Now in the context of simplicial $\mathbb{Z}[G]$-modules the concept of duality can be defined in complete analogy to the 'non-linear' case, and $K(\mathbb{Z}[G])$ can be constructed from a larger category of $\mathbb{Z}[G]$-modules by including duality data. This again leads to an involution on $K(\mathbb{Z}[G])$, which by its very construction is compatible with that on A(X) via the linearization map.
The composition of the linearization map $A(X) \longrightarrow K(\mathbb{Z}[G])$ with the map $K(\mathbb{Z}[G]) \longrightarrow K(\mathbb{Z}[\pi_0 G]) = K(\mathbb{Z}[\pi_1 X])$ induced from the connected component map $G \longrightarrow \pi_0 G$ is identical with the map of proposition 2.11.

§ 3. The splitting theorem

In this section we apply the concept of duality developed in the previous sections to give another proof of the splitting theorem, [17], [18]:

Theorem: The canonical map $\Omega^\infty S^\infty(X_+) \longrightarrow A(X)$ is a coretraction up to weak homotopy.

The theorem will be proved by constructing a splitting map $A(X) \longrightarrow \Omega^\infty S^\infty(X_+)$. To make the proof more transparent we give an informal preview of the argument.

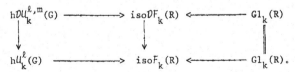

Recall the category $h\mathcal{DR}^n(X)$ from § 1. We agree that duality is taken with respect to the trivial spherical fibration $\varepsilon = X \times S^0$ if no spherical fibration is mentioned explicitly. This category approximates $A(X)$ in a sense which was made precise there. To define the splitting map, the category $h\mathcal{DR}^n(X)$ will have to be replaced by a certain simplicial topological space. This is done in two steps. First a simplicial set $DR^n(X)_\circ$ is constructed together with a chain of homotopy equivalences

$$h\mathcal{DR}^n(X) \longrightarrow h\mathcal{DR}^n(X)_\circ \longleftarrow DR^n(X)_\circ$$

where $h\mathcal{DR}^n(X)_\circ$ is a certain simplicial category combining both $h\mathcal{DR}^n(X)$ and $DR^n(X)_\circ$. In a second step each simplex of $DR^n(X)_\circ$ is replaced by a certain contractible space. This gives a simplicial topological space, which is denoted $\underline{DR}^n(X)_\circ$. It is on this space that the splitting map is defined.

To show that the map constructed is a retraction up to homotopy we consider the following diagram: (For simplicity let $X = *$)

$$\begin{array}{ccccc} & & \underline{DR}^n(*)_\circ & \longrightarrow & \Omega^\infty S^\infty \\ & \nearrow & \Big\downarrow \simeq & & \\ B\Sigma_\infty & \longrightarrow & DR^n(*)_\circ & & \end{array}$$

The lower horizontal arrow represents the inclusion $\Omega^\infty S^\infty \longrightarrow A(*)$ before the + construction. It is shown that this arrow may be lifted as indicated by the broken arrow, and further that the composite $B\Sigma_\infty \longrightarrow \Omega^\infty S^\infty$ agrees up to homotopy with a map described by Segal, [12]. Hence, after performing the + construction it gives a weak homotopy equivalence.

To make precise the way these spaces approximate $A(*)$ (resp. $A(X)$) one has to stabilize them in various ways.

Finally let us mention that from the description of the splitting map given here it is not clear how this map is related to the splittings constructed in [17] and [18].

We start by giving the precise definitions now. Define a simplicial set $R(X)_\circ$ by stipulating that a p-simplex be given by an object $Y \underset{s}{\overset{r}{\rightleftarrows}} X \times \Delta^p$ of $R_{hf}(X \times \Delta^p)$ such that for each face inclusion $\alpha\colon \Delta^q \subset \Delta^p$ the following conditions are satisfied:

(i) $\alpha^*Y := a^{-1}(\Delta^q)$ is an object of $R_{hf}(X \times \Delta^q)$, where a is the composite $Y \xrightarrow{r} X \times \Delta^p \xrightarrow{pr} \Delta^p$;

(ii) the following diagram commutes

$$\begin{array}{ccc} X \times \Delta^q & \xrightarrow{\ id \times \alpha\ } & X \times \Delta^p \\ s\Big\downarrow & & \Big\downarrow s \\ \alpha^* Y & \longrightarrow & Y \\ r\Big\downarrow & & \Big\downarrow r \\ X \times \Delta^q & \xrightarrow{\ id \times \alpha\ } & X \times \Delta^p \end{array} ;$$

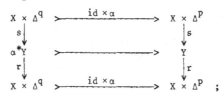

(iii) $\alpha^*Y \rightarrowtail Y$ is a weak homotopy equivalence.

The face maps are given by the obvious restriction maps. This definition can be modified to include duality data. Concretely, let $DR(X).$ denote the simplicial set a p-simplex of which is given by a tuple (Y,Y',u), where Y and Y' are objects of $\mathcal{R}_{hf}(X)_p$ and

$$u\colon (Y \cup_{X\times\Delta^p} X) \wedge (Y' \cup_{X\times\Delta^p} X) \longrightarrow Th_n(\varepsilon)$$

is an n-duality map in $\mathcal{R}(X)$. (Hence, for each $\alpha\colon \Delta^q \subset \Delta^p$ the restriction of u to $Y_\alpha := \alpha^*Y \cup_{X\times\Delta^q} X$ is also an n-duality map.)

The simplicial set $DR^n(X).$ and the category $h\mathcal{DR}^n(X)$ described earlier can be combined into a simplicial category $h\mathcal{DR}^n(X).$. By definition the objects of $h\mathcal{DR}^n(X)_p$ are given by $DR^n(X)_p$. Let (Y,Y',u) (resp. (Z,Z',v)) denote a simplex of $DR^n(X)_p$. A morphism $(Y,Y',u) \longrightarrow (Z,Z',v)$ in $h\mathcal{DR}^n(X)_p$ is a pair of weak homotopy equivalences $f\colon Y \longrightarrow Z$, $f'\colon Z' \longrightarrow Y'$ satisfying that the following diagrams commute

(i)

$$\begin{array}{ccc} & X\times\Delta^p & \\ \swarrow & & \searrow \\ Y & \xrightarrow{f} & Z \\ \searrow & & \swarrow \\ & X & \end{array} \quad , \quad \begin{array}{ccc} & X\times\Delta^p & \\ \swarrow & & \searrow \\ Z' & \xrightarrow{f'} & Y' \\ \searrow & & \swarrow \\ & X & \end{array}$$

(ii) for each $:\Delta^q \subset \Delta^p$

$$\begin{array}{ccc} \alpha^*Y & \xrightarrow{f|\alpha^*Y} & \alpha^*Z \\ \downarrow & & \downarrow \\ Y & \xrightarrow{f} & Z \end{array} \qquad \begin{array}{ccc} \alpha^*Z' & \xrightarrow{f'|\alpha^*Z'} & \alpha^*Y' \\ \downarrow & & \downarrow \\ Z' & \xrightarrow{f'} & Y' \end{array}$$

$$\begin{array}{ccc} Y_\alpha \wedge Z'_\alpha & \xrightarrow{f_\alpha \wedge id} & Z_\alpha \wedge Z'_\alpha \\ {\scriptstyle id\wedge f'_\alpha}\downarrow & & \downarrow{\scriptstyle v_\alpha} \\ Y_\alpha \wedge Y'_\alpha & \xrightarrow[u_\alpha]{} & Th_n(\varepsilon) \end{array} .$$

The simplicial category $h\mathcal{DR}^n(X).$ contains both $h\mathcal{DR}^n(X) = h\mathcal{DR}^n(X)_0$ and $DR^n(X).$ being its simplicial set of objects.

Lemma 3.1. The inclusions

$$DR^n(X). \longrightarrow h\mathcal{DR}^n(X). \longleftarrow h\mathcal{DR}^n(X)$$

are weak homotopy equivalences.

Proof: Let $s\colon h\mathcal{DR}^n(X) \longrightarrow h\mathcal{DR}^n(X)_m$ denote the degeneracy map; let d be the map in the other direction given by restriction to the m-th vertex of Δ^m. Define a functor f from $h\mathcal{DR}^n(X)_m$ to itself by mapping $(Y \xrightarrow{p} \Delta^m,\ Y' \xrightarrow{p'} \Delta^m,\ \ldots)$ to

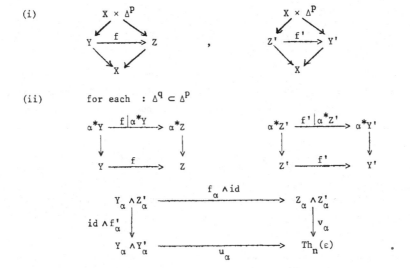

$(p^{-1}(v_m) \times \Delta^m \xrightarrow{pr_2} \Delta^m,\ Y' \xrightarrow{p'} \Delta^m, \dots)$. (Here v_m denotes the m-th vertex of Δ^m).
Similarly f' is the endofunctor of $h\mathcal{DR}^n(X)_m$ given by

$$(Y \xrightarrow{p} \Delta^m,\ Y' \xrightarrow{p'} \Delta^m, \dots) \longmapsto (Y \xrightarrow{p} \Delta^m,\ p^{-1}(v_m) \times \Delta^m \xrightarrow{pr_2} \Delta^m, \dots).$$

Clearly sd = f'f. Define another functor g: $h\mathcal{DR}^n(X)_m \longrightarrow h\mathcal{DR}^n(X)_m$ by
$(Y \longrightarrow \Delta^m,\ Y' \longrightarrow \Delta^m, \dots) \longmapsto (Y \times \Delta^m \xrightarrow{pr} \Delta^m,\ Y' \longrightarrow \Delta^m)$.
There are natural transformations given by the inclusion $p^{-1}(v_m) \times \Delta^m \longrightarrow Y \times \Delta^m$, resp. by the map $Y \xrightarrow{(id,p)} Y \times \Delta^m$. This proves that f is homotopic to the identity. By a similar argument the functor f' is homotopic to the identity. Since ds = id anyway this proves that s is a homotopy equivalence. This is true for every m, so by the realization lemma (cf. e.g. [16]) the right arrow in the lemma is a homotopy equivalence.

To show that the left arrow is a homotopy equivalence we employ a variant of Quillen's theorem A, [10]. Let i. denote the arrow in question. We show that the left fibre $i./((Z,Z',v);[m])$ over a fixed object in degree m is contractible. Since $DR^n(X).$ is a simplicial set the fibre will be a simplicial set, too, rather than a simplicial category. A p-simplex of the fibre is given by

$$(a: [p] \to [m];\ (Y,Y',u) \in h\mathcal{DR}^n(X)_p,\ (\alpha,\alpha'): (Y,Y'u) \xrightarrow{\simeq} a^*(Z,Z',v)).$$

There is a simplicial subset F. of the fibre defined by the condition that $Y' = a^*Z'$, and the structure map $a^*Z' \to Y'$ is the identity. In fact, F. is a deformation retract of $i./((Z,Z',v);[m])$. To see this let $j: F. \longrightarrow i./((Z,Z',v);[m])$ denote the inclusion map. There is an obvious retraction

$$k: i./((Z,Z',v);[m]) \longrightarrow F.$$

given by

$$(a;\ (Y,Y',u),\ (\alpha,\alpha'): (Y,Y') \xrightarrow{\simeq} (a^*Z,a^*Z'))$$
$$\longmapsto (a;\ (Y,a^*Z'),\ (\alpha,id): (Y,a^*Z') \xrightarrow{\simeq} (a^*Z,a^*Z')).$$

We describe a simplicial homotopy from the identity map on $i./((Z,Z',v);[m])$ to the composite jk by specifying a family of maps $h_q: i./((Z,Z',v);[m])_p \longrightarrow i./((Z,Z',v);[m])_{p+1}$, $q = 0,\dots,p$.

Let x be a p-simplex of $i./((Z,Z',v);[m])$ as described above. Let $T(\alpha')$ denote the mapping cylinder of the map $\alpha': a^*Z' \longrightarrow Y'$. There are canonical maps over $\Delta^p \times \Delta^1$ $\beta: T(\alpha') \longrightarrow Y' \times \Delta^1$, and $\gamma: a^*Z' \times \Delta^1 \longrightarrow T(\alpha')$. The map h_q is defined to take the p-simplex x to the (p+1)-simplex of $i./((Z,Z',v);[m])$ given by

$$(a\sigma_q: [p+1] \to [m], (\varphi_q^*(Y \times \Delta^1), \varphi_q^*(T(\alpha')), u_q),$$
$$(\alpha_q,\alpha_q'): (\varphi_q^*(Y \times \Delta^1), \varphi_q^*(T(\alpha')), u_q) \longrightarrow (a\sigma_q)^*(Z,Z',v)),$$

where (i) $\varphi_q: \Delta^{p+1} \to \Delta^p \times \Delta^1$ are the characteristic maps of the non-degenerate (p+1)-simplices of $\Delta^p \times \Delta^1$, $q = 0,\dots,p$.

(ii) σ_q are the surjective maps $[p+1] \longrightarrow [p]$, $q = 0,\ldots,p$;

(iii) u_q is the composite $\varphi_q^*(Y \times \Delta^1) \wedge \varphi_q^*(T(\alpha')) \xrightarrow{\simeq}$
$(Y \times \Delta^1) \wedge T(\alpha') \xrightarrow{id \wedge \beta} (Y \times \Delta^1) \wedge (Y' \times \Delta^1) \longrightarrow$
$Y \wedge Y' \xrightarrow{u} Th_{n-d}(\varepsilon)$;

(iv) $\alpha_q : \varphi_q^*(Y \times \Delta^1) \xrightarrow{\varphi_q^*(\beta \times id)} \varphi_q^*(a^*Z \times \Delta^1) = (a\sigma_q)^*Z$

$\alpha_q' : (a\sigma_q)^*(Z') = \varphi_q^*(Z' \times \Delta^1) \xrightarrow{\varphi_q^*(\gamma \times id)} \varphi_q^*(T(\alpha'))$.

One checks that the maps h_q assemble to a simplicial homotopy from the identity map on $i./((Z,Z',v);[m])$ to the map kj. Hence $F.$ is a deformation retract of $i./((Z,Z',v);[m])$. By an argument which is very similar, one proves that the canonical map $F. \to \Delta^m$ is a homotopy equivalence. Hence $i./((Z,Z',v);[m])$ is contractible, and by an application of theorem A we conclude that $i.$ is a homotopy equivalence as asserted. □

Remark: The homotopy equivalences of the lemma restrict to homotopy equivalences of the subcategories $DR_k^{\ell,m}(X).$ (resp. $h\mathcal{DR}_k^{\ell,m}(X).$) defined by restricting the homotopy type of the spaces involved.
There is a (left) stabilization map

$$DR_k^{\ell,m}(X). \longrightarrow DR_k^{\ell+1,m}(X).$$

given by

$$(Y,Y',u) \longmapsto ((S^1 \times \Delta^p) \wedge_{\Delta^p} Y, Y', \ldots).$$
$${}_{X\times\Delta^p}$$

Similarly, there is a right stabilization map, by suspending Y', and finally, in the k-variable, one stabilizes by taking the wedge sum with an ℓ-sphere (resp. m-sphere).

In view of this remark, the algebraic K-theory of X may now be described using the simplicial sets $DR_k^{\ell,m}(X).$ in the following way:

Corollary 3.2. $$A(X) \simeq \mathbf{Z} \times |\varinjlim_{k,\ell,m} DR_k^{\ell,m}(X).|^+$$ □

Let us now specialize our arguments to the case X = point. The general case will be delt with afterwards.

Recall that an n-duality map was defined to be a pointed map $u: Y \wedge Y' \to S^n$ satisfying a certain non-singularity condition. It is also possible to describe this duality by a certain map $v: S^n \to Y \wedge Y'$. Namely, given the map u, define

$$w: Y \wedge Y' \wedge Y \wedge Y' \longrightarrow S^n \wedge S^n$$

by

$$(a \wedge a' \wedge b \wedge b') \longmapsto (u(a \wedge b') \wedge u(b \wedge a')) .$$

It is easy to check that this defines a 2n-duality pairing. Define v to be the dual of u with respect to the duality w. Equivalently, v is characterized by the condition that the following diagram commutes up to homotopy

$$(*)\qquad \begin{array}{ccc} S^n & \xrightarrow{\quad v \quad} & Y \wedge Y' \\ {\scriptstyle id^{\#}}\downarrow & & \downarrow{\scriptstyle w^{\#}} \\ \mathrm{Map}(S^n,S^{2n}) & \xrightarrow{\quad u^{*} \quad} & \mathrm{Map}(Y \wedge Y',S^{2n}) \end{array} .$$

Here Map(A,B) denotes the space of pointed maps from A to B. The vertical arrows in the diagram are given by the adjoint of the identity (resp. the adjoint of w).

Let $*$ denote the one-point space. We are going to construct a certain simplicial *space* from the simplicial set $DR(*)$. by including further duality data.
Let x be a 0-simplex of $DR^n(*)$, which is represented by the n-duality map $u\colon Y \wedge Y' \longrightarrow S^n$. Let E_x denote the space of pointed maps

$$S^n \longrightarrow \underset{\leftarrow}{\mathrm{holim}}(Y \wedge Y' \xrightarrow{w^{\#}} \mathrm{Map}(Y \wedge Y',S^{2n}))$$

satisfying that the following diagram commutes

$$\begin{array}{ccc} S^n & \longrightarrow & \underset{\leftarrow}{\mathrm{holim}}(Y \wedge Y' \xrightarrow{w^{\#}} \mathrm{Map}(Y \wedge Y',S^{2n})) \\ \downarrow & & \downarrow \\ \mathrm{Map}(S^n,S^{2n}) & \longrightarrow & \mathrm{Map}(Y \wedge Y',S^{2n}) \end{array} .$$

In other words, a point of E_x is given by a map $v\colon S^n \longrightarrow Y \wedge Y'$ making diagram $(*)$ commute up to homotopy, together with a specific homotopy commutativity $h\colon S^n \wedge Y \wedge Y' \wedge [0,1]_+ \longrightarrow S^{2n}$ between the maps

$$S^n \wedge Y \wedge Y' \xrightarrow{v \wedge id} Y \wedge Y' \wedge Y \wedge Y' \xrightarrow{\ w\ } S^{2n}$$

and

$$S^n \wedge Y \wedge Y' \xrightarrow{id \wedge u} S^n \wedge S^n =\!=\!=\!= S^{2n} .$$

The space E_x is the same up to homotopy as

$$\Omega^n(\mathrm{fibre}(Y \wedge Y' \longrightarrow \mathrm{Map}(Y \wedge Y',S^{2n}))) .$$

The map $w^{\#}$ is (2n-1)-connected. Therefore, E_x is (n-1)-connected. Similarly, if x denotes a p-simplex of $DR^n(*)_p$ represented by Y,Y' and the n-duality map $u\colon Y/\Delta^p \wedge Y'/\Delta^p \longrightarrow S^n$, we let E_x denote the space of pointed maps

$$S^n \wedge \Delta^p_+ \longrightarrow \underset{\leftarrow}{\mathrm{holim}}(Y/\Delta^p \wedge Y/\Delta^p \xrightarrow{w^{\#}} \mathrm{Map}(Y/\Delta^p \wedge Y'/\Lambda^p,S^{2n}))$$

satisfying that

(i) the following diagram commutes

$$\begin{array}{ccc}
S^n \wedge \Delta^p_+ & \longrightarrow & \underleftarrow{\text{holim}}(Y/\Delta^p \wedge Y'/\Delta^p \xrightarrow{w^\#} \text{Map}(Y/\Delta^p \wedge Y'/\Delta^p, S^{2n})) \\
\downarrow{\scriptstyle pr} & & \\
S^n & & \downarrow \\
\downarrow{\scriptstyle id^\#} & & \\
\text{Map}(S^n, S^{2n}) & \xrightarrow{u^*} & \text{Map}(Y/\Delta^p \wedge Y'/\Lambda^p, S^{2n})
\end{array}$$

(ii) the map $S^n \wedge \Delta^p_+ \longrightarrow \underleftarrow{\text{holim}}(Y/\Delta^p \wedge Y'/\Delta^p \xrightarrow{w^\#} \text{Map}(Y/\Delta^p \wedge Y'/\Delta^p, S^{2n})$ $\longrightarrow Y/\Delta^p \wedge Y'/\Delta^p$ is a map in the category $R(*)_\circ$, i.e. for each face inclusion $\alpha: \Delta^q \subset \Delta^p$ there is a commutative diagram

$$\begin{array}{ccc}
S^n \wedge \Delta^q_+ & \longrightarrow & Y_\alpha \wedge Y'_\alpha \\
{\scriptstyle id \wedge \alpha}\downarrow & & \downarrow \\
S^n \wedge \Delta^p_+ & \longrightarrow & Y/\Delta^p \wedge Y'/\Delta^p
\end{array} \quad .$$

Again E_x is an (n-1)-connected space. For every p let $\underline{DR}^n(*)_p$ be the disjoint union of the spaces E_x for all $x \in DR^n(*)_p$. In view of condition (ii) this defines a simplicial space $DR^n(*)_\circ$. There is a canonical map

$$\underline{DR}^n(*)_\circ \longrightarrow DR^n(*)_\circ$$

$$E_x \longmapsto x \quad .$$

This map is (n-1)-connected, since it is (n-1)-connected in each simplicial degree.

Of course, one can again restrict the homotopy type of the spaces involved in the construction of these simplicial sets. Thus one obtains simplicial spaces $\underline{DR}_k^{\ell,m}(*)_\circ$. There are three stabilization maps. For example, stabilization with respect to ℓ is given by

$$\underline{DR}_k^{\ell,m}(*)_p \longrightarrow \underline{DR}_k^{\ell+1,m}(*)_p$$

$$(S^n \wedge \Delta^p_+ \to Y/\Delta^p \wedge Y'/\Delta^p \to S^n) \longmapsto (S^{n+1} \wedge \Delta^p_+ \to \Sigma(Y)/\Delta^p \wedge Y'/\Delta^p \to S^{n+1}).$$

One therefore has a well-defined map

$$\varinjlim_{\ell,m} \underline{DR}_k^{\ell,m}(*)_\circ \longrightarrow \varinjlim_{\ell,m} DR_k^{\ell,m}(*)_\circ \quad .$$

This map is a weak homotopy equivalence since by the above it is the limit of $(\ell+m-1)$-connected maps. Hence one obtains still another description of $A(*)$:

Proposition 3.3. $A(*) \simeq \mathbb{Z} \times |\varinjlim_{k,\ell,m} \underline{DR}_k^{\ell,m}(*)_\circ|^+$ □

We now give a description of the map $\Omega^\infty S^\infty \longrightarrow A(*)$. By the theorem of Barratt-Priddy-Quillen-Segal, cf. [13], there is a weak homotopy equivalence $B\Sigma_\infty^+ \simeq \Omega^\infty S^\infty_{(o)}$, where Σ_∞ denotes the infinite symmetric group, and $\Omega^\infty S^\infty_{(o)}$ is the 0-component of stable homotopy. We define a map $B\Sigma_\infty \longrightarrow \lim\limits_{\overrightarrow{n}} DR^n(*)$, which induces the map $\Omega^\infty S^\infty \longrightarrow A(*)$ in view of the description of $A(*)$ afforded by cor. 3.2.. We need a suitable model of $B\Sigma_\infty$.

Consider the following *configuration space*.

Let $C_k'^{\ell,m} = ((\mathbf{R}^\ell)^k$ - (fat) diagonal) $\times$ $((\mathbf{R}^m)^k$ - (fat) diagonal). (Recall that the fat diagonal is defined by the condition that at least two vectors of a k-tuple of vectors are identical.) This space is $(\min(\ell,m)-2)$-connected. The symmetric group Σ_k acts freely on $C_k'^{\ell,m}$ via the diagonal action. Let $C_k^{\ell,m}$ be the orbit space of this action. It follows that the spaces $C_k^{\ell,m}$ approximate $B\Sigma_k$. The space $C_k^{\ell,m}$ contains as a deformation retract the space $D_k^{\ell,m}$ defined by thickening the points of a configuration in $C_k^{\ell,m}$ to unit discs, one in $\mathbf{R}^\ell$, the other one in $\mathbf{R}^m$.

Let $c = (\alpha_i: D^\ell \longrightarrow \mathbf{R}^\ell,\ \alpha_i^!: D^m \longrightarrow \mathbf{R}^m)_{i\in I}$ be a point in $D_k^{\ell,m}$. (I denotes an index set of cardinality k.) To this point there can be associated a map $\varphi_c: S^{\ell+m} = \mathbf{R}^{\ell+m} \cup \{\infty\} \longrightarrow S^{\ell+m}$ of degree k as follows. Choose a fixed degree 1 map $f:(D^\ell \times D^m, \partial) \longrightarrow S^{\ell+m}$. Define

$$\varphi_c(x) = \begin{cases} f((\alpha_i \times \alpha_i^!)^{-1}(x)) & \text{if } x \in (\alpha_i \times \alpha_i^!)(D^\ell \times D^m) \\ * & \text{otherwise.} \end{cases}$$

Taking c to φ_c defines a map $\varphi: D_k^{\ell,m} \longrightarrow \Omega^{\ell+m} S^{\ell+m}{}_{(k)}$. The subscript 'k' on the right refers to the component of degree k maps.

Lemma 3.4. The map φ induces a map

$$(\lim_{\overrightarrow{k}} D_k^{\ell,m})^+ \longrightarrow \Omega^{\ell+m} S^{\ell+m}{}_{(o)}$$

which is $(\min(\ell,m)-2)$-connected.

Proof: Consider the usual configuration space of k disjoint particles in $\mathbf{R}^\ell$. Denote this space by the symbol D_k^ℓ. There is a canonical diagonal map $D_k^\ell \longrightarrow D_k^{\ell,\ell}$, which is $(\ell-2)$-connected. More generally one can define a map $D_k^\ell \longrightarrow D_k^{\ell,m}$ if $m \geqq \ell$. Further there is a map $D_k^\ell \longrightarrow \Omega^\ell S^\ell{}_{(k)}$ and a commutative diagram $(m \geqq \ell)$

$$\begin{array}{ccc} D_k^\ell & \longrightarrow & \Omega^\ell S^\ell{}_{(k)} \\ \downarrow & & \downarrow \\ D_k^{\ell,m} & \longrightarrow & \Omega^{\ell+m} S^{\ell+m}{}_{(k)} \end{array} .$$

It is proved in [12] that the upper horizontal map induces a weak homotopy equivalence $(\lim\limits_{\overrightarrow{k}} D_k^\ell)^+ \longrightarrow \lim\limits_{\overrightarrow{k}} \Omega^\ell S^\ell{}_{(k)} \longrightarrow \Omega^\ell S^\ell{}_{(o)}$. The vertical maps in the diagram are $(\ell-2)$-connected in the case that $m \geqq \ell$. The other case follows since $D_k^{\ell,m} \simeq D_k^{m,\ell}$. □

Let $S(D_k^{\ell,m})$ denote the singular complex of $D_k^{\ell,m}$. Define a map

$$S(D_k^{\ell,m}) \longrightarrow DR_k^{\ell,m}(*).$$

in the following way. Let $c = (\alpha_i: \Delta^p \times D^\ell \longrightarrow \mathbb{R}^\ell,\ \alpha_i^!: \Delta^p \times D^m \longrightarrow \mathbb{R}^m)_{i \in I}$ represent a p-simplex of $S(D_k^{\ell,m})$. Consider the spaces $S^\ell \wedge I_+$ (resp. $S^m \wedge I_+$), where I is considered as a discrete topological space. There is an obvious duality pairing

$$u_c: (S^\ell \wedge I_+) \wedge (S^m \wedge I_+) \longrightarrow S^{\ell+m}$$

which is induced by the map $I \times I \longrightarrow S^0$ which takes exactly the complement of the diagonal to the base point of S^0. Associating to the configuration c the tuple consisting of the spaces $(S^\ell \wedge I_+) \times \Delta^p$ (resp. $(S^m \wedge I_+) \times \Delta^p$) and the canonical duality of these spaces induced from u_c defines the required map. We would like to lift this map to the simplicial space $\underline{DR}^{\ell+m}(*)$. . Let $f: (D^\ell \times D^m, \partial) \longrightarrow (S^\ell \wedge S^m, *)$ be as before, and let

$$v_c': S^{\ell+m} \wedge \Delta_+^p \longrightarrow S^\ell \wedge S^m \wedge I_+ \wedge \Delta_+^p$$

$$x \wedge s \longmapsto \begin{cases} f(\alpha_i \times \alpha_i^!)^{-1}(x) \wedge i \wedge s & \text{if } x \in (\alpha_i \times \alpha_i^!)(\{s\} \times D^\ell \times \{s\} \times D^m) \\ * & \text{otherwise} \end{cases} .$$

(Again, $S^{\ell+m}$ is regarded as $\mathbb{R}^\ell \times \mathbb{R}^m \cup \{\infty\}$.)
Let v_c denote the composite of v_c' with the diagonal map

$$S^\ell \wedge S^m \wedge I_+ \wedge \Delta_+^p \longrightarrow (S^\ell \wedge I_+ \wedge \Delta_+^p) \wedge (S^m \wedge I_+ \wedge \Delta_+^p).$$

It is clear that v_c is a duality map and that v_c is dual to u_c in the sense defined above, at least up to a sign depending on the parity of m. Since we eventually have to pass to the limit with respect to ℓ and m anyway, we may assume m even without essential loss of generality.

Next we have to define a certain homotopy

$$h_c: (S^{\ell+m} \wedge \Delta_+^p) \wedge (S^\ell \wedge I_+ \wedge \Delta_+^p) \wedge (S^m \wedge I_+ \wedge \Delta_+^p) \wedge [0,1]_+ \longrightarrow S^{2(\ell+m)}$$

which is part of the data of a point in $\underline{DR}^{\ell+m}(*)$. .
We proceed as follows. A point $(\alpha_i, \alpha_i^!)$ of a configuration c determines a map $\alpha_i \times \alpha_i^!: D^\ell \times \Delta^p \times D^m \times \Delta^p \longrightarrow \mathbb{R}^{\ell+m}$ and hence a map of degree 1 $S^\ell \wedge \Delta_+^p \wedge S^m \wedge \Delta_+^p \to S^{\ell+m}$. Letting the radius of the disc D^ℓ (resp. D^m) grow to infinity defines a canonical homotopy between this map and the projection $S^\ell \wedge \Delta_+^p \wedge S^m \wedge \Delta_+^p \to S^{\ell+m}$. For each $i \in I$ define

$$h_i: S^{\ell+m} \wedge \Delta_+^p \wedge S^\ell \wedge \Delta_+^p \wedge S^m \wedge \Delta_+^p \wedge [0,1]_+ \longrightarrow S^{\ell+m} \wedge S^{\ell+m}$$

to be the projection on the first two factors, and on the other factors the homotopy determined by $\alpha_i, \alpha_i^!$ just described. Define the map

$$h_c': S^{\ell+m} \wedge \Delta_+^p \wedge S^\ell \wedge I_+ \wedge \Delta_+^p \wedge S^m \wedge I_+ \wedge \Delta_+^p \wedge [0,1]_+ \longrightarrow S^{\ell+m} \wedge S^{\ell+m}$$

$$(x \wedge s \wedge x' \wedge i \wedge s' \wedge x'' \wedge j \wedge s'' \wedge t) \longmapsto \begin{cases} * & \text{if } i \neq j \\ h_i(x \wedge s \wedge x' \wedge s' \wedge x'' \wedge s'' \wedge t) & \text{if } i = j. \end{cases}$$

Compose the homotopy h'_c with some standard homotopy between the map

$$S^{\ell} \wedge S^{m} \wedge S^{\ell} \wedge S^{m} \longrightarrow S^{\ell} \wedge S^{m} \wedge S^{\ell} \wedge S^{m}$$

$$(x \wedge x' \wedge y \wedge y') \longmapsto (x \wedge y' \wedge y \wedge x')$$

and the identity map. Such a homotopy exists because of our assumption on the parity of m. This defines the required homotopy h_c .

Taking the configuration c to (u_c, v_c, h_c) gives a map

$$(\dagger) \qquad s\colon S(D_k^{\ell,m}) \longrightarrow \underline{DR}_k^{\ell,m}(*)_{\cdot} \quad .$$

Let $D_k := \lim\limits_{\substack{\to \\ \ell,m}} D_k^{\ell,m}$. By lemma 3.4. this is a classifying space for Σ_k. Let $S(D_k)$ denote the singular complex of D_k. Passing to the limit with respect to ℓ and m in (†) hence gives a map

$$B\Sigma_k \longrightarrow \lim_{\substack{\to \\ \ell,m}} \underline{DR}_k^{\ell,m}(*)_{\cdot}$$

which, after passing to the limit in k and performing the + construction gives the map $\Omega^\infty S^\infty \longrightarrow A(*)$.

We now describe a splitting of this map. Associate to any p-simplex (u_c, v_c, h_c) of $DR_k^{\ell,m}(*)_{\cdot}$ the composite

$$u_c\, v_c\colon S^{\ell+m} \wedge \Delta_+^p \longrightarrow S^{\ell+m} \quad .$$

This defines a map

$$DR_k^{\ell,m}(*)_p \longrightarrow \mathrm{Map}(S^{\ell+m} \wedge \Delta_+^p,\ S^{\ell+m})_{(k)}$$

resp.

$$r\colon DR_k^{\ell,m}(*)_{\cdot} \longrightarrow \mathrm{Map}(S^{\ell+m} \wedge \Delta_+^{\cdot},\ S^{\ell+m})_{(k)} \quad .$$

(Again, $\mathrm{Map}(\ldots)_{(k)}$ denotes the component of degree k maps).

The simplicial space on the right is the singular complex of $\Omega^{\ell+m}S^{\ell+m} = \mathrm{Map}(S^{\ell+m}, S^{\ell+m})$ considered as a simplicial *space* in the natural way. Checking the restriction of r to the subspace $S(D_k^{\ell,m})$ of $\underline{DR}_k^{\ell,m}(*)_{\cdot}$ immediately reveals that this is nothing else but the map φ described above. Hence one obtains:

Lemma 3.5. There is a commutative diagram

$$\begin{array}{ccc}
S(D_k^{\ell,m}) & \xrightarrow{\quad\varphi\quad} & S(\Omega^{\ell+m}S^{\ell+m}{}_{(k)}) \\
{\scriptstyle s}\big\downarrow & & \big\downarrow \\
\underline{DR}_k^{\ell,m}(*)_{\cdot} & \xrightarrow{\quad r\quad} & S(\Omega^{\ell+m}S^{\ell+m}{}_{(k)})_{\cdot}
\end{array} \quad ,$$

where $S(\Omega^{\ell+m}S^{\ell+m})_{\cdot}$ denotes the singular complex considered as a simplicial space, and the vertical map on the right is the natural inclusion.

□

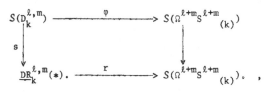

After passing to the limit with respect to k, ℓ, and m and applying the + construction, the map φ becomes a homotopy equivalence by lemma 3.4. The vertical arrow on the right is a homotopy equivalence anyway, and the + construction does not change the terms on the right. This proves that the map

$$|\lim_{\substack{\to\\ k,\ell,m}} \underline{DR}_k^{\ell,m}(*).|^+ \longrightarrow \lim_{\vec{k}} S(\Omega^\infty S^\infty_{(k)}).| \simeq \Omega^\infty S^\infty_{(o)}$$

is a retraction up to weak homotopy, and hence the theorem of this paragraph in the case $X = *$.

The modifications required for the general case are straightforward: To a duality

$$(Y \cup_{X\times\Delta^p} X) \wedge (Y' \cup_{X\times\Delta^p} X) \longrightarrow Th_n(\varepsilon) \quad \text{in } DR^n(X)_p$$

there is associated another duality in $DR^n(*)_p$, which is given by

$$Y/X \times \Delta^p) \wedge Y'(X \times \Delta^p) \longrightarrow Th_n(\varepsilon)/X^2 = S^n \wedge X_+ \longrightarrow S^n$$

(cf. prop. 1.17.). This defines a map $DR^n(X). \longrightarrow DR^n(*).$. Define the simplicial space $\underline{DR}^n(X).$ as the pull-back of the following diagram.

$$\begin{array}{ccc}
\underline{DR}^n(X). & \longrightarrow & \underline{DR}^n(*). \\
\downarrow & & \downarrow \\
DR^n(X). & \longrightarrow & DR^n(*).
\end{array}$$

Similarly define $\underline{DR}_k^{\ell,m}(X).$. Hence a p-simplex of $\underline{DR}^n(X).$ (resp. $\underline{DR}_k^{\ell,m}(X).$) consists of a certain n-duality (resp. $(\ell+m)$-duality)

$$(Y \cup_{X\times\Delta^p} X) \wedge (Y' \cup_{X\times\Delta^p} X) \longrightarrow Th_n(\varepsilon)$$

over Δ^p, together with a map $S^n \wedge \Delta^p_+ \to Y/X \times \Delta^p \wedge Y'/X \times \Delta^p$, and additional data. Associating to such a p-simplex the composite

$$S^n \wedge \Delta^p_+ \longrightarrow Y/X \times \Delta^p \wedge Y'/X \times \Delta^p \longrightarrow Th(\varepsilon)/X^2 = S^n \wedge X_+$$

defines a map

$$\underline{DR}_k^{\ell,m}(X). \longrightarrow Map(S^{\ell+m} \wedge \Delta^\bullet_+, S^{\ell+m} \wedge X_+) = S(\Omega^{\ell+m} S^{\ell+m}(X_+)).$$

By the same argument as in the case $X = *$ this map is shown to be a retraction up to homotopy after passing to the limit with respect to k,ℓ,m, and performing the + construction.

This ends the proof of the theorem. □

References

[1] D. Burghelea, *Automorphisms of manifolds*, Proc. Symp. Pure Math. vol. 32, part I, Am. Math. Soc., 1978, 347-372

[2] D. Burghelea, *The rational homotopy groups of Diff(M) and Homeo(M) in the stability range*, Proc. Conf. Alg. Topology, Aarhus, 1978, Lecture Notes in Mathematics, vol. 763, Springer, Berlin-Heidelberg-New York, 1979, 604-626

[3] D. Burghelea, Z. Fiedorowicz, *Hermitian algebraic K-theory of topological spaces*, Algebraic K-theory, Number Theory, Geometry and Analysis, Proceedings, Bielefeld 1982, Lecture Notes in Mathematics, vol. 1046, Springer, Berlin-Heidelberg-New York, 1984

[4] A. Hatcher, *Concordance spaces, higher simple homotopy theory, and applications*, Proc. Symp. Pure Math. vol. 32, part I, Am. Math. Soc., 1978, 3-22

[5] W.C. Hsiang, B.Jahren, *A note on the homotopy groups of the diffeomorphism groups of spherical space forms*, Alg. K-theory proceedings, Oberwolfach 1980, Lecture Notes in Mathematics, vol. 967, Springer, Berlin-Heidelberg-New York, 1983, 132-145

[6] D.M. Kan, *On c.s.s. complexes*, Amer. J. of Math. 79, 1957, 449-476

[7] D.M. Kan, *A combinatorial definition of homotopy groups*, Ann. of Math. (2), 67, 1958, 282-312

[8] J. Lemaire, *Le transfert dans les espaces fibrês (d'après J. Becker et D. Gottlieb)*, Séminaire Bourbaki, 23e année, n° 472, 1975

[9] D. Puppe, *Duality in monoidal categories and applications*, Game theory and related topics, North-Holland, Amsterdam-New York-Oxford, 1979, 173-185

[10] D. Quillen, *Higher algebraic K-theory I*, Lecture Notes in Mathematics, vol. 341, Springer, Berlin-Heidelberg-New York, 1973, 85-147

[11] G. Segal, *Classifying spaces and spectral sequences*, Publ. Math. IHES, 34, 1968, 105-112

[12] G. Segal, *Configuration spaces and iterated loop spaces*, Invent. Math. 21, 1973, 213-221

[13] G. Segal, *Categories and cohomology theories*, Topology 13, 1974, 293-312

[14] E. Spanier, *Function spaces and duality*, Ann. of Math. (2), 70, 1958, 338-378

[15] W. Vogell, *The canonical involution on the algebraic K-theory of spaces*, Proc. Conf. Alg. Topology, Aarhus 1982, Lecture Notes in Mathematics, vol. 1051, Springer, Berlin-Heidelberg-New York, 1984, 156-172

[16] F. Waldhausen, *Algebraic K-theory of topological spaces I*, Proc. Symp. Pure Math. vol. 32, part I, Am. Math. Soc., 1978, 35-60

[17] F. Waldhausen, *Algebraic K-theory of topological spaces II*, Proc. Conf. Alg. Topology, Aarhus 1978, Lecture Notes in Mathematics vol. 763, Springer, Berlin-Heidelberg-New York, 1979, 356-394

[18] F. Waldhausen, *Algebraic K-theory of spaces, a manifold approach*, Current trends in topology, Canad. Math. Soc. proc. vol. 2, part I, 1982, 141-186

[19] F. Waldhausen, *Algebraic K-theory of spaces*, these proceedings

[20] J.F. Adams, *Prerequisites*, Proc. Conf. Alg. Topology, Aarhus 1982, Lecture Notes in Mathematics, vol. 1051, Springer, Berlin-Heidelberg-New York, 1984, 483-532

Universität Bielefeld

Fakultät für Mathematik

4800 Bielefeld 1, FRG

ALGEBRAIC K-THEORY OF SPACES.

Friedhelm Waldhausen

This is an account of foundational material on the algebraic K-theory of spaces functor $X \mapsto A(X)$.

The paper is in three parts which are entitled "Abstract K-theory", "A(X)", and "Relation of A(X) to $Wh^{PL}(X)$", respectively.

The main result of the paper is in the second part. It says that several definitions of A(X) are in fact equivalent to each other, up to homotopy. The proof uses most of the results of the first part. An introduction to this circle of ideas can be obtained from looking at the sections entitled "Review of A(X)" and "Review of algebraic K-theory" in the papers [17] and [18] (these two sections were written with that purpose in mind).

The third part of the paper is devoted to an abstract version of the relation of the A-functor to concordance theory. The content of the *parametrized h-cobordism theorem* in the sense of Hatcher is that PL concordance theory, stabilized with respect to dimension, can be re-expressed in terms of non-manifold data. A detailed account of the translation is given elsewhere [16], in particular the relevant results of Hatcher's are (re-)proved there. The result of the translation (after a dimension shift) is a functor $X \mapsto Wh^{PL}(X)$. It is shown here that there is a map $A(X) \to Wh^{PL}(X)$ and that the homotopy fibre of that map is a homology theory (i.e., that, as a functor of X , the homotopy fibre satisfies the excision property).

The first part of the paper, on which everything else depends, may perhaps look a little frightening because of the abstract language that it uses throughout. This is unfortunate, but there is no way out. It is not the purpose of the abstract language to strive for great generality. The purpose is rather to simplify proofs, and indeed to make some proofs understandable at all. The reader is invited to run the following test: take theorem 2.2.1 (this is about the worst case), translate the complete proof into not using the abstract language, and then try to communicate it to somebody else.

Contents.

1. ABSTRACT K-THEORY.

1.1. Categories with cofibrations, and the language of filtered objects.

A category $\mathcal{C}$ is called *pointed* if it is equipped with a distinguished zero object $*$, i.e. an object which is both initial and terminal.

A *category with cofibrations* shall mean a pointed category $\mathcal{C}$ together with a subcategory $co\mathcal{C}$ satisfying the axioms Cof 1 - Cof 3 below. The feathered arrows ' $\rightarrowtail$ ' will be used to denote the morphisms in $co\mathcal{C}$. Informally the morphisms in $co\mathcal{C}$ will simply be referred to as the *cofibrations in* $\mathcal{C}$.

Cof 1. The isomorphisms in $\mathcal{C}$ are cofibrations (in particular $co\mathcal{C}$ contains all the objects of $\mathcal{C}$).

Cof 2. For every $A \in \mathcal{C}$, the arrow $* \to A$ is a cofibration.

Cof 3. Cofibrations admit cobase changes. This means the following two things. If $A \rightarrowtail B$ is a cofibration, and $A \to C$ any arrow, then firstly the pushout $C \cup_A B$ exists in $\mathcal{C}$, and secondly the canonical arrow $C \to C \cup_A B$ is a cofibration again.

Here is some more language. If $A \rightarrowtail B$ is a cofibration then B/A will denote any representative of $* \cup_A B$. We think of it as the quotient of B by A . The canonical map $B \to B/A$ will be referred to as a *quotient map*. The double headed arrows ' $\twoheadrightarrow$ ' are reserved to denote quotient maps. (Note that it is neither asked, nor asserted, that the quotient maps form a category, i.e. that the composite of two quotient maps is always a quotient map again.)

Our usage of the term *cofibration sequence* conforms to the usage in homotopy theory. It refers to a sequence $A \rightarrowtail B \twoheadrightarrow B/A$ where $B \twoheadrightarrow B/A$ is the quotient map associated to $A \rightarrowtail B$.

Beware that we will also be using the term *sequence of cofibrations* which of course refers to a sequence of the type $A_1 \rightarrowtail A_2 \rightarrowtail \ldots \rightarrowtail A_n$.

The most important example of a category with cofibrations, for our purposes, is that of the spaces having a given space X as a retract. We will denote this category by $\mathcal{R}(X)$. As a technical point, there will be several cases to consider depending on whether *space* means simplicial set, or cell complex, or whatever, and perhaps with a finiteness condition imposed. In any case the term cofibration has essentially its usual meaning here. (As a technical point again, note that the axiom Cof 2 may force us to put a condition on one of the structural maps of an object of $R(X)$ - the section should be a cofibration).

Another important example, though of less concern to us here, is that of an *exact category* in the sense of Quillen. Any exact category can be considered as a category with cofibrations by choosing a zero object, and declaring the admissible monomorphisms to be the cofibrations. The re-interpretation involves a loss of structure: one ignores that pullbacks used to play a role, too (the base change by admissible epimorphisms).

Since our axioms are so primitive it will not be surprising that they admit examples which are not important at all, and perhaps even embarrassing. Here is a particularly bad case. Consider a category having a zero object and finite colimits. It can be made into a category with cofibrations by declaring *all* morphisms to be cofibrations.

Here is some more language. A functor between categories with cofibrations is called *exact* if it preserves all the relevant structure: it takes $*$ to $*$, cofibrations to cofibrations, and it preserves the pushout diagrams of axiom Cof 3.

For example, a map $X \to X'$ induces an exact functor $R(X) \to R(X')$. On total spaces it is given by pushout of $X \to X'$ with the structural sections.

Another example of an exact functor is the linearization functor (or Hurewicz map) which takes an object of $R(X)$ to the abelian-group-object in $R(X)$ which it generates.

There is a concept slightly stronger than that of an *exact inclusion functor* which we will have to consider. We say that C' is a *subcategory with cofibrations of* C if in addition to the exactness of the inclusion functor the following condition is satisfied: an arrow in C' is a cofibration in C' if it is a cofibration in C and the quotient is in C' (up to isomorphism).

An example of a subcategory-with-cofibrations arises if we consider a subcategory of $R(X)$ defined by a finiteness condition.

Here is a more interesting example. For $n \geq 2$ let $R^n(X)$ denote the full subcategory of $R(X)$ whose objects are obtainable from X by attaching of n-cells (up to homotopy). It can be considered as a subcategory with cofibrations of $R(X)$.

In the remainder of the section we will check that certain elementary constructions with categories do not lead one out of the framework of categories with cofibrations. In particular we will be interested in *filtered objects*; that is, sequences of cofibrations. (Despite the fact, exemplified above, that cofibrations need not be monomorphic at all, we shall let ourselves be guided by the more relevant examples to justify using this terminology). The arguments below will not go beyond trivial manipulation with colimits. There is, however, one idea involved. The idea is that the notion of *bifiltered object* (or *lattice*) can be formulated without pullbacks. Namely if the diagram

$$\begin{array}{ccc} A & \longrightarrow & B \\ \downarrow & & \downarrow \\ C & \longrightarrow & D \end{array}$$

is to be a 'lattice' we are inclined to ask this in the form of two conditions: firstly, that all the arrows be cofibrations, and secondly, that the 'images' in D satisfy $\mathrm{Im}(A) \supset \mathrm{Im}(B) \cap \mathrm{Im}(C)$. The latter does not make sense in our context, in general, but we can substitute it with the condition that the arrow $B\cup_A C \to D$ be a cofibration.

For any category $\mathcal{C}$ we let $\mathrm{Ar}\mathcal{C}$ denote the category whose objects are the arrows of $\mathcal{C}$ and whose morphisms are the commutative squares

$$\begin{array}{ccc} \cdot & \longrightarrow & \cdot \\ \downarrow & & \downarrow \\ \cdot & \longrightarrow & \cdot \end{array}$$

in $\mathcal{C}$. If $\mathcal{C}$ is a category with cofibrations then so is $\mathrm{Ar}\mathcal{C}$ in an obvious way: a map is in $\mathrm{coAr}\mathcal{C}$ if and only if the two associated maps in $\mathcal{C}$ are in $\mathrm{co}\mathcal{C}$.

Definition. $F_1\mathcal{C}$ is the full subcategory of $\mathrm{Ar}\mathcal{C}$ whose objects are the cofibrations in $\mathcal{C}$, and $\mathrm{co}F_1\mathcal{C}$ is the class of the maps $(A \rightarrowtail B) \to (A' \rightarrowtail B')$ in $F_1\mathcal{C}$ having the property that both $A \to A'$ and $A'\cup_A B \to B'$ are cofibrations in $\mathcal{C}$.

<u>Lemma</u> 1.1.1. $\mathrm{co}F_1\mathcal{C}$ makes $F_1\mathcal{C}$ a category with cofibrations.

Proof. There are two points that require proof: that $\mathrm{co}F_1\mathcal{C}$ is a category, and that the axiom Cof 3 is satisfied.

As to the first, let $(A \rightarrowtail B) \rightarrowtail (A' \rightarrowtail B')$ and $(A' \rightarrowtail B') \rightarrowtail (A'' \rightarrowtail B'')$ be in $\mathrm{co}F_1\mathcal{C}$. Then $A \rightarrowtail A''$ since $\mathrm{co}\mathcal{C}$ is a category. By assumption about the second map $A''\cup_{A'} B' \rightarrowtail B''$; and by assumption about the first map and by axioms Cof 1 and Cof 3 for $\mathrm{co}\mathcal{C}$, all the following terms are defined and the composed map

$$A''\cup_A B \xrightarrow{\approx} A''\cup_{A'} A'\cup_A B \longrightarrow A''\cup_{A'} A'\cup_A B\cup_{(A'\cup_A B)} B' \xrightarrow{\approx} A''\cup_{A'} B'$$

is also in $\mathrm{co}\mathcal{C}$. Taking the composition of the two maps we obtain that $A''\cup_A B \to B''$ is in $\mathrm{co}\mathcal{C}$, as was to be shown.

As to the second, let $(A \rightarrowtail B) \rightarrowtail (A' \rightarrowtail B')$ and $(A \rightarrowtail B) \to (C \rightarrowtail D)$ be maps in $\mathrm{co}F_1\mathcal{C}$, resp. $F_1\mathcal{C}$. Their pushout exists in $\mathrm{Ar}\mathcal{C}$ by Cof 3 for $\mathcal{C}$ (because $A \rightarrowtail A'$ and $A'\cup_A B \rightarrowtail B'$ implies $B \rightarrowtail B'$) where it is represented by

$$A'\cup_A C \longrightarrow B'\cup_B D \ .$$

We show below that this is an object of (and consequently also a pushout in) $\mathcal{F}_1\mathcal{C}$. We must in addition show that the canonical map $(C \rightarrowtail D) \to (A'\cup_A C \to B'\cup_B D)$ is in $co\mathcal{F}_1\mathcal{C}$. This amounts to the two assertions that $C \rightarrowtail A'\cup_A C$, which is clear, and that $(A'\cup_A C)\cup_C D \rightarrowtail (B'\cup_B D)$. The latter map is isomorphic to $A'\cup_A D \to B'\cup_B D$ which in turn is isomorphic to the composed map

$$(A'\cup_A B)\cup_B D \longrightarrow B'\cup_{(A'\cup_A B)}(A'\cup_A B)\cup_B D \xrightarrow{\approx} B'\cup_B D$$

and this is a cofibration since $A'\cup_A B \to B'$ is one. Finally $A'\cup_A C \to (A'\cup_A C)\cup_C D$ is a cofibration since $C \to D$ is one. Composing it with the cofibration $(A'\cup_A C)\cup_C D \to B'\cup_B D$ (above) we obtain the map $A'\cup_A C \to B'\cup_B D$. This proves the postponed claim that the latter map is a cofibration. □

Definition. $\mathcal{F}_1^+\mathcal{C}$ is the category equivalent to $\mathcal{F}_1\mathcal{C}$ in which an object consists of an object $A \rightarrowtail B$ of $\mathcal{F}_1\mathcal{C}$ together with the choice of a quotient B/A; in other words, $\mathcal{F}_1^+\mathcal{C}$ is the category of cofibration sequences $A \rightarrowtail B \twoheadrightarrow B/A$ in $\mathcal{C}$. It is made into a category with cofibrations by means of the equivalence $\mathcal{F}_1^+\mathcal{C} \to \mathcal{F}_1\mathcal{C}$.

<u>Lemma</u> 1.1.2. The three functors $s, t, q\colon \mathcal{F}_1^+\mathcal{C} \to \mathcal{C}$ sending $A \rightarrowtail B \twoheadrightarrow B/A$ to A, B, and B/A, respectively, are exact.

Proof. For s this holds by definition, and for t almost so. The case of q requires proof. We must show that q takes $co\mathcal{F}_1^+\mathcal{C}$ to $co\mathcal{C}$, and that q preserves the pushout diagrams of axiom Cof 3.

As to the first, if $(A \rightarrowtail B) \to (A' \rightarrowtail B')$ is in $co\mathcal{F}_1^+\mathcal{C}$ then, by definition, $A'\cup_A B \to B'$ is in $co\mathcal{C}$. Hence so is

$$B/A \xrightarrow{\approx} *\cup_{A'} A'\cup_A B \longrightarrow *\cup_{A'} A'\cup_A B\cup_{(A'\cup_A B)} B' \xrightarrow{\approx} B'/A'$$

as claimed.

As to the second, let such a pushout diagram in $\mathcal{F}_1^+\mathcal{C}$ be given by the diagram

$$\begin{array}{ccc}
(A \rightarrowtail B \twoheadrightarrow B/A) & \longrightarrow & (C \rightarrowtail D \twoheadrightarrow D/C) \\
\big\downarrow & & \big\downarrow \\
(A' \rightarrowtail B' \twoheadrightarrow B'/A') & \longrightarrow & (A'\cup_A C \rightarrowtail B'\cup_B D \twoheadrightarrow (B'\cup_B D)/(A'\cup_A C))\ .
\end{array}$$

Then the assertion means that

$$(B'\cup_B D)/(A'\cup_A C) \qquad \text{and} \qquad B'/A'\cup_{B/A} D/C$$

are canonically isomorphic. But this is clear from the fact that an iterated colimit may be computed in any way desired provided only that all the colimits involved exist. In particular the two objects at hand are canonically isomorphic because both represent the colimit of the diagram

$$\begin{array}{ccccc} D & \leftarrowtail & C & \longrightarrow & * \\ \uparrow & & \uparrow & & \uparrow \\ B & \leftarrowtail & A & \longrightarrow & * \\ \downarrow & & \downarrow & & \downarrow \\ B' & \leftarrowtail & A' & \longrightarrow & * \end{array}$$

when this colimit is computed in the two obvious ways. □

Definition. $F_m C$ is the category in which an object is a sequence of cofibrations

$$A_0 \rightarrowtail A_1 \rightarrowtail \dots \rightarrowtail A_m$$

in C, and where a morphism is a natural transformation of diagrams. $F_m^+ C$ is the category equivalent to $F_m C$ in which an object consists of one of $F_m C$ together with a choice, for every $0 \leqslant i < j \leqslant m$, of a quotient $A_{i,j} = A_j/A_i$.

Lemma 1.1.3. Let $A \to A'$ be a map in $F_m C$, resp. $F_m^+ C$. Suppose that the maps

$$A_j \longrightarrow A'_j \quad , \qquad A'_j \cup_{A_j} A_{j+1} \longrightarrow A'_{j+1}$$

are cofibrations in C. Then

for every pair $j < k$ the map $A'_j \cup_{A_j} A_k \to A'_k$ is a cofibration, and

for every triple $i < j < k$ the map $A'_{i,j} \cup_{A_{i,j}} A_{i,k} \to A'_{i,k}$ is a cofibration.

Proof. The first results inductively by considering the compositions

$$A'_j \cup_{A_j} A_k \cup_{A_k} A_{k+1} \longrightarrow A'_k \cup_{A_k} A_{k+1} \longrightarrow A'_{k+1}$$

and the second follows from the first by the preceding lemma applied to the cofibration in $F_1 C$,

$$(A'_i \cup_{A_i} A_i \rightarrowtail A'_i) \rightarrowtail (A'_j \cup_{A_j} A_k \rightarrowtail A'_k) \ . \qquad □$$

Proposition 1.1.4. $F_m C$ and $F_m^+ C$ are categories with cofibrations in a natural way. The forgetful map $F_m^+ C \to F_m C$ is an exact equivalence. The 'subquotient' maps

$$q_j : F_m C \longrightarrow C \ , \qquad q_{i,j} : F_m^+ C \longrightarrow C$$
$$A \longmapsto A_j \qquad\qquad A \longmapsto A_j/A_i$$

are exact.

In fact, a map in $F_m C$, resp. $F_m^+ C$, is defined to be a cofibration if it satisfies the hypothesis of lemma 1.1.3, and the assertions of the proposition just summarize the preceding lemmas. □

Iterating the construction one can obtain categories with cofibrations $F_n F_m C$ and $F_n^+ F_m^+ C$.

Lemma 1.1.5. There are natural isomorphisms of categories with cofibrations

$$F_nF_mC \approx F_mF_nC \ , \qquad F_n^+F_m^+C \approx F_m^+F_n^+C \ .$$

Proof. It suffices to remark that an object of F_nF_mC can be more symmetrically defined as a rectangular array of squares each of which consists of cofibrations only and satisfies the condition in the definition of a cofibration in F_1C ; the point is that the condition is symmetric with respect to *horizontal* and *vertical*. Similarly, a cofibration in F_nF_mC , or sequence of such, may be identified to a 3-dimensional diagram satisfying conditions with respect to which none of the three directions is preferred. □

We will want to know that categories with cofibrations reproduce under certain other simple constructions. By the *fibre product* of a pair of functors $f\colon A \to C$, $g\colon B \to C$ is meant the category $\Pi(f,g)$ whose objects are the triples

$$(A,c,B) \ , \quad A \in A \ , \quad B \in B \ , \quad c\colon f(A) \xrightarrow{\approx} g(B) \ ,$$

and where a morphism from (A,c,B) to (A',c',B') is a pair of morphisms (a,b) compatible with the isomorphisms c and c' . In some special cases the fibre product category is equivalent to the pullback category $A\times_C B$; notably this is so if either f or g is a retraction. (If the two are not the same, up to equivalence, the pullback should be regarded as pathological.)

Lemma 1.1.6. If $f\colon A \to C$ and $g\colon B \to C$ are exact functors of categories with cofibrations then $\Pi(f,g)$ can be made into a category with cofibrations by letting

$$co(\Pi(f,g)) = \Pi(co(f),co(g)) \ ,$$

and the projection functors from $\Pi(f,g)$ to A and B are exact.

Similarly, if $j \to C_j$, $j \in J$, is a direct system of categories with cofibrations and exact functors then $\varinjlim C_j$ is a category with cofibrations, with

$$co(\varinjlim C_j) = \varinjlim coC_j \ ,$$

and the functors $C_j \to \varinjlim C_j$ are exact. □

Definition and corollary. Let A, B, C be categories with cofibrations and let A and B be subcategories of C in such a way that the inclusion functors are exact. Define $E(A,C,B)$ as the category of the cofibration sequences in C ,

$$A \rightarrowtail C \twoheadrightarrow B \ , \quad A \in A \ , \quad B \in B \ .$$

Then $E(A,C,B)$ is a category with cofibrations, and the projections to A, C, B are exact.

Indeed, $E(A,C,B)$ is the pullback of a diagram $F_1^+C \longrightarrow C \times C \longleftarrow A \times B$; the pullback is not pathological since the first arrow has a section. □

1.2. Categories with cofibrations and weak equivalences.

Let $\mathcal{C}$ be a category with cofibrations in the sense of section 1.1 (we will from now on drop explicit mentioning of the category of cofibrations $co\mathcal{C}$ from the notation). A *category of weak equivalences* in $\mathcal{C}$ shall mean a subcategory $w\mathcal{C}$ of $\mathcal{C}$ satisfying the following two axioms.

Weq 1. The isomorphisms in $\mathcal{C}$ are contained in $w\mathcal{C}$ (and in particular therefore the category $w\mathcal{C}$ contains all the objects of $\mathcal{C}$).

Weq 2. (*Gluing lemma*). If in the commutative diagram

$$\begin{array}{ccccc} B & \leftarrowtail & A & \longrightarrow & C \\ \downarrow & & \downarrow & & \downarrow \\ B' & \leftarrowtail & A' & \longrightarrow & C' \end{array}$$

the horizontal arrows on the left are cofibrations, and all three vertical arrows are in $w\mathcal{C}$, then the induced map

$$B \cup_A C \longrightarrow B' \cup_{A'} C'$$

is also in $w\mathcal{C}$.

Here are some examples. Any category with cofibrations can be equipped with a category of weak equivalences in at least two ways: the minimal choice is to let $w\mathcal{C}$ be the category of isomorphisms in $\mathcal{C}$, while the maximal choice is to let $w\mathcal{C}$ be equal to $\mathcal{C}$ itself.

To obtain an example of a category of weak equivalences on the category $\mathcal{R}(X)$ (the preceding section) choose a homology theory and define $w\mathcal{R}(X)$ to be the category of those maps which induce isomorphisms of that homology theory.

To obtain another example define $h\mathcal{R}(X)$ to be the category of the *weak homotopy equivalences*.

To obtain yet another example define $s\mathcal{R}(X)$ to be the category of the *simple maps*, i.e. the maps whose point inverses have the shape (or Cech homotopy type) of a point. (We shall consider simple maps in the simplicial setting only in which case the definition simplifies to asking that the point inverses in the geometric realization of the map are contractible.) Neither the fact that $s\mathcal{R}(X)$ is a category nor the gluing lemma are trivial to prove.

The following two further axioms may, or may not, be satisfied by a given category of weak equivalences.

Saturation axiom. If a, b are composable maps in C and if two of a, b, ab are in wC then so is the third.

For example the simple maps do not satisfy the saturation axiom. E.g. consider the two maps a, b in $R(*)$ given by the inclusion of the basepoint in a 1-simplex and by the projection of that 1-simplex to the basepoint, respectively.

Extension axiom. Let

$$\begin{array}{ccccc} A & \rightarrowtail & B & \twoheadrightarrow & B/A \\ \downarrow & & \downarrow & & \downarrow \\ A' & \rightarrowtail & B' & \twoheadrightarrow & B'/A' \end{array}$$

be a map of cofibration sequences. If the arrows $A \to A'$ and $B/A \to B'/A'$ are in wC then it follows that $B \to B'$ is in wC, too.

For example the weak homotopy equivalences do not satisfy the extension axiom. E.g. consider the diagram in $R(*)$

$$\begin{array}{ccccc} BZ & \overset{=}{\rightarrowtail} & BZ & \twoheadrightarrow & * \\ {\scriptstyle =}\downarrow & & \downarrow & & \downarrow \\ BZ & \rightarrowtail & BG & \twoheadrightarrow & BG/BZ \end{array}$$

where BZ is the classifying space of the infinite cyclic group and BG the classifying space of a suitable non-abelian group which is normally generated by a subgroup Z, for example a classical knot group.

As the examples show there may be a great profusion of categories of weak equivalences on a given category with cofibrations. Also, we will have occasion to consider a category with cofibrations equipped with *two* categories of weak equivalences at the same time, one finer than the other, and study their interplay. We must therefore exercise some care with the notation, and in general the category of weak equivalences will be explicitly mentioned.

Still there are some situations where there is no danger of confusion. On those occasions we will allow ourselves the abuse of referring to the maps in wC as the *weak equivalences in* C, and denote them by the decorated arrows '$\xrightarrow{\sim}$'.

By a *category with cofibrations and weak equivalences* will be meant a category with cofibrations equipped with one (and only one) category of weak equivalences. A functor between such is called *exact* if it preserves all the relevant structure.

As in the preceding section, the notion of an exact inclusion functor may be sharpened to that of a *subcategory with cofibrations and weak equivalences*.

Finally we note that categories of weak equivalences are inherited by diagram categories. There are lemmas similar to, but easier than, those of the preceding section. We omit their formulation.

1.3. The K-theory of a category with cofibrations and weak equivalences.

Consider the partially ordered set of pairs (i,j) , $0 \leq i \leq j \leq n$, where $(i,j) \leq (i',j')$ if and only if $i \leq i'$ and $j \leq j'$. Regarded as a category it may be identified to the arrow category $Ar[n]$ where as usual $[n]$ denotes the ordered set $(0 < 1 < \ldots < n)$ (considered as a category).

Let C be a category with cofibrations. We consider the functors

$$A: Ar[n] \longrightarrow C$$
$$(i,j) \longmapsto A_{i,j}$$

having the property that for every j ,

$$A_{j,j} = * ,$$

and that for every triple $i \leq j \leq k$, the map

$$A_{i,j} \longrightarrow A_{i,k}$$

is a cofibration, and the diagram

$$\begin{array}{ccc} A_{i,j} & \longrightarrow & A_{i,k} \\ \downarrow & & \downarrow \\ A_{j,j} & \longrightarrow & A_{j,k} \end{array}$$

is a pushout; in other words,

$$A_{i,j} \rightarrowtail A_{i,k} \twoheadrightarrow A_{j,k}$$

is a cofibration sequence. We denote the category of these functors and their natural transformations by S_nC .

To give an object $A \in S_n$ is really the same thing as to give a sequence of cofibrations

$$A_{o,1} \rightarrowtail A_{o,2} \rightarrowtail \ldots \rightarrowtail A_{o,n}$$

together with a choice of subquotients

$$A_{i,j} = A_{o,j}/A_{o,i} .$$

It results that the category S_nC can be identified with one of the categories of filtered objects considered in section 1.1 (namely F^+_{n-1}) and in particular therefore S_nC can be regarded as a category with cofibrations in a natural way.

The definition of S_nC given here has the advantage of making it clear that $[n] \mapsto Ar[n] \mapsto S_nC$ is contravariantly functorial on the category Δ of the ordered sets $[0], [1], \ldots$. We therefore have a simplicial category

$$S.C : \Delta^{op} \longrightarrow (cat)$$
$$[n] \longmapsto S_nC \quad .$$

In fact, we have a *simplicial category with cofibrations*; that is, a simplicial object in the category whose objects are the categories with cofibrations and whose morphisms are the exact functors between those. This results from the lemmas of section 1.1 upon inspection of what the face and degeneracy maps are. For example the face map $d_i : S_nC \to S_{n-1}C$ corresponds, for $i > 0$, to the forgetful map which drops $A_{o,i}$ from the sequence $A_{o,1} \rightarrowtail \ldots \rightarrowtail A_{o,n}$; and for $i = 0$ it corresponds to the map "quotient by $A_{o,1}$ " which replaces that sequence by $A_{1,2} \rightarrowtail \ldots \rightarrowtail A_{1,n}$.

If C is equipped with a category of weak equivalences, wC , then S_nC comes naturally equipped with a category of weak equivalences, wS_nC . By definition here an arrow $A \to A'$ of S_nC is in wS_nC if and only if the arrow $A_{i,j} \to A'_{i,j}$ is in wC for every pair $i \leqslant j$; or what amounts to the same in view of the assumed gluing lemma, if this is so for $i = 0$. It results that $S.C$ is a *simplicial category with cofibrations and weak equivalences* in this case.

Let us take a look at the simplicial category of weak equivalences

$$wS.C : \Delta^{op} \longrightarrow (cat)$$
$$[n] \longmapsto wS_nC \quad .$$

The category S_oC , and therefore also its subcategory wS_oC , is the trivial category with one object and one morphism. Hence the geometric realization $|wS_oC|$ is the one-point space.

The category S_1C is the category of diagrams

$$* = A_{o,o} \rightarrowtail A_{o,1} \twoheadrightarrow A_{1,1} = *$$

and is thus isomorphic to C . Hence the category of weak equivalences may be identified to wC .

Consider $|wS.C|$, the geometric realization of the simplicial category $wS.C$. The '1-skeleton' in the $S.$-direction is obtained from the '0-skeleton' (which is $|wS_oC|$) by attaching of $|wS_1C| \times |\Delta^1|$ (where $|\Delta^1|$ denotes the topological space 1-*simplex*). It results that the '1-skeleton' is naturally isomorphic to the suspension $S^1 \wedge |wC|$. As a consequence we obtain an inclusion $S^1 \wedge |wC| \to |wS.C|$, and by adjointness therefore an inclusion of $|wC|$ into the loop space of $|wS.C|$,

$$|wC| \longrightarrow \Omega|wS.C| \quad .$$

The passage from $|wC|$ to $\Omega|wS.C|$ is reminiscent of the 'group completion' process of Segal [11] (by which it was originally motivated, to some extent). We will have occasion to make an actual comparison later (in section 1.8).

Definition. The *algebraic K-theory* of the category with cofibrations C, with respect to the category of weak equivalences wC, is given by the pointed space

$$\Omega|wS.C| .$$

To pursue the analogy with Segal's version of group completion a little further, one can actually describe K-theory as a spectrum rather than just a space. Namely the $S.$-construction extends, by naturality, to simplicial categories with cofibrations and weak equivalences. In particular therefore it applies to $S.C$ to produce a bisimplicial category with cofibrations and weak equivalences, $S.S.C$. Again the construction extends to bisimplicial categories with cofibrations and weak equivalences; and so on. There results a spectrum

$$n \longmapsto |\underbrace{wS. \ldots S.}_{\longleftarrow n \longrightarrow}C|$$

whose structural maps are defined just as the map $|wC| \to \Omega|wS.C|$ above.

It turns out that the spectrum is a Ω-spectrum beyond the first term (the additivity theorem is needed to prove this, below). As the spectrum is connective (the n-th term is (n-1)-connected) an equivalent assertion is that in the sequence

$$|wC| \longrightarrow \Omega|wS.C| \longrightarrow \Omega\Omega|wS.S.C| \longrightarrow \ldots$$

all maps except the first are homotopy equivalences. It results that the K-theory of (C,wC) could equivalently be defined as the space

$$\Omega^\infty|wS.^{(\infty)}C| = \lim_{\vec{n}} \Omega^n|wS.^{(n)}C| , \qquad wS.^{(n)}C = \underbrace{wS. \ldots S.}_{\longleftarrow n \longrightarrow}C .$$

There is another way of making K-theory into a spectrum. Namely the pushout of the cofibrations $* \to A$ induces a sum in C and therefore a composition law in the sense of Segal on wC, $wS.C$, $wS.^{(2)}C$, and so on. As $\Omega|wS.C|$ is 'group-like' Segal's machine produces a connective Ω-spectrum from it. To see that the spectrum is equivalent to the former it suffices to note that the two spectra can be combined into a connective *bi-spectrum.* (A more direct relationship can also be established.)

The definition of K-theory is natural for categories with cofibrations and weak equivalences: an exact functor $F: C' \to C$ induces maps $wS.F: wS.C' \to wS.C$, etc.

Let a *weak equivalence* of exact functors $F, F': C' \to C$ mean a natural transformation $F \to F'$ having the property that for every $A \in C'$ the map $F(A) \to F'(A)$ is a weak equivalence in C.

Proposition 1.3.1. A weak equivalence from F to F' induces a homotopy between $wS.F$ and $wS.F'$.

Proof. The weak equivalence from F to F' restricts to a natural transformation of the restricted functors $F, F'\colon wC' \to wC$ and thereby induces a homotopy between these by a well known remark due to Segal [10]. Similarly there is what may be called a simplicial natural transformation from $wS.F$ to $wS.F'$. It gives rise to a homotopy in the same way. □

Let a *cofibration sequence* of exact functors $C' \to C$ mean a sequence of natural transformations $F' \to F \to F''$ having the following two properties: (i) for every $A \in C'$ the sequence $F'(A) \to F(A) \to F''(A)$ is a cofibration sequence, and (ii) for every cofibration $A' \to A$ in C' the square of cofibrations

$$\begin{array}{ccc} F'(A') & \longrightarrow & F'(A) \\ \downarrow & & \downarrow \\ F(A') & \longrightarrow & F(A) \end{array}$$

is *admissible* in the sense that $F(A') \cup_{F'(A')} F'(A) \to F(A)$ is also a cofibration.

Recall the category $E(A,C,B)$ (section 1.1), and let $E(C) = E(C,C,C)$.

<u>Proposition</u> 1.3.2. (*Equivalent formulations of the additivity theorem*). Each of the following four assertions implies all the three others.

(1) The following projection is a homotopy equivalence,

$$\begin{array}{ccc} wS.E(A,C,B) & \longrightarrow & wS.A \times wS.B \\ A \rightarrowtail C \twoheadrightarrow B & \longmapsto & A\ ,\ B\ . \end{array}$$

(2) The following projection is a homotopy equivalence,

$$\begin{array}{ccc} wS.E(C) & \longrightarrow & wS.C \times wS.C \\ A \rightarrowtail C \twoheadrightarrow B & \longmapsto & A\ ,\ B\ . \end{array}$$

(3) The following two maps are homotopic (resp. weakly homotopic),

$$\begin{array}{ccc} wS.E(C) & \longrightarrow & wS.C \\ A \rightarrowtail C \twoheadrightarrow B & \longmapsto & C\ ,\ \text{resp. } A \vee B\ . \end{array}$$

(4) If $F' \to F \to F''$ is a cofibration sequence of exact functors $C' \to C$ then there exists a homotopy

$$|wS.F| \simeq |wS.F'| \vee |wS.F''| \qquad (= |wS.(F' \vee F'')|\ )\ .$$

Proof. (2) is a special case of (1), and (3) is a special case of (4). So it will suffice to show the implications (2) ⇒ (3) ⇒ (4) and (4) ⇒ (1).

Ad (3)⇒(4). To give a cofibration sequence of functors $F' \rightarrowtail F \twoheadrightarrow F''$ from C' to C is equivalent to giving an exact functor $G\colon C' \to E(C)$, with $F' = sG$, $F = tG$, and $F'' = qG$, where s, t, q are the maps $A \rightarrowtail C \twoheadrightarrow B \longmapsto A, C, B,$ respectively (which are exact by proposition 1.1.4). Thus (4) follows from (3) by naturality.

Ad (2)⇒(3). The desired homotopy $|wS.t| \simeq |wS.(s \vee q)|$ is certainly valid upon restriction along the map

$$\begin{array}{ccc} |wS.\mathcal{C}| \times |wS.\mathcal{C}| & \longrightarrow & |wS.\mathcal{E}(\mathcal{C})| \\ A\ ,\ B & \longmapsto & A \rightarrowtail A \vee B \twoheadrightarrow B\ , \end{array}$$

so it will suffice to know that this map is a homotopy equivalence. But the map is a section to the map in (2) and therefore is a homotopy equivalence if that is one.

Ad (4)⇒(1). The map $p\colon wS.\mathcal{E}(\mathcal{A},\mathcal{C},\mathcal{B}) \to wS.\mathcal{A} \times wS.\mathcal{B}$ is a retraction, with section σ given by $A,B \longmapsto A \rightarrowtail A\vee B \twoheadrightarrow B$. To show p is a homotopy equivalence it therefore suffices to show that the identity map on $wS.\mathcal{E}(\mathcal{A},\mathcal{C},\mathcal{B})$ is homotopic to the map σp. (In fact, it would suffice to know that the two maps are *weakly homotopic*, that is, homotopic upon restriction to any compactum, for that would still imply that the map σ is surjective, and hence bijective, on homotopy groups.) The desired homotopy results from (4) applied to a suitable cofibration sequence of endofunctors on $\mathcal{E}(\mathcal{A},\mathcal{C},\mathcal{B})$. The cofibration sequence is shown by the following diagram which depicts the functors (the rows) applied to an object $A \rightarrowtail C \twoheadrightarrow B$,

$$\begin{array}{c} (A \overset{=}{\rightarrow} A \rightarrow *) \\ \downarrow \\ (A \rightarrowtail C \twoheadrightarrow B) \\ \downarrow \\ (* \rightarrow B \underset{=}{\rightarrow} B) \end{array}$$

This completes the proof. □

The actual proof of the additivity theorem is rather long and it will be given later (it occupies the next section). We will now convince ourselves that a considerable short cut to the proof is possible if the definition of K-theory is adjusted somewhat. We begin with the

Observation 1.3.3. Let s, t, q denote the maps from $\mathcal{E}(\mathcal{C})$ to $\mathcal{C}$ given by $A \rightarrowtail C \twoheadrightarrow B \longmapsto A,\ C,\ B$, respectively, and let $s\vee q$ denote the sum of s and q. Then the following two composite maps are homotopic,

$$|w\mathcal{E}(\mathcal{C})| \underset{s\vee q}{\overset{t}{\rightrightarrows}} |w\mathcal{C}| \longrightarrow \Omega|wS.\mathcal{C}|\ .$$

This results from an inspection of $|wS.\mathcal{C}|_{(2)}$, the '2-skeleton' of $|wS.\mathcal{C}|$ in the $S.$-direction. Let us identify $w\mathcal{C}$ to $wS_1\mathcal{C}$, as before, and let us identify $w\mathcal{E}(\mathcal{C})$ to $wS_2\mathcal{C}$ whose objects are the cofibration sequences $A_{0,1} \rightarrowtail A_{0,2} \twoheadrightarrow A_{1,2}$. The face maps from $wS_2\mathcal{C}$ to $wS_1\mathcal{C}$ then correspond to the three maps s, t, q, respectively, and which is which can be seen from the diagram

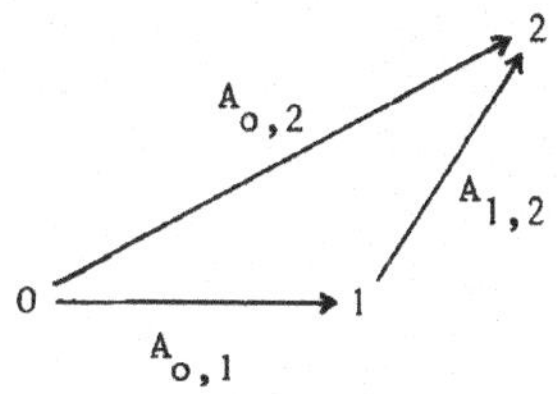

Let us consider the canonical map $|wS_2C| \times |\Delta^2| \to |wS.C|_{(2)}$. Regarding the 2-simplex $|\Delta^2|$ as a homotopy from the edge (0,2) to the edge path (0,1)(1,2) we obtain a homotopy from the composite map jt ,

$$|wE(C)| \xrightarrow{t} |wC| \xrightarrow{j} \Omega|wS.C|_{(2)} ,$$

to the *loop product* of the two composite maps js and jq . But in $\Omega|wS.C|$ the loop product is homotopic to the composition law, by a well known fact about loop spaces of H-spaces, whence the observation as stated.

The same consideration shows, more generally,

Observation 1.3.4. For every $n \geq 0$ the two composite maps

$$|wS.^{(n)}E(C)| \underset{svq}{\overset{t}{\rightrightarrows}} |wS.^{(n)}C| \longrightarrow \Omega|wS.^{(n+1)}C|$$

are homotopic, where $wS.^{(n)}C = \underbrace{wS.\ \dots\ S.}_{n}C$.

Corollary 1.3.5. The additivity theorem (proposition 1.3.2) is valid if the definition of K-theory as $\Omega|wS.C|$ is substituted with $\Omega^\infty|wS.^{(\infty)}C| = \varinjlim \Omega^n|wS.^{(n)}C|$.

Proof. First, proposition 1.3.2 is formal in the sense that it applies to the present definition of K-theory just as well. Second, by the preceding observation the two composite maps

$$\Omega^\infty|wS.^{(\infty)}E(C)| \underset{svq}{\overset{t}{\rightrightarrows}} \Omega^\infty|wS.^{(\infty)}C| \longrightarrow \Omega^\infty|wS.^{(\infty)}C|$$

are weakly homotopic. Since the arrow on the right is an isomorphism this is one of the equivalent formulations of the additivity theorem (proposition 1.3.2). □

Remark. As a consequence of the corollary we could add yet another reformulation of the additivity theorem to the list of proposition 1.3.2. Namely the additivity theorem as stated there implies (section 1.5) that the maps $|wS.^{(n)}C| \to \Omega|wS.^{(n+1)}C|$ are homotopy equivalences for $n \geq 1$. Conversely if these maps are homotopy equivalences then so is $\Omega|wS.C| \to \Omega^\infty|wS.^{(\infty)}C|$, and thus the additivity theorem is provided by the corollary.

To conclude this section we describe a modification of the simplicial category $wS.C$ which was suggested by Thomason. It is a simplicial category $wT.C$. By definition wT_nC is a subcategory of the functor category $C^{[n]}$. The objects of wT_nC are the sequences of cofibrations

$$C_o \rightarrowtail C_1 \rightarrowtail \ldots \rightarrowtail C_n$$

and the morphisms are the natural transformations $C \to C'$ satisfying the condition that for every $i \leq j$ the induced map

$$C_i' \cup_{C_i} C_j \longrightarrow C_j'$$

is a map in wC.

$wT.C$ is 'better' than $wS.C$ insofar as it may be regarded as the horizontal nerve of a *bicategory*.

In order to compare the two we have to modify $wT.C$ a little, by including choices. Namely let wT_n^+C be defined just as wT_nC except that in the data of an object we include a choice of quotients $C_{ij} = C_j/C_i$ for every $i \leq j$; the choice is to be arbitrary except if $i = j$ where we insist that $C_{ii} = *$, the basepoint. The forgetful map $wT_.^+C \to wT.C$ is an equivalence of categories in each degree, and therefore a homotopy equivalence. The comparison is now made by means of a map of simplicial categories $wT_.^+C \to wS.C$ which we show to be a homotopy equivalence. The map is defined as the forgetful map which forgets the C_i and remembers only the subquotients C_{ij}.

To show the map is a homotopy equivalence it suffices to show $wT_n^+C \to wS_nC$ is a homotopy equivalence for every n. For fixed n now wS_nC may be regarded as a retract of wT_n^+C; the section is the map which defines C_i as $C_{o,i}$ (the section is *not* induced by a simplicial map). We show the retraction is a deformation retraction by exhibiting a homotopy explicitly. There is a natural transformation from the identity functor to the composed map $wT_n^+C \to wS_nC \to wT_n^+C$, it is given on an object $C_o \rightarrowtail \ldots \rightarrowtail C_n$ by the quotient map to $C_{o,o} \rightarrowtail \ldots \rightarrowtail C_{o,n}$ which is a map in wT_nC in view of the definition of what this means. The natural transformation gives the desired homotopy.

1.4. The additivity theorem.

The proof of the additivity theorem involves only the cofibration structure, not the weak equivalences. It will therefore be convenient to explicitly concentrate on the cofibrations, a kind of 'separation of variables'.

If $\mathcal{C}$ is a (small) category with cofibrations we let $s_n\mathcal{C} = \mathrm{Ob}(S_n\mathcal{C})$, the set of objects of $S_n\mathcal{C}$, and $s.\mathcal{C}$ the simplicial set $[n] \mapsto s_n\mathcal{C}$.

Lemma 1.4.1. An exact functor of categories with cofibrations $f\colon \mathcal{C} \to \mathcal{C}'$ induces a map $s.f\colon s.\mathcal{C} \to s.\mathcal{C}'$. An isomorphism between two such functors f and f' induces a homotopy between $s.f$ and $s.f'$.

Before proving this we note the following consequence.

Corollary. (1) An exact equivalence of categories with cofibrations $\mathcal{C} \to \mathcal{C}'$ induces a homotopy equivalence $s.\mathcal{C} \to s.\mathcal{C}'$.

(2) Let $\mathcal{C}$ be made into a category with cofibrations and weak equivalences by means of the category $i\mathcal{C}$ of isomorphisms in $\mathcal{C}$. Then there is a homotopy equivalence $s.\mathcal{C} \to iS.\mathcal{C}$.

Indeed, (1) is clear, and (2) results by considering the simplicial object $[m] \mapsto i_mS.\mathcal{C}$, the nerve of $iS.\mathcal{C}$ in the i-direction, and noting that $i_oS.\mathcal{C} = s.\mathcal{C}$ and that the face and degeneracy maps are homotopy equivalences by (1).

Proof of lemma. The first part is clear. To prove the second part we will explicitly write down a simplicial homotopy. This is best done in categorical language. It is quite well known that simplicial objects in a category $\mathcal{D}$ can be regarded as functors $X\colon \Delta^{op} \to \mathcal{D}$, $[n] \mapsto X[n]$; and maps of simplicial objects as natural transformations of such functors. It seems to be less well known that simplicial homotopies can be described in similar fashion. Namely let $\Delta/[1]$ denote the category of objects over $[1]$ in Δ ; the objects are the maps $[n] \to [1]$. For any $X\colon \Delta^{op} \to \mathcal{D}$ let X^* denote the composed functor

$$\begin{array}{ccccc} (\Delta/[1])^{op} & \longrightarrow & \Delta^{op} & \xrightarrow{X} & \mathcal{D} \\ ([n] \to [1]) & \longmapsto & [n] & \longmapsto & X[n] . \end{array}$$

Then a *simplicial homotopy of maps from* X *to* Y may be identified with a natural transformation $X^* \to Y^*$.

In the case at hand suppose that a functor isomorphism from f to f' is given and write it as a functor $F: \mathcal{C} \times [1] \to \mathcal{C}'$. The required simplicial homotopy then is the map from $([n]\to[1]) \mapsto s_n\mathcal{C}$ to $([n]\to[1]) \mapsto s_n\mathcal{C}'$ given by

$$(a: [n] \to [1]) \longmapsto (\ (A: Ar[n] \to \mathcal{C}) \longmapsto (A': Ar[n] \to \mathcal{C}')\ )$$

where A' is defined as the composition

$$Ar[n] \xrightarrow{(A,a_*)} \mathcal{C} \times Ar[1] \xrightarrow{id \times p} \mathcal{C} \times [1] \xrightarrow{F} \mathcal{C}'$$

and $p: Ar[1] \to [1]$ is given by $(0,0) \mapsto 0$, $(1,1) \mapsto 1$, and $(0,1) \mapsto 1$. □

Recall the equivalent formulations of the additivity theorem given in proposition 1.3.2. We will now prove one of them.

Theorem 1.4.2. *(Additivity theorem)*. Let $\mathcal{C}$ be a category with cofibrations and weak equivalences. Then the following map is a homotopy equivalence,

$$wS.\mathcal{E}(\mathcal{C}) \longrightarrow wS.\mathcal{C} \times wS.\mathcal{C}$$
$$A \rightarrowtail C \twoheadrightarrow B \longmapsto A\ ,\ B\ .$$

We deduce this from

Lemma 1.4.3. The map $s.\mathcal{E}(\mathcal{C}) \to s.\mathcal{C} \times s.\mathcal{C}$ is a homotopy equivalence.

The lemma may be regarded as a special case of the theorem, namely the case of the map $iS.\mathcal{E}(\mathcal{C}) \to iS.\mathcal{C} \times iS.\mathcal{C}$, in view of lemma 1.4.1. Conversely,

Proof of theorem from lemma 1.4.3. Define $\mathcal{C}(m,w)$ to be the full subcategory of the functor category $\mathcal{C}^{[m]}$ of those functors which take values in $w\mathcal{C}$. Then $\mathcal{C}(m,w)$ is a subcategory-with-cofibrations of $\mathcal{C}^{[m]}$, and $[m] \mapsto \mathcal{C}(m,w)$ defines a simplicial category with cofibrations. Applying the lemma we obtain that each of the maps $s.\mathcal{E}(\mathcal{C}(m,w)) \to s.\mathcal{C}(m,w) \times s.\mathcal{C}(m,w)$ is a homotopy equivalence. It follows, by the realization lemma, that the map of simplicial objects

$$(\ [m] \mapsto s.\mathcal{E}(\mathcal{C}(m,w))\ ) \longrightarrow (\ [m] \mapsto s.\mathcal{C}(m,w)\ ) \times (\ [m] \mapsto s.\mathcal{C}(m,w)\ )$$

is a homotopy equivalence. But this is equivalent to the assertion of the theorem in view of the natural isomorphism of $[m],[n] \mapsto s_n\mathcal{C}(m,w)$ with the bisimplicial set $[m],[n] \mapsto w_mS_n\mathcal{C}$, the nerve of the simplicial category $wS.\mathcal{C}$. □

In the proof of lemma 1.4.3 we will need a version of the fibration criterion, theorem B of Quillen [8], in the framework of simplicial sets. We proceed to formulate this.

Let Δ^n denote the simplicial set *standard* n-*simplex*, $[m] \mapsto Hom_\Delta([m],[n])$. If Y is any simplicial set then its set of n-simplices may be identified with the set of maps $\Delta^n \to Y$ (a case of the Yoneda lemma). Let $f: X \to Y$ be a map of simplicial sets and let y be a n-simplex of Y . Define a simplicial set $f/(n,y)$

as the pullback

$$\begin{array}{ccc} f/(n,y) & \longrightarrow & X \\ \downarrow & & \downarrow f \\ \Delta^n & \xrightarrow{\;y\;} & Y \end{array} .$$

<u>Lemma</u> 1.4.A. If $f/(n,y)$ is contractible for every (n,y) then f is a homotopy equivalence.

<u>Lemma</u> 1.4.B. If for every $a\colon [m] \to [n]$, and every $y \in Y_n$, the induced map from $f/(m,a^*y)$ to $f/(n,y)$ is a homotopy equivalence then for every (n,y) the pullback diagram above is homotopy cartesian.

These two lemmas follow at once from theorems A and B of Quillen [8]. For let $\mathrm{simp}(Y)$ denote the category whose objects are the (n,y) and where a morphism from (n',y') to (n,y) is a morphism $a\colon [n'] \to [n]$ in Δ such that $a^*y = y'$. By applying $\mathrm{simp}(-)$ to everything in sight we obtain a translation of lemmas A and B into cases of theorems A and B, respectively. This uses that $\mathrm{simp}(f/(n,y))$ is naturally isomorphic with $\mathrm{simp}(f)/(n,y)$, the left fibre over (n,y) of the map of categories $\mathrm{simp}(f)$. And it uses further that, if N denotes the nerve functor, there is a natural transformation $N\mathrm{simp}(Y) \to Y$ which is a homotopy equivalence (cf. the end of section 1.6).

Proof of lemma 1.4.3. We defer till later the proof of the following

<u>Sublemma</u>. The map $f\colon s.\mathcal{E}(\mathcal{C}) \to s.\mathcal{C}$, $A \rightarrowtail C \twoheadrightarrow B \longmapsto A$, satisfies the hypothesis of lemma B above.

Applying lemma B we obtain a certain homotopy cartesian square for each simplex (n,y) of $s.\mathcal{C}$. In particular we obtain such a square for the unique 0-simplex $*$ of $s.\mathcal{C}$ in which case the homotopy cartesian square may be rewritten as a fibration up to homotopy $f/(0,*) \to s.\mathcal{E}(\mathcal{C}) \to s.\mathcal{C}$. The term $f/(0,*)$ can be identified with $s.\mathcal{E}'(\mathcal{C})$ where $\mathcal{E}'(\mathcal{C})$ denotes the subcategory with cofibrations of $\mathcal{E}(\mathcal{C})$ whose objects are the cofibration sequences $* \rightarrowtail C \twoheadrightarrow B$. As the quotient map in those cofibration sequences is necessarily an isomorphism, $\mathcal{E}'(\mathcal{C})$ is equivalent to $\mathcal{C}$, and by lemma 1.4.1 therefore $s.\mathcal{E}'(\mathcal{C})$ is homotopy equivalent to $s.\mathcal{C}$. We conclude that the sequence

$$\begin{array}{ccccc} s.\mathcal{C} & \longrightarrow & s.\mathcal{E}(\mathcal{C}) & \longrightarrow & s.\mathcal{C} \\ & & A \rightarrowtail C \twoheadrightarrow B & \longmapsto & A \\ B & \longmapsto & * \rightarrowtail B \twoheadrightarrow B & & \end{array}$$

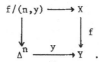

is a fibration up to homotopy. There is a map to this fibration sequence from the product fibration sequence. The map is the identity on the fibre and on the base, and on total spaces it is given by the *split cofibration sequences*, i.e. it is the map $s.\mathcal{C} \times s.\mathcal{C} \to s.\mathcal{E}(\mathcal{C})$, $(A,B) \mapsto (A \rightarrowtail A \vee B \twoheadrightarrow B)$. It follows that this map is a homotopy equivalence. The map is a section to the map of lemma 1.4.3, so that map must be a homotopy equivalence, too. □

Proof of sublemma. The assertion is that for every $y \in s_n\mathcal{C}$ and $w\colon [m] \to [n]$ in Δ , the map $w_*\colon f/(m,w^*y) \to f/(n,y)$ is a homotopy equivalence.

It will suffice to consider the special case of maps $[0] \to [n]$. For any map $w\colon [m] \to [n]$ can be embedded in some commutative triangle

$$\begin{array}{ccc} [m] & \xrightarrow{\;w\;} & [n] \\ & {\scriptstyle u}\nwarrow \quad \nearrow{\scriptstyle v} & \\ & [0] & \end{array}$$

and if we know that u_* and v_* are both homotopy equivalences then it follows that w_* is a homotopy equivalence, too.

We are thus reduced to proving this: let A' be a n-simplex of $s.\mathcal{C}$, for some n , and $*$ the unique 0-simplex of $s.\mathcal{C}$. Let $v_i\colon [0] \to [n]$ denote the map which takes 0 to i . Then for every i the map

$$v_{i*} : f/(0,*) \longrightarrow f/(n,A')$$

is a homotopy equivalence.

A m-simplex of $s.\mathcal{E}(\mathcal{C})$ may be identified to an object of $\mathcal{E}(S_m\mathcal{C})$, that is, a cofibration sequence $A \rightarrowtail C \twoheadrightarrow B$ in the category $S_m\mathcal{C}$.

A m-simplex of $f/(n,A')$ now consists of such a m-simplex $A \rightarrowtail C \twoheadrightarrow B$ together with a map $u\colon [m] \to [n]$, and these data are subject to the condition that A is equal to the composite

$$\mathrm{Ar}[m] \xrightarrow{\;u_*\;} \mathrm{Ar}[n] \xrightarrow{\;A'\;} \mathcal{C} \ .$$

The quotient projection $A \rightarrowtail C \twoheadrightarrow B \longmapsto B$ induces a map $p\colon f/(n,A') \to s.\mathcal{C}$. It will suffice to show that p is a homotopy equivalence. Indeed, p is left inverse to each of the composed maps

$$s.\mathcal{C} \xrightarrow{\;j_*\;} f/(0,*) \xrightarrow{\;v_{i*}\;} f/(n,A') \ ,$$

therefore if p is a homotopy equivalence then so is $v_{i*}j_*$, and hence also v_{i*} , since j_* certainly is a homotopy equivalence, being induced by the equivalence $\mathcal{C} \to f/(0,*)$, $B \mapsto (* \rightarrowtail B \overset{=}{\twoheadrightarrow} B, -)$.

Finally, in order to show p is a homotopy equivalence, it suffices to show that the particular map $v_{n*}j_*p\colon f/(n,A') \to f/(n,A')$ is homotopic to the identity map on $f/(n,A')$. We will construct such a homotopy explicitly.

The homotopy to be constructed will be a lifting of the simplicial homotopy that contracts Δ^n to its last vertex. In categorical language, this simplicial homotopy is given by a map of the composed functor

$$(\Delta/[1])^{op} \longrightarrow \Delta^{op} \longrightarrow (\text{sets})$$
$$([m] \to [1]) \longmapsto [m] \longmapsto \mathrm{Hom}([m],[n])$$

to itself, namely by

$$(v\colon [m] \to [1]) \longmapsto (\ (u\colon [m] \to [n]) \longmapsto (\bar{u}\colon [m] \to [n])\ )$$

where $\bar{u}$ is defined as the composite

$$[m] \xrightarrow{(u,v)} [n] \times [1] \xrightarrow{w} [n]$$

and where $w(j,0) = j$, $w(j,1) = n$.

A lifting of this homotopy to one on $f/(n,A')$ will be a map taking

$$(v\colon [m] \to [1])$$

to

$$(A \rightarrowtail C \twoheadrightarrow B\ ,\ \ u\colon [m] \to [n]) \longmapsto (\bar{A} \rightarrowtail \bar{C} \twoheadrightarrow \bar{B}\ ,\ \ \bar{u}\colon [m] \to [n])$$

where $\bar{u}$ is obtained from (v,u) as before and where certain compatibility conditions must be satisfied. In particular $\bar{A}$ must be equal to the composite

$$\mathrm{Ar}[m] \xrightarrow{\bar{u}_*} \mathrm{Ar}[n] \xrightarrow{A'} \mathcal{C}$$

and is thus entirely forced.

To see that the rest of the data can be found in the required way we note that for every $j \in [m]$ we have

$$u(j) \leqslant \bar{u}(j)\ .$$

This may be expressed by saying that there is a map of functors

$$(u\colon [m] \to [n]) \longrightarrow (\bar{u}\colon [m] \to [n])\ .$$

Consequently there is also a map of functors

$$(u_*\colon \mathrm{Ar}[m] \to \mathrm{Ar}[n]) \longrightarrow (\bar{u}_*\colon \mathrm{Ar}[m] \to \mathrm{Ar}[n])\ ,$$

and the latter induces a map of the composed functors

$$\mathrm{Ar}[m] \longrightarrow \mathrm{Ar}[n] \longrightarrow \mathcal{C}\ ,$$

that is, a map from A to $\bar{A}$ in $S_m\mathcal{C}$.

For later reference we record that a map $A \to \bar{A}$ obtained in this fashion is necessarily unique. Indeed, $A \to \bar{A}$ is induced by a map of functors $\mathrm{Ar}[m] \to \mathrm{Ar}[n]$ and the latter map, if it exists at all, is unique because $\mathrm{Ar}[n]$ is a partially ordered set.

We now *define* a cofibration sequence $\overline{A} \rightarrowtail \overline{C} \twoheadrightarrow \overline{B}$ as being obtained from $A \rightarrowtail C \twoheadrightarrow B$ by cobase change, in $S_m\mathcal{C}$, with the map $A \to \overline{A}$. Thus

$$\begin{array}{ccccc} A & \rightarrowtail & C & \twoheadrightarrow & B \\ \downarrow & & \downarrow & & \downarrow \| \\ \overline{A} & \rightarrowtail & \overline{C} & \twoheadrightarrow & \overline{B} \end{array} .$$

The definition involves a choice of pushouts; that is, given $\overline{A} \leftarrow A \rightarrowtail C$ we must complete it to a pushout diagram, with pushout $\overline{C}$, in some definite way. We insist at this point that those choices shall be made in $\mathcal{C}$ rather than in $S_m\mathcal{C}$. Because of the way pushouts in $S_m\mathcal{C}$ are computed (proposition 1.1.4) this gives the required choices in $S_m\mathcal{C}$ as well.

We are left to verify that the construction of $\overline{A} \rightarrowtail \overline{C} \twoheadrightarrow \overline{B}$ is compatible with the structure maps of the category $\Delta/[1]$; that is, if in our data we replace $[m]$ by $[m']$ throughout, by means of some map $[m'] \to [m]$, then the structure map in $s_{\cdot}\mathcal{E}(\mathcal{C})$ induced by $[m'] \to [m]$ takes the one cofibration sequence to the other.

To see this we review the steps of the construction. The first step was the definition of the map $A \to \overline{A}$. The definition is compatible with structure maps because of the uniqueness property pointed out above.

The second step was the choice of actual pushout diagrams. But this choice was made in $\mathcal{C}$, and an element of $S_m\mathcal{C}$ is a certain kind of diagram in $\mathcal{C}$ on which the simplicial structure maps operate by omission and/or reduplication of data. So again there is the required compatibility.

With a little extra care we can arrange the choices so that the homotopy starts from the identity map (namely if $A \to \overline{A}$ is an identity map we insist that $C \to \overline{C}$ is also an identity map); and that the image of $v_{n*}j_*$ is fixed under the homotopy (namely if $\overline{A} = *$ we insist that $\overline{C} \to \overline{B}$ is the identity map on $\overline{B}$). We have now constructed the desired homotopy. This completes the proof of the sublemma and hence that of the additivity theorem. □

1.5. Applications of the additivity theorem to relative K-theory, de-looping, and cofinality.

Let $X\colon \Delta^{op} \to \mathcal{D}$ be a simplicial object in a category $\mathcal{D}$. The associated *path object* PX is defined as the composition of X with the *shift functor* $\Delta \to \Delta$ which takes $[n]$ to $[n+1]$ (by 'sending i to i+1' - this fixes the behaviour on morphisms). The fact that a path space deforms into the subspace of constant paths has the following well known analogue here, e.g. [11], which we record in detail because we need to know the homotopy.

Lemma 1.5.1. PX is simplicially homotopy equivalent to the constant simplicial object $[n] \mapsto X_0$.

Proof. We show there is a simplicial homotopy between the identity on PX and the composite map $PX \to X_0 \to PX$ induced from

$$[n] \longmapsto ([n+1] \to [0] \to [n+1])$$
$$0 \longmapsto 0 \quad .$$

The homotopy is given by the natural transformation

$$(a\colon [n] \to [1]) \longmapsto (\varphi_a^*\colon X_{n+1} \to X_{n+1})$$

induced from $(a\colon [n] \to [1]) \mapsto (\varphi_a\colon [n+1] \to [n+1])$ where $\varphi_a(0) = 0$ and

$$\varphi_a(j+1) = \begin{cases} j+1 & \text{if } a(j) = 1 \\ 0 & \text{if } a(j) = 0 \ . \end{cases}$$ □

PX comes equipped with a projection $PX \to X$ (it is induced by the 0-face map of X which is not otherwise used in PX) and there is an inclusion of X_1 considered as a constant simplicial object (because $(PX)_0 = X_1$). There results a sequence $X_1 \to PX \to X$.

In particular if $\mathcal{C}$ is a category with cofibrations and weak equivalences we obtain a sequence $wS_1\mathcal{C} \to P(wS.\mathcal{C}) \to wS.\mathcal{C}$ which in view of the isomorphism of $wS_1\mathcal{C}$ with $w\mathcal{C}$ we may rewrite as

$$w\mathcal{C} \longrightarrow P(wS.\mathcal{C}) \longrightarrow wS.\mathcal{C} \ .$$

The composite map is constant, and $|P(wS.\mathcal{C})|$ is contractible (for by the preceding lemma it is homotopy equivalent to the one-point space $|wS_0\mathcal{C}|$), so we obtain a map, well defined up to homotopy,

$$|w\mathcal{C}| \longrightarrow \Omega|wS.\mathcal{C}| \ .$$

Lemma 1.5.2. The map can be chosen to agree with the corresponding map in the preceding section.

Proof. From the explicit homotopy of the preceding lemma one actually obtains an explicit choice of the map. This *is* the map in question. □

By naturality we can substitute $\mathcal{C}$ with the simplicial category $S.\mathcal{C}$ in the above sequence. We obtain a sequence

$$wS.\mathcal{C} \longrightarrow P(wS.S.\mathcal{C}) \longrightarrow wS.S.\mathcal{C}$$

(where the ' P ' refers to the first $S.$-direction, say).

Proposition 1.5.3. The sequence is a fibration up to homotopy. That is, the map from $|wS.\mathcal{C}|$ to the homotopy fibre of $|P(wS.S.\mathcal{C})| \to |wS.S.\mathcal{C}|$ is a homotopy equivalence.

Proof. This is a special case of proposition 1.5.5 below. □

Thus $|wS.\mathcal{C}| \to \Omega|wS.S.\mathcal{C}|$ is a homotopy equivalence and more generally therefore, in view of the realization lemma, also the map $|wS.^{(n)}\mathcal{C}| \to \Omega|wS.^{(n+1)}\mathcal{C}|$ for every $n \geqslant 1$, proving the postponed claim (section 1.3) that the spectrum $n \mapsto |wS.^{(n)}\mathcal{C}|$ is a Ω-spectrum beyond the first term.

We digress to indicate in which way the twice de-looped K-theory $wS.S.\mathcal{C}$ is used in defining *products*; or better, *external pairings* (products are induced from those). The ingredient that one needs is a *bi-exact functor* of categories with cofibrations and weak equivalences. This is a functor $\mathcal{A}\times\mathcal{B} \to \mathcal{C}$, $(A,B) \mapsto A\wedge B$, having the property that for every $A \in \mathcal{A}$ and $B \in \mathcal{B}$ the partial functors $A \wedge ?$ and $? \wedge B$ are exact, and where in addition the following more technical condition must also be satisfied; namely for every pair of cofibrations $A \rightarrowtail A'$ and $B \rightarrowtail B'$ in $\mathcal{A}$ and $\mathcal{B}$, respectively, the induced square of cofibrations in $\mathcal{C}$ must be *admissible* in the sense that the map $A'\wedge B \cup_{A\wedge B} A\wedge B' \to B\wedge B'$ is a cofibration.
A bi-exact functor induces a map, of bisimplicial bicategories,

$$wS.\mathcal{A} \times wS.\mathcal{B} \longrightarrow wwS.S.\mathcal{C}$$

which upon passage to geometric realization factors through the smash product

$$|wS.\mathcal{A}| \wedge |wS.\mathcal{B}| \longrightarrow |wwS.S.\mathcal{C}|$$

and in turn induces

$$\Omega|wS.\mathcal{A}| \wedge \Omega|wS.\mathcal{B}| \longrightarrow \Omega\Omega|wwS.S.\mathcal{C}| \ .$$

This is the desired pairing in K-theory in view of the homotopy equivalence of $|wS.\mathcal{C}|$ with $\Omega|wS.S.\mathcal{C}|$, and a (much more innocent) homotopy equivalence of $wS.S.\mathcal{C}$ with $wwS.S.\mathcal{C}$ which we will have occasion later on to consider in detail (the 'swallowing lemma' in section 1.6).

Definition 1.5.4. Let $f: A \to B$ be an exact functor of categories with cofibrations and weak equivalences. Then $S.(f:A\to B)$ is the pullback of the diagram

$$S.A \longrightarrow S.B \longleftarrow PS.B \ .$$

Thus for every n we have a pullback diagram

$$\begin{array}{ccc} S_n(f:A\to B) & \longrightarrow & (PS.B)_n = S_{n+1}B \\ \downarrow & & \downarrow \\ S_nA & \longrightarrow & S_nB \end{array} \ .$$

The vertical map on the right has a section (it is not compatible with face maps), so the pullback category is equivalent to the fibre product category and in any case is not pathological. It results (sections 1.1 and 1.2) that $S.(f:A\to B)$ is a simplicial category with cofibrations and weak equivalences in a natural way, and all the maps in the defining diagram (definition 1.5.4) are exact.

Considering B as a simplicial category in a trivial way we have an inclusion $B \to P(S.B)$ whose composition with the projection to $S.B$ is trivial (cf. above). Lifting the inclusion to the pullback, and combining with the other projection, we then obtain a sequence

$$B \longrightarrow S.(f:A\to B) \longrightarrow S.A$$

in which the composed map is trivial. The sequence is formally very similar to the sequence describing the homotopy fibration associated to a map of spaces. The following result says that in fact the sequence serves a similar purpose.

<u>Proposition</u> 1.5.5. The sequence

$$wS.B \longrightarrow wS.S.(f:A\to B) \longrightarrow wS.S.A$$

is a fibration up to homotopy.

Proof. There is a fibration criterion which says that it is enough to show that for every n the sequence $wS.B \to wS.S_n(f:A\to B) \to wS.S_nA$ is a fibration up to homotopy (e.g. since the base term $wS.S_nA$ is connected for every n, the criterion given by lemma 5.2 of [13] will do). Using the additivity theorem we will show that, in fact, the sequence is the same, up to homotopy, as the trivial fibration sequence associated to the product $wS.B \times wS.S_nA$.

Neglecting choices to simplify the notation, we can identify an object of $S_n(f:A\to B)$ to a pair of filtered objects in A and B, respectively, say $A_{o,1} \rightarrowtail \dots \rightarrowtail A_{o,n}$ and $B_o \rightarrowtail B_1 \rightarrowtail \dots \rightarrowtail B_n$, together with an isomorphism of filtered objects,

$$f(A_{o,1}) \rightarrowtail \dots \rightarrowtail f(A_{o,n}) \approx B_1/B_o \rightarrowtail \dots \rightarrowtail B_n/B_o \ .$$

Let C' denote the subcategory of the objects where all the maps $B_o \to B_1 \to \ldots \to B_n$ are identities and all the $A_{o,i}$ are equal to the basepoint; then C' is isomorphic to $\mathcal{B}$. Let C'' denote the subcategory where B_o is equal to the basepoint; then C'' is isomorphic to $S_n A$. There is an obvious cofibration sequence of endofunctors

$$j' \rightarrowtail id \twoheadrightarrow j''$$

where j' and j'' take values in C' and C'', respectively. Applying the additivity theorem (in formulation (4) of proposition 1.3.2) we obtain that the identity map on $wS.S_n(f{:}A{\to}\mathcal{B})$ is homotopic to the sum of $wS.j'$ and $wS.j''$. It results that the map, given by the split cofibration sequences,

$$wS.\mathcal{B} \times wS.S_n A \longrightarrow wS.S_n(f{:}A{\to}\mathcal{B})$$

is a retraction, up to homotopy. On the other hand the map is obviously also a coretraction. It is therefore a homotopy equivalence. We conclude with the remark that the homotopy equivalence can be induced by a map from the product fibration sequence to the sequence in question (i.e. the degree n part of the sequence of the proposition). It follows that the two sequences are the same, up to homotopy. This completes the proof of the proposition. □

In a special situation we can modify the definition of $S.(f{:}A{\to}\mathcal{B})$ to obtain a variant which is technically a little more convenient. Namely suppose that A is a *subcategory with cofibrations and weak equivalences* of $\mathcal{B}$ as defined in sections 1.1 and 1.2. Then we define

$$F_n(\mathcal{B},A)$$

as the category whose objects are the sequences of cofibrations in $\mathcal{B}$,

$$B_o \rightarrowtail B_1 \rightarrowtail \ldots \rightarrowtail B_n$$

subject to the condition that for every pair $i \leq j$ the object B_j/B_i is isomorphic to some object of A. There is a forgetful map

$$S_n(A{\to}\mathcal{B}) \longrightarrow F_n(\mathcal{B},A)$$

(forget choices of quotients B_j/B_i in A). It is an equivalence of categories with cofibrations and weak equivalences. Further the $F_n(\mathcal{B},A)$ may be assembled to a simplicial category with cofibrations and weak equivalences $F.(\mathcal{B},A)$. By the realization lemma then the forgetful map

$$wS.S.(A{\to}\mathcal{B}) \longrightarrow wS.F.(\mathcal{B},A)$$

is a homotopy equivalence. Thus $F.(\mathcal{B},A)$ may be used interchangeably with $S.(A{\to}\mathcal{B})$ if A is a subcategory with cofibrations and weak equivalences of $\mathcal{B}$.

Corollary 1.5.6. If $\mathcal{A} \to \mathcal{B} \to \mathcal{C}$ are exact functors of categories with cofibrations and weak equivalences then the square

$$\begin{array}{ccc} wS.\mathcal{B} & \longrightarrow & wS.\mathcal{S}.(\mathcal{A}\to\mathcal{B}) \\ \downarrow & & \downarrow \\ wS.\mathcal{C} & \longrightarrow & wS.\mathcal{S}.(\mathcal{A}\to\mathcal{C}) \end{array}$$

is homotopy cartesian. Similarly the square

$$\begin{array}{ccc} wS.\mathcal{B} & \longrightarrow & wS.\mathcal{F}.(\mathcal{B},\mathcal{A}) \\ \downarrow & & \downarrow \\ wS.\mathcal{C} & \longrightarrow & wS.\mathcal{F}.(\mathcal{C},\mathcal{A}) \end{array}$$

is homotopy cartesian if the terms on the right are defined.

Proof. There is a commutative diagram

$$\begin{array}{ccccc} wS.\mathcal{B} & \longrightarrow & wS.\mathcal{S}.(\mathcal{A}\to\mathcal{B}) & \longrightarrow & wS.\mathcal{S}.\mathcal{A} \\ \downarrow & & \downarrow & & \downarrow \\ wS.\mathcal{C} & \longrightarrow & wS.\mathcal{S}.(\mathcal{A}\to\mathcal{C}) & \longrightarrow & wS.\mathcal{S}.\mathcal{A} \end{array}$$

in which the vertical map on the right is an identity map and where the rows are fibrations up to homotopy, by the preceding proposition. It results that the square on the left is homotopy cartesian.

Concerning the second square, if that is defined, there is a natural transformation between the two squares in which all the maps are homotopy equivalences. The second assertion is just a rewriting of the first. □

Corollary 1.5.7. To an exact functor $\mathcal{B} \to \mathcal{C}$ there is associated a sequence of the homotopy type of a fibration (with a preferred null-homotopy of the composed map)

$$wS.\mathcal{B} \longrightarrow wS.\mathcal{C} \longrightarrow wS.\mathcal{S}.(\mathcal{B}\to\mathcal{C}) \ .$$

Indeed, this is the case $\mathcal{A} = \mathcal{B}$ of corollary 1.5.6 since $wS.\mathcal{S}.(\mathcal{A} \xrightarrow{=} \mathcal{A})$ is contractible.

Corollary 1.5.8. If $\mathcal{C}$ is a retract of $\mathcal{B}$ (by exact functors) there is a splitting

$$wS.\mathcal{B} \;\simeq\; wS.\mathcal{C} \times wS.\mathcal{S}.(\mathcal{C}\to\mathcal{B}) \ .$$

Indeed, this is the case of corollary 1.5.6 where the composed map $\mathcal{A} \to \mathcal{B} \to \mathcal{C}$ is an identity map (or more generally, an exact equivalence) since $wS.\mathcal{S}.(\mathcal{A}\to\mathcal{C})$ is contractible in that case.

Let A be a subcategory with cofibrations and weak equivalences of B. We say that A is *strictly cofinal* in B if for every $B \in B$ there exists a $A \in A$ such that $B \vee A$ is isomorphic to an object of A.

For example the category of free modules over a ring qualifies as strictly cofinal in the category of stably free modules, but not in the category of projective modules.

Proposition 1.5.9. If A is strictly cofinal in B then $wS.A \to wS.B$ is a homotopy equivalence.

Proof. It will be convenient to assume that A is *saturated* in B in the sense that every object of B isomorphic to one of A is actually contained in A. Since A can be enlarged to an equivalent category which is saturated in B and since such an enlargement does not affect any homotopy types, this assumption is not a loss of generality.

By corollary 1.5.7 or 1.5.6 the map $wS.A \to wS.B$ will be a homotopy equivalence if the bisimplicial category $wS.F.(B,A)$ is contractible. By the realization lemma this follows if $wS_nF.(B,A)$ is contractible for every n. We can rewrite

$$wS_nF.(B,A) \approx wF.(S_nB,S_nA) .$$

Assertion 1. If A is strictly cofinal in B then, for every n, S_nA is strictly cofinal in S_nB.

The assertion will be proved later. It reduces us to showing that $wF.(B,A)$ is contractible if A is strictly cofinal in B. By the realization lemma again this follows if the simplicial set $w_mF.(B,A)$, i.e. the degree-m-part of the nerve in the w-direction, is contractible for every m. Let, as before, $B(m,w)$ denote the category of the diagrams $B_0 \to B_1 \to \dots \to B_n$ in B in which the arrows are weak equivalences; and similarly with $A(m,w)$. Let $\mathfrak{f}.(B,A)$ denote the simplicial set of objects of $F.(B,A)$. We can rewrite

$$w_mF.(B,A) \approx \mathfrak{f}.(B(m,w),A(m,w)) .$$

Assertion 2. If A is strictly cofinal in B then, for every m, $A(m,w)$ is strictly cofinal in $B(m,w)$.

The assertion reduces us to proving

Assertion 3. If A is strictly cofinal in B then $\mathfrak{f}.(B,A)$ is contractible.

It remains to prove the assertions.

Proof of assertion 1. Let $B \in S_n\mathcal{B}$. We think of it as a filtration $B_{o,1} \rightarrowtail B_{o,2} \rightarrowtail \dots \rightarrowtail B_{o,n}$, plus a choice of subquotients $B_{i,j}$. By applying the cofinality hypothesis for $\mathcal{A} \subset \mathcal{B}$ we can find objects $A'_{i,j}$ in $\mathcal{A}$ (not subquotients of a filtration) so that $B_{i,j} \vee A'_{i,j}$ is in $\mathcal{A}$ for every (i,j) . Let A' be the sum of all the $A'_{i,j}$. Then $B_{i,j} \vee A'$ is in $\mathcal{A}$ for every (i,j) . We can define an object A of $S_n\mathcal{A}$ where, for every $i < j$, $A_{i,j}$ involves at least one summand A' ; briefly, $A_{o,i}$ is the i-fold sum of A' with itself. Then $B \vee A$ is in $S_n\mathcal{B}$, and all the objects involved in it are in $\mathcal{A}$; it is therefore in $S_n\mathcal{A}$ in view of the definition of what it means for $\mathcal{A}$ to be a subcategory with cofibrations of $\mathcal{B}$.

Proof of assertion 2. This is similar, but easier.

Proof of assertion 3. A n-simplex of $\mathcal{S}.(\mathcal{B},\mathcal{A})$ is a sequence of cofibrations in $\mathcal{B}$, $B_o \rightarrowtail \dots \rightarrowtail B_n$, subject to the condition that every subquotient B_j/B_i is isomorphic to some object of $\mathcal{A}$ (in fact, equal to an object of $\mathcal{A}$, for any choice whatsoever, in view of the assumed fact that $\mathcal{A}$ is saturated in $\mathcal{B}$). We apply the cofinality hypothesis to each of the B_i and then add all the objects of $\mathcal{A}$ obtained. This gives an object A in $\mathcal{A}$ with the property that $B_i \vee A$ is in $\mathcal{A}$ for every i ; the sequence $B_o \vee A \rightarrowtail \dots \rightarrowtail B_n \vee A$ is thus a sequence of cofibrations in $\mathcal{A}$ (since $\mathcal{A}$ is a subcategory with cofibrations of $\mathcal{B}$). We refer to this situation by saying that the object A *moves* the simplex $B_o \rightarrowtail \dots \rightarrowtail B_n$.

More generally, given finitely many simplices, not necessarily of the same dimension, we can find objects as before and add them all up to obtain a single object A which moves every one of these simplices.

The simplicial set $\mathcal{S}.(\mathcal{A},\mathcal{A})$ is contractible (it is the nerve of the category of cofibrations in $\mathcal{A}$, which has an initial object). To show $\mathcal{S}.(\mathcal{B},\mathcal{A})$ is contractible it suffices therefore to show that the inclusion $\mathcal{S}.(\mathcal{A},\mathcal{A}) \to \mathcal{S}.(\mathcal{B},\mathcal{A})$ is a homotopy equivalence. This follows if we can show that for every *finite* pair of simplicial subsets $(L,K) \subset (\mathcal{S}.(\mathcal{B},\mathcal{A}),\mathcal{S}.(\mathcal{A},\mathcal{A}))$ there is a homotopy, of pairs, from the inclusion map to some map with image in $\mathcal{S}.(\mathcal{A},\mathcal{A})$.

The simplicial set L has only finitely many non-degenerate simplices. So there is an object $A \in \mathcal{A}$ which moves every one of these simplices. But then A moves every other simplex of L as well.

$\mathcal{S}.(\mathcal{B},\mathcal{A})$ is a simplicial subset of the nerve of the category of cofibrations in $\mathcal{B}$. The sum with A induces a natural transformation of that category, and in turn a homotopy of the identity map on $\mathcal{S}.(\mathcal{B},\mathcal{A})$. The restriction of that homotopy to L , resp. K , is entirely in $\mathcal{S}.(\mathcal{B},\mathcal{A})$, resp. $\mathcal{S}.(\mathcal{A},\mathcal{A})$, and the homotopy terminates at a map which takes L into $\mathcal{S}.(\mathcal{A},\mathcal{A})$. This gives the required homotopy of pairs. The proof is complete. □

1.6. Cylinder functors, the generic fibration, and the approximation theorem.

Let $\mathcal{C}$ be a category with cofibrations and weak equivalences. By a *cylinder functor* on $\mathcal{C}$ is meant a functor from $Ar\mathcal{C}$ to the category of diagrams in $\mathcal{C}$ taking $f: A \to B$ to a diagram

$$\begin{array}{ccccc} A & \xrightarrow{j_1} & T(f) & \xleftarrow{j_2} & B \\ & {\scriptstyle f}\searrow & \downarrow{\scriptstyle p} & \swarrow{\scriptstyle =} & \\ & & B & & \end{array}$$

The functor is required to satisfy the axioms Cyl 1 - Cyl 3 below. The object $T(f)$ will be referred to as the *cylinder of* f, and the maps j_1, j_2, p as the *front inclusion*, *back inclusion*, and *projection*, respectively.

Cyl 1. The front and back inclusions assemble to an exact functor

$$\begin{array}{ccc} Ar\mathcal{C} & \longrightarrow & F_1\mathcal{C} \\ (A \xrightarrow{f} B) & \longmapsto & (A \vee B \xrightarrow{j_1 \vee j_2} T(f)) . \end{array}$$

Cyl 2. $T(* \to A) = A$, for every $A \in \mathcal{C}$, and the projection and back inclusion are the identity map on A.

Cyl 3.

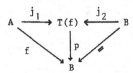

(fool's morning song [9], the tune replaces an unnecessary axiom)

Consider, for example, the category $\mathcal{R}(X)$ of the spaces having X as retract. It has a cylinder functor where $T(Y \to Y') = X \cup_{X\times[0,1]} Y\times[0,1] \cup_{Y\times 1} Y'$.

The following axiom may, or may not, be satisfied by a particular category of weak equivalences $w\mathcal{C}$.

Cylinder axiom. The projection $p: T(f) \to B$ is in $w\mathcal{C}$ for every $f: A \to B$ in $\mathcal{C}$.

Note. If in addition to the cylinder axiom $w\mathcal{C}$ also satisfies the saturation axiom (section 1.2) it follows that the back inclusion j_2 is always in $w\mathcal{C}$, and the front inclusion j_1 is in $w\mathcal{C}$ whenever f is.

For example in $\mathcal{R}(X)$ the weak homotopy equivalences and the simple maps satisfy the cylinder axiom while the isomorphisms do not. However the simple maps do not satisfy the saturation axiom, and in fact j_1 and j_2 are not, in general, simple maps.

Lemma 1.6.1. Cylinder functors are inherited by filtered objects. That is, a cylinder functor on $\mathcal{C}$ induces one on $\mathcal{S}_n\mathcal{C}$, for every n . If the weak equivalences in $\mathcal{C}$ satisfy the cylinder axiom then so do those in $\mathcal{S}_n\mathcal{C}$.

Proof. The required functor on $\mathrm{Ar}\mathcal{S}_n\mathcal{C}$ is defined as the induced map

$$\mathrm{Ar}\mathcal{S}_n\mathcal{C} \approx \mathcal{S}_n\mathrm{Ar}\mathcal{C} \longrightarrow \mathcal{S}_n(\text{diagrams in } \mathcal{C}) \approx (\text{diagrams in } \mathcal{S}_n\mathcal{C}) .$$

The only non-trivial point to check is the exactness of the functor $\mathrm{Ar}\mathcal{S}_n\mathcal{C} \to \mathcal{F}_1\mathcal{S}_n\mathcal{C}$ of axiom Cyl 1 . But this functor may be identified to the composite

$$\mathrm{Ar}\mathcal{S}_n\mathcal{C} \approx \mathcal{S}_n\mathrm{Ar}\mathcal{C} \longrightarrow \mathcal{S}_n\mathcal{F}_1\mathcal{C} \approx \mathcal{F}_1\mathcal{S}_n\mathcal{C}$$

and hence is exact since $\mathrm{Ar}\mathcal{C} \to \mathcal{F}_1\mathcal{C}$ is exact by axiom Cyl 1 in $\mathcal{C}$. □

Definition. The *cone functor* $A \mapsto cA$ is defined by

$$cA = T(A \to *) ,$$

and the *suspension functor* is defined as the quotient of the cone by the front inclusion $A \rightarrowtail T(A \to *)$,

$$\Sigma A = cA/A .$$

Proposition 1.6.2. If $\mathcal{C}$ has a cylinder functor and the weak equivalences satisfy the cylinder axiom then the suspension map

$$\Sigma : wS.\mathcal{C} \longrightarrow wS.\mathcal{C}$$

represents a homotopy inverse with respect to the H-space structure on $wS.\mathcal{C}$ given by the sum.

Proof. By the additivity theorem the cofibration sequence of functors $id \rightarrowtail c \twoheadrightarrow \Sigma$ implies a homotopy of self-maps on $wS.\mathcal{C}$, $id \vee \Sigma \simeq c$. The natural transformation $cA \to *$ is a weak equivalence in view of the assumed cylinder axiom. By lemma 1.3.1 therefore c, and hence $id \vee \Sigma$, is null-homotopic. □

Define $\bar{w}\mathcal{C}$ to be the subcategory of $w\mathcal{C}$ of those weak equivalences which are also cofibrations. (This is *not*, in general, a category of weak equivalences in the sense of section 1.2.)

Lemma 1.6.3. If $\mathcal{C}$ has a cylinder functor, and the weak equivalences in $\mathcal{C}$ satisfy the cylinder axiom and saturation axiom, then the inclusion $\bar{w}\mathcal{C} \to w\mathcal{C}$ is a homotopy equivalence.

Proof. Calling the inclusion i, it suffices to show by theorem A [8] that for every $B \in w\mathcal{C}$ the left fibre i/B is contractible. An object of i/B is a pair (A,f) where $f: A \to B$ is a map in $w\mathcal{C}$. Since the cylinder projection $p: T(f) \to B$ is in $w\mathcal{C}$ (by the cylinder axiom) we can define a functor $t: i/B \to i/B$ by letting $t(A,f) = (T(f),p)$. The front inclusion $j_1: A \to T(f)$ and back inclusion $j_2: B \to T(f)$ are weak equivalences as well as cofibrations (by the cylinder axiom and saturation axiom), so they define natural transformations to the functor t, one from the identity functor (using that $p\, j_1 = f$) and one from the constant functor with value (B, id_B) (using that $p\, j_2 = id_B$). It results that t is homotopic to both the identity map on i/B and the trivial map (B, id_B). Hence the latter two are homotopic, and i/B is contractible. □

To formulate the next result suppose that $\mathcal{C}$ is a category with cofibrations and that $\mathcal{C}$ is equipped with *two* categories of weak equivalences, one finer than the other, $v\mathcal{C} \subset w\mathcal{C}$. Let $\mathcal{C}^w$ denote the subcategory with cofibrations of $\mathcal{C}$ given by the objects A in $\mathcal{C}$ having the property that the map $* \to A$ is in $w\mathcal{C}$. It inherits categories of weak equivalences $v\mathcal{C}^w = \mathcal{C}^w \cap v\mathcal{C}$ and $w\mathcal{C}^w = \mathcal{C}^w \cap w\mathcal{C}$.

Theorem 1.6.4. (*Fibration theorem*). If $\mathcal{C}$ has a cylinder functor, and the coarse category of weak equivalences $w\mathcal{C}$ satisfies the cylinder axiom, saturation axiom, and extension axiom, then the square

$$\begin{array}{ccc} vS.\mathcal{C}^w & \longrightarrow & wS.\mathcal{C}^w \quad (\simeq *) \\ \downarrow & & \downarrow \\ vS.\mathcal{C} & \longrightarrow & wS.\mathcal{C} \end{array}$$

is homotopy cartesian, and the upper right term is contractible.

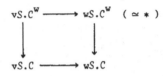

Proof. Define vwC to be the bicategory of the commutative squares

$$\begin{array}{ccc} \cdot & \longrightarrow & \cdot \\ \downarrow & & \downarrow \\ \cdot & \longrightarrow & \cdot \end{array}$$

in C in which the vertical and horizontal arrows are in vC and wC, respectively. Considering wC as a bicategory in a trivial way we have an inclusion $wC \to vwC$ which is a homotopy equivalence (lemma 1.6.5 below). There is a map in the other direction. The map exists only after passing to nerves, and diagonalizing (briefly, the map takes each square to its diagonal arrow), but to simplify the notation we will allow ourselves the abuse of writing the map as $vwC \to wC$. The map is left inverse to the former map, hence is a homotopy equivalence itself.

We can similarly define a simplicial bicategory $vwS.C$. By the realization lemma it results from the above that the maps $wS.C \to vwS.C$ and $vwS.C \to wS.C$ are homotopy equivalences as well (again the second map exists only after passing to nerves and diagonalizing the v- and w-directions).

Let $v\bar{w}C$ denote the sub-bicategory of vwC of the squares in which the horizontal arrows are in $\bar{w}C$ rather than just wC. Then the inclusion $v\bar{w}C \to vwC$ is a homotopy equivalence by lemma 1.6.3, which applies in view of the assumed cylinder axiom and saturation axiom. (In detail, by the realization lemma we can reduce to passing to nerves in the v-direction and showing that $v_n\bar{w}C \to v_nwC$ is a homotopy equivalence for every n. The map may be rewritten, in a way we have used before, as $\bar{w}C(v,n) \to wC(v,n)$, and lemma 1.6.3 now applies to the latter). Similarly there is a simplicial bicategory $v\bar{w}S.C$, and the inclusion $v\bar{w}S.C \to vwS.C$ is a homotopy equivalence. (For by the realization lemma we can reduce to showing that $v\bar{w}S_nC \to vwS_nC$ is a homotopy equivalence for every n. As S_nC inherits a cylinder functor from C (lemma 1.6.1) the above considerations apply to it.)

The square of the theorem may be identified to the large square in the following diagram

$$\begin{array}{ccccccc} vS.C^w & \longrightarrow & v\bar{w}S.C^w & \longrightarrow & vwS.C^w & \longrightarrow & wS.C^w \\ \downarrow & & \downarrow & & \downarrow & & \downarrow \\ vS.C & \longrightarrow & v\bar{w}S.C & \longrightarrow & vwS.C & \longrightarrow & wS.C \end{array}$$

As the preceding discussion shows, the horizontal maps in the middle and on the right are homotopy equivalences. So the square will be homotopy cartesian if and only if the square on the left is. After passing to nerves in the $\bar{w}$-direction we can identify the square on the left to one of the squares of corollary 1.5.6 associated to the categories at hand, namely

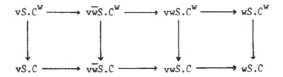

$$\begin{array}{ccc} vS.C^w & \longrightarrow & vS.F.(C^w,C^w) \\ \downarrow & & \downarrow \\ vS.C & \longrightarrow & vS.F.(C,C^w) \end{array} \quad ;$$

the point is that a map in $\overline{w}C$ can be characterized as a cofibration in C whose quotient is in C^w (this uses the assumed fact that wC satisfies the extension axiom). The square is thus homotopy cartesian by corollary 1.5.6.

Finally the simplicial category $wS.C^w$ is contractible because in each degree it has an initial object. □

The following lemma was used in the preceding argument; cf. [13] for some generalities on *bicategories*.

Lemma 1.6.5. (*Swallowing lemma*). Let A be a subcategory of B , and AB the bicategory of the commutative squares with vertical and horizontal arrows in A and B , respectively. The inclusion $B \to AB$ is a homotopy equivalence.

Proof. By the realization lemma it will suffice to take the nerve in the A-direction and show that for every n the map $B \to A_nB$ is a homotopy equivalence. For fixed n we can define a map $A_nB \to B$ by taking the sequence $A_o \to \ldots \to A_n$ to A_o . This is left inverse to the inclusion of B . Composing the other way we obtain the map which takes $A_o \to \ldots \to A_n$ to the appropriate sequence of identity maps on A_o . There is a natural transformation of this map to the identity map; it is given by the diagram

$$\begin{array}{ccccccc} A_o & \xrightarrow{=} & A_o & \xrightarrow{=} & \ldots & \xrightarrow{=} & A_o \\ \downarrow & & \downarrow{\scriptstyle a_1} & & & & \downarrow{\scriptstyle a_n \ldots a_2 a_1} \\ A_o & \xrightarrow[a_1]{} & A_1 & \xrightarrow[a_2]{} & \ldots & \xrightarrow[a_n]{} & A_n \end{array}$$

This shows that B is a deformation retract of A_nB . □

In order to formulate the next result it is convenient to introduce the following notion. Let $F: A \to B$ be an exact functor of categories with cofibrations and weak equivalences. We say it has the *approximation property* if it satisfies the conditions App 1 and App 2 below.

App 1. An arrow in A is a weak equivalence in A if (and only if) its image in B is a weak equivalence in B .

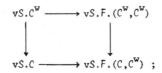

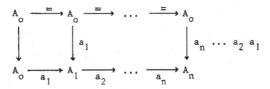

App 2. Given any object A in $\mathcal{A}$ and any map $x\colon F(A) \to B$ in $\mathcal{B}$ there exist a cofibration $a\colon A \rightarrowtail A'$ in $\mathcal{A}$ and a weak equivalence $x'\colon F(A') \to B$ in $\mathcal{B}$ so that the following triangle commutes,

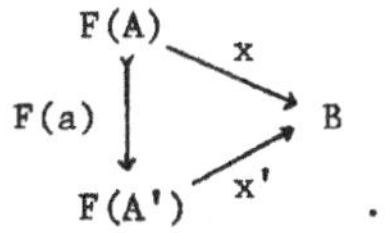

Lemma 1.6.6. If $F\colon \mathcal{A} \to \mathcal{B}$ has the approximation property then so does $S_n\mathcal{A} \to S_n\mathcal{B}$.

Proof. The non-trivial thing to verify is the condition App 2 for the map S_nF . We think of an object of $S_n\mathcal{A}$ as a filtration $A_{o,1} \rightarrowtail A_{o,2} \rightarrowtail \dots \rightarrowtail A_{o,n}$, plus a choice of subquotients. Proceeding by induction on n we suppose we have found already a sequence $A'_{o,1} \rightarrowtail \dots \rightarrowtail A'_{o,n-1}$ together with maps as required. From these data we obtain an object in $\mathcal{A}$,

$$A_{o,n} \cup_{A_{o,n-1}} A'_{o,n-1} \quad ,$$

and a map in $\mathcal{B}$,

$$F(\, A_{o,n} \cup_{A_{o,n-1}} A'_{o,n-1}\,) \longrightarrow B_{o,n} \quad ,$$

to which the hypothesis App 2 for F may be applied. This gives a cofibration

$$A_{o,n} \cup_{A_{o,n-1}} A'_{o,n-1} \rightarrowtail A'_{o,n}$$

and a weak equivalence $F(A'_{o,n}) \to B_{o,n}$ so that the following diagram commutes (where the broken arrow $A_{o,n} \rightarrowtail\!\!-\!\!\to A'_{o,n}$ is defined as the composite)

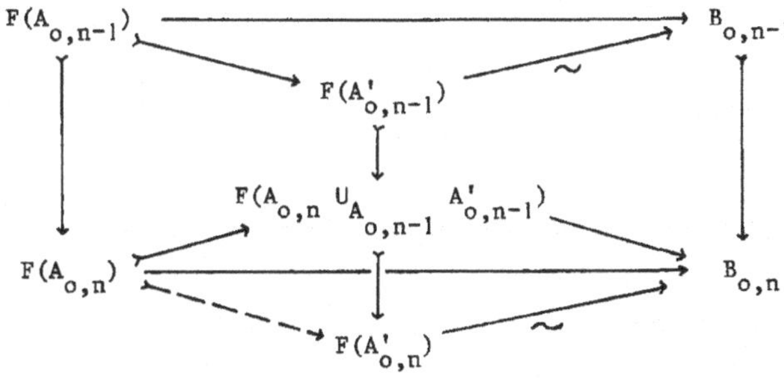

We are done. □

<u>Theorem</u> 1.6.7. (*Approximation theorem*). Let $\mathcal{A}$ and $\mathcal{B}$ be categories with cofibrations and weak equivalences. Suppose the weak equivalences in $\mathcal{A}$ and $\mathcal{B}$ satisfy the saturation axiom. Suppose further that $\mathcal{A}$ has a cylinder functor and the weak equivalences in $\mathcal{A}$ satisfy the cylinder axiom. Let $F\colon \mathcal{A} \to \mathcal{B}$ be an exact functor. Suppose F has the approximation property. Then the induced maps $w\mathcal{A} \to w\mathcal{B}$ and $wS.\mathcal{A} \to wS.\mathcal{B}$ are homotopy equivalences.

Proof. It will suffice to show that $w\mathcal{A} \to w\mathcal{B}$ is a homotopy equivalence. For this implies, in view of the preceding lemma, that $wS_n\mathcal{A} \to wS_n\mathcal{B}$ is a homotopy equivalence for every n, and hence, by the realization lemma, that $wS.\mathcal{A} \to wS.\mathcal{B}$ is a homotopy equivalence.

The proof that $w\mathcal{A} \to w\mathcal{B}$ is a homotopy equivalence, is quite long. It occupies the rest of this section. Calling the map f, it suffices to show, by theorem A [8], that for every $B \in w\mathcal{B}$ the left fibre f/B is contractible, and this is what we shall prove.

The idea for the proof of contractibility of f/B is in the following observation which says that certain diagrams $\mathcal{D}$ in f/B admit extensions to their cones and are thus contractible in f/B; by the *cone* on $\mathcal{D}$ is meant here the diagram $\mathcal{D}$ together with an added terminal vertex.

Observation. Let $\mathcal{D}$ be a diagram in f/B. Suppose that *as a diagram in* F/B it extends to the cone (for example, this is the case if the colimit of $\mathcal{D}$ exists in F/B). Then $\mathcal{D} \to f/B$ also extends to the cone.

Indeed, suppose that $\mathcal{D} \to f/B \subset F/B$ extends to the cone. Let the cone point be represented by $(A', F(A') \to B)$ in F/B. Applying the approximation property of F we find a cofibration $A' \rightarrowtail A''$ in $\mathcal{A}$ and a weak equivalence $F(A'') \to B$ in $\mathcal{B}$ so that the triangle

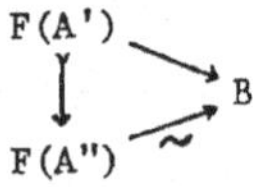

commutes. Then $(A'', F(A'') \to B)$ may be regarded as a terminal vertex to $\mathcal{D}$ in f/B rather than just F/B as we see by checking that certain maps are weak equivalences. Namely let $(A, F(A) \to B)$ represent any vertex of $\mathcal{D}$. Then there is a triangle

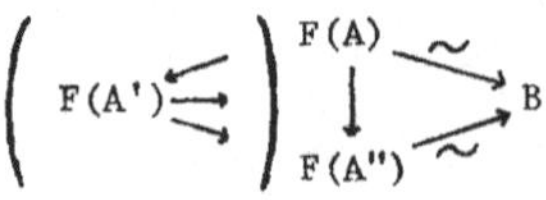

in which both of the maps going to B are weak equivalences. Applying the saturation axiom we obtain that $F(A) \to F(A'')$ is a weak equivalence in $\mathcal{B}$. From this we deduce in turn, using property App 1 of F, that $A \to A''$ is a weak equivalence, as required.

For example the empty diagram in f/B has a colimit in F/B provided by the initial object of $\mathcal{A}$. In view of the observation we conclude that f/B is non-empty.

Similarly any discrete two-point-diagram in f/B has a colimit in F/B provided by the sum in $\mathcal{A}$. In view of the observation this shows that f/B is connected.

To show that f/B is contractible it remains to find sufficiently many diagrams to which the observation applies. The sublemma below claims that this can be done. But we must first explain what 'sufficiently many' means in this context.

Let a *non-singular* simplicial set mean one where for every n and every non-degenerate n-simplex, the representing map from Δ^n is an embedding. For example *ordered simplicial complexes* may be regarded as simplicial sets and as such are non-singular.

In order to show the simplicial set $N(f/B)$, the nerve of f/B , is contractible it will suffice to show that for every non-singular X and every map from X to $N(f/B)$, this map is null-homotopic. (E.g. think of X as running through iterated subdivisions of spheres. There are sufficiently many maps from such X to represent all the elements of the homotopy groups of $N(f/B)$. If they are all trivial $N(f/B)$ is thus contractible by the Whitehead theorem).

To any simplicial set Y we can associate its category of simplices $\mathrm{simp}(Y)$, and there is a natural transformation $N(\mathrm{simp}(Y)) \to Y$ (the *last vertex map*) which is a homotopy equivalence (this will be recalled at the end of this section). If Y happens to be the nerve of a category then the natural transformation is the nerve of a map of categories. In particular we have a map $\mathrm{simp}(N(f/B)) \to f/B$.

If Y is non-singular then the category $\mathrm{simp}(Y)$ has a subcategory which is given by the non-degenerate simplices (it is a partially ordered set really). The inclusion $\mathrm{simp}^{n.d.}(Y) \to \mathrm{simp}(Y)$ is a homotopy equivalence (cf. the end of the section).

The map $X \to N(f/B)$ now gives rise to a sequence of maps

$$\mathrm{simp}^{n.d}(X) \xrightarrow{\sim} \mathrm{simp}(X) \longrightarrow \mathrm{simp}(N(f/B)) \xrightarrow{\sim} f/B$$

as well as a diagram

$$\begin{array}{ccc} N\,\mathrm{simp}(X) & \longrightarrow & N\,\mathrm{simp}(N(f/B)) \\ \downarrow\wr & & \downarrow\wr \\ X & \longrightarrow & N(f/B) \end{array} .$$

This shows that the map $X \to N(f/B)$ will be null-homotopic as soon as the induced map $\mathrm{simp}^{n.d.}(X) \to f/B$ is. The proof of the theorem has thus been reduced to the

Assertion. Let X be a non-singular finite simplicial set and $q: X \to N(f/B)$ a map. Then the induced map $q_*: \text{simp}^{n.d.}(X) \to f/B$ is null-homotopic.

We prove below

Sublemma. In this situation there exists a functor

$$T_q : \text{simp}^{n.d.}(X) \longrightarrow f/B$$

with the following two properties.

(1) There is a natural transformation from T_q to q_*.

(2) The composite functor

$$\text{simp}^{n.d.}(X) \xrightarrow{T_q} f/B \subset F/B$$

extends to a functor on $s(X)$, the partially ordered set of the simplicial subsets of X.

The sublemma implies the assertion and hence the theorem. For the partially ordered set $s(X)$ has a maximal element, therefore part (2) of the sublemma implies that $\text{simp}^{n.d.}(X) \to F/B$ extends to the cone on $\text{simp}^{n.d.}(X)$. In view of the observation therefore $T_q: \text{simp}^{n.d.}(X) \to f/B$ extends to the cone, too, thus T_q is null-homotopic. By part (1) of the sublemma T_q is homotopic to q_*. It results that q_* is null-homotopic.

Proof of sublemma. In order to define T_q we need the notion of *iterated mapping cylinder*, a notion derived from the cylinder functor on A. Let $A_o \to \ldots \to A_n$ be a sequence of maps in A. We will associate to this sequence the following data

(1) the (iterated) cylinder object $T(A_o \to \ldots \to A_n)$,

(2) a map $\partial_i: T(A_o \to \ldots \to \hat{A}_i \to \ldots \to A_n) \to T(A_o \to \ldots \to A_n)$ for every $0 \leqslant i \leqslant n$, where the hat indicates the omission of A_i from the sequence,

(3) a map $p: T(A_o \to \ldots \to A_n) \to A_n$.

Proceeding inductively we define $T(A_o \to \ldots \to A_n)$ as $T(T(A_o \to \ldots \to A_{n-1}) \to A_n)$, the cylinder of the composed map

$$T(A_o \to \ldots \to A_{n-1}) \xrightarrow{p} A_{n-1} \longrightarrow A_n ,$$

and $p: T(A_o \to \ldots \to A_n) \to A_n$ as the cylinder projection.

The definition of ∂_i requires a case distinction. The map

$$\partial_n : T(A_o \to \ldots \to A_{n-1}) \longrightarrow T(A_o \to \ldots \to A_n)$$

is defined as the front inclusion of the cylinder. If $n = 1$ the map

$$\partial_o : A_1 \longrightarrow T(A_o \to A_1)$$

is the back inclusion. And in general, finally, if $i < n$ and $n > 1$ then the map

$$\partial_i \;:\; T(A_o\to\ldots\to\hat{A}_i\to\ldots\to A_n) \longrightarrow T(A_o\to\ldots\to A_n)$$

is defined inductively as $T(\partial_i^!)$ where $\partial_i^!$ is the (vertical) map of diagrams

$$\begin{array}{ccc} T(A_o\to\ldots\to\hat{A}_i\to\ldots\to A_{n-1}) & \longrightarrow & A_n \\ \downarrow{\scriptstyle\partial_i} & & \Vert \\ T(A_o\to\ldots\to A_{n-1}) & \longrightarrow & A_n \end{array} .$$

From the particular sequence $A_o \to \ldots \to A_n$ we can obtain a functor

$$\mathrm{simp}^{n.d.}(\Delta^n) \longrightarrow A$$

taking each face of Δ^n to the iterated cylinder of the subsequence indexed by that face. On morphisms the functor is given by the maps ∂_i and their composites. To justify this we must check that the maps ∂_i satisfy the identities for iterated face maps. But for the identities not involving ∂_n this follows inductively from the case $n-1$, and for the identities which do involve ∂_n it follows from the fact that the front inclusion is a natural transformation.

The desired functor T_q is obtained by a slight modification, and generalization, of this construction. Namely let X be a non-singular simplicial set, and q a map from X to the nerve of f/B. Then the image of q on a n-simplex x of X is given by a sequence of weak equivalences in A, over $B \in B$,

$$A_o(x) \longrightarrow \ldots \longrightarrow A_n(x) .$$

Assuming now that x is a non-degenerate n-simplex of X we define $T_q(x)$ to be the iterated cylinder of that sequence, making it an object of f/B by means of the composite map $F(T(A_o(x)\to\ldots\to A_n(x))) \to F(A_n(x)) \to B$ (the first map here is induced from the projection p by the functor F, it is a weak equivalence in view of the assumed cylinder axiom). On morphisms T_q is defined by the maps ∂_i and their iterates (the morphisms are in f/B rather than just F/B in view of the assumed cylinder axiom and saturation axiom). It was checked above that the rule for morphisms is compatible with the identities for iterated face maps. There are no other identities in $\mathrm{simp}^{n.d.}(X)$, so T_q is a functor on it.

The desired natural transformation from T_q to q_* is given by the projection

$$p \;:\; T(A_o(x)\to\ldots\to A_n(x)) \xrightarrow{\;\sim\;} A_n(x) .$$

This completes the argument for part (1) of the sublemma.

In defining the proposed extension t of the composed functor

$$\mathrm{simp}^{n.d.}(X) \xrightarrow{T_q} f/B \subset F/B$$

we will insist on the following two properties of t

(1) t takes maps in $s(X)$ to cofibrations (as maps in A, after neglect of the

structure maps to B, that is),

(2) $t(X_1 \cup_{X_o} X_2) = t(X_1) \cup_{t(X_o)} t(X_2)$.

Given its restriction to $\mathrm{simp}^{n.d.}(X)$ provided by T_q, the functor t is uniquely determined by these conditions, up to isomorphism.

To establish the existence of t we proceed by induction, assuming in the inductive step that t does exist on the (n-1)-skeleton of X. Our aim is to establish the existence of t on the n-skeleton. There is only one thing that could conceivably go wrong with the inductive step. Namely if x is a n-simplex of X and ∂x its boundary (the union of the proper faces) then $t(\partial x)$ and $t(x)$ are both defined, and a map $t(\partial x) \to t(x)$ is also defined. The problem now is if this map is a cofibration.

Let $\Lambda^n x$ be the n-th horn of x, the union of all the proper faces except $d_n x$; so

$$x = \Lambda^n x \cup_{\partial d_n x} d_n x .$$

Condition (2) above expresses $t(\Lambda^n x)$ in terms of values of t on faces of x. Since a similar formula is valid for the cylinder functor, in view of its exactness, we conclude that

$$t(\Lambda^n x) \approx T(t(\partial d_n x) \to A_n)$$

where A_n denotes the value of t on the n-th vertex of x (and where, for ease of notation, we are ignoring the structure maps of objects in F/B). Applying condition (2) again we obtain that the map $t(\partial x) \to t(x)$ can be identified to the map

$$t(d_n x) \cup_{t(\partial d_n x)} T(t(\partial d_n x) \to A_n) \longrightarrow T(t(d_n x) \to A_n) .$$

That the latter map is a cofibration, is one of the conditions that must be satisfied for the following map in $F_1 A$ to be a cofibration in $F_1 A$,

$$(\, t(\partial d_n x) \to T(t(\partial d_n x) \to A_n) \,) \longrightarrow (\, t(d_n x) \to T(t(d_n x) \to A_n) \,) ,$$

so it will suffice to know that. The map is the image, with respect to

(†) $$(\, A' \to A'' \,) \longmapsto (\, A' \xrightarrow{j_1} T(A' \to A'') \,) ,$$

of the following map in ArA,

$$(\, t(\partial d_n x) \to A_n \,) \longrightarrow (\, t(d_n x) \to A_n \,) ,$$

which is a cofibration in ArA because $t(\partial d_n x) \to t(d_n x)$ is a cofibration by condition (1) above and the inductive hypothesis. We conclude by recalling that a cylinder functor has certain exactness properties, as specified in the axiom Cyl 1. In particular therefore the map (†) preserves cofibrations. This completes the proof of the sublemma and hence also that of the theorem. □

It remains to say a few words, as promised, about the map $\mathrm{Nsimp}(Y) \to Y$. In view of the natural isomorphisms $\mathrm{Nsimp}(Y) \approx \mathrm{colim}_{\mathrm{simp}(Y)}(([n],y) \mapsto \mathrm{Nsimp}(\Delta^n))$ and $Y \approx \mathrm{colim}_{\mathrm{simp}(Y)}(([n],y) \mapsto \Delta^n)$, the map is fully described once one knows the special case of simplices Δ^n. A m-simplex of $\mathrm{Nsimp}(\Delta^n)$ is a sequence of maps in Δ,

$$[n_0] \xrightarrow{a_0} [n_1] \xrightarrow{a_1} \dots \longrightarrow [n_m] \xrightarrow{a_m} [n] ,$$

and one associates to it the m-simplex $b\colon [m] \to [n]$ in Δ^n given by the last vertices, i.e.

$$b(i) = a_m a_{m-1} \dots a_i(n_i) .$$

$\mathrm{Nsimp}(\Delta^n)$ is contractible since $\mathrm{simp}(\Delta^n)$ has a terminal object. Therefore the map $\mathrm{Nsimp}(\Delta^n) \to \Delta^n$ is a homotopy equivalence. In view of the gluing lemma it results from this that $\mathrm{Nsimp}(Y) \to Y$ is a homotopy equivalence in general (cf. the appendix A to [11]).

Suppose now that Y is the nerve of a category C. Then $\mathrm{simp}(NC)$ is the category of pairs $([m],x)$, $x\colon [m] \to C$, and we can define a natural transformation $\mathrm{simp}(NC) \to C$ by $([m],x) \mapsto x(m)$. On passing to nerves this induces the above natural transformation in the case when $C = [n]$, and consequently also in general.

We conclude with

Lemma. If X is non-singular there is a functor $\mathrm{simp}(X) \to \mathrm{simp}^{n.d.}(X)$ which is left adjoint, and left inverse, to the inclusion functor.

Proof. The functor associates to each simplex of X the unique non-degenerate simplex of which the simplex is a degenerate. It is clear that this works in the special case where X is Δ^n. The general case reduces to this special case in view of the non-singularity of X. □

1.7. Spherical objects and cell filtrations.

By a *homology theory* on a category with cofibrations $\mathcal{C}$, with values in an abelian category $\mathcal{A}$, will be meant a sequence of functors $H_i : \mathcal{C} \to \mathcal{A}$, $i = 0,1,..,$ together with connecting maps $(A \rightarrowtail B) \longmapsto (H_{i+1}(B/A) \to H_i(A))$ such that the long sequence resulting from a cofibration sequence $A \rightarrowtail B \twoheadrightarrow B/A$ is exact and terminates in a surjection $H_o(B) \twoheadrightarrow H_o(B/A)$.

Given such a homology theory, $\mathcal{C}$ may be regarded as a category with cofibrations and weak equivalences where the latter are defined as the maps inducing isomorphisms in homology. The category of weak equivalences will be denoted $w\mathcal{C}$. It satisfies the saturation axiom and extension axiom.

Suppose given a full subcategory $\mathcal{E}$ of the abelian category $\mathcal{A}$ which is closed under the formation of extensions and kernels; that is, if $E' \rightarrowtail E \twoheadrightarrow E''$ is short exact then $E', E'' \in \mathcal{E}$ implies $E \in \mathcal{E}$, and $E, E'' \in \mathcal{E}$ implies $E' \in \mathcal{E}$. For example $\mathcal{A}$ itself will do.

Definition. An object $A \in \mathcal{C}$ is $(H_*,\mathcal{E})$-*spherical of dimension* n if

$$H_i(A) = 0 \quad \text{if} \quad i \neq n\,, \qquad \text{and} \quad H_n(A) \in \mathcal{E}\,.$$

With H_* and $\mathcal{E}$ being understood, such an A will also be simply referred to as n-*spherical*.

We denote the category of the n-spherical objects by $\mathcal{C}^n$. It is a subcategory with cofibrations and weak equivalences of $\mathcal{C}$ (section 1.1).

Example. On the category $\mathcal{R}(X)$ of the spaces having X as a retract there is a homology theory with values in the category of $Z[\pi_1 X]$-modules, $H_i(Y,r,s) = H_i(Y,s(X),r^*(Z[\pi_1 X]))$. For $\mathcal{E}$ one can take the category of projective $Z[\pi_1 X]$-modules, or even the subcategory of the stably free ones. The n-spherical objects include the objects (Y,r,s) where Y is obtainable, up to homotopy, by attaching n-cells to X.

We assume that $\mathcal{C}$ has a cylinder functor and that the weak equivalences satisfy the cylinder axiom. Any map $f: A \to B$ then gives rise to a long exact sequence $\ldots \to H_i(A) \to H_i(B) \to H_i(f) \to H_{i-1}(A) \to \ldots$ where

$$H_i(f) \;=\; H_i(T(f)/A)\,.$$

We say the map f is k-*connected* if $H_i(f) = 0$ for $i \leqslant k$.

The following hypothesis will be needed in the theorem below.

Hypothesis. For every m-connected map $X_m \to Y$ in $\mathcal{C}$ there is a factorization

$$\begin{array}{ccccccccc} X_m & \rightarrowtail & X_{m+1} & \rightarrowtail & \cdots & \rightarrowtail & X_n & \xrightarrow{\sim} & Y \\ & & \searrow & & & & \searrow & & \\ & & & X_{m+1}/X_m \in \mathcal{C}^{m+1} & \cdots & & & X_n/X_{n-1} \in \mathcal{C}^n & . \end{array}$$

Recall (proposition 1.6.2) that the suspension induces an exact functor $\Sigma\colon \mathcal{C} \to \mathcal{C}$ and a homotopy equivalence $wS.\mathcal{C} \to wS.\mathcal{C}$. As a consequence if we denote by $\varinjlim_{(\Sigma)} wS.\mathcal{C}$ the direct limit of the system $n \mapsto wS.\mathcal{C}$ in which the maps are given by suspension then

$$wS.\mathcal{C} \longrightarrow \varinjlim_{(\Sigma)} wS.\mathcal{C}$$

is a homotopy equivalence.

The suspension also induces an exact functor $\mathcal{C}^n \to \mathcal{C}^{n+1}$, so we can form $\varinjlim_{n} \mathcal{C}^n$.

<u>Theorem</u> 1.7.1. The map

$$\varinjlim_{n} wS.\mathcal{C}^n \longrightarrow \varinjlim_{(\Sigma)} wS.\mathcal{C}$$

is a homotopy equivalence, provided that the hypothesis is satisfied.

The proof of the theorem occupies all of this section. The strategy of the proof is to replace $\mathcal{C}$ by a category of *cell filtrations*, and to study two notions of weak equivalence, as well as their interplay, on that category.

Definition. A *cell filtration* in $\mathcal{C}$ is an eventually stationary sequence of cofibrations

$$* = A_{-1} \rightarrowtail A_o \rightarrowtail \cdots \rightarrowtail A_n \rightarrowtail \cdots$$

such that

$$A_n/A_{n-1} \in \mathcal{C}^n$$

for every n . The object to which the sequence stabilizes is denoted A_∞ .

For example, given any object $A \in \mathcal{C}$ one can find a cell filtration $\{A_i\}$ together with a weak equivalence $A_\infty \to A$. This results from the hypothesis of the theorem applied to the map $* \to A$ in $\mathcal{C}$.

The category of cell filtrations will be denoted $\hat{\mathcal{C}}$. It is a category with cofibrations where, by definition, a map $\{A_i\} \to \{A'_i\}$ is a cofibration if, and only if, for all n the map

$$A'_{n-1} \cup_{A_{n-1}} A_n \longrightarrow A'_n$$

is a cofibration in C with quotient in C^n. (Note this implies that the maps $A_n/A_{n-1} \to A'_n/A'_{n-1}$ and $A'_{n-1}/A_{n-1} \to A'_n/A_n$ are cofibrations, with the same quotient as the above. It also implies that for all n the map $A_n \to A'_n$ is a cofibration.)

The cylinder functor on C induces one on $\hat{C}$ where

$$T(\ \{f_i\}: \{A_i\} \to \{A'_i\}\ ) \quad = \quad \{\ A_i \cup_{A_{i-1}} T(f_{i-1}) \cup_{A'_{i-1}} A'_i\ \}\ .$$

As usual the cylinder functor induces functors *cone* and *suspension*. The suspension functor on $\hat{C}$ relates very simply to that on C, namely it is given by

$$\Sigma\{A_i\} \quad = \quad \{\Sigma A_{i-1}\}\ .$$

Of the two categories of weak equivalences in $\hat{C}$ to be considered, the coarse one is the category $w\hat{C}$ of the maps $\{A_i\} \to \{A'_i\}$ having the property that $A_\infty \to A'_\infty$ is in wC. The category $w\hat{C}$ satisfies the saturation axiom and extension axiom, and also the cylinder axiom.

Lemma 1.7.2. The map $wS.\hat{C} \to wS.C$, $\{A_i\} \mapsto A_\infty$, is a homotopy equivalence.

Proof. This is where the hypothesis of the theorem is used. We verify that the approximation theorem 1.6.7 applies to the forgetful map $\hat{C} \to C$. The non-trivial thing to prove is that given $\{A_i\} \in \hat{C}$ and a map $x: A_\infty \to B$ in C, we can find a cofibration $\{a_i\}: \{A_i\} \rightarrowtail \{A'_i\}$ in $\hat{C}$ and a weak equivalence $x': A'_\infty \to B$ in C so that $x = x'a_\infty$.

Let, for definiteness, $A_\infty = A_m$. If $m = -1$ then $\{A_i\} = *$, and $\{A'_i\}$ can be found by applying the hypothesis of the theorem to the map $* \to B$.

For $m \geq 0$ we proceed by induction. Truncating $\{A_i\}$ at level $m-1$ we can apply the inductive hypothesis to find $\{A''_i\}$, a cofibration $\{A_i\}_{(m-1)} \rightarrowtail \{A''_i\}$, and a weak equivalence $A''_\infty \to B$ so that the resulting triangle commutes.

A homology computation (downward induction on i) shows that $A''_i \to B$ is i-connected for every i, in particular $A''_{m-1} \to B$ is $(m-1)$-connected. By another homology computation we deduce from this that

$$A_m \cup_{A_{m-1}} A''_{m-1} \longrightarrow B$$

is also $(m-1)$-connected. We can now apply the hypothesis of the theorem to factor the map as

$$A_m \cup_{A_{m-1}} A''_{m-1} \rightarrowtail A'_m \rightarrowtail \ldots \rightarrowtail A'_n \xrightarrow{\sim} B$$

where the quotients of the cofibrations are spherical of the appropriate dimensions. We define $A'_i = A''_i$ for $i \leq m-1$. Then everything has been proved already except for the fact that $A'_m/A'_{m-1} \in C^m$. To see this we consider the sequence

$$A'_{m-1} = A''_{m-1} \rightarrowtail A_m \cup_{A_{m-1}} A''_{m-1} \rightarrowtail A'_m .$$

The associated cofibration sequence

$$A_m/A_{m-1} \rightarrowtail A'_m/A'_{m-1} \twoheadrightarrow A'_m/(A_m \cup_{A_{m-1}} A''_{m-1})$$

has both its 'subobject' and quotient in C^m. Since C^m is extension closed in C we conclude that $A'_m/A'_{m-1} \in C^m$. The lemma is proved. □

Let the fine category of weak equivalences in $\hat{C}$ be defined as the category $v\hat{C}$ of the maps $\{A_i\} \to \{A'_i\}$ having the property that $A_i \to A'_i$ is in wC for every i.

Let $\hat{C}_m$ denote the category of the cell filtrations in dimensions $\leqslant m$, i.e. the full subcategory of the $\{A_i\}$ in $\hat{C}$ with $A_m = A_\infty$. We consider $\hat{C}_m$ as a subcategory-with-cofibrations-and-weak-equivalences (sections 1.1 and 1.2) of $(\hat{C}, v\hat{C})$.

<u>Lemma</u> 1.7.3. The map

$$\begin{array}{ccc} vS.\hat{C}_m & \longrightarrow & wS.C^o \times wS.C^1 \times \ldots \times wS.C^m \\ (A_o \rightarrowtail A_1 \rightarrowtail . \rightarrowtail A_m) & \longmapsto & A_o\ ,\ A_1/A_o\ ,\ \ldots\ ,\ A_m/A_{m-1} \end{array}$$

is a homotopy equivalence.

Proof. By induction it suffices to show that the map

$$\begin{array}{ccc} vS.\hat{C}_m & \longrightarrow & vS.\hat{C}_{m-1} \times vS.C^m \\ (A_o \rightarrowtail . \rightarrowtail A_m) & \longmapsto & (A_o \rightarrowtail . \rightarrowtail A_{m-1})\ ,\ A_m/A_{m-1} \end{array}$$

is a homotopy equivalence. The map is a retraction. We show that it is also a coretraction, up to homotopy. The desired homotopy is given by the additivity theorem applied to the cofibration sequence of functors $f' \rightarrowtail \mathrm{id} \twoheadrightarrow f''$ on $\hat{C}_m$ where f' and f'' take $(A_o \rightarrowtail . \rightarrowtail A_m)$ to $(A_o \rightarrowtail . \rightarrowtail A_{m-1} \xrightarrow{=} A_{m-1})$ and $(* \rightarrowtail . \rightarrowtail * \rightarrowtail A_m/A_{m-1})$ respectively. □

Let, as usual, $\hat{C}^w$ denote the subcategory of the $\{A_i\}$ in $\hat{C}$ where $* \to \{A_i\}$ is in $w\hat{C}$. Let $\hat{C}^w_m = \hat{C}^w \cap \hat{C}_m$; it is the category of the cell filtrations $(A_o \rightarrowtail . \rightarrowtail A_{m-1} \rightarrowtail A_m)$ having the property that A_m is acyclic. We consider $\hat{C}^w_m$ as a subcategory-with-cofibrations-and-weak-equivalences of $(\hat{C}, v\hat{C})$.

<u>Lemma</u> 1.7.4. If $\{A_i\} \in \hat{C}^w$ then $A_n \in C^n$ for all n.

Proof. Using suitable long exact sequences we obtain

if $k > n$ then $H_k(A_n) \xleftarrow{\approx} H_k(A_{n-1}) \xleftarrow{\approx} \ldots \xleftarrow{\approx} H_k(A_{-1}) = 0$, and

if $k < n$ then $H_k(A_n) \xrightarrow{\approx} H_k(A_{n+1}) \xrightarrow{\approx} \ldots \xrightarrow{\approx} H_k(A_\infty) = 0$, thus $H_k(A_n) = 0$ if $k \neq n$. There is a short exact sequence

$$H_n(A_n) \rightarrowtail H_n(A_n/A_{n-1}) \twoheadrightarrow H_{n-1}(A_{n-1}) .$$

By induction we may assume $H_{n-1}(A_{n-1}) \in E$, and by definition of a cell filtration we have $H_n(A_n/A_{n-1}) \in E$. It follows that $H_n(A_n) \in E$ in view of the assumed fact that the category E is closed under taking kernels. □

Lemma 1.7.5. The map

$$\begin{array}{ccc} vS.\hat{C}^w_m & \longrightarrow & wS.C^o \times wS.C^1 \times \ldots \times wS.C^{m-1} \\ (A_o \rightarrowtail A_1 \rightarrowtail . . \rightarrowtail A_m) & \longmapsto & A_o , \quad A_1 , \quad \ldots \quad A_{m-1} \end{array}$$

is a homotopy equivalence.

Proof. The map exists by the preceding lemma. To show it is a homotopy equivalence it suffices, by induction, to show that the map p,

$$\begin{array}{ccc} vS.\hat{C}^w_m & \longrightarrow & vS.\hat{C}^w_{m-1} \times wS.C^{m-1} \\ (A_o \rightarrowtail . . \rightarrowtail A_m) & \longmapsto & (A_o \rightarrowtail . . \rightarrowtail A_{m-2} \rightarrowtail A_m) , \quad A_{m-1} , \end{array}$$

is a homotopy equivalence (p exists by the preceding lemma since $H_i(A_m/A_{m-2}) \approx H_{i-1}(A_{m-2})$). We show that the map s in the other direction,

$$(B_o \rightarrowtail . . \rightarrowtail B_{m-1}) , B \longmapsto (B_o \rightarrowtail . . \rightarrowtail B_{m-2} \rightarrowtail B_{m-1} \vee B \rightarrowtail B_{m-1} \vee cB) ,$$

is homotopy inverse to p where, as usual, cB denotes the cone on B.

The composite sp is given by

$$(B_o \rightarrowtail . . \rightarrowtail B_{m-1}) , B \longmapsto (B_o \rightarrowtail . . \rightarrowtail B_{m-2} \rightarrowtail B_{m-1} \vee cB) , B_{m-1} \vee B .$$

There is a natural transformation from the identity map to sp. It is a weak equivalence since both $B_{m-1} \to B_{m-1} \vee cB$ and $B \to B_{m-1} \vee B$ are weak equivalences. Hence it induces a homotopy (lemma 1.3.1), showing that s is left inverse to p.

To show that s is right inverse to p we construct a homotopy by applying the additivity theorem to a cofibration sequence of maps on $\hat{C}^w_m$. We can write $ps = f'vf''$ where f' and f'' are the self-maps of $\hat{C}^w_m$ taking $(A_o \rightarrowtail . . \rightarrowtail A_m)$ to $(* \to . . \to * \rightarrowtail A_{m-1} \rightarrowtail cA_{m-1})$ and $(A_o \rightarrowtail . . \rightarrowtail A_{m-2} \rightarrowtail A_m \xrightarrow{=} A_m)$, respectively. If we could find a cofibration sequence $f' \rightarrowtail f \twoheadrightarrow f''$, where f denotes the identity map on $\hat{C}^w_m$, it would follow by the additivity theorem that there is a homotopy between f and $f'vf''$, and we would be done.

The desired cofibration sequence does not exist directly, but it exists after the maps f and f'' have been modified a little. The modified maps are related to the original maps by chains of weak equivalences.

In a first step we replace the identity map f by a map f_1 taking $(A_o \rightarrowtail . . \rightarrowtail A_m)$ to $(A_o \rightarrowtail . . \rightarrowtail A_{m-1} \rightarrowtail c(A_m \cup_{A_{m-1}} cA_{m-1}))$. There is a weak equivalence $f \to f_1$

and we can define a map $f' \to f_1$ now. In a second step we blow up f_1 to a weakly equivalent f_2 so that the map $f' \to f_1$ can be replaced by a cofibration $f' \rightarrowtail f_2$. By definition, f_2 takes $(A_o \rightarrowtail . \rightarrowtail A_m)$ to

$$(A_o \rightarrowtail \dots \rightarrowtail A_{m-2} \rightarrowtail TA_{m-1} \rightarrowtail Tc(A_m \cup_{A_{m-1}} cA_{m-1}))$$

where TA is defined as $T(id_A)$, the cylinder of the identity map on A.

Let f''_3 be defined as the quotient f_2/f'. There is a weak equivalence to it from f''_2,

$$(A_o \rightarrowtail \dots \rightarrowtail A_{m-2} \rightarrowtail TA_{m-1}/A_{m-1} \xrightarrow{=} TA_{m-1}/A_{m-1}) ,$$

the latter maps by weak equivalence to f''_1,

$$(A_o \rightarrowtail \dots \rightarrowtail A_{m-2} \rightarrowtail TA_m/A_m \xrightarrow{=} TA_m/A_m) ,$$

and, to conclude, we have a weak equivalence $f'' \to f''_1$. We are done. □

Lemma 1.7.6. The map

$$vS.\hat{C}^w_m \times wS.C^m \longrightarrow vS.\hat{C}_m$$

is a homotopy equivalence.

Proof. The map

$$wS.C^o \times \dots \times wS.C^{m-1} \longrightarrow vS.\hat{C}^w_m$$

$$A_o , \dots , A_{m-1} \longmapsto A_o \rightarrowtail cA_o vA_1 \rightarrowtail \dots \rightarrowtail cA_o v..vcA_{m-2}vA_{m-1} \rightarrowtail cA_o v..vcA_{m-1}$$

is a homotopy equivalence. For by composing it with the homotopy equivalence of the preceding lemma we obtain a map induced by a self-map of $C^o \times \dots \times C^{m-1}$ weakly equivalent to the identity map. As a result it will suffice to show that the composite map

$$(C^o \times \dots \times C^{m-1}) \times C^m \longrightarrow \hat{C}^w_m \times C^m \longrightarrow \hat{C}_m \longrightarrow C^o \times \dots \times C^m ,$$

where the right hand map is that of lemma 1.7.3, induces a homotopy equivalence of $wS.C^o \times .. \times wS.C^m$ to itself. The composite map is given by

$$(A_o , \dots , A_m) \longmapsto (A_o , \Sigma A_o vA_1 , \Sigma A_1 vA_2 , \dots , \Sigma A_{m-1} vA_m) .$$

This is clearly a homotopy equivalence. □

Lemma 1.7.7. The map

$$\varinjlim_m wS.C^m \times \varinjlim_{(\Sigma)} vS.\hat{C}^w \longrightarrow \varinjlim_{(\Sigma)} vS.\hat{C}$$

(limits by suspension) is a homotopy equivalence.

Proof. The desired homotopy equivalence results by direct limit once it is known that the maps $\varphi_k : \varinjlim_m wS.C^m \times \varinjlim_m vS.\hat{C}^w_{m+k} \longrightarrow \varinjlim_m vS.\hat{C}_{m+k}$ are homotopy equivalences. The case $k = 0$ follows from the preceding lemma by direct limit. We deduce the case $k = 1$ from the case $k = 0$. Namely the two maps

$$\varinjlim_m C^m \xrightarrow[\approx]{\Sigma} \varinjlim_m C^{m+1} \xrightarrow{\psi_o} \varinjlim_m \hat{C}_{m+1}\ , \qquad \varinjlim_m C^m \xrightarrow{\psi_1} \varinjlim_m \hat{C}_{m+1}$$

are related by a cofibration sequence of functors $\psi_1 \rightarrowtail \Theta \twoheadrightarrow \psi_o\Sigma$ where Θ is the composite map

$$\begin{array}{ccccc} \varinjlim C^m & \longrightarrow & \varinjlim \hat{C}^w_{m+1} & \longrightarrow & \varinjlim \hat{C}_{m+1} \\ A & \longmapsto & (.. \overset{=}{\to} * \rightarrowtail A \rightarrowtail cA \overset{=}{\to} ..) & & \end{array}$$

By the additivity theorem there results a homotopy of the induced maps, $\psi_1 \vee \psi_o\Sigma \simeq \Theta$, showing that, modulo $\varinjlim \hat{C}^w_{m+1}$, the maps ψ_1 and $\psi_o\Sigma$ are the same up to sign. We conclude that φ_1 is a homotopy equivalence since φ_o is. Similarly it follows that φ_2 is a homotopy equivalence since φ_1 is. And so on. □

Proof of theorem 1.7.1. By the fibration theorem 1.6.4 there is a homotopy cartesian square

$$\begin{array}{ccc} vS.\hat{C}^w & \longrightarrow & vS.\hat{C} \\ \downarrow & & \downarrow \\ wS.\hat{C}^w & \longrightarrow & wS.\hat{C} \end{array} .$$

Suspension induces a self-map of the square, and hence a direct system. Passing to the direct limit we obtain a square which is homotopy cartesian again. It is the large square in the following diagram

$$\begin{array}{ccccc} \varinjlim vS.\hat{C}^w & \longrightarrow & \varinjlim (\ vS.\hat{C}^w \times wS.C^m\ ) & \longrightarrow & \varinjlim vS.\hat{C} \\ \downarrow & & \downarrow & & \downarrow \\ \varinjlim wS.\hat{C}^w & \longrightarrow & \varinjlim (\ wS.\hat{C}^w \times wS.C^m\ ) & \longrightarrow & \varinjlim wS.\hat{C} \end{array} .$$

By comparing the vertical homotopy fibres we see that the left square in the diagram is also homotopy cartesian. It follows that the square on the right is homotopy cartesian. By the preceding lemma the upper horizontal map in the right hand square is a homotopy equivalence. We conclude that the lower horizontal map is a homotopy equivalence. Discarding the contractible factor $\varinjlim wS.\hat{C}^w$ we obtain the map

$$\varinjlim wS.C^m \longrightarrow \varinjlim wS.\hat{C}$$

which is therefore a homotopy equivalence. In view of the homotopy equivalence

$$\varinjlim wS.\hat{C} \longrightarrow \varinjlim wS.C$$

of lemma 1.7.2 this completes the proof of the theorem. □

1.8. Split cofibrations, and K-theory via group completion.

Let $\mathcal{A}$ be a category with *sum* (categorical coproduct), and let $\mathcal{A}$ be pointed by an initial object $*$. There is an associated simplicial category

$$\begin{aligned} \mathcal{N}.\mathcal{A} \ : \ \Delta^{op} &\longrightarrow (cat) \\ [n] &\longmapsto \mathcal{N}_n\mathcal{A} \ , \end{aligned}$$

the *nerve with respect to the composition law*. By definition $\mathcal{N}_n\mathcal{A}$ is the category equivalent to $\mathcal{A}^n$ in which an object consists of a tuple $A_1,\ldots,A_n$ together with appropriate sum diagrams, one for each subset of $\{1,\ldots,n\}$; these choices are to be compatible, and for the subsets of cardinality $\leqslant 1$ they are to be given by the objects $A_1,\ldots,A_n$ themselves and by the initial object $*$, respectively.

By a *category of weak equivalences* in $\mathcal{A}$ will be meant any subcategory $w\mathcal{A}$ which contains the isomorphisms and is closed under sum formation; that is, if $A_1 \to A_1'$ and $A_2 \to A_2'$ are in $w\mathcal{A}$ then so is $A_1 \vee A_2 \to A_1' \vee A_2'$.

If $\mathcal{A}$ is a *category with sum and weak equivalences* let $w\mathcal{N}_n\mathcal{A}$ be defined as the subcategory of $\mathcal{N}_n\mathcal{A}$ whose morphisms are the natural transformations with values in $w\mathcal{A}$. It is a category of weak equivalences in $\mathcal{N}_n\mathcal{A}$, and it is equivalent to $w\mathcal{A}^n$ by the forgetful map. $\mathcal{N}.\mathcal{A}$ may be regarded as a *simplicial category with sum and weak equivalences*, and the simplicial category of weak equivalences is

$$\begin{aligned} w\mathcal{N}.\mathcal{A} \ : \ \Delta^{op} &\longrightarrow (cat) \\ [n] &\longmapsto w\mathcal{N}_n\mathcal{A} \ . \end{aligned}$$

The construction is a special case of Segal's construction of *Γ-categories* [11]. The present notation has been chosen to conform to that of section 1.3.

Let $\mathcal{C}$ be a category with cofibrations and weak equivalences. By neglect of structure $\mathcal{C}$ is a category with sum and weak equivalences, $A \vee B = A \cup_* B$. There is a map of simplicial categories

$$w\mathcal{N}.\mathcal{C} \longrightarrow wS.\mathcal{C} \ ,$$

it takes

$$(\ A_1,\ldots,A_n \ , \ \text{choices} \ )$$

to

$$(\ A_1 \rightarrowtail A_1 \vee A_2 \rightarrowtail \ldots \rightarrowtail A_1 \vee \ldots \vee A_n \ , \ \text{(fewer) choices} \ ) \ .$$

The theorem to be formulated below says that the map is a homotopy equivalence in certain cases.

Suppose that $\mathcal{C}$, a category with cofibrations and weak equivalences, has a cylinder functor and that the weak equivalences in $\mathcal{C}$ satisfy the cylinder axiom, saturation axiom, and extension axiom.

Suppose given a sequence of subcategories-with-cofibrations-and-weak-equivalences $\mathcal{C}^n$ in $\mathcal{C}$ subject to the condition that suspension takes $\mathcal{C}^n$ into $\mathcal{C}^{n+1}$ for all n. The example to be kept in mind is that of a sequence of categories of spherical objects in the sense of the preceding section.

Let us say that a cofibration $A \rightarrowtail B$ in $\mathcal{C}^n$ is *splittable up to weak equivalence* if there is a chain of weak equivalences, relative to A, relating $A \rightarrowtail B$ to $A \rightarrowtail B'$ where $B' \approx A \vee B'/A$.

Theorem 1.8.1. The map

$$\varinjlim_n wN.\mathcal{C}^n \longrightarrow \varinjlim_n wS.\mathcal{C}^n$$

is a homotopy equivalence, provided that, for every n, all cofibrations in $\mathcal{C}^n$ are splittable up to weak equivalence.

The proof of the theorem occupies the present section. The argument will be summarized at the end of the section. The splittability condition actually used is slightly weaker than the one formulated here.

For any $X \in \mathcal{C}$ let $\mathcal{C}_X$ denote the *category of the cofibrant objects under* X; the objects of $\mathcal{C}_X$ are the cofibrations $X \rightarrowtail A$ in $\mathcal{C}$, and the morphisms are the maps $A \to A'$ restricting to the identity map on X. $\mathcal{C}_X$ is a category with sum,

$$(X \rightarrowtail A) \vee (X \rightarrowtail A') \quad = \quad (X \rightarrowtail A \cup_X A') ,$$

and it comes equipped with a category of weak equivalences $w\mathcal{C}_X$, the pre-image of $w\mathcal{C}$ under the projection $\mathcal{C}_X \to \mathcal{C}$, $(X \rightarrowtail A) \mapsto A$.

Let as usual c denote the cone functor derived from the cylinder functor ($cA = T(A \rightarrowtail *)$) and Σ the suspension functor, $\Sigma A = cA/A = cA \cup_A *$.

Lemma 1.8.2. To $X \rightarrowtail A$ in $\mathcal{C}_X$ there is naturally associated a chain of weak equivalences in $\mathcal{C}_{\Sigma X}$,

$$(\Sigma X \rightarrowtail \Sigma A \cup_{\Sigma X} \Sigma A) \quad \sim \quad (\Sigma X \rightarrowtail \Sigma A \cup_* \Sigma A/\Sigma X) \quad .$$

Proof. The chain consists of two maps. These are given by the two diagonal arrows in the following diagram

$$\begin{array}{ccc}
 & \nearrow & \Sigma A \cup_{\Sigma X} \Sigma A \cup_{*} * \\
 & & \uparrow \\
\Sigma A \cup_{\Sigma X} (cA/X \cup_{A/X} cA/cX) & \longrightarrow & \Sigma A \cup_{\Sigma X} \Sigma A \cup_{*} \Sigma A/\Sigma X \\
 & & \downarrow \\
 & \searrow & \Sigma A \cup_{*} \Sigma A/\Sigma X
\end{array}$$

By definition, the horizontal arrow is given by pushout with the map $A/X \to *$, and the downward vertical arrow is induced by the folding map $\Sigma A \cup_{\Sigma X} \Sigma A \to \Sigma A$. The upper diagonal arrow is a weak equivalence since it is given by pushout with the weak equivalence $cA/cX \to *$. The lower diagonal arrow is a weak equivalence in view of the assumed extension axiom. For by cobase change with the map $\Sigma A \to *$ one obtains from it the weak equivalence $cA/cX \cup_{A/X} cA/cX \to \Sigma A/\Sigma X$. □

Remark. If C happens to be an *additive* category the lemma is true without suspension, one can define a weak equivalence $A \cup_X A \to A \cup_* A/X$ as a map whose restriction to the second A is the sum of the identity $A \to A \cup_* *$ and the projection $A \to * \cup_* A/X$. In the additive case the argument leading to the theorem, and the theorem itself, can thus be simplified. □

If $X \in C^m$ we can form C^m_X . There are maps, of categories with sum and weak equivalences,

$$q : C^m_X \longrightarrow C^m , \quad X \rightarrowtail A \longmapsto A/X \qquad\qquad j : C^m \longrightarrow C^m_X , \quad B \longmapsto X \rightarrowtail X \cup_* B$$

and q is left inverse to j , up to natural isomorphism of $q\, j$ to the identity on C^m .

<u>Proposition</u> 1.8.3. The map

$$\varinjlim_n wN.C^{m+n} \longrightarrow \varinjlim_n wN.C^{m+n}_{\Sigma^n X}$$

(limits by suspension) is a homotopy equivalence.

Proof. It will suffice to know that for each n the composite $j\, q$ becomes homotopic to the identity upon suspension. The next lemma provides this; upon re-indexing it will suffice to formulate the lemma for the case $n = 0$. □

<u>Lemma</u> 1.8.4. The geometric realizations of the two maps

$$\Sigma \;,\quad \Sigma\, j\, q \;:\quad wN.C^m_X \longrightarrow wN.C^{m+1}_{\Sigma X}$$

are homotopic.

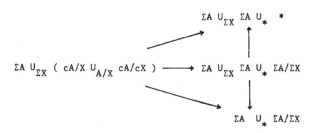

Proof. The natural transformations of lemma 1.8.1 provide a homotopy between the two maps $wN.C^m_X \to wN.C^{m+1}_{\Sigma X}$ which take $X \rightarrowtail A$ to

$$\Sigma X \rightarrowtail \Sigma A \cup_{\Sigma X} \Sigma A \quad \text{and} \quad \Sigma X \rightarrowtail \Sigma A \cup_* \Sigma A/\Sigma X \ ,$$

respectively; that is, the maps

$$\Sigma \vee \Sigma \quad \text{and} \quad \Sigma \vee \Sigma jq \ .$$

The geometric realization of $wN.C^{m+1}_{\Sigma X}$ is an H-space (by $\vee$) which is connected and hence group-like. So we can cancel the left Σ to obtain the desired homotopy. □

The following is the analogue of definition 1.5.4 with the $S.$ construction replaced by the $N.$ construction. In particular the letter P refers to the *simplicial path object* construction whose elementary properties have been recalled in the beginning of section 1.5.

Definition 1.8.5. Let $f\colon A \to B$ be a map of categories with sum and weak equivalences. Then $N.(f\colon A\to B)$ is the simplicial category with sum and weak equivalences given by the pullback of the diagram

$$N.A \longrightarrow N.B \longleftarrow PN.B \ .$$

$N.(f\colon A\to B)$ represents a *one-sided bar construction* of A acting on B by the sum via f . In fact, notice that in particular for every n there is a pullback diagram

$$\begin{array}{ccc} N_n(f\colon A\to B) & \longrightarrow & (PN.B)_n = N_{n+1}B \\ \downarrow & & \downarrow \\ N_nA & \longrightarrow & N_nB \end{array}$$

and the vertical map on the right corresponds, under the equivalence of N_mB with the product category B^m , to the projection map $B^{n+1} \to B^n$, the projection away from the first factor; and $N_n(f\colon A\to B)$ is equivalent to the product category $B\times A^n$.

Considering B as a simplicial category in a trivial way we have a sequence of simplicial categories with sum and weak equivalences

$$B \longrightarrow N.(f\colon A\to B) \longrightarrow N.A \ .$$

We would like this sequence to represent a fibration, up to homotopy, of the associated simplicial categories of weak equivalences, but we cannot expect this to be true in general since A need not act invertibly on B . We circumvent the difficulty by introducing another simplicial direction, using either the $S.$ or the $N.$ construction (we need both cases), as follows.

If $f\colon A \to B$ is a map of categories with cofibrations and weak equivalences then $N.(f\colon A\to B)$ is a simplicial category with cofibrations and weak equivalences, so we can form $S.N.(f:A\to B)$. Alternatively we could apply the definition 1.8.5 to the map $S.f: S.A \to S.B$ to obtain $N.(S.f:S.A\to S.B)$, and the two bisimplicial categories are naturally isomorphic. There is a sequence, of bisimplicial categories with cofibrations and weak equivalences,

$$S.B \longrightarrow S.N.(f:A\to B) \longrightarrow S.N.A \quad ;$$

alternatively we could rewrite it, up to isomorphism, as

$$S.B \longrightarrow N.(S.f:S.A\to S.B) \longrightarrow N.S.A \quad .$$

In general we can apply the $N.$ construction to the simplicial category with sum and weak equivalences $N.(f:A\to B)$ to obtain $N.N.(f:A\to B)$. Alternatively we could apply the definition 1.8.5 to the map $N.f: N.A \to N.B$ to obtain $N.(N.f:N.A\to N.B)$, and the two bisimplicial categories are naturally isomorphic (the isomorphism involves a switch of the two $N.$ directions). There is a sequence, of bisimplicial categories with sum and weak equivalences,

$$N.B \longrightarrow N.N.(f:A\to B) \longrightarrow N.N.A \quad ;$$

alternatively we could rewrite it, up to isomorphism, as

$$N.B \longrightarrow N.(N.f:N.A\to N.B) \longrightarrow N.N.A \quad .$$

Lemma 1.8.6. The sequence

$$wN.B \longrightarrow wN.N.(f:A\to B) \longrightarrow wN.N.A$$

is a fibration up to homotopy. Similarly so is the sequence

$$wS.B \longrightarrow wS.N.(f:A\to B) \longrightarrow wS.N.A$$

if that is defined. In either case, if f is an identity map then the middle term $wN.N.(f:A\to B)$, resp. $wS.N.(f:A\to B)$, is contractible.

Proof. This is a special case of a result of Segal [11]. Essentially the same proof results if the argument of proposition 1.5.5 is adapted to the present situation. That is, one observes that (in the second case, say) for every n one has a fibration

$$wS.B \longrightarrow wN_n(S.f:S.A\to S.B) \longrightarrow wN_nS.A$$

namely a product fibration, and one draws the desired conclusion from this, using a suitable fibration criterion for simplicial objects. □

Let $\mathcal{D}$ be a category with cofibrations and weak equivalences. The example to be kept in mind is that of the category $\varinjlim C^n$ of the theorem. Our next result is of a formal nature. It gives a sufficient condition for the conclusion of the theorem to be valid.

Proposition 1.8.7. If for every $X \in \mathcal{D}$ the simplicial category $wN.(j:\mathcal{D} \to \mathcal{D}_X)$ is contractible then the map $wN.\mathcal{D} \to wS.\mathcal{D}$ is a homotopy equivalence.

Proof. Applying lemma 1.8.6 we obtain that the map of the proposition de-loops to $wN.N.\mathcal{D} \to wN.S.\mathcal{D}$, so it will suffice to show that the latter map is a homotopy equivalence. By the realization lemma this follows if for every n the map

$$wN.N_n\mathcal{D} \longrightarrow wN.S_n\mathcal{D}$$

is a homotopy equivalence, and this is what we shall show.

The simplicial category on the left is equivalent to the product $(wN.\mathcal{D})^n$, so our task is to show that the simplicial category on the right is homotopy equivalent to that same product by the subquotient map. In other words, our task is to establish a case of the additivity theorem for the $N.$ construction rather than the $S.$ construction.

By induction it will suffice to show that the map

$$\begin{array}{ccc} wN.S_n\mathcal{D} & \longrightarrow & wN.S_{n-1}\mathcal{D} \times wN.\mathcal{D} \\ (A_1 \rightarrowtail \ldots \rightarrowtail A_n \text{ , choices }) & \longmapsto & (A_1 \rightarrowtail \ldots \rightarrowtail A_{n-1} \text{ , choices; } A_n/A_{n-1}) \end{array}$$

is a homotopy equivalence. To reduce further we consider the map

$$\begin{array}{rccc} j_n : & \mathcal{D} & \longrightarrow & S_n\mathcal{D} \\ & A & \longmapsto & * \rightarrowtail \ldots \rightarrowtail * \rightarrowtail A \end{array} .$$

By combining these two maps, and using lemma 1.8.6, we obtain a diagram of homotopy fibrations

$$\begin{array}{ccccc} wN.S_n\mathcal{D} & \longrightarrow & wN.N.(j_n:\mathcal{D} \to S_n\mathcal{D}) & \longrightarrow & wN.N.\mathcal{D} \\ \downarrow & & \downarrow & & \downarrow \| \\ wN.(S_{n-1}\mathcal{D} \times \mathcal{D}) & \longrightarrow & wN.N.(\mathcal{D} \to S_{n-1}\mathcal{D} \times \mathcal{D}) & \longrightarrow & wN.N.\mathcal{D} \end{array} .$$

So our task of showing that the vertical map on the left is a homotopy equivalence, translates into the task of showing that the vertical map in the middle is one. By the realization lemma this will follow if we can show that

$$wN.(j_n:\mathcal{D} \to S_n\mathcal{D}) \longrightarrow wN.(\mathcal{D} \to S_{n-1}\mathcal{D} \times \mathcal{D})$$

is a homotopy equivalence. Now

$$wN.(\mathcal{D} \to S_{n-1}\mathcal{D} \times \mathcal{D}) \approx wS_{n-1}\mathcal{D} \times wN.(\mathcal{D} \xrightarrow{=} \mathcal{D})$$

and the factor $wN.(\mathcal{D} \to \mathcal{D})$ is contractible. So the proof of the proposition has been reduced to proving the following lemma:

Lemma 1.8.8. If for every $X \in \mathcal{D}$ the simplicial category $wN.(j:\mathcal{D} \to \mathcal{D}_X)$ is contractible then the map $p\colon wN.(j_n:\mathcal{D} \to S_n\mathcal{D}) \to wS_{n-1}\mathcal{D}$ is a homotopy equivalence.

Proof. There is a variant of theorem A [8] for simplicial categories. A special case, sufficient for the present application, has been described in [13, prop. 6.5] in great detail. A neater, and more general, version may be found in [15, section 4] with a sketch proof. In any case, the criterion says that for the map p to be a homotopy equivalence it suffices that for every object

$$B = (B_1 \rightarrowtail \ldots \rightarrowtail B_{n-1} , \text{ choices }) \in wS_{n-1}\mathcal{D}$$

the left fibre $(p/B).$ is contractible.

Capitalizing on the special feature that $wS_{n-1}\mathcal{B}$, the target of p , is only a simplicial category in a trivial way, we can re-express $(p/B).$ in terms of left fibres of maps of categories, namely

$$(p/B)_m = p_m/B .$$

An object of p_m/B consists of a diagram

$$\begin{array}{ccccccc} A_1 & \rightarrowtail & \ldots & \rightarrowtail & A_{n-1} & \rightarrowtail & A_n \\ \downarrow\wr & & & & \downarrow\wr & & \\ B_1 & \rightarrowtail & \ldots & \rightarrowtail & B_{n-1} & & \end{array}$$

plus a m-tuple of objects in $\mathcal{D}$, plus certain sum diagrams formed from this m-tuple and A_n (plus, as usual, certain other choices).

There is a natural transformation of the identity map on p_m/B , it is given by pushout with the vertical map(s) in the diagram. For varying m the natural transformations are compatible, so they combine to give a homotopy of the identity map of $(p/B).$; namely a deformation retraction into the simplicial subcategory defined by the condition that the vertical map(s) be the identity.

That subcategory is isomorphic to $wN.(j:\mathcal{D} \to \mathcal{D}_X)$ where $X = B_{n-1}$, it is thus contractible by assumption. We are done. □

Let $\mathcal{D}$ be a category with cofibrations and weak equivalences, and $X \in \mathcal{D}$. It turns out that the contractibility of $wN.(\mathcal{D} \to \mathcal{D}_X)$ may be re-expressed in terms of two other conditions which appear to be rather independent of each other.

<u>Proposition</u> 1.8.9. $wN.(\mathcal{D} \to \mathcal{D}_X)$ is contractible if and only if the following two conditions are satisfied:

(1) $wN.(\mathcal{D} \to \mathcal{D}_X)$ is connected,

(2) the map $wN.\mathcal{D} \to wN.\mathcal{D}_X$ is a homotopy equivalence.

Proof. If $wN.(\mathcal{D} \to \mathcal{D}_X)$ is connected it has $wN.N.(\mathcal{D} \to \mathcal{D}_X)$ as a de-loop (by [11] or a variant of lemma 1.8.6). Therefore, provided it is connected, it is contractible if and only if $wN.N.(\mathcal{D} \to \mathcal{D}_X)$ is contractible. By lemma 1.8.6 we have a diagram of homotopy fibrations

$$\begin{array}{ccccc}
wN.\mathcal{D} & \longrightarrow & wN.N.(\mathcal{D} \overset{=}{\to} \mathcal{D}) & \longrightarrow & wN.N.\mathcal{D} \\
\downarrow & & \downarrow & & \downarrow \| \\
wN.\mathcal{D}_X & \longrightarrow & wN.N.(\mathcal{D} \to \mathcal{D}_X) & \longrightarrow & wN.N.\mathcal{D}
\end{array}$$

and the middle term in the upper row is contractible. Therefore $wN.N.(\mathcal{D} \to \mathcal{D}_X)$ is contractible if and only if the vertical map on the left is a homotopy equivalence. □

Proof of theorem 1.8.1. The nerve of the simplicial category $wN.(\mathcal{D} \to \mathcal{D}_X)$ is a bisimplicial set whose vertices are the objects $X \rightarrowtail A$ in $\mathcal{D}_X$. There are two kinds of 1-simplices, corresponding to the morphisms of $w\mathcal{D}_X$ on the one hand, and to the 'operation' of the objects of $\mathcal{D}$ on those of $\mathcal{D}_X$ on the other. It results that the set of connected components is the set of equivalence classes of the $X \rightarrowtail A$ under the equivalence relation generated by

(i) $(X \rightarrowtail A) \sim (X \rightarrowtail A')$ if there is a map $(X \rightarrowtail A) \to (X \rightarrowtail A')$ in $w\mathcal{D}_X$

(ii) $(X \rightarrowtail A) \sim (X \rightarrowtail A \cup_* A'')$ if $A'' \in \mathcal{D}$.

The condition referred to in the theorem, that *cofibrations in* $\mathcal{D}$ *are splittable up to weak equivalence*, implies that every object of $\mathcal{D}_X$ can be related (in a special way, in fact) to the trivial object $X \overset{=}{\rightarrowtail} X$, thus $wN.(\mathcal{D} \to \mathcal{D}_X)$ is connected.

Let $\mathcal{D} = \varinjlim C^n$ now. Then, as just observed, $wN.(\mathcal{D} \to \mathcal{D}_X)$ is connected for every X, and, by proposition 1.8.3, the map $wN.\mathcal{D} \to wN.\mathcal{D}_X$ is a homotopy equivalence. By proposition 1.8.9 these two properties imply that $wN.(\mathcal{D} \to \mathcal{D}_X)$ is contractible for every X which in turn, by proposition 1.8.7, implies that

$$wN.\mathcal{D} \longrightarrow wS.\mathcal{D}$$

is a homotopy equivalence, as desired. □

1.9. Appendix: Relation with the Q construction.

Let A be an *exact category* in the sense of Quillen [8]. One can make A into a category with cofibrations and weak equivalences by choosing a zero object and by defining the cofibrations and the weak equivalences to be the admissible monomorphisms and the isomorphisms, respectively. So a simplicial category $iS.A$ is defined. It turns out that $iS.A$ is naturally homotopy equivalent to the category QA of Quillen.

To see this we first replace QA by a homotopy equivalent simplicial category $iQ.A$. Namely let iQA be the bicategory of the commutative squares in QA in which the vertical arrows are the isomorphisms (in either A or QA — those are the same). Then QA and iQA are homotopy equivalent (lemma 1.6.5), and we let $iQ.A$ be a partial nerve of iQA , namely the nerve in the Q direction.

Next we replace $iS.A$ by a homotopy equivalent simplicial category $iS^e_.A$. We use the *edgewise subdivision* functor [12] which to any simplicial object $X.$, say $X. : \Delta^{op} \to K$, associates another $X^e_. : \Delta^{op} \to K$, namely the composite

$$X^e_. \;=\; X.\, d^{op}$$

where $d\colon \Delta \to \Delta$ is the *doubling map* which takes $[n]$ to $[2n+1]$ and whose behaviour on maps may be described by saying that it takes

$$(0 < 1 < \ldots < n) \quad \text{to} \quad (n' < \ldots < 1' < 0' < 0 < 1 < \ldots < n) \ .$$

If $X.$ is a simplicial space then the geometric realizations $|X.|$ and $|X^e_.|$ are naturally homeomorphic [12, prop. (A.1)]. Applying this fact to the simplicial space $[n] \mapsto |iS_nA|$ we obtain that $iS.A$ and its edgewise subdivision $iS^e_.A$, or rather their geometric realizations, are homotopy equivalent.

There is a map of simplicial categories

$$iS^e_.A \longrightarrow iQ.A$$

which is an equivalence of categories in each degree, and therefore a homotopy equivalence. The map is best explained by drawing a diagram to illustrate the situation for $n = 3$.

An object of iS^e_3A ($\approx iS_7A$) is a sequence of cofibrations

$$A_{(3',2')} \rightarrowtail A_{(3',1')} \rightarrowtail A_{(3',0')} \rightarrowtail A_{(3',0)} \rightarrowtail A_{(3',1)} \rightarrowtail A_{(3',2)} \rightarrowtail A_{(3',3)}$$

together with a choice of quotients

$$A_{(i,j)} \;=\; A_{(3',j)}/A_{(3',i)} \ .$$

By dropping some of the choices while retaining others we can associate to the object the following diagram

$$\begin{array}{ccccccc}
A_{(3',0)} & \rightarrowtail & A_{(3',1)} & \rightarrowtail & A_{(3',2)} & \rightarrowtail & A_{(3',3)} \\
\downarrow & & \downarrow & & \downarrow & & \\
A_{(2',0)} & \rightarrowtail & A_{(2',1)} & \rightarrowtail & A_{(2',2)} & & \\
\downarrow & & \downarrow & & & & \\
A_{(1',0)} & \rightarrowtail & A_{(1',1)} & & & & \\
\downarrow & & & & & & \\
A_{(0',0)} & & & & & &
\end{array}$$

The diagram describes a sequence of three composable morphisms in $Q\mathcal{A}$ as well as the different ways in which the actual composition can be performed. In particular the diagram defines an object of $iQ_3\mathcal{A}$. The object in question is not identical to the diagram itself, rather it is an equivalence class of diagrams; two diagrams are considered equivalent if they are isomorphic by an isomorphism which restricts to the identity on each of the diagonal objects $A_{(j',j)}$.

To conclude we note a variant of the homotopy equivalence. Let $s.\mathcal{A}$ denote the simplicial set of objects of $S.\mathcal{A}$. Considering $s.\mathcal{A}$ as a simplicial category in a trivial way we have an inclusion $s.\mathcal{A} \rightarrow iS.\mathcal{A}$ which is a homotopy equivalence by lemma 1.4.1. Let $Q.\mathcal{A}$ denote the nerve of the category $Q\mathcal{A}$. Above we have described a map

$$s^e_. \mathcal{A} \longrightarrow Q.\mathcal{A} \quad .$$

This map is a homotopy equivalence. For it fits into a diagram

$$\begin{array}{ccc}
s^e_. \mathcal{A} & \longrightarrow & Q.\mathcal{A} \\
\downarrow & & \downarrow \\
iS^e_. \mathcal{A} & \longrightarrow & iQ.\mathcal{A}
\end{array}$$

and we know already that the three other maps in the diagram are homotopy equivalences.

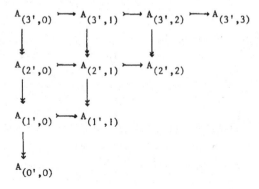

2. THE FUNCTOR A(X) .

2.1. Equivariant homotopy theory, and the definition of A(X) .

Let X be a space. A(X) is defined as the K-theory, in the sense of the preceding chapter, of an equivariant homotopy theory associated to X .

There are several ways of making this precise. The main purpose of this section is to describe a few of those ways in detail and to show that they all lead to the same result, up to homotopy.

The various cases arise from the fact that we want to keep the option of interpreting each of the terms *space*, *equivariant*, and *finite type* in two different ways. Namely we will want to work either with topological spaces or with simplicial sets. We want to use spaces over X on the one hand or spaces with an action of G(X) , the loop group of X , on the other. And finally we want to be free to impose a condition of strict finiteness on the objects of the category or to be content with a condition of finiteness up to homotopy.

We begin with a construction that combines the two equivariant points of view. We will be mainly interested, eventually, in the two special cases where one of G and W below is trivial and the other one is X , resp. a loop group of X .

Let G be a simplicial monoid and W a simplicial set on which G acts (by a monoid is meant an associative semigroup with 1). We define

$$\mathcal{R}(W,G)$$

to be the category of the G-simplicial sets having W as a retract. In detail, the objects of $\mathcal{R}(W,G)$ are the triples (Y,r,s) where Y is a simplicial set with G-action and $s: W \to Y$ and $r: Y \to W$ are G-maps so that $rs = \mathrm{Id}_W$, and the morphisms from (Y,r,s) to (Y',r',s') are the G-maps $f: Y \to Y'$ so that $r'f = r$ and $fs = s'$.

If G is the trivial monoid we omit it from the notation. In other words, we let $\mathcal{R}(X)$ denote the category of the simplicial sets having X as a retract.

There are similar constructions in the topological case, and geometric realization induces a functor $\mathcal{R}(W,G) \to \mathcal{R}(|W|,|G|)$.

We define our finite type conditions now. We proceed in the following order:

1. finiteness in the simplicial case,
2. finiteness in the topological case,
3. homotopy finiteness in the topological case,
4. homotopy finiteness in the simplicial case.

1. *Finiteness in the simplicial case.* An object (Y,r,s) of $\mathcal{R}(X)$ is called *finite* if the simplicial set Y is generated by the simplices of $s(X)$ together with finitely many other simplices. An equivalent condition is that the geometric realization $|Y|$ is a finite CW complex relative to the subspace $|s(X)|$. The full subcategory of the finite objects is denoted $\mathcal{R}_f(X)$.

In the general case of $\mathcal{R}(W,G)$ we must combine the finite generation condition with a freeness condition. *Finite generation* of (Y,r,s) means that Y is generated, as a G-simplicial set, by the simplices of $s(W)$ together with finitely many other simplices. *Freeness* means that, for every k, the action of G_k on Y_k is free away from W_k; precisely, the condition is that Y may be obtained from W by *attaching of free* G-*cells*, that is, by direct limit and the formation of pushouts of diagrams of the kind $Y' \leftarrow \partial\Delta^n \times G \rightarrow \Delta^n \times G$ where Δ^n denotes the simplicial set n-*simplex*, and $\partial\Delta^n$ the simplicial subset *boundary*. We denote $\mathcal{R}_f(W,G)$ the full subcategory of $\mathcal{R}(W,G)$ given by the objects which are both finitely generated and free; the objects (Y,r,s), in other words, where Y can be obtained from W by attaching of finitely many free G-cells. $\mathcal{R}_f(W,G)$ is a category with cofibrations and weak equivalences in the sense of sections 1.1 and 1.2, the cofibrations are the injective maps, and the weak (homotopy) equivalences are the maps $(Y,r,s) \rightarrow (Z,t,u)$ whose underlying maps $Y \rightarrow Z$ are weak homotopy equivalences in the usual sense (that is, induce isomorphisms of homotopy groups upon geometric realization). We denote the category of the weak homotopy equivalences by $h\mathcal{R}_f(W,G)$.

2. *Finiteness in the topological case.* Let $|X|$ be a topological space, not necessarily the geometric realization of a simplicial set X. An object (Y,r,s) of $\mathcal{R}(|X|)$ is called *finite* if Y is equipped with the structure of a finite CW complex relative to the subspace $s(|X|)$. We let $\mathcal{R}_f(|X|)$ denote the category of these objects and their *cellular* maps (it is not, of course, a full subcategory of $\mathcal{R}(|X|)$). We consider $\mathcal{R}_f(|X|)$ as a category with cofibrations and weak (homotopy) equivalences; by definition, a map in $\mathcal{R}_f(|X|)$ is a cofibration if it is isomorphic to a cellular inclusion.

More generally, in the case of $\mathcal{R}(|W|,|G|)$, we define $\mathcal{R}_f(|W|,|G|)$ to be the category of the finite $|G|$-free CW complexes, relative to $|W|$, and their cellular maps.

3. *Homotopy finiteness in the topological case.* We define $R_{hf}(|W|,|G|)$ as the full subcategory of $R(|W|,|G|)$ given by the (Y,r,s) where (Y,s) has the $|G|$-homotopy type, in the strong sense, of a finite $|G|$-free CW complex relative to $|W|$. This is a category with cofibrations and weak (homotopy) equivalences, where *cofibration* has its usual meaning as a map having the $|G|$-homotopy extension property (after neglect of structural retractions, that is). To see that cobase change by cofibrations does not take one out of the category, i.e. preserves homotopy finiteness, it suffices to note that weak homotopy equivalences have homotopy inverses, after neglect of structural retractions (the Whitehead theorem for $|G|$-free CW complexes).

Remark. On the face of it there are set theoretical difficulties in the construction of K-theory from $R_{hf}(|X|)$. For $hS.R_{hf}(|X|)$ is not a 'small' simplicial category, nor even equivalent to one (in the sense of category theory). Here are a few ways of dealing with this matter, each with its own virtues and drawbacks: (a) one can pick an explicit small category $R'_{hf}(|X|)$ with which to work (for example, have all one's spaces embedded in $|X| \times R^{\infty}$), (b) one may postulate the existence of a universe, in the sense of Grothendieck, work in a fixed one, and check that an enlargement of the universe does not alter the homotopy type, (c) one may regard the notion of a 'large' space as just as legitimate as that of a 'large' category, provided only that certain constructions are avoided (this is the naive version of the preceding). Which one of these or other alternatives to adopt seems a matter of taste. We will not pursue the matter further.

4. *Homotopy finiteness in the simplicial case.* We reduce to the topological case. That is, we define $R_{hf}(W,G)$ as the full subcategory of $R(W,G)$ given by the (Y,r,s) whose geometric realizations are homotopy finite in the sense of the preceding case.

Recall that the *approximation theorem* 1.6.7 describes sufficient conditions for an exact functor $C \to C'$ to induce a homotopy equivalence $hS.C \to hS.C'$.

Proposition 2.1.1. The approximation theorem applies to the map

$$R_f(W,G) \longrightarrow R_{hf}(W,G) ,$$

resp. its topological analogue.

Proof. The non-trivial thing to verify is the following assertion (the part App 2 of the *approximation property*).

Assertion. Let $(Y,r,s) \in R_f(W,G)$, and let $(Y,r,s) \to (Y',r',s')$ be any map in $R_{hf}(W,G)$. Then the map can be factored as $(Y,r,s) \to (Y_1,r_1,s_1) \to (Y',r',s')$ where $(Y_1,r_1,s_1) \in R_f(W,G)$, the first map is a cofibration in $R_f(W,G)$, and the second map is a weak equivalence in $R_{hf}(W,G)$.

To prove the assertion it will suffice to find a factorization

$$(Y,s) \longrightarrow (Y_1,s_1) \longrightarrow (Y',s') .$$

For it is then possible to *define* the structural retraction r_1 as the composite of $Y_1 \to Y'$ with $r'_1 : Y' \to W$.

We treat the topological case first. The Whitehead theorem for $|G|$-free CW complexes relative to $|W|$ is available here, so we can find a finite (Y_o,s_o) together with homotopy equivalences $(Y_o,s_o) \to (Y',s')$ and $(Y',s') \to (Y_o,s_o)$, homotopy inverse to each other. Choose a cellular map $(Y,s) \to (Y_o,s_o)$ homotopic to the composition $(Y,s) \to (Y',s') \to (Y_o,s_o)$, and define (Y_1,s_1) as its mapping cylinder. Then there exists a map $(Y_1,s_1) \to (Y',s')$ extending the given maps on (Y,s) and (Y_o,s_o) . This has the required properties.

In the simplicial case we know, by the topological case, that there exists some factorization

$$(|Y|,|s|) \longrightarrow (Y_1,s_1) \longrightarrow (|Y'|,|s'|) .$$

We show that, by perturbing (Y_1,s_1) a little, we may lift it back to the simplicial framework.

Proceeding by induction on the cells of Y_1 not in $|Y|$ we suppose that we have found a subcomplex $|Z|$ of Y_1 which does arise by geometric realization, and so that the map $|Z| \to |Y'|$ is a geometric realization, too. To add another one of the cells of Y_1 to $|Z|$, means that we form the pushout of a diagram of the kind

$$|Z| \longleftarrow |\partial\Delta^n| \times |G| \longrightarrow |\Delta^n| \times |G| .$$

We use simplicial approximation to rigidify this. Namely let Sd denote the *subdivision functor* for simplicial sets [4], and Sd_k its k-fold iteration. Then if k is large enough one knows [4] that there is a map of simplicial sets,

$$Sd_k\, \partial\Delta^n \longrightarrow Z ,$$

whose geometric realization is homotopic to the map

$$|Sd_k\, \partial\Delta^n| \approx |\partial\Delta^n| \times 1 \longrightarrow |Z| , \qquad 1 \in |G| ,$$

and, again if k is large enough, the composite map $Sd_k\partial\Delta^n \to Z \to Y'$ extends to $Sd_k\Delta^n$, in the preferred homotopy class. We now define

$$Z' = Z \cup_{Sd_k\partial\Delta^n \times G} Sd_k\Delta^n \times G .$$

Then $Z \to Y'$ extends to a map $Z' \to Y'$ in the preferred homotopy class. By the $|G|$-homotopy extension theorem $|Z'|$ in turn may be extended, by induction on the remaining cells, to a $|G|$-CW complex Y'_1 mapping to Y_1 by homotopy equivalence. This completes the inductive step, and hence the proof of the proposition. □

Proposition 2.1.2. The approximation theorem applies to the geometric realization map

$$R_f(W,G) \longrightarrow R_f(|W|,|G|) .$$

Proof. The non-trivial thing to verify is the following assertion.

Assertion. Let $(Y,r,s) \in R_f(W,G)$, and let $(|Y|,|r|,|s|) \to (Y',r',s')$ be any map in $R_f(|W|,|G|)$. Then the map can be factored as

$$(|Y|,|r|,|s|) \longrightarrow (|Y''|,|r''|,|s''|) \longrightarrow (Y',r',s')$$

where the first map is the geometric realization of a cofibration in $R_f(W,G)$, and the second map is a weak equivalence in $R_f(|W|,|G|)$.

As before (the preceding proof) it suffices to find a factorization

$$(|Y|,|s|) \longrightarrow (|Y''|,|s''|) \longrightarrow (Y',s') .$$

Define (Y_1,s_1) as the mapping cylinder of $(|Y|,|s|) \to (Y',s')$. Then (Y'',s'') is obtained from (Y_1,s_1) by rigidifying, one after the other, the cells of Y_1 not in $|Y|$. The argument is the same as that in the second part of the preceding proof. □

Let G be a simplicial group now, not just monoid, and X a simplicial set. By a *principal* G-*bundle with base* X is meant a free G-simplicial set P together with an isomorphism of X with $P\times^G *$, the simplicial set of orbits.

Lemma 2.1.3. There is an equivalence of categories $R(X) \sim R(P,G)$.

Proof. We can define functors between these categories by pullback with $P \to X$ and by the orbit map, respectively. If $(Y,r,s) \in R(X)$ then $(Y\times_X P)\times^G * \approx Y$. And if $(Y',r',s') \in R(P,G)$ then the diagram

$$\begin{array}{ccc} Y' & \longrightarrow & P \\ \downarrow & & \downarrow \\ Y'\times^G * & \longrightarrow & P\times^G * \end{array}$$

is a pullback, thanks to the freeness of the G-action on P and the fact that G is a simplicial group, not just monoid. Hence $Y' \approx (Y'\times^G *)\times_X P$, and the two functors are inverse to each other, up to isomorphism. □

By a *universal* G-*bundle with base* X will be meant a principal bundle whose total space P is contractible (in the weak sense). In this situation it is necessarily the case that G represents the loop space of X, but apart from this restriction one knows that universal bundles exist in great profusion. Specifically there is a functor, due to Kan, which to connected pointed X associates a universal $G(X)$-bundle where $G(X)$ is a certain free simplicial group, the *loop group* of X . Conversely it is also possible, in any of several functorial ways, to associate to a simplicial group G a universal bundle over a *classifying space*.

Given a universal G-bundle over X we can define a functor

$$\begin{array}{ccc} \mathcal{R}(X) & \longrightarrow & \mathcal{R}(*,G) \\ (Y,r,s) & \longmapsto & (\ (Y\times_X P)/(X\times_X P),\bar{r},\bar{s}\ ) \end{array} .$$

The functor respects the notion of finiteness, resp. homotopy finiteness, and it is *exact* (sections 1.1 and 1.2), so it induces a map in K-theory. In a similar way we can also use P to define a map $\mathcal{R}(|X|) \to \mathcal{R}(*,|G|)$.

<u>Proposition</u> 2.1.4. The map $hS.\mathcal{R}_{hf}(X) \to hS.\mathcal{R}_{hf}(*,G)$ is a homotopy equivalence.

Proof. In view of its definition, the map arises as the composite of the equivalence $\mathcal{R}_{hf}(X) \to \mathcal{R}_{hf}(P,G)$ of lemma 2.1.3 with the map $\mathcal{R}_{hf}(P,G) \to \mathcal{R}_{hf}(*,G)$ given by pushout with $P \to *$. It therefore suffices to show that the latter map induces a homotopy equivalence. We show this by providing a homotopy inverse. Consider the map $\mathcal{R}(*,G) \to \mathcal{R}(P,G)$ given by product with P , using the diagonal action of G . The map respects the notion of homotopy finiteness, in view of the contractibility of P , and it is exact, so it induces a map in K-theory. The composite map on $\mathcal{R}(*,G)$ admits a natural transformation to the identity,

$$Y\times P \cup_{*\times P} * \longrightarrow Y ,$$

and the composite map on $\mathcal{R}(P,G)$ admits a natural transformation from the identity,

$$Y \longrightarrow Y\times P \cup_{P\times P} P .$$

In view of the contractibility of P each of these two natural transformations is a weak equivalence. Using proposition 1.3.1 now we are done. □

<u>Theorem</u> 2.1.5. If X is a simplicial set (resp. if G is a simplicial monoid) there is a 2×2 diagram of homotopy equivalences, namely the left one (resp. right one) of the following two squares

$$\begin{array}{ccc} hS.\mathcal{R}_f(X) & \longrightarrow & hS.\mathcal{R}_{hf}(X) \\ \downarrow & & \downarrow \\ hS.\mathcal{R}_f(|X|) & \longrightarrow & hS.\mathcal{R}_{hf}(|X|) \end{array} \qquad \begin{array}{ccc} hS.\mathcal{R}_f(*,G) & \longrightarrow & hS.\mathcal{R}_{hf}(*,G) \\ \downarrow & & \downarrow \\ hS.\mathcal{R}_f(*,|G|) & \longrightarrow & hS.\mathcal{R}_{hf}(*,|G|) \end{array} .$$

If G is a loop group of X , and if a universal G-bundle with base X is given, there is a natural transformation from the left square to the one on the right, and all the arrows in the resulting 2×2×2 diagram are homotopy equivalences.

Proof. This results from propositions 2.1.1, 2.1.2, and 2.1.4. □

Picking one of the choices offered by the theorem we now make the definition

$$A(X) \quad = \quad \Omega|hS.R_f(X)|$$

if X is a simplicial set.

A map $x: X \to X'$ induces $x_*: R(X) \to R(X')$ by pushout with x, and hence a map in K-theory. In this way $A(X)$ becomes a covariant functor. Below we give an argument to show that this functor is a homotopy functor (proposition 2.1.7).

We have to consider functorial behaviour in a slightly more general situation. Namely let $g: G \to G'$ be a group map, and $w: W \to W'$ a map under g. These induce a map $(g,w)_*: R(W,G) \to R(W',G')$ as the composite

$$R(W,G) \longrightarrow R(W\times^G G',G') \longrightarrow R(W',G')$$

where the first map is given by product with G' under G, and the second map by pushout with $W\times^G G' \to W'$.

Let a *map of universal bundles* mean a triple of maps

$$(x,p,g) \;:\; (X,P,G) \longrightarrow (X',P',G')$$

where p is a map under g, and over x. We note that $X\times_{X'}P' \approx P\times^G G'$ in this situation.

<u>Lemma</u> 2.1.6. To such a map there is associated a commutative diagram

$$\begin{array}{ccc} R(X) & \xrightarrow{\;x_*\;} & R(X') \\ \downarrow & & \downarrow \\ R(*,G) & \xrightarrow[(*,g)_*]{} & R(*,G') \end{array} \quad .$$

Proof. This results from the definition of the maps and the commutativity of the diagram

$$\begin{array}{ccccc} R(X) & \xrightarrow{\;=\;} & R(X) & \longrightarrow & R(X') \\ \downarrow\wr & & \downarrow\wr & & \downarrow\wr \\ R(P,G) & \xrightarrow{\;\sim\;} & R(P\times^G G',G') & \longrightarrow & R(P',G') \\ \downarrow & & \downarrow & & \downarrow \\ R(*,G) & \longrightarrow & R(*,G') & \xrightarrow{\;=\;} & R(*,G') \end{array}$$

where the arrows $\xrightarrow{\;\sim\;}$ denote equivalences of categories (lemma 2.1.3). □

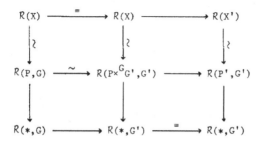

Proposition 2.1.7. If $x: X \to X'$ is a weak homotopy equivalence then so is the induced map $x_*: A(X) \to A(X')$.

Proof. The functor $X \mapsto hS.\mathcal{R}_f(X)$ commutes with direct limit, and it takes finite disjoint unions to products. As a result it suffices to prove the proposition in the case where X and X' are connected. We may further replace 'h' by 'hf' . Our task is then to show that $x_*: hS.\mathcal{R}_{hf}(X) \to hS.\mathcal{R}_{hf}(X')$ is a homotopy equivalence in that special case.

Choose a universal bundle over X' , say a universal G'-bundle P' . Since $x: X \to X'$ is a weak homotopy equivalence, pullback with it defines a universal G'-bundle $P = X \times_{X'} P'$ over X . There is a map of universal bundles now,

$$(x, pr_2, Id_{G'}) : (X,P,G') \longrightarrow (X',P',G') .$$

Hence (the preceding lemma) there is a commutative diagram

$$\begin{array}{ccc} hS.\mathcal{R}_{hf}(X) & \xrightarrow{x_*} & hS.\mathcal{R}_{hf}(X') \\ \downarrow & & \downarrow \\ hS.\mathcal{R}_{hf}(*,G') & \xrightarrow{=} & hS.\mathcal{R}_{hf}(*,G') \end{array}$$

and the vertical arrows are homotopy equivalences by proposition 2.1.4. It follows that x_* is a homotopy equivalence. □

Remark. For simplicial monoids in general, as opposed to simplicial groups, it does not follow in the same way that $G \mapsto \Omega|hS.\mathcal{R}_f(*,G)|$ is a homotopy functor. The result is still true, however. For example it follows from theorem 2.2.1 below.

2.2. A(X) via spaces of matrices.

Let G be a simplicial monoid. We consider the free pointed $|G|$-CW complex with k $|G|$-cells in dimension n and no other cells; or what is the same thing,

$$V^k S^n \wedge |G|_+ ,$$

the half-smash product of $|G|$ with a wedge of k spheres of dimension n .

Let

$$\mathcal{H}^n_k(G) = \mathcal{H}_{|G|}(V^k S^n \wedge |G|_+)$$

denote the simplicial monoid of pointed $|G|$-equivariant (weak) homotopy equivalences, and let $B\mathcal{H}^n_k(G)$ denote its classifying space. There are stabilization maps

$$B\mathcal{H}^n_k(G) \longrightarrow B\mathcal{H}^{n+1}_k(G) \quad , \qquad B\mathcal{H}^n_k(G) \longrightarrow B\mathcal{H}^n_{k+1}(G)$$

given by suspension and by the addition of an identity map, respectively.

The purpose of this section is to show that the K-theory of the preceding section can be re-expressed in terms of the + construction of Quillen, as follows.

Theorem 2.2.1. There is a natural chain of homotopy equivalences

$$\Omega|hS.\mathcal{R}_f(*,G)| \simeq Z \times \varinjlim_{n,k} B\mathcal{H}^n_k(G)^+ .$$

By combining with theorem 2.1.5 we obtain that, in particular, A(X) may be so re-expressed for connected X ,

$$A(X) \simeq Z \times \varinjlim_{n,k} B\mathcal{H}^n_k(G(X))^+ .$$

This may be regarded as a description of A(X) in terms of *spaces of matrices*, analogous to the definition of the algebraic K-theory of a ring in terms of matrices and the + construction, as follows.

In the case at hand, the 'ring' in question is the *ring up to homotopy*

$$\Omega^\infty S^\infty |G|_+ = \varinjlim_{n} \mathit{Map}(S^n, S^n \wedge |G|_+) .$$

Let $M_{k\times k}(\Omega^\infty S^\infty |G|_+)$ denote the product of $k\times k$ copies of this space, considered as a multiplicative H-space by means of matrix multiplication. We denote

$$\widehat{GL}_k(\Omega^\infty S^\infty |G|_+)$$

the sub-H-space of the homotopy-invertible matrices; it is the union of those connected components which are invertible in the monoid of connected components. The point now is simply that

$$\lim_{\overrightarrow{n}} BH_k^n(G)$$

provides a classifying space for the H-space $\widehat{GL}_k(\Omega^\infty S^\infty|G|_+)$. Indeed, there is a homotopy equivalence of H-spaces

$$\lim_{\overrightarrow{n}} H_k^n(G) \simeq \widehat{GL}_k(\Omega^\infty S^\infty|G|_+) .$$

It is given, in the limit, by the (n-1)-connected map

$$\begin{aligned} Map_{|G|}(V^k S^n \wedge |G|_+, V^k S^n \wedge |G|_+) &\approx Map(V^k S^n, V^k S^n \wedge |G|_+) \\ \longrightarrow Map(V^k S^n, \Pi^k S^n \wedge |G|_+) &\approx Map(S^n, S^n \wedge |G|_+)^{k \times k} . \end{aligned}$$

Proof of theorem. Define $R_k^n(*,G)$ to be the full subcategory of $R_f(*,G)$ given by the objects which are n-*spherical of rank* k . By definition, these are the objects weakly equivalent to

$$* \cup_{\amalg^k \partial\Delta^n \times G} \amalg^k \Delta^n \times G ,$$

that is, the objects which are in the same connected component, in $hR_f(*,G)$, as that particular object.

It is plausible, and will be shown below (proposition 2.2.5), that there is a natural chain of homotopy equivalences

$$BH_k^n(G) \simeq |hR_k^n(*,G)| .$$

Define $R^n(*,G)$ to be the subcategory of $R_f(*,G)$ of the objects which are n-spherical of unspecified rank; that is, the union of the categories $R_k^n(*,G)$. This is a category with sum and weak equivalences (section 1.8), so the group completion in the sense of Segal is defined; in the language of section 1.8 this is the simplicial category $hN.R^n(*,G)$. By a theorem of Segal [11] there is a homotopy equivalence, well defined up to weak homotopy (homotopy on compacta),

$$\Omega|hN.R^n(*,G)| \simeq Z \times \lim_{\overrightarrow{k}} |hR_k^n(*,G)|^+ .$$

Combining with the homotopy equivalence above, and passing to the limit with respect to n , we obtain now a homotopy equivalence

$$\lim_{\overrightarrow{n}} \Omega|hN.R^n(*,G)| \simeq Z \times \lim_{\overrightarrow{n,k}} BH_k^n(G)^+ .$$

This reduces the proof of the theorem to the following proposition.

<u>Proposition</u> 2.2.2. There is a natural chain of homotopy equivalences

$$\lim_{\overrightarrow{n}} hN.R^n(*,G) \simeq hS.R_f(*,G) .$$

The proof of the proposition is an application of theorems 1.7.1 and 1.8.1. To make these theorems applicable we have to check some things first. Let us define

$$h_*(Y) = \tilde{H}_*(Y \times^G \pi_0 G)$$

for $Y \in R_f(*,G)$, where $\tilde{H}_*$ denotes the reduced integral homology of pointed spaces.

<u>Lemma</u> 2.2.3. If $\tilde{H}_i(Y) = 0$ for $i < m$ then $\tilde{H}_m(Y) \to h_m(Y)$ is an isomorphism.

Proof. We give two proofs. The first applies to the special case where G is a simplicial group, not just monoid. In this case $Y \times^G \pi_0 G \approx Y \times^F *$ where F is the connected component of $1 \in G$. Choose a universal F-bundle E and form the associated bundle over $E \times^F *$, i.e. $(Y \times E) \times^F *$. Then $Y \times^F *$ may be identified, up to homotopy, to the quotient $(Y \times E) \times^F * / E \times^F *$, and the lemma results from the Serre spectral sequence of the fibration.

In the general case one notices that the lemma is really a special case of one in the next section (lemma 2.3.4) which concerns simplicial modules over a simplicial ring and whose proof depends on a spectral sequence of Quillen's on (derived) tensor products. □

Let $R_f^{(2)}(*,G)$ denote the subcategory of $R_f(*,G)$ of the objects which are 1-connected.

<u>Lemma</u> 2.2.4. The inclusion $hS.R_f^{(2)}(*,G) \to hS.R_f(*,G)$ is a homotopy equivalence.

Proof. Double suspension defines an endomorphism of each of these which is homotopic to the identity map (proposition 1.6.2). On the other hand, double suspension takes $hS.R_f(*,G)$ into $hS.R_f^{(2)}(*,G)$, so it gives a deformation retraction. □

Proof of proposition 2.2.2. The functor $Y \mapsto h_*(Y)$ defines a homology theory on $R_f(*,G)$, in the sense of section 1.7, with values in the category of $Z[\pi_0 G]$-modules.

Restricting attention to 1-connected objects, as we may by lemma 2.2.4, we obtain from lemma 2.2.3 together with the Hurewicz theorem that the weak equivalences are *homologically defined*: a map is a weak equivalence if and only if it induces an isomorphism on h_* .

The objects of $R^n(*,G)$ have the property that $h_i(Y)$ is 0 for $i \neq n$, and free over $Z[\pi_0 G]$ for $i = n$. Conversely they are characterized by this property. To see this it suffices to construct a map from a standard object inducing an isomorphism on h_* . Such a map is obtained by mapping each generating cell $\Delta^n \times 1$, suitably subdivided, so as to represent an appropriate generating element of the module $\pi_n|Y| \approx H_n(Y) \approx h_n(Y)$.

We show next that the hypothesis of section 1.7 is satisfied: if $Y_p \to Y$ is any p-connected map then it is possible to construct a factorization

$$Y_p \longrightarrow Y_{p+1} \longrightarrow \ldots \longrightarrow Y_q \longrightarrow Y$$

where each Y_{n+1} is obtained from Y_n by attaching (n+1)-cells and where the map $Y_q \to Y$ is a weak homotopy equivalence. First, the inductive construction of Y_{n+1} from Y_n is done as follows. The module $h_{n+1}(Y_n \to Y) \approx \pi_{n+1}(|Y_n| \to |Y|)$ is finitely generated over $Z[\pi_0 G]$, and each element may be represented by mapping a (suitably subdivided) pair $(\Delta^{n+1}, \partial\Delta^{n+1})$. Picking a generating set, we can use these maps to attach (n+1)-cells to Y_n and to extend the map to Y to the cells. Next, the construction can terminate. For suppose that q is at least as large as the dimension of Y . Then $h_q(Y_{q-1} \to Y)$ is computed from a finitely generated free chain complex which is both (q-1)-connected and q-dimensional. It follows that h_q is the only non-vanishing homology, and that it is stably free. After attaching some more (q-1)-cells to Y_{q-1} , if necessary, we may suppose the homology is actually free, so that in a last step, finally, we can attach q-cells to kill the homology without introducing new homology in the next dimension.

We have verified most of the hypotheses of theorem 1.7.1 now. The one exception is the condition that the category E , in the definition of spherical objects in section 1.7, should be closed under the operation of taking kernels of surjections. Our E so far is the category of finitely generated free modules over $Z[\pi_0 G]$. This does not satisfy the condition, in general, so we must enlarge it. We therefore replace $R^n(*,G)$ by $\tilde{R}^n(*,G)$ which we define as follows. It is the subcategory of $R_f(*,G)$ of the objects which are n-spherical in the following sense: $h_i(Y)$ is 0 for $i \neq n$, and it is stably free for $i = n$.

Theorem 1.7.1 now applies to give homotopy equivalences

$$\varinjlim_n hS.\tilde{R}^n(*,G) \longrightarrow \varinjlim_{(\Sigma)} hS.R_f(*,G) \longleftarrow hS.R_f(*,G)$$

(we have used lemma 2.2.4 to suppress the superscript (2) on R_f again).

It is plain from the preceding discussion, on the other hand, that $R^n(*,G)$ is *strictly cofinal* in $\tilde{R}^n(*,G)$ in the sense of proposition 1.5.9, so the inclusion

$$hS.R^n(*,G) \longrightarrow hS.\tilde{R}^n(*,G)$$

is a homotopy equivalence.

Finally it is also plain that the cofibrations in $R^n(*,G)$ are *splittable up to weak equivalence* in the sense of theorem 1.8.1, so the map

$$\varinjlim_n hN.R^n(*,G) \longrightarrow \varinjlim_n hS.R^n(*,G)$$

is a homotopy equivalence.

The proof of the proposition is now complete. □

Remark. The preceding argument can be varied a little. Namely instead of replacing $\widetilde{R}^n(*,G)$ by $R^n(*,G)$ as we have just done, we could also argue directly that

$$\lim_{\vec{n}} hN.\widetilde{R}^n(*,G) \longrightarrow \lim_{\vec{n}} hS.\widetilde{R}^n(*,G)$$

is a homotopy equivalence. Segal's theorem used elsewhere in the proof of the theorem then applies in the form of giving a homotopy equivalence

$$\Omega|hN.\widetilde{R}^n(*,G)| \quad \simeq \quad K_o'(Z[\pi_o G]) \times \lim_{\vec{k}} |hR_k^n(*,G)|$$

where $K_o'(Z[\pi_o G])$ denotes the subgroup of the class group given by the stably free modules (that subgroup is of course Z again).

The theorem itself can also be varied. Namely the category $R_f(*,G)$ may be enlarged to the category $R_{df}(*,G)$ of the objects *dominated by finite ones* (these are the objects which are retracts, up to homotopy, of finite ones). The theorem then goes through unchanged except that the restricted class group $K_o'(Z[\pi_o G])$ has to be replaced by the full class group $K_o(Z[\pi_o G])$. □

To complete the proof of the theorem we are still left to compare $BH_k^n(G)$ with $hR_k^n(*,G)$.

Let C denote any of the categories $hR_{hf}(*,G)$, $hR_f(*,|G|)$, $hR_{hf}(*,|G|)$. We blow it up to a simplicial category $C.$, $[m] \mapsto C_m$, where C_m is defined as the category whose objects are the same as those of C and whose morphisms are the m-parameter families of morphisms in C . That is, a morphism in C_m from Y to Z is a map

$$Y \longrightarrow Z^{\Delta^m}$$

in C (resp. similarly with Δ^m replaced by $|\Delta^m|$ in the topological case) or, what is the same, a map $Y \times \Delta^m / * \times \Delta^m \to Z$. Considering C as a simplicial category in a trivial way, we have a map $C \to C.$.

If $Y \in C$ we let C_Y , resp. $C._Y$, denote the connected component of C , resp. $C.$, containing Y , and $C.(Y)$ the simplicial subcategory of self-maps of Y in $C.$.

Proposition 2.2.5. In the topological case, the maps

$$C_Y \longrightarrow C._Y \longleftarrow C.(Y)$$

are homotopy equivalences. The same is true in the simplicial case provided that Y satisfies the Kan extension condition.

Corollary. There is a natural chain of homotopy equivalences

$$BH_k^n(G) \quad \simeq \quad |hR_k^n(*,G)| \ .$$

Proof. Let $C = hR_f(*,|G|)$ in the proposition, and $Y = \vee^k S^n \wedge |G|_+$. Then $|C.(Y)|$ is the same as $BH^n_k(G)$, by definition of the latter, and it is homotopy equivalent to $hR^n_k(*,|G|)$, by application of the proposition. On the other hand, the geometric realization map $hR^n_k(*,G) \to hR^n_k(*,|G|)$ is a homotopy equivalence by proposition 2.1.2. □

Proof of proposition. By lemma 2.2.6 below, each of the (degeneracy) maps $C \to C_m$ is a homotopy equivalence. It follows (the realization lemma) that $C \to C.$ is a homotopy equivalence. Consequently, $C_Y \to C._Y$ is one, too.

In the topological case, the inclusion $C._Y \leftarrow C.(Y)$ is a homotopy equivalence by lemma 2.2.7 below.

In the simplicial case, that lemma does not apply to $C.$ directly, it only applies to the simplicial subcategory $C'.$ of the objects which satisfy the Kan extension condition. It remains to see that the inclusion $C'. \to C.$ is a homotopy equivalence. By the first part of the proposition we can reduce to showing that $C' \to C$ is a homotopy equivalence. This follows if we can find a functor $C \to C'$ together with a natural transformation from the identity functor. The desired functor is given by one of the standard devices of forcing the extension condition, namely the process of *filling horns* (which may be arranged in a G-equivariant way). □

Lemma 2.2.6. The map $C \to C_m$ is a homotopy equivalence.

Proof. Call this map j . We define a map $p\colon C_m \to C$. It is the identity on objects, and it takes a morphism $Y{\times}\Delta^m/{*}{\times}\Delta^m \to Z$ to the map $Y \to Z$ given by restriction to the last vertex of Δ^m . Then pj is the identity map on C . We will show that jp is homotopic to the identity map on C_m .

To construct the homotopy we use an auxiliary functor $F\colon C_m \to C_m$ which on objects is given by

$$Y \longmapsto Y{\times}\Delta^1/{*}{\times}\Delta^1 .$$

To define F on morphisms we use the standard contraction of Δ^m , that is, the map $f\colon \Delta^m{\times}\Delta^1 \to \Delta^m$ whose restrictions to $\Delta^m{\times}0$ and $\Delta^m{\times}1$ are the identity map on Δ^m , and the projection of Δ^m into its last vertex, respectively. By definition now F takes a map $Y{\times}\Delta^m/{*}{\times}\Delta^m \to Z$ to the map given by

$$Y{\times}\Delta^1{\times}\Delta^m \approx Y{\times}\Delta^m{\times}\Delta^1 \xrightarrow{(a,b)} Z{\times}\Delta^1$$

(or rather the induced map of quotients) where b is the projection $Y{\times}\Delta^m{\times}\Delta^1 \to \Delta^1$, and a is the composite map

$$Y \times (\Delta^m{\times}\Delta^1) \xrightarrow{\text{Id} \times f} Y \times \Delta^m \longrightarrow Z .$$

The point of considering F is that there are natural transformations $\text{Id} \to F$

and $jp \to F$. They are induced by the inclusions $Y \to Y\times\Delta^1/*\times\Delta^1$ taking Y to $Y\times 0$ and $Y\times 1$, respectively. In view of these natural transformations, each of the functors Id and jp is homotopic to F . Hence they are homotopic to each other. □

In order to formulate the next lemma we need a little preparation. Let $C.$ be a simplicial category. We say it is *special* if all the categories C_m have the same objects, and the face and degeneracy maps are the identity on objects. By abuse we can then speak of the objects of $C.$, rather than objects in some fixed degree, and for any two objects Y and Z we have a simplicial set of morphisms, which we denote $C.(Y,Z)$.

As before we let $C.(Y)$ denote the simplicial category of endomorphisms of Y . We must carefully distinguish between $C.(Y)$ and $C.(Y,Y)$. For they have different geometric realizations (the geometric realization of the former takes the composition law into account, whereas that of the latter does not).

We will say that two objects Y and Z are *strictly homotopy equivalent* if there exist $f \in C_o(Y,Z)$ and $g \in C_o(Z,Y)$ so that the composite gf is homotopic, in the simplicial set $C.(Y,Y)$, to the identity map on Y , and so that similarly the composite fg is homotopic in $C.(Z,Z)$ to the identity map on Z .

Lemma 2.2.7. Let $C.$ be a special simplicial category in which all objects are strictly homotopy equivalent to each other. Then for every object Y the inclusion $C.(Y) \to C.$ is a homotopy equivalence.

We deduce the lemma from a version of Quillen's theorem A for simplicial categories. In the case of special simplicial categories it takes the following form, cf. [15].

Criterion. Let $F\colon \mathcal{D}. \to C.$ be a map of special simplicial categories. A sufficient condition for F to be a homotopy equivalence is that for every object Z of $C.$ the simplicial category $F./Z : [m] \mapsto F_m/Z$ is contractible.

Proof of lemma. By the criterion applied to the inclusion $F\colon C.(Y) \to C.$ it suffices to show that for every Z the simplicial category $F./Z$ is contractible.

Suppose that $f \in C_o(Z,Z')$. It induces a map $f_*\colon F./Z \to F./Z'$,

$$(u \in C_m(Y,Z)) \longmapsto (d^*(f)\, u \in C_m(Y,Z'))$$

where d^* denotes the (degeneracy) map induced by $d\colon [m] \to [0]$.

Suppose next that $f_1 \in C_1(Z,Z')$, and let f and f' be its faces in $C_o(Z,Z')$. Then we claim that f_* and f'_* are homotopic. Indeed, a simplicial homotopy from f_* to f'_* is given (cf. the proof of lemma 1.4.1 for a discussion of simplicial homotopies) by the natural transformation which takes $a: [m] \to [1]$ to the map $F_m/Z \to F_m/Z'$,

$$(u \in C_m(Y,Z)) \longmapsto (a^*(f_1) u \in C_m(Y,Z')) .$$

By induction we conclude that if f and f'' are in the same connected component of $C.(Z,Z')$ then they induce homotopic maps $F./Z \to F./Z'$.

In turn we conclude that if Z_0 and Z_1 are strictly homotopy equivalent to each other, then $F./Z_0$ and $F./Z_1$ are homotopy equivalent.

Applying the hypothesis of the lemma now we obtain that, for every Z, $F./Z$ is homotopy equivalent to $F./Y$.

But $F./Y$ is the same as $\mathrm{Id}_{C.}/Y : [m] \to \mathrm{Id}_{C_m}/Y$. This is a simplicial object of contractible categories (each has a terminal object). Hence it is contractible. We are done. □

2.3. K-theory of simplicial rings, and linearization of A(X) .

The theme of this section is that much of the material of the preceding two sections can be redone in a 'linearized' setting. This leads to considering a K-theory of simplicial rings, and specifically, to comparing several definitions of it. In the case of discrete rings the K-theory is the same as Quillen's.

There is a natural transformation, *linearization*, from the 'non-linear' to the 'linear' setting. We record the plausible fact that, up to homotopy, the induced map in K-theory does not depend on which particular definition of K-theory is used.

Let R be a simplicial ring (with 1). By a *module* over R is meant a simplicial abelian group A together with a (unital and associative) action of R , that is, a map $A \otimes R \to A$ (degreewise tensor product). We let $M(R)$ denote the category of these modules and their R-linear maps.

A simplicial set Y gives rise to a module R[Y] where $(R[Y])_n = R_n[Y_n]$, the free R_n-module generated by Y_n . By the *attaching of a* n-*cell* to a module A is meant the formation of a pushout of the kind

$$A \longleftarrow R[\partial\Delta^n] \longrightarrow R[\Delta^n] \ .$$

We say that B is *obtainable from* A *by attaching of cells* if it can be built up by this process together with, perhaps, direct limit; we will also refer to this situation by saying that $A \to B$ is a *free map* (the notion is the same as that of a free map in [6]).

We define $M_f(R)$ to be the full subcategory of the modules which are obtainable from the zero module by attaching of finitely many cells. This is a category with cofibrations (free maps) and weak (homotopy) equivalences.

More generally, we define $M_{hf}(R)$ as the category given by the modules obtainable from 0 by attaching of perhaps infinitely many cells, but homotopy equivalent to some module in $M_f(R)$. Again this is a category with cofibrations and weak equivalences, in the same way.

$M_f(R)$ and $M_{hf}(R)$ give rise to the same K-theory, that is, the map

$$\Omega|hS.M_f(R)| \longrightarrow \Omega|hS.M_{hf}(R)|$$

is a homotopy equivalence. This results from

Proposition 2.3.1. The approximation theorem applies to the map $M_f(R) \to M_{hf}(R)$.

Proof. The argument is the same as that in the first part of the proof of proposition 2.1.1. The point is that the Whitehead theorem is available for objects in $M_f(R)$ or $M_{hf}(R)$ (one just constructs any desired map by induction on the generating simplices $\Delta^n \cdot 1$, $1 \in R$; it is not even necessary to subdivide Δ^n in the process since simplicial abelian groups satisfy the Kan extension condition). □

Let $M_{k\times k}(R)$ denote the simplicial ring of the $k\times k$ matrices in R . We define $\hat{GL}_k(R)$ to be the multiplicative simplicial monoid given by the matrices in $M_{k\times k}(R)$ which are invertible *up to homotopy*. Let $B\hat{GL}_k(R)$ denote the classifying space.

Theorem 2.3.2. There is a natural chain of homotopy equivalences

$$\Omega|hS.M_f(R)| \simeq K_0'(\pi_0 R) \times \lim_{\vec{k}} B\hat{GL}_k(R)^+ .$$

Here $K_0'(\pi_0 R)$ denotes the subgroup of the class group of the ring $\pi_0 R$ given by the free modules (it is cyclic, and in cases of interest it is usually Z).

Remark. There is a variant of the theorem where the category $M_f(R)$ is replaced by the larger category $M_{df}(R)$ of the objects *dominated by finite ones*; that is, the objects which are retracts of such in $M_{hf}(R)$. In that case the restricted class group $K_0'(\pi_0 R)$ in the theorem has to be replaced by the full class group $K_0(\pi_0 R)$.

Proof of theorem. Define $M_k^n(R)$ to be the full subcategory of $M_f(R)$ given by the objects which are *n-spherical of rank* k ; that is, the objects weakly equivalent to $R[\amalg^k \Delta^n]/R[\amalg^k \partial\Delta^n]$.

It will be shown below (proposition 2.3.5) that there is a natural homotopy equivalence

$$B\hat{GL}_k(R) \simeq |hM_k^n(R)|$$

compatible with suspension (the passage from n to $n+1$ on the right hand side).

Define $M^n(R)$ as the union of the categories $M_k^n(R)$. According to Segal [11] we have a homotopy equivalence

$$\Omega|hN.M^n(R)| \simeq K_0'(\pi_0 R) \times \lim_{\vec{k}} |hM_k^n(R)|^+ .$$

Combining with the former homotopy equivalence we obtain one

$$\Omega|hN.M^n(R)| \simeq K_0'(\pi_0 R) \times \lim_{\vec{k}} B\hat{GL}_k(R)^+ ,$$

compatible with suspension. The proof of the theorem has thus been reduced to the following proposition.

Proposition 2.3.3. There is a natural chain of homotopy equivalences

$$\lim_{\vec{n}} hN.M^n(R) \simeq hS.M_f(R) .$$

The proposition is actually true without passage to the limit on the left, but the limit makes for easier quoting of the general results (which were designed for different applications).

The proof is an application of theorems 1.7.1 and 1.8.1. To make these theorems applicable we have to check some things first. Let us define

$$h_*M = \pi_*(M \otimes_R \pi_o R) .$$

Lemma 2.3.4. Let $M \in M_{hf}(R)$. If $\pi_i M = 0$ for $i < n$ then the map $\pi_n M \to h_n M$ is an isomorphism.

Proof. If M and M' are right and left R-modules, respectively, there is a *derived tensor product* $M \overset{L}{\otimes}_R M'$, well defined up to homotopy [6 ,p.6.8]. If the module M happens to be 'free' (in the sense that $0 \to M$ is a *free map* — the objects of $M_{hf}(R)$ have that property, by definition) then the derived tensor product is represented by the actual tensor product $M \otimes_R M'$, by the corollary [6 ,p.6.10]. Therefore the spectral sequence (b) of theorem 6 [6 ,p.6.8] gives, in the case at hand, a first quadrant spectral sequence

$$E^2_{p,q} = \mathrm{Tor}_p^{\pi_* R}(\pi_* M, \pi_o R)_q \implies \pi_{p+q}(M \otimes_R \pi_o R)$$

where $\mathrm{Tor}_p(..)_q$ denotes the degree q part of the graded abelian group $\mathrm{Tor}_p(..)$. Now $\pi_i M = 0$ for $i < n$, so $E^2_{p,q} = 0$ for $q < n$, and we obtain an isomorphism $\pi_n(M \otimes_R \pi_o R) \approx E^2_{o,n}$, proving the lemma. □

Proof of proposition. The argument is precisely the same as that of the proof of proposition 2.2.2. Here is a brief account.

The objects of $M^n(R)$ may be characterized by the property that $h_i M$ is 0 for $i \neq n$, and free of finite rank over $\pi_o R$ for $i = n$. Let $\widetilde{M}^n(R)$ be the corresponding category with *free* replaced by *stably free*. Then all the hypotheses of section 1.7 are satisfied, so by theorem 1.7.1 we have homotopy equivalences

$$\lim_{\vec{n}} hS.\widetilde{M}^n(R) \longrightarrow \lim_{\to(\Sigma)} hS.M_f(R) \longleftarrow hS.M_f(R) .$$

On the other hand, $M^n(R)$ is strictly cofinal in $\widetilde{M}^n(R)$, so the inclusion

$$hS.M^n(R) \longrightarrow hS.\widetilde{M}^n(R)$$

is a homotopy equivalence by proposition 1.5.9. And finally the cofibrations in $M^n(R)$ are splittable up to weak equivalence, so theorem 1.8.1 applies to show that

$$\lim_{\vec{n}} hN.M^n(R) \longrightarrow \lim_{\vec{n}} hS.M^n(R)$$

is a homotopy equivalence. By combining the homotopy equivalences we obtain the proposition. □

To complete the proof of the theorem we are now left to compare $|hM_k^n(R)|$ and $B\widehat{GL}_k(R)$.

Let us write C instead of $hM_f(R)$, for short. We blow up C to a simplicial category $C.$, $[m] \mapsto C_m$. The objects of C_m are the same as those of C , and the morphisms in C_m are the m-parameter families of morphisms in C . That is, a morphism in C_m from A to B is a map $A[\Delta^m] \approx A \otimes Z[\Delta^m] \to B$. Considering C as a simplicial category in a trivial way we have a map $C \to C.$.

If $A \in C$ we let C_A , resp. $C._A$, denote the connected component of C , resp. $C.$, containing A , and $C.(A)$ the simplicial category of self-maps of A in $C.$.

Proposition 2.3.5. For every $A \in hM_f(R)$ there are homotopy equivalences

$$C_A \longrightarrow C._A \longleftarrow C.(A) .$$

Proof. The argument is similar to that of proposition 2.2.5. □

Corollary. There is a natural chain of homotopy equivalences, $B\widehat{GL}_k(R) \simeq |hM_k^n(R)|$, compatible with suspension.

Proof. Let $A = A_k^n$ denote the module obtained by attaching k n-cells to zero,

$$A_k^n = R[\amalg^k \Delta^n]/R[\amalg^k \partial\Delta^n] .$$

We claim that the simplicial ring of self-maps of A_k^n is homotopy equivalent to $M_{k\times k}(R)$, independently of n . To see this we can reduce, by a direct sum argument, to the special case $k = 1$. Restricting to the generating simplex we then obtain an isomorphism

$$Map_R(A_1^n, A_1^n) \approx Map(\Delta^n/\partial\Delta^n, R[\Delta^n]/R[\partial\Delta^n]) .$$

But it is well known, and easy to prove, that the n-fold loop space of the simplicial abelian group $R[\Delta^n]/R[\partial\Delta^n]$ is R again, up to homotopy. For example consider the *horn* Λ^n , the union of all the faces of Δ^n except the last. Then $R[\Delta^n]/R[\Lambda^n]$ is contractible. Hence the short exact sequence

$$R[\Delta^{n-1}]/R[\partial\Delta^{n-1}] \longrightarrow R[\Delta^n]/R[\Lambda^n] \longrightarrow R[\Delta^n]/R[\partial\Delta^n]$$

gives a looping fibration. It follows from the claim that the simplicial monoid of self-equivalences of A_k^n is homotopy equivalent, as monoid, to $\widehat{GL}_k(R)$. Hence $B\widehat{GL}_k(R) \simeq |C.(A_k^n)|$. Applying the proposition now we obtain that the latter is homotopy equivalent to $|C_A| = |hM_k^n(R)|$. The corollary results. □

Remark. The theorem includes a description of the Quillen K-theory of a discrete ring in terms of chain complexes over that ring. For if R is discrete then a 'module' in the sense used above is really the same thing as a *simplicial module over* R . In view of the Dold-Kan theorem there is therefore an equivalence (it is given by the normalized chain complex functor) of the category $M_f(R)$ with a category of chain complexes over R . □

Below, in the context of linearization, it will be convenient to know that the foregoing material can be redone topologically rather than simplicially. We record this now.

As a technical point, we will want to know that the geometric realization functor commutes with finite products. Therefore products should be formed in the category of compactly generated spaces. As a result we will restrict ourselves to working in that category. For example, if we mention a topological abelian group it will be tacitly understood that the underlying topological space is compactly generated.

Let $|A|$ be a topological abelian group, not necessarily the geometric realization of a simplicial abelian group A , and $|X|$ a topological space, not necessarily the geometric realization of a simplicial set X either. In this situation we can form $|A|[|X|]$, the topological abelian group freely generated by $|X|$ over $|A|$. The underlying space is the space of linear combinations of the kind

$$a_1x_1 + \dots + a_kx_k ,$$

subject to a suitable equivalence relation, and topologized accordingly. In detail, one forms

$$\coprod_k |A|^k \times |X|^k / \sim$$

where the equivalence relation is generated by the rule that for every map of finite sets, $\theta\colon \underline{m} \to \underline{n}$, the two maps

$$|A|^n \times |X|^n \xleftarrow{\theta_* \times \mathrm{Id}} |A|^m \times |X|^n \xrightarrow{\mathrm{Id} \times \theta^*} |A|^m \times |X|^m$$

are to be equalized.

If, in particular, $|R|$ is a topological ring, and $|X|$ a topological space, we can in this way obtain $|R|[|X|]$, the free $|R|$-module generated by $|X|$. The construction is compatible with geometric realization in the sense that if R is a simplicial ring, and X a simplicial set, then $|R|[|X|] \approx |R[X]|$.

We have the means now of defining the notion of the *attaching of a* n*-cell* to a $|R|$-module M . Namely this is the formation of a pushout of the kind

$$M \longleftarrow |R|[|\partial\Delta^n|] \longrightarrow |R|[|\Delta^n|] .$$

Starting from this notion we can proceed as in section 2.1 to carry over the definitions of $M_f(R)$ and $M_{hf}(R)$ to the topological context to obtain definitions of $M_f(|R|)$ and $M_{hf}(|R|)$.

Proposition 2.3.6. Let R be a simplicial ring. The approximation theorem applies to the geometric realization map $M_f(R) \to M_f(|R|)$.

Proof. The argument is similar to that of proposition 2.1.2. □

Define $\widehat{GL}_k(|R|)$ as in the simplicial case; that is, it is the simplicial monoid of the homotopy-invertible matrices over $|R|$.

Corollary 2.3.7. Let R be a simplicial ring. There is a natural chain of homotopy equivalences

$$\Omega|hS.M_f(|R|)| \simeq K_o'(\pi_o R) \times \varinjlim_k B\widehat{GL}_k(|R|)^+ ,$$

and the chain is compatible, via geometric realization, to that of theorem 2.3.2.

Proof. We consider the chain of maps in theorem 2.3.2 as consisting of three parts. The first part is the chain of maps between $\varinjlim hN.M^n(R)$ and $hS.M_f(R)$ in proposition 2.3.3. The preceding proposition applies to each map in the transformation from this chain to its topological analogue, so these maps are homotopy equivalences. As a result, since the maps in the former chain are homotopy equivalences, it follows that so are those in the latter.

The second part of the chain is Segal's homotopy equivalence of $\Omega|hN.M^n(R)|$ with $K_o'(\pi_o R) \times \varinjlim |hM_k^n(R)|^+$. This is certainly compatible with its topological analogue.

The third part of the chain, finally, is given by the maps in proposition 2.3.5, resp. its corollary. There is a compatible chain of maps in the topological case, and the maps are homotopy equivalences by the version of proposition 2.3.5 in the topological case. □

Suppose now that G is a simplicial monoid. Let Z be the ring of integers. There is an exact functor

$$\begin{aligned} R(*,G) &\longrightarrow M(Z[G]) \\ Y &\longmapsto \tilde{Z}[Y] = Z[Y]/Z[*] \end{aligned}$$

and hence an induced map in K-theory, the *linearization map*

$$\Omega|hS.R_f(*,G)| \longrightarrow \Omega|hS.M_f(Z[G])| .$$

On the other hand, the map of *rings up to homotopy* $\Omega^\infty S^\infty |G|_+ \to Z[|G|]$ induces, by matrix multiplication, a map of H-spaces

$$\hat{GL}(\Omega^\infty S^\infty|G|_+) \longrightarrow \hat{GL}(Z[|G|]) .$$

This is de-loopable to a map of classifying spaces $B\hat{GL}(\Omega^\infty S^\infty|G|_+) \to B\hat{GL}(Z[|G|])$, well defined up to homotopy. Namely the latter is obtained by composing, in the limit with respect to n and k , the map

$$B\,Aut_{|G|}(V^kS^n\wedge|G|_+) \longrightarrow B\,Aut_{Z[|G|]}(\tilde{Z}[V^kS^n\wedge|G|_+])$$

with a homotopy inverse to the homotopy equivalence

$$B\hat{GL}_k(Z[|G|]) \approx B\,Aut_{Z[|G|]}(\tilde{Z}[V^kS^o\wedge|G|_+]) \longrightarrow B\,Aut_{Z[|G|]}(\tilde{Z}[V^kS^n\wedge|G|_+]) .$$

We can further compose with an inverse to the homotopy equivalence

$$B\hat{GL}_k(Z[G]) \longrightarrow B\hat{GL}_k(Z[|G|]) .$$

<u>Corollary</u> 2.3.8. The linearization map corresponds, under the homotopy equivalences of theorems 2.2.1 and 2.3.2, to the map

$$Z \times B\hat{GL}(\Omega^\infty S^\infty|G|_+)^+ \longrightarrow Z \times B\hat{GL}(Z[G])^+ .$$

As indicated in [14], this result can be used to obtain numerical information. For example, as a consequence of the fact that the map $\Omega^\infty S^\infty|G|_+ \to Z[|G|]$ is a rational homotopy equivalence as well as an isomorphism on π_o , it follows that the map of the corollary is a rational homotopy equivalence.

Proof of corollary. This is a matter of checking, similar to the preceding corollary. We regard the chain of homotopy equivalences in theorem 2.2.1 as consisting of three parts. The first part is the chain of maps between $\varinjlim hN.R^n(*,G)$ and $hS.R_f(*,G)$ in proposition 2.2.2. This is compatible, by linearization, to the corresponding chain of maps between $\varinjlim hN.M^n(Z[G])$ and $hS.M_f(Z[G])$ in proposition 2.3.3.

The second part of the chain is Segal's homotopy equivalence of $\Omega|hN.R^n(*,G)|$ with $Z \times \varinjlim |hR^n_k(*,G)|^+$. This is compatible to its linear analogue, the homotopy equivalence between $\Omega|hN.M^n(Z[G])|$ and $Z \times \varinjlim |hM^n_k(Z[G])|^+$.

The third part, finally, is the commutative diagram of homotopy equivalences, with the notation as in proposition 2.2.5, and Y the simplicial version of $V^kS^n\wedge|G|_+$,

$$\begin{array}{ccccc} C_Y & \longrightarrow & C._Y & \leftarrow - - - & \\ \downarrow & & \downarrow & & \vdots \\ C_{|Y|} & \longrightarrow & C._{|Y|} & \longleftarrow & C.(|Y|) \end{array} .$$

The notation of the broken arrows here simply means that these arrows are missing.

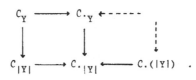

For we have not tried to put anything into the upper right corner. Such a Y would have to satisfy the Kan extension condition (proposition 2.2.5) and it would also have to fit into a sequence of Y's related to each other by some kind of suspension.

At any rate, the diagram is compatible, by linearization, to one

$$\begin{array}{ccccc} C_A & \longrightarrow & C._A & \longleftarrow & C.(A) \\ \downarrow & & \downarrow & & \downarrow \\ C_{|A|} & \longrightarrow & C._{|A|} & \longleftarrow & C.(|A|) \end{array}$$

where the upper row is that of proposition 2.3.5, with $A = Z[Y]$, and the lower row is the topological analogue of it. □

To conclude the topic of linearization let us briefly mention that, in the case of $A(X)$, there is a description of the linearization map which uses only spaces over X , not the loop group of X . The map is defined in terms of an exact functor $R(X) \to R^{ab}(X)$ where $R^{ab}(X)$ denotes the category of abelian group objects in $R(X)$.

In particular this means that, for connected X , there is a description of $K(Z[G(X)])$ in terms of $R^{ab}(X)$. To obtain that description, one *defines* a notion of weak equivalence in $R^{ab}(X)$ so that the map $R^{ab}(X) \to R^{ab}(*,G) \approx M(Z[G])$ corresponding to that of proposition 2.1.4, respects *and detects* weak equivalences. The argument of proposition 2.1.4 may then be adapted.

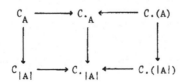

3. THE WHITEHEAD SPACE $Wh^{PL}(X)$, AND ITS RELATION TO $A(X)$.

3.1. Simple maps and the Whitehead space.

A map of simplicial sets is called *simple* if its geometric realization has contractible point inverses. We will admit here that simple maps form a category, that is, that a composite of simple maps is simple again, and that the gluing lemma is valid for simple maps. Proofs of these facts may be found e.g. in [16] where also a few other characterizations of simple maps are given.

If X is a simplicial set we denote by $C(X)$ the category of the cofibrant objects under X ; the objects are the pairs (Y,s) , $s\colon X \rightarrowtail Y$, and the morphisms from (Y,s) to (Y',s') are the maps $f\colon Y \to Y'$ with $fs = s'$.

As before we let $\mathcal{R}(X)$ denote the category of the triples (Y,r,s) , $rs = \mathrm{Id}_X$.

In either case, the subscript 'f' will denote the subcategory of the *finite* objects (where Y is generated, as simplicial set, by the simplices of $s(X)$ together with finitely many other simplices) and the superscript 'h' will denote the subcategory of the *homotopically trivial* objects (where $s\colon X \to Y$ is a weak homotopy equivalence). Finally the prefix 's' will denote the subcategory of the *simple maps*.

The category $sC_f^h(X)$ is of interest because of its role in the classification of PL manifolds and their automorphisms [2] [3] [16]; cf. also [15] and especially the proof of proposition 5.5 in that paper.

By the *Whitehead space* (the PL Whitehead space, to be precise) is meant a space whose fundamental group turns out to be the Whitehead group (the Whitehead group of $\pi_1 X$, that is, if X is connected) and which can be obtained from the (classifying space of the) category $sC_f^h(X)$ by de-looping, as follows.

In the language of section 1.8, the category $C_f^h(X)$ may be regarded as a category with sum (gluing at X) and weak equivalences (simple maps). Hence the group completion in the sense of Segal, the simplicial category $s\mathcal{N}.C_f^h(X)$, is defined.

<u>Proposition</u> 3.1.1. There is a natural homotopy equivalence

$$|sC_f^h(X)| \simeq \Omega|s\mathcal{N}.C_f^h(X)| .$$

Proof. Thanks to Segal [11] one knows that the canonical map from $|sC^h_f(X)|$ to $\Omega|sN.C^h_f(X)|$ is a homotopy equivalence if the H-space $|sC^h_f(X)|$ is *group-like* or, what amounts to the same thing, if the monoid $\pi_o|sC^h_f(X)|$ is a group. But it is well known that this is the case, cf. e.g. [16] for a proof. □

The main goal of this section is to prove the result (theorem 3.1.7 below) that the *sum construction* in $sN.C^h_f(X)$ can be traded for the *cofibration construction*; that is, that '$N.$' can be replaced by '$S.$'. In order for this replacement to make sense it is necessary to trade 'C' for 'R' first, that is, to impose structural retractions throughout. We also need an auxiliary construction; its purpose is to prevent the homotopy property of the functor $X \mapsto sN.C^h_f(X)$ from being lost upon transition from 'C' to 'R'.

Let F be a functor defined on the category of simplicial sets, with values in a category $\mathcal{B}$, say. We associate to it another functor $\check{F}$, with values in the category of simplicial objects in $\mathcal{B}$,

$$\check{F}(X) \quad = \quad ([n] \mapsto F(X^{\Delta^n}))$$

where X^{Δ^n} denotes the simplicial set of maps $\Delta^n \to X$.

Remark. In cases where the name of the functor is not F but something lengthy, such as for example $sN.C^h_f$, the notation $\check{F}(X)$ would be awkward. We will therefore use instead the notation $F(X^{\Delta^\cdot})$ on such occasions. □

Using the identification of $F(X)$ with $F(X^{\Delta^o})$, and considering objects of $\mathcal{B}$ as simplicial objects in a trivial way, we can define a natural transformation from F to $\check{F}$.

Supposing now that in the receiving category $\mathcal{B}$ it makes sense to speak of weak homotopy equivalences, we will say that the functor F *respects weak homotopy equivalences* if $X \xrightarrow{\sim} X'$ always implies $F(X) \xrightarrow{\sim} F(X')$.

<u>Lemma</u> 3.1.2. If F respects weak homotopy equivalences then the natural transformation $F \to \check{F}$ is a weak homotopy equivalence.

Proof. The (degeneracy) map $X^{\Delta^o} \to X^{\Delta^n}$ is a weak homotopy equivalence and therefore so is $F(X^{\Delta^o}) \to F(X^{\Delta^n})$, by assumption about F. We conclude with the realization lemma. □

<u>Lemma</u> 3.1.3. For any F, the functor $\check{F}$ preserves simplicial homotopies.

Proof. Let $X \to Y^{\Delta^1}$ be a simplicial homotopy. The claim is that one can naturally associate to it a simplicial homotopy of maps $\check{F}(X) \to \check{F}(Y)$. Such a simplicial homotopy may be identified to a natural transformation of functors on the category $\Delta/[1]$,

$$(a: [n] \to [1]) \longmapsto (\check{F}(X)_n \to \check{F}(Y)_n) .$$

The desired map on the right is defined as the composite map

$$F(X^{\Delta^n}) \longrightarrow F(Y^{\Delta^1 \times \Delta^n}) \longrightarrow F(Y^{\Delta^n})$$

where the first and second map are induced, respectively, by the homotopy $X \to Y^{\Delta^1}$, and by the map

$$\Delta^n \xrightarrow{(a_*, \mathrm{Id})} \Delta^1 \times \Delta^n .$$

□

Lemma 3.1.4. Let $F(X) = sR^h_f(X)$. Then the functor $\check{F}$ respects weak homotopy equivalences. Similarly with the functors $sN.R^h_f(X)$ and $sS.R^h_f(X)$.

Proof. By a well known argument (which e.g. may be found in [16]) it suffices to show that $\check{F}(X) \to \check{F}(X')$ is a weak homotopy equivalence if X' is obtained from X by *filling a horn*, that is, if it is the pushout of a diagram $X \leftarrow \Lambda^n_i \to \Delta^n$ where Λ^n_i is the i-*th horn* in Δ^n, the union of all the faces except the i-th. The idea of the following argument is to construct, in this situation, a deformation retraction of $\check{F}(X')$ to $\check{F}(X)$ by using the preceding lemma. Since it is not true, in general, that X is a deformation retract of X' by a simplicial homotopy, we must subdivide first.

Let Sd denote the subdivision functor for simplicial sets, and Sd_k its k-fold iteration. One knows that the subdivision of a simple map is simple again, cf. [16], so we can use Sd_2, say, to define a map

$$\Phi : sR^h_f(X') \longrightarrow sR^h_f(\mathrm{Sd}_2X') .$$

We compose with the map $f_*: sR^h_f(\mathrm{Sd}_2X') \to sR^h_f(X')$ induced by pushout with $f: \mathrm{Sd}_2X' \to X'$ (the composite of the 'last vertex map' $\mathrm{Sd}(X'') \to X''$ with itself). The composite map on $sR^h_f(X')$ then is homotopic to the identity. For, it takes (Y,r,s) to

$$\mathrm{Sd}_2Y \cup_{\mathrm{Sd}_2X'} X' ,$$

with the appropriate structure maps, and the desired homotopy is given by the natural transformation to the identity functor induced from $\mathrm{Sd}_2Y \to Y$, which is a simple map, cf. [16].

As shown below, $f: \mathrm{Sd}_2X' \to X'$ is simplicially homotopic, relative to Sd_2X, to a map into X. Applying the preceding lemma we thus obtain a simplicial homotopy of the map $\check{f}_*$. We conclude that there is a map homotopic to the identity on $sR^h_f(X'^{\Delta^\cdot})$, namely $\check{f}_*\Phi$, which is also homotopic to a map into $sR^h_f(X^{\Delta^\cdot})$. The latter homotopy is relative to the 'identity' on $sR^h_f(X^{\Delta^\cdot})$; more precisely, the homotopy is constant on the analogue of the map $\check{f}_*\Phi$ constructed from X instead of X'. So we can draw the desired conclusion that the map $sR^h_f(X^{\Delta^\cdot}) \to sR^h_f(X'^{\Delta^\cdot})$

is a weak homotopy equivalence.

We are left to show that $Sd_2X' \to X'$ is simplicially homotopic, relative to Sd_2X, to a map into X. Since the subdivision functor commutes with pushouts, this reduces to the following special case.

Assertion. The map $Sd_2\Delta^n \to \Delta^n$ is simplicially homotopic, relative to $Sd_2\Lambda^n_i$, to a map into Λ^n_i.

To see this we note that there is a homotopy of maps $|Sd_1\Delta^n| \to |\Delta^n|$ which has all the asserted properties except that it is not quite the geometric realization of a simplicial homotopy; it is only a linear homotopy of *unordered* simplicial complexes. We can get the ordering right by subdividing once more. This gives a simplicial homotopy of maps $Sd_2\Delta^n \to Sd_1\Delta^n$. Composing with the map $Sd_1\Delta^n \to \Delta^n$ we obtain the desired homotopy from it.

The other cases of the lemma are handled similarly. □

Lemma 3.1.5. If X satisfies the Kan condition, the map $sR^h_f(X^{\Delta^\cdot}) \to sC^h_f(X^{\Delta^\cdot})$ is a homotopy equivalence.

Proof. We define a simplicial category $[m] \mapsto sR^h_f(X)_m$ in which an object is one of $sC^h_f(X)$, say (Y,y), together with a map $Y\times\Delta^m \to X$ extending the projection $X\times\Delta^m \to X$. Since y is a weak homotopy equivalence, and X satisfies the extension condition, the simplicial set of those objects of $sR^h_f(X)_\cdot$ which arise from any particular (Y,y), is contractible. In other words, the simplicial set of objects of $sR^h_f(X)_\cdot$ maps by homotopy equivalence to the set of objects of $sC^h_f(X)$. Similarly, the simplicial set of morphisms of $sR^h_f(X)_\cdot$ maps by homotopy equivalence to the set of morphisms of $sC^h_f(X)$; and so on. It follows (the realization lemma) that the forgetful map $sR^h_f(X)_\cdot \to sC^h_f(X)$ is a homotopy equivalence.

Next we define a bisimplicial category $[m],[n] \mapsto sR^h_f(X)_{m,n} = sR^h_f(X^{\Delta^n})_m$. In view of the homotopy equivalence just established it follows, by the realization lemma, that the map $sR^h_f(X)_{\cdot\cdot} \to sC^h_f(X^{\Delta^\cdot})$ is a homotopy equivalence. Passing to the diagonal simplicial category of the bisimplicial category on the left (it has the same geometric realization, up to isomorphism) we obtain

$$\operatorname{diag} sR^h_f(X)_{\cdot\cdot} \longrightarrow sC^h_f(X^{\Delta^\cdot}) .$$

The lemma now results by checking that $\operatorname{diag} sR^h_f(X)_{\cdot\cdot}$ contains $sR^h_f(X^{\Delta^\cdot})$ as a deformation retract, and that the map of the lemma is the restriction of the latter homotopy equivalence.

An object of $sR^h_f(X)_{n,n}$ consists of an injective map $X^{\Delta^n} \to Y$ (with a finiteness condition) together with a map $Y\times\Delta^n \to X^{\Delta^n}$ which on $X^{\Delta^n}\times\Delta^n$ restricts to the projection. The object is in the subcategory $sR^h_f(X^{\Delta^n})$ if the map on $Y\times\Delta^n$ itself factors through the projection.

Passing to the adjoint, we can rewrite the map as $Y \to X^{\Delta^n \times \Delta^n}$. The desired simplicial homotopy now is induced by a simplicial deformation retraction of $[n] \mapsto X^{\Delta^n \times \Delta^n}$ to $[n] \mapsto X^{\Delta^n}$. Cf. e.g. [16] for a description of the homotopy. □

Lemma 3.1.6. If X satisfies the Kan condition, the forgetful map

$$\begin{array}{ccc} sS_n\mathcal{R}_f^h(X^{\Delta^\cdot}) & \longrightarrow & sS_{n-1}\mathcal{R}_f^h(X^{\Delta^\cdot}) \times s\mathcal{R}_f^h(X^{\Delta^\cdot}) \\ (\, Y_1 \rightarrowtail \dots \rightarrowtail Y_{n-1} \rightarrowtail Y_n \,) & \longmapsto & (\, Y_1 \rightarrowtail \dots \rightarrowtail Y_{n-1} \,,\, Y_n/Y_{n-1} \,) \end{array}$$

is a homotopy equivalence.

Proof. Define a category $s\widetilde{S}_n\mathcal{R}_f^h(X)$ just as $sS_n\mathcal{R}_f^h(X)$ except that there is no structural retraction on the object Y_n in the filtration $Y_o \rightarrowtail \dots \rightarrowtail Y_{n-1} \rightarrowtail Y_n$. There is a forgetful map

$$sS_n\mathcal{R}_f^h(X^{\Delta^\cdot}) \longrightarrow s\widetilde{S}_n\mathcal{R}_f^h(X^{\Delta^\cdot})$$

which forgets the structural retraction in question. This forgetful map is a homotopy equivalence as one sees by a straightforward adaption of the argument of the preceding lemma. Consequently (and in view of the preceding lemma) the assertion of the lemma is equivalent to the assertion that the map

$$s\widetilde{S}_n\mathcal{R}_f^h(X^{\Delta^\cdot}) \longrightarrow sS_{n-1}\mathcal{R}_f^h(X^{\Delta^\cdot}) \times s\mathcal{C}_f^h(X^{\Delta^\cdot})$$

is a homotopy equivalence. By the realization lemma this follows if we can show it degreewise, for fixed m . Writing X instead of X^{Δ^m} now, we are reduced to showing that the map

$$s\widetilde{S}_n\mathcal{R}_f^h(X) \longrightarrow sS_{n-1}\mathcal{R}_f^h(X) \times s\mathcal{C}_f^h(X)$$

is a homotopy equivalence.

Let us denote the components of this map by p and q , respectively, and the section of the map q by i . In order to show that (p,q) is a homotopy equivalence, it will suffice to show that the sequence

$$s\mathcal{C}_f^h(X) \xrightarrow{\ i\ } s\widetilde{S}_n\mathcal{R}_f^h(X) \xrightarrow{\ p\ } sS_{n-1}\mathcal{R}_f^h(X)$$

is a fibration, up to homotopy. We use Quillen's theorem B [8] to prove this. We proceed to show that the theorem applies, in its version for left fibres, to the map p .

Let $(Y_1 \to .. \to Y_{n-1})$ be an object of $sS_{n-1}\mathcal{R}_f^h(X)$. An object of the category $p/(Y_1 \to .. \to Y_{n-1})$ consists of an object $(Y'_o \to .. \to Y'_{n-1} \to Y'_n)$ of $s\widetilde{S}_n\mathcal{R}_f^h(X)$ together with a map g , say, in $sS_{n-1}\mathcal{R}_f^h(X)$, the (vertical) transformation

$$\begin{array}{ccc} Y'_o \to & \dots & \to Y'_{n-1} \\ \downarrow & & \downarrow \\ Y_o \to & \dots & \to Y_{n-1} \end{array} .$$

Let $p/(Y_o \to .. \to Y_{n-1})'$ denote the subcategory of the objects for which the structural map g is the identity map. It is a deformation retract of $p/(Y_o \to .. \to Y_{n-1})$; in fact, a deformation retraction is given by pushout with g .

On the other hand, $p/(Y_o \to .. \to Y_{n-1})'$ is isomorphic to $sC_f^h(Y_{n-1})$. As shown in [16], the functor $X \mapsto sC_f^h(X)$ respects weak homotopy equivalences. Hence the structural inclusion $X \to Y_{n-1}$ induces a homotopy equivalence $sC_f^h(X) \to sC_f^h(Y_{n-1})$. It results that the maps in $sS_{n-1}R_f^h(X)$ induce homotopy equivalences of the left fibres. Thus theorem B applies, showing that for every $(Y_o \to .. \to Y_{n-1})$ the square

$$\begin{array}{ccc} p/(Y_o \to .. \to Y_{n-1}) & \longrightarrow & s\widetilde{S}_n R_f^h(X) \\ \downarrow & & \downarrow \\ Id/(Y_o \to .. \to Y_{n-1}) & \longrightarrow & sS_{n-1}R_f^h(X) \end{array}$$

is homotopy cartesian. In particular this is so for the distinguished object $(X \to .. \to X)$. We saw above that $p/(X \to .. \to X)$ contains as a deformation retract a subcategory isomorphic to $sC_f^h(X)$. Under the horizontal map in the square this subcategory projects to the image of the inclusion map i , and under the vertical map it projects trivially into the contractible category $Id/(X \to .. \to X)$. We obtain that the maps i and p form a homotopy fibration, as claimed. □

Theorem 3.1.7. Let X be a simplicial set. There are homotopy equivalences

$$sN.C_f^h(X) \longrightarrow sN.C_f^h(X^{\Delta^\cdot}) \longleftarrow sN.R_f^h(X^{\Delta^\cdot}) \longrightarrow sS.R_f^h(X^{\Delta^\cdot}) .$$

Proof. It is shown in [16] that the functor $X \mapsto sC_f^h(X)$ respects weak homotopy equivalences. By lemma 3.1.2 therefore the map from $sC_f^h(X)$ to $sC_f^h(X^{\Delta^\cdot})$ is a homotopy equivalence, and consequently also $sN.C_f^h(X) \to sN.C_f^h(X^{\Delta^\cdot})$, in view of the realization lemma. To proceed we choose a weak equivalence $X \to X'$ where X' is a simplicial set satisfying the Kan condition. Then all maps in the transformation of the chain of the theorem to the corresponding chain with X replaced by X' are weak equivalences by lemma 3.1.4. Thus we can reduce to proving the theorem for simplicial sets which actually satisfy the Kan condition. Applying lemmas 3.1.5 and 3.1.6 now to the second and third map, respectively, we obtain that these maps are homotopy equivalences degreewise in the $N.$, resp. $S.$, directions. We conclude with the realization lemma. □

3.2. The homology theory associated to A(*) .

Let F be a functor defined on the category of simplicial sets, with values in some category of spaces. We say F is *excisive* if it satisfies the following two axioms.

(Limit). F commutes with direct limit.

(Excision). If $X_o \to X_1$ is a cofibration, and $X_o \to X_2$ any map, then the square

$$\begin{array}{ccc} F(X_o) & \longrightarrow & F(X_2) \\ \downarrow & & \downarrow \\ F(X_1) & \longrightarrow & F(X_1 \cup_{X_o} X_2) \end{array}$$

is homotopy cartesian.

We say F is a *homological functor* (or a *homology theory*) if, in addition to being excisive, it also satisfies

(Homotopy). If $X \to X'$ is a weak homotopy equivalence then so is $F(X) \to F(X')$.

Recall (the preceding section) that $\check{F}(X) = F(X^{\Delta^\cdot})$ denotes the functor

$$X \longmapsto ([n] \mapsto F(X^{\Delta^n})) .$$

The purpose of this section is to prove the following result.

Theorem 3.2.1. The functor $X \mapsto sS.R_f(X^{\Delta^\cdot})$ is a homology theory.

Addendum 3.2.2. The functor $X \mapsto \Omega|sS.R_f(X^{\Delta^\cdot})|$ may be identified, up to a natural chain of maps, to the homology theory associated to A(*) .

In fact, the chain is given by the maps (of loop spaces of)

$$([n] \mapsto sS.R_f(X^{\Delta^n})) \longleftarrow ([n] \mapsto sS.R_f(X_n)) \longrightarrow ([n] \mapsto hS.R_f(X_n))$$

where $X = ([n] \mapsto X_n)$ and where the first map is induced by the identification $X_n = (X^{\Delta^n})_o$. Each of the three terms is a homology theory. In the first case this is so by the theorem, and in the second and third cases, the terms are the homology theories associated to the Γ-spaces with underlying spaces $sS.R_f(*)$ and $hS.R_f(*)$, respectively (cf. e.g. [13] for a detailed description of the homology theory associated to a (special) Γ-space). Given the fact that the three terms are homology theories, and connected, the proof that the maps are homotopy equivalences can be reduced to checking the case $X = *$. In that case, the first map is an isomorphism, while the second map is the inclusion $sS.R_f(*) \to hS.R_f(*)$. There does not seem to exist a direct proof that the latter map is a homotopy equivalence, but an indirect proof is provided by theorem 3.3.1, below, together with the fact that $sS.R_f^h(*)$ is contractible (which, e.g., follows from proposition 1.3.1).

In order to prove the theorem it will suffice to prove the following two propositions 3.2.3 and 3.2.4.

Proposition 3.2.3. The functor $X \mapsto sS.R_f(X)$ is excisive.

Proof. First, it is clear that the functor commutes with direct limit (up to isomorphism).

Next, suppose that $X_o \to X_1$ is an injective map. Pullback with it defines a map $R_f(X_1) \to R_f(X_o)$ which respects simple maps. The inclusion-induced map $R_f(X_o) \to R_f(X_1)$ also respects simple maps. Composing the two we therefore obtain a subfunctor f of the identity functor on $R_f(X_1)$ which is *exact*, and hence a cofibration sequence of exact functors $f \to \mathrm{Id} \to f'$ where f' is defined as the quotient $f' = \mathrm{Id}/f$. Let $R_f(X_1,X_o)$ be defined as the category of the objects (Y,r,s) in $R_f(X_1)$ having *support away from* X_o ; that is, having the property that the pullback

$$X_o \times_{X_1} Y$$

is not bigger than X_o . Then f' takes values in $R_f(X_1,X_o)$, and it restricts to the identity map on that subcategory. Applying the additivity theorem to the cofibration sequence $f \to \mathrm{Id} \to f'$ now, we obtain a homotopy equivalence of $sS.R_f(X_1)$ with the product $sS.R_f(X_o) \times sS.R_f(X_1,X_o)$. In particular, therefore, the sequence

$$sS.R_f(X_o) \longrightarrow sS.R_f(X_1) \longrightarrow sS.R_f(X_1,X_o)$$

is a fibration, up to homotopy.

Applying this consideration in the situation of the excision axiom, we obtain a diagram of homotopy fibrations

$$\begin{array}{ccccc} sS.R_f(X_o) & \longrightarrow & sS.R_f(X_1) & \longrightarrow & sS.R_f(X_1,X_o) \\ \downarrow & & \downarrow & & \downarrow \\ sS.R_f(X_2) & \longrightarrow & sS.R_f(X_1 \cup_{X_o} X_2) & \longrightarrow & sS.R_f(X_1 \cup_{X_o} X_2, X_2) \end{array} \quad .$$

The vertical map on the right is an isomorphism (an inverse is induced by pullback). It follows that the square on the left is homotopy cartesian, as asserted by the excision axiom. This completes the proof. □

Proposition 3.2.4. Let F be an excisive functor, and suppose that $F(X)$ is connected for every X. Then the associated functor $\check{F}$ is a homology theory.

The proof will be given at the end of this section. Together with the preparatory material, it occupies the rest of the section.

Remark. The artificial looking connectivity assumption comes from the fact that our proof of the proposition uses the following lemma 3.2.5. Some auxiliary condition, such as connectivity, is definitely needed in that lemma.

Lemma 3.2.5. Let

$$\begin{array}{ccc} W_{\cdot\cdot} & \longrightarrow & X_{\cdot\cdot} \\ \downarrow & & \downarrow \\ Y_{\cdot\cdot} & \longrightarrow & Z_{\cdot\cdot} \end{array}$$

be a commutative diagram of bisimplicial sets. Suppose that for every m the diagram of simplicial sets

$$\begin{array}{ccc} W_{m\cdot} & \longrightarrow & X_{m\cdot} \\ \downarrow & & \downarrow \\ Y_{m\cdot} & \longrightarrow & Z_{m\cdot} \end{array}$$

is homotopy cartesian. Suppose further that for every m the simplicial sets $Y_{m\cdot}$ and $Z_{m\cdot}$ are connected. Then the diagram of bisimplicial sets is also homotopy cartesian.

Remark. There are easy examples to show that the connectivity assumption cannot be dropped without replacing it by something else. Here is a particularly bad case. Take any pullback diagram of simplicial sets, and consider it as a diagram of bisimplicial sets in a trivial way. Then in each degree m we have a pullback diagram of sets, and certainly therefore a homotopy cartesian square (of sets !). But it rarely happens, on the other hand, that a pullback diagram of simplicial sets is also homotopy cartesian.

Proof of lemma. We deduce the lemma from a corresponding result for homotopy fibrations which we refer to as the *fibre realization lemma.* A proof may be found in [13]; for convenience we recall the statement here. By a *fibration up to homotopy* is meant here a sequence of maps of 'spaces' of some sort, $X \rightarrow Y \rightarrow Z$, having the property that, firstly, the composite map $X \rightarrow Z$ is a trivial map, with image $*$ say, and, secondly, the map from X to the homotopy fibre of $Y \rightarrow Z$ at $*$ is a weak homotopy equivalence. The fibre realization lemma says the following. Let $X_{\cdot\cdot} \rightarrow Y_{\cdot\cdot} \rightarrow Z_{\cdot\cdot}$ be a sequence of maps of bisimplicial sets so that the composite map $X_{\cdot\cdot} \to Z_{\cdot\cdot}$ is a trivial map. Suppose that, for every m, the sequence of maps of simplicial sets $X_{m\cdot} \to Y_{m\cdot} \to Z_{m\cdot}$ is a fibration up to homotopy. Suppose further that for every m the simplicial set $Z_{m\cdot}$ is connected. Then the sequence of bisimplicial sets, $X_{\cdot\cdot} \to Y_{\cdot\cdot} \to Z_{\cdot\cdot}$, is itself a fibration up to homotopy.

The idea for proving the present lemma comes from the fact that a homotopy cartesian square with connected bases can be characterized as a commutative square in which the homotopy fibres of the vertical maps are mapped to each other by homotopy equivalence. Using this one hopes to obtain a translation of the assertion which follows from the fibre realization lemma.

To get the details right, it is convenient to replace homotopy fibres by actual fibres in a systematic way. We need to know that there is a functorial way of turning a map of simplicial sets into a Kan fibration; e.g., the process of *filling horns* [1] will do. Using it we replace, for every m, the square of the lemma by a square

$$\begin{array}{ccc} W'_{m\cdot} & \longrightarrow & X'_{m\cdot} \\ \downarrow & & \downarrow \\ Y'_{m\cdot} & \longrightarrow & Z'_{m\cdot} \end{array}$$

in which the vertical maps are Kan fibrations. In view of the naturality of the construction, these squares still assemble to a square of bisimplicial sets

$$\begin{array}{ccc} W'_{\cdot\cdot} & \longrightarrow & X'_{\cdot\cdot} \\ \downarrow & & \downarrow \\ Y'_{\cdot\cdot} & \longrightarrow & Z'_{\cdot\cdot} \end{array} \quad .$$

There is a natural transformation from the old square to the new, and the maps $W_{\cdot\cdot} \to W'_{\cdot\cdot}$, etc., are homotopy equivalences by the realization lemma. To prove the lemma it will therefore suffice to show that the new square is homotopy cartesian.

Choose any point of $Y'_{\cdot\cdot}$ (i.e., a compatible family of points in the $Y'_{m\cdot}$) as a basepoint; denote it $*$. Let $\mathrm{fibre}(W'_{m\cdot} \to Y'_{m\cdot})_{(*)}$ denote the actual fibre at $*$. Since $W'_{m\cdot} \to Y'_{m\cdot}$ is a Kan fibration, it is certainly true that the sequence

$$\mathrm{fibre}(W'_{m\cdot} \to Y'_{m\cdot})_{(*)} \longrightarrow W'_{m\cdot} \longrightarrow Y'_{m\cdot}$$

is a fibration up to homotopy, for every m. In view of the fibre realization lemma we deduce from this that the sequence

$$\mathrm{fibre}(W'_{\cdot\cdot} \to Y'_{\cdot\cdot})_{(*)} \longrightarrow W'_{\cdot\cdot} \longrightarrow Y'_{\cdot\cdot}$$

is also a fibration up to homotopy, where the term on the left denotes the actual fibre again; the point is that $\mathrm{fibre}(W'_{\cdot\cdot} \to Y'_{\cdot\cdot})_{(*)} \approx ([m] \mapsto \mathrm{fibre}(W'_{m\cdot} \to Y'_{m\cdot})_{(*)})$.

There are similar fibrations if W' and Y' are replaced by X' and Z'.

We can now complete the proof of the lemma as follows. In view of the assumption of homotopy cartesianness we have, for every m, a homotopy equivalence

$$\text{fibre}(W'_m.\to Y'_m.)_{(*)} \longrightarrow \text{fibre}(X'_m.\to Z'_m.)_{(\text{Im}(*))} .$$

By the realization lemma this implies a homotopy equivalence

$$\text{fibre}(W'_!.\to Y'_!.)_{(*)} \longrightarrow \text{fibre}(X'_!.\to Z'_!.)_{(\text{Im}(*))} ,$$

and therefore, in view of the preceding, a homotopy equivalence of the vertical homotopy fibres in the $W'_!.$-$X'_!.$-$Y'_!.$-$Z'_!.$ square. Thus that square is homotopy cartesian, as was to be shown. □

The lemma enters into the proof of proposition 3.2.4 through the following consequence.

Proposition 3.2.6. Let $[m] \mapsto F_m$ be a simplicial object of functors. Suppose that $F_m(X)$ is connected for every m and every X. Then if the F_m are excisive, it follows that so is F, where $F(X) = ([m] \mapsto F_m(X))$.

Proof. The validity of the limit axiom for F is automatic. The validity of the excision axiom for F follows from its validity for the F_m by application of the preceding lemma. □

For later use we record the following here.

Lemma 3.2.7. Let F^1 and F^2 be excisive functors so that $F^1(X)$ and $F^2(X)$ are connected for every X. Let $F^1 \to F^2$ be a natural transformation. If the natural transformation is a weak equivalence in the cases $X = \Delta^n$, $n = 0, 1, 2, \ldots$, then it is a weak equivalence in general.

Proof. By the limit axiom we can reduce to showing that $F^1(X) \to F^2(X)$ is a weak equivalence for finite X. Let X be obtained by attaching a 'last' simplex Δ^n to a simplicial set Y. In other words, choose an isomorphism of X to the push-out in a diagram

$$\begin{array}{ccc} \partial\Delta^n & \longrightarrow & \Delta^n \\ \downarrow & & \downarrow \\ Y & \longrightarrow & Y \cup_{\partial\Delta^n} \Delta^n . \end{array}$$

Applying F^1 to the diagram we obtain a homotopy cartesian square, in view of excision, and applying F^2 we obtain another. The map $F^1 \to F^2$ gives a map of the first homotopy cartesian square to the second. Since $F^1(X)$ and $F^2(X)$ are connected we conclude that, in order for $F^1(X) \to F^2(X)$ to be a homotopy equivalence, it suffices that the map is a homotopy equivalence in the other three cases. But in the case of Δ^n this is true by hypothesis, and in the cases of $\partial\Delta^n$ and Y it may be assumed true by induction. □

The crucial step in the proof of proposition 3.2.4 is the construction given in the following two definitions.

Definition 3.2.8. Let X be a simplicial set. Define $[k] \mapsto \mathit{Cov}(X)_k$ to be the simplicial object, in the category of simplicial sets, given by

$$\mathit{Cov}(X)_k \quad = \quad \coprod_{m,n} \Delta^m \times N_k(m,n) \times X_n$$

where $N_k(m,n)$ denotes the set of sequences in Δ ,

$$[m] \to [m_1] \to \dots \to [m_{k-1}] \to [n] \qquad (\text{ k arrows}) \;.$$

To describe the simplicial structure one rewrites $\mathit{Cov}(X)$ as the bisimplicial set where a bisimplex in bidegree (q,k) consists of a sequence

$$[q] \to [m_o] \to [m_1] \to \dots \to [m_{k-1}] \to [m_k]$$

together with an element $x \in X[m_k]$. By definition now the i-th face map with respect to the k-direction is given by omitting $[m_i]$ from the sequence; except if $i = k$ in which case, in addition, the element $x \in X[m_k]$ must be taken to the appropriate element of $X[m_{k-1}]$. The degeneracy maps are given by the insertion of identity maps in the sequence.

Definition 3.2.9. Let F be a functor on the category of simplicial sets. Then

$$F^{\times}(X) \quad = \quad ([k] \mapsto F(\mathit{Cov}(X)_k)) \;.$$

Considering the simplicial set X as a simplicial object in a trivial way, we can define a natural transformation

$$\mathit{Cov}(X). \longrightarrow X \;;$$

by definition, its restriction to $(\Delta^m, [m]\to\dots\to[n], x)$ is the composite map

$$\Delta^m \xrightarrow{([m]\to[n])_*} \Delta^n \xrightarrow{\;x\;} X \;.$$

Lemma 3.2.10. If X is a simplex Δ^p or, more generally, a disjoint union of simplices, then this map is the retraction in a simplicial deformation retraction from the simplicial object $[k] \mapsto \mathit{Cov}(X)_k$ to the trivial simplicial object $[k] \mapsto X$

Proof. In the case $X = \Delta^p$, the simplicial homotopy is defined as the natural transformation on the category $\Delta/[1]$ taking $a\colon [k] \to [1]$ to the map of $\mathit{Cov}(\Delta^p)_k$ to Δ^p defined in the following way. The map a_* takes the sequence

$$[q] \to [m_o] \to [m_1] \to \dots \to [m_k] \to [p]$$

to the sequence

$$[q] \to [m_o] \to \dots \to [m_{i(a)}] \to [p] \overset{=}{\to} \dots \overset{=}{\to} [p]$$

where $i(a)$ is the largest of the $i \in [k]$ which are in the pre-image of $0 \in [1]$; if a takes $[k]$ entirely into $1 \in [1]$ then the image sequence is

$$[q] \to [p] \overset{=}{\to} \dots \overset{=}{\to} [p] .$$

The homotopy is similarly defined in the more general case where X is a disjoint union of simplices. □

Considering the objects of the receiving category of the functor F as simplicial objects in a trivial way, we can define a natural transformation

$$F^{\times}(X) \longrightarrow F(X)$$

as the map which in degree k takes $F(\mathit{Cov}(X)_k)$ into $F(X)$ by the map induced from $\mathit{Cov}(X)_k \to X$.

Lemma 3.2.11. In the case where X is a simplex, or a disjoint union of such, the map $F^{\times}(X) \to F(X)$ is a (simplicial) homotopy equivalence.

Proof. The functor $F^{\times}$ has been defined by means of degreewise extension in the k-variable, so it preserves simplicial homotopies in the k-variable. The present lemma thus results from the preceding lemma. □

Remark. It is not difficult to show that $\mathit{Cov}(X). \to X$ is a *weak* homotopy equivalence for all X . On the other hand there seems little reason to suppose, in general, that the natural transformation $F^{\times}(X) \to F(X)$ is a weak equivalence for X which are not just disjoint unions of simplices.

Proposition 3.2.12. Suppose that $F(X)$ is connected for all X , and that F is excisive. Then $F^{\times}(X) \to F(X)$ is a weak homotopy equivalence for all X .

Proof. The functor

$$X \longmapsto \mathit{Cov}(X)_k = \coprod_{m,n} \Delta^m \times N_k(m,n) \times X_n$$

preserves monomorphisms and pushouts. As a result, the functor

$$X \longmapsto F(\mathit{Cov}(X)_k)$$

is excisive since F is. Applying proposition 3.2.6 now we obtain that

$$X \longmapsto ([k] \mapsto F(\mathit{Cov}(X)_k))$$

is an excisive functor, too.

Thus $F^{\times}(X) \to F(X)$ is a map of excisive functors. By lemma 3.2.11 the map is a weak equivalence in the case $X = \Delta^n$. Consequently, by lemma 3.2.7, it is a weak equivalence in general. □

Proposition 3.2.13. Let G be a functor satisfying that $G(X)$ is connected for all X. Suppose that G commutes with direct limit, and that it takes finite disjoint unions to products (up to homotopy); e.g., suppose that G is excisive. Then the functor $\check{G}^{\times}$ is excisive.

Proof. Let $X = ([j] \mapsto X_j)$. Then the functor $X \mapsto G(X_j)$ is excisive by hypothesis about G. By proposition 3.2.6 therefore the functor $X \mapsto ([j] \mapsto G(X_j))$ is excisive, too. We will show that the latter functor is weakly equivalent to $\check{G}^{\times}$. We show this by constructing an intermediate functor H and relating it to both.

Recalling the definitions

$$F^{\times}(X) \;=\; ([k] \mapsto F(\mathit{Cov}(X)_k)) \qquad \text{and} \qquad \check{F}(X) \;=\; ([j] \mapsto F(X^{\Delta^j}))$$

we unravel the definition of $\check{G}^{\times}$ as

$$\begin{aligned} \check{G}^{\times}(X) &= ([j] \mapsto ([k] \mapsto G(\mathit{Cov}(X^{\Delta^j})_k))) \\ &= ([j] \mapsto ([k] \mapsto G(\textstyle\coprod_{m,n} \Delta^m \times N_k(m,n) \times (X^{\Delta^j})_n))) \\ &\approx ([k] \mapsto ([j] \mapsto G(\textstyle\coprod_{m,n} \Delta^m \times N_k(m,n) \times (X^{\Delta^n})_j))) . \end{aligned}$$

We define the intermediate functor H by replacing X^{Δ^n} by X in the latter term,

$$H(X) \;=\; ([k] \mapsto ([j] \mapsto G(\textstyle\coprod_{m,n} \Delta^m \times N_k(m,n) \times X_j))) .$$

The projection $\Delta^n \to \Delta^o$ induces an inclusion $X \to X^{\Delta^n}$ and hence a map of $H(X)$ to $\check{G}^{\times}(X)$. We claim this map is a homotopy equivalence.

In fact, the map $([j] \mapsto X_j) \to ([j] \mapsto (X^{\Delta^n})_j)$ is a *simplicial* homotopy equivalence. The process of applying functors degreewise preserves simplicial homotopies. Hence the map

$$([j] \mapsto G(\textstyle\coprod_{m,n} \Delta^m \times N_k(m,n) \times X_j)) \longrightarrow ([j] \mapsto G(\textstyle\coprod_{m,n} \Delta^m \times N_k(m,n) \times (X^{\Delta^n})_j))$$

is a (simplicial) homotopy equivalence still. Applying the realization lemma with respect to the k-variable now, we conclude that $H(X) \to \check{G}^{\times}(X)$ is a (weak) homotopy equivalence.

To proceed, we rewrite $H(X)$ as

$$([j] \mapsto ([k] \mapsto G(\textstyle\coprod_{m,n} \Delta^m \times N_k(m,n) \times X_j))) .$$

The map

$$([k] \mapsto G(\textstyle\coprod_{m,n} \Delta^m \times N_k(m,n) \times X_j)) \longrightarrow ([k] \mapsto G(X_j))$$

is a (simplicial) homotopy equivalence by lemma 3.2.11. Applying the realization lemma with respect to the j-variable now we conclude that the map

$$H(X) \longrightarrow (\ [j] \mapsto (\ [k] \mapsto G(X_j)\ )\ )$$

is a (weak) homotopy equivalence. The target of this map is the simplicial object $[j] \to G(X_j)$ considered as a bisimplicial object in a trivial way. We are done. □

Proof of proposition 3.2.4. Recall, the claim is that if F is an excisive functor such that $F(X)$ is connected for every X, then the functor $\check{F}$ is a homology theory.

The main problem is to show that $\check{F}$ is excisive again. To see this we introduce the functor $F^{\times}$ (definition 3.2.9). The natural transformation $F^{\times} \to F$ is a weak homotopy equivalence in the situation at hand (proposition 3.2.12). By the realization lemma it follows that the natural transformation $\check{F}^{\times} \to \check{F}$ is a weak homotopy equivalence as well. Thus we can reduce to showing that the functor $\check{F}^{\times}$ is excisive. This was shown in proposition 3.2.13.

We are left to show now that the functor $\check{F}$ respects weak homotopy equivalences. By a well known argument (which e.g. may be found in [1]) it suffices to show that $\check{F}(X) \to \check{F}(X')$ is a homotopy equivalence if X' is obtained from X by filling a horn, that is, if there is a pushout diagram

$$\begin{array}{ccc} \Lambda^n_i & \longrightarrow & \Delta^n \\ \downarrow & & \downarrow \\ X & \longrightarrow & X' \end{array} .$$

$\check{F}$ applied to this diagram gives a homotopy cartesian square, by excision, so we can reduce further to showing that $\check{F}(\Lambda^n_i) \to \check{F}(\Delta^n)$ is a homotopy equivalence.

Now Δ^n is contractible to its i-th vertex by simplicial homotopy (if $i = 0$ or n, a single homotopy will do; otherwise one needs a chain of two) and the contraction restricts to one of Λ^n_i. Since $\check{F}$ preserves simplicial homotopies (lemma 3.1.3) we conclude that indeed $\check{F}(\Lambda^n_i) \to \check{F}(\Delta^n)$ is a homotopy equivalence. The proof is now complete. □

3.3. The fibration relating $Wh^{PL}(X)$ and $A(X)$.

The fibration arises from the interplay of two notions of weak equivalence on the category $\mathcal{R}_f(X)$, where X is a simplicial set. The two notions are given by the *simple maps* on the one hand and by the *weak homotopy equivalences* on the other.

Let the superscript 'h' denote the subcategory of the objects which are homotopically trivial; that is, the (Y,r,s) where s is a weak homotopy equivalence. As before (the preceding two sections) let $\mathcal{R}_f(X^{\Delta^\bullet})$ denote the simplicial category $[n] \mapsto \mathcal{R}_f(X^{\Delta^n})$.

Theorem 3.3.1. The square

$$\begin{array}{ccc} sS.\mathcal{R}^h_f(X^{\Delta^\bullet}) & \longrightarrow & hS.\mathcal{R}^h_f(X^{\Delta^\bullet}) \\ \downarrow & & \downarrow \\ sS.\mathcal{R}_f(X^{\Delta^\bullet}) & \longrightarrow & hS.\mathcal{R}_f(X^{\Delta^\bullet}) \end{array}$$

is homotopy cartesian, and the term on the upper right is contractible. The other terms are as follows,

$$\Omega|hS.\mathcal{R}_f(X^{\Delta^\bullet})| \simeq A(X) ,$$

$$X \mapsto sS.\mathcal{R}_f(X^{\Delta^\bullet}) \quad \text{is a homology theory,}$$

$$sS.\mathcal{R}^h_f(X^{\Delta^\bullet}) \simeq Wh^{PL}(X) ,$$

and each of the homotopy equivalences can be described by a natural chain of maps.

Proof. In order to show that the square is homotopy cartesian it will suffice to show, by lemma 3.2.5, that for each n the square with $X^{\Delta^\bullet}$ replaced by X^{Δ^n} is homotopy cartesian. Writing X instead of X^{Δ^n} now we have reduced to showing that the square

$$\begin{array}{ccc} sS.\mathcal{R}^h_f(X) & \longrightarrow & hS.\mathcal{R}^h_f(X) \\ \downarrow & & \downarrow \\ sS.\mathcal{R}_f(X) & \longrightarrow & hS.\mathcal{R}_f(X) \end{array}$$

is homotopy cartesian. The desired fact is essentially a special case of theorem 1.6.4. There is a little technical point. Namely the category of weak homotopy equivalences on $R_f(X)$ does not satisfy the *extension axiom* as required for a direct application of theorem 1.6.4. For this reason we compare with the square

$$\begin{array}{ccc} sS.R_f^h(X) & \longrightarrow & hS.R_f^h(X) \\ \downarrow & & \downarrow \\ sS.R_f^{(2)}(X) & \longrightarrow & hS.R_f^{(2)}(X) \end{array}$$

where $R_f^{(2)}(X)$ denotes the subcategory of $R_f(X)$ of the (Y,r,s) where $s: X \to Y$ is a 1-connected map. The weak homotopy equivalences in $R_f^{(2)}(X)$ may alternatively be characterized as the maps inducing isomorphisms in homology (the Whitehead theorem), consequently they do satisfy the extension axiom. Hence theorem 1.6.4 applies to show the latter square is homotopy cartesian. We conclude by noting that the map to the former square is a homotopy equivalence on each of the four corners. In fact, double suspension induces an endomorphism of each of the terms, the endomorphism is homotopic to the identity map (proposition 1.6.2), and it takes $R_f(X)$ into $R_f^{(2)}(X)$.

The upper right term $hS.R_f^h(X^{\Delta^\cdot})$ is contractible since it is a bisimplicial object of categories with initial objects.

The term $hS.R_f(X^{\Delta^\cdot})$ is a de-loop of $A(X)$ since $hS.R_f(X) \to hS.R_f(X^{\Delta^\cdot})$ is a homotopy equivalence (by lemma 3.1.2) in view of the fact that $X \mapsto hS.R_f(X)$ respects weak homotopy equivalences (proposition 2.1.7).

The homotopy equivalence $sS.R_f^h(X^{\Delta^\cdot}) \simeq Wh^{PL}(X)$ is given in theorem 3.1.7.

The fact that $X \mapsto sS.R_f(X^{\Delta^\cdot})$ is a homology theory, finally, is provided by theorem 3.2.1. □

The theorem may be reformulated a little by defining the auxiliary simplicial structure in a slightly different way. Namely define a simplicial category $R_f(X)_\cdot$ as follows. $R_f(X)_n$ is the subcategory of $R_f(X\times\Delta^n)$ given by the objects (Y,r,s) which have the property that the composite map

$$Y \xrightarrow{\ r\ } X \times \Delta^n \xrightarrow{\ pr_2\ } \Delta^n$$

is locally fibre homotopy trivial.

<u>Proposition</u> 3.3.2. There is a homotopy cartesian square

$$\begin{array}{ccc} sS.R_f^h(X)_\cdot & \longrightarrow & hS.R_f^h(X)_\cdot \\ \downarrow & & \downarrow \\ sS.R_f(X)_\cdot & \longrightarrow & hS.R_f(X)_\cdot \end{array}$$

and it is homotopy equivalent to the square of the theorem by a natural map.

Proof. The homotopy cartesianness of the square is established in the same way as in the theorem. There is a map from the square of the theorem to that of the proposition. It is induced from the map of simplicial categories $\mathcal{R}_f(X^{\Delta^\cdot}) \to \mathcal{R}_f(X)_\cdot$ defined as follows. The map in degree n is the composite map

$$\mathcal{R}_f(X^{\Delta^n}) \longrightarrow \mathcal{R}_f(X^{\Delta^n} \times \Delta^n) \longrightarrow \mathcal{R}_f(X \times \Delta^n)$$

where the first map is given by product with Δ^n, and the second map is induced from a map

$$X^{\Delta^n} \times \Delta^n \longrightarrow X \times \Delta^n ,$$

namely the map whose second and first components are the projection map pr_2 and the evaluation map

$$X^{\Delta^n} \times \Delta^n \longrightarrow X ,$$

respectively.

In order to show that the transformation of squares is a homotopy equivalence it suffices, in view of the homotopy cartesianness of the two squares, to show that the map is a homotopy equivalence on three of the four corners.

This is automatic in the case of the upper right corner as both terms are contractible.

It is still easy in the case of the lower right corner. Namely in view of the homotopy equivalence $hS.\mathcal{R}_f(X) \to hS.\mathcal{R}_f(X^{\Delta^\cdot})$ (the theorem) it suffices to know that the map $hS.\mathcal{R}_f(X) \to hS.\mathcal{R}_f(X)_\cdot$ is a homotopy equivalence. This follows from the fact (by the argument of lemma 2.2.6) that for every n the map $hS.\mathcal{R}_f(X) \to hS.\mathcal{R}_f(X)_n$ is a homotopy equivalence.

As our third case we take that of the upper left corner. That case is less easy. We consider the diagram

$$\begin{array}{ccccccc}
sN.\mathcal{C}^h_f(X) & \longrightarrow & sN.\mathcal{C}^h_f(X^{\Delta^\cdot}) & \longleftarrow & sN.\mathcal{R}^h_f(X^{\Delta^\cdot}) & \longrightarrow & sS.\mathcal{R}^h_f(X^{\Delta^\cdot}) \\
\Big\| & & \Big\downarrow & & \Big\downarrow & & \Big\downarrow \\
sN.\mathcal{C}^h_f(X) & \longrightarrow & sN.\mathcal{C}^h_f(X)_\cdot & \longleftarrow & sN.\mathcal{R}^h_f(X)_\cdot & \longrightarrow & sS.\mathcal{R}^h_f(X)_\cdot
\end{array}$$

where the upper row is the chain of maps of theorem 3.1.7, and the lower row is an analogue of that chain for the other auxiliary simplicial structure. The maps in the upper row are homotopy equivalences (theorem 3.1.7), so it will suffice to know that the maps in the lower row are homotopy equivalences, too. The second and third maps in the chain now are handled as before (lemmas 3.1.5 and 3.1.6). In the case of the first map one can reduce (by the realization lemma) to showing that the map $s\mathcal{C}^h_f(X) \to s\mathcal{C}^h_f(X)_\cdot$ is a homotopy equivalence; or in fact, that $s\mathcal{C}^h_f(X) \to s\mathcal{C}^h_f(X)_n$ is, for every n. But this has been proved in [16]. □

References.

1. P. Gabriel and M. Zisman, *Calculus of fractions and homotopy theory*, Ergebnisse der Mathematik und ihrer Grenzgebiete, Band 35, Springer (1967).

2. A. Hatcher, *Higher simple homotopy theory*, Ann. of Math. 102 (1975), 101-137.

3. ————, *Concordance spaces, higher simple homotopy theory, and applications*, Proc. Symp. Pure Math. vol. 32, part I, A.M.S. (1978), 3-21.

4. D.M. Kan, *On c.s.s. complexes*, Amer. J. Math. 79 (1957), 449-476.

5. J.L. Loday, *Homotopie des espaces de concordances*, Séminaire Bourbaki, 30e année, 1977/78, n° 516.

6. D.G. Quillen, *Homotopical algebra*, Springer Lecture Notes in Math. 43.

7. ————, *Cohomology of groups*, Actes, Congrès Intern. Math. 1970, tom 2, 47-51.

8. ————, *Higher Algebraic K-theory. I*, Springer Lecture Notes in Math. 341 (1973), 85-147.

9. M. Ravel, *Alborada del gracioso*, Editions Max Eschig, Paris.

10. G. Segal, *Classifying spaces and spectral sequences*, Publ. Math. I.H.E.S. 34 (1968), 105-112.

11. ————, *Categories and cohomology theories*, Topology 13 (1974), 293-312.

12. ————, *Configuration spaces and iterated loop spaces*, Invent. math. 21 (1973), 213-221.

13. F. Waldhausen, *Algebraic K-theory of generalized free products*, Ann. of Math. 108 (1978), 135-256.

14. ————, *Algebraic K-theory of topological spaces. I*, Proc. Symp. Pure Math. vol. 32, part I, A.M.S. (1978), 35-60.

15. ————, *Algebraic K-theory of spaces, a manifold approach*, Canadian Math. Soc., Conf. Proc., vol. 2, part 1, A.M.S. (1982), 141-184.

16. ————, *Spaces of PL manifolds and categories of simple maps.*

17. ————, *Operations in the algebraic K-theory of spaces*, Springer Lecture Notes in Math. 967 (1982), 390-409.

18. ————, *Algebraic K-theory of spaces, localization, and the chromatic filtration of stable homotopy*, Springer Lecture Notes in Math. 1051 (1984), 173-195.

FAKULTÄT FÜR MATHEMATIK
UNIVERSITÄT BIELEFELD
4800 BIELEFELD, FRG.

Oliver's Formula and Minkowski's Theorem

Shmuel Weinberger

In this note we give an elementary verification of an unpublished formula of Bob Oliver. This leads to a three line proof of Minkowski's theorem, that a finite group acting effectively on a surface of genus at least two is represented faithfully by its action on homology.

Theorem. (Oliver)

If $\mathbb{Z}_n$ acts cellularly on a finite complex X then

$$\chi(X^{\mathbb{Z}_n}) = L(g)$$

where L(g) is the Lefshetz number of a generator.

Proof.

Examine the equivariant chain complex of X

$$C_*(X) \approx C_*(X^{\mathbb{Z}_n}) \oplus \bar{C}_*(X)$$

where $\bar{C}_*(X)$ is freely generated by cells not in $X^{\mathbb{Z}_n}$. Thus

$$L(g) = \Sigma\ (-1)^i \operatorname{Tr} g_{i_*}$$

$$= \Sigma\ (-1)^i \operatorname{Tr} g|C_i(X)$$

$$= \Sigma\ (-1)^i \operatorname{Tr} g|C_i(X^{\mathbb{Z}_n}) + \Sigma\ (-1)^i \operatorname{Tr} g\ \bar{C}_i(X)$$

$$= \chi(X^{\mathbb{Z}_n})$$

since $g|C_i(X^{\mathbb{Z}_n})$ is the identity and $g|\bar{C}_i(X)$ has only zeroes along the diagonal. □

Corollary. (Minkowski)

If π acts effectively on a surface M of genus at least two then $\pi \to \operatorname{Aut} H_1(M;\mathbb{Q})$ is injective.

Proof.

Let $\mathbb{Z}_n \subset$ Kernel. Since the action is effective $M^{\mathbb{Z}_n}$ is a union of circles and points so that $\chi(M^{\mathbb{Z}_n}) \geq 0$. On the other hand, by assumption $g_* = 1$ so $L(g) = \chi(M) < 0$. □

An easy consequence is that only finitely many groups act effectively on any fixed surface of genus larger than one.

Princeton University

Some Nilpotent Complexes

Shmuel Weinberger

In this note we construct some nilpotent complexes whose existences were unknown. (A CW complex X is nilpotent if $\pi_1(X)$ is, and $\pi_i(X)$ are nilpotent $\pi_1 X$ modules.

Proposition 1: For any finite nilpotent group π, there is a three dimensional nilpotent complex with fundamental group π.

Proposition 2: There are nilpotent finite Poincaré complexes with $\pi_1 = \mathbb{Z}_{15}$ which are not simple Poincaré complexes in the sense of (Wa) Wall.

For Proposition 1, it is not hard to see that $\dim X \geq 3$ (if π is nontrivial) and P. Kahn asked if one can arrange for $\dim X = 3$, for then by crossing with a torus one obtains for abelian groups nilpotent spaces of smallest possible dimension with the given group as π_1 ((BK)).

For Proposition 2, note that in [CW] it is shown that nilpotent finite Poincaré complexes with π_1 an odd p-group are simple.

Lemma. If $\pi_1 X$ is nilpotent and acts nilpotently on $H_i(\tilde{X})$, then X is nilpotent.

Proof. Quite easy from localization theory. See eg. [We].

Proof of Proposition 1: Nilpotent groups are products of their p-Sylow subgroups, $\pi = \Pi\pi_p$. The version of Wall finiteness obstruction theory given in [We] shows that $S^3 \times K(\pi_p, 1)$ has the $\mathbb{Z}[\frac{1}{p}][\pi_p]$ homology type of a finite 3-dimensional complex M_p. Now just Zabrodsky mix the M_p's. This produces a finitely dominated three dimensional homologically nilpotent complex as desired. Q.E.D.

Problem: Can this complex be taken finite?

Proof of Proposition 2: Recall from [KM] that $\tilde{K}_0(\mathbb{Z}_{15}) = \mathbb{Z}_2$. As a result

$$\begin{array}{ccccc} L^p_{odd}(\mathbb{Z}_{15}) & \to & H(\mathbb{Z}_2;K_0(\mathbb{Z}_{15})) & \to & L^h_{even}(\mathbb{Z}_{15}) \\ \| & & \| & & \\ 0 & & \mathbb{Z}_2 & & \end{array}$$

$L^h_{even}(\mathbb{Z}_{15})$ has an element of order 2 which we will construct shortly. Since L^s is torsion free [Wa 2] this element is detected by $H(\mathbb{Z}_2;Wh(\mathbb{Z}_{15}))$ in the Rothenberg sequence.

Let M be the nontrivial module $\mathbb{Z}_9$ over $\mathbb{Z}_{15}$. M is nilpotent, and has a resolution $0 \to P \to \mathbb{Z}(\mathbb{Z}_{15}) \to M \to 0$ where P is the nontrivial element of $K_0(\mathbb{Z}_{15})$. $P \oplus P^*$ admits a hyperbolic form which is the desired element of L^h. Better yet, consider the form (= as an element of L^h) on $N \oplus N^*$ where $N = [\mathbb{Z}_{15}] \oplus P^*$. This has a free hyperbolic pair in it based on $P \oplus P^* \subset N$.

Apply the proof of the Wall realization theorem to $S^2 \times L^3_{15}$ and the form on $N \oplus N^*$ and glue in copies of $D^3 \times L^3_{15}$. It is easy to see that the result of surgering the geometric spheres corresponding to $P \oplus P^*$ produces the desired complex. Q.E.D.

Remark: Mislin has also used the module M in his work on finiteness obstructions for nilpotent complexes.

References

[BK] K. Brown and P. Kahn, Homotopy dimension and simple cohomological dimension of spaces, Comm. Math. Helv. 52(1977) 111-127.

[CW] S. Cappell and S. Weinberger, Homology Propagation of Group Actions (preprint)

[KM] M. Kervaire and M. Murthy, On the Projective Class Group of cyclic groups of prime power, Comm. Math. Helv. 52(1977) 415-452.

[M] G. Mislin, The geometric realization of Wall obstructions by nilpotent and simple spaces, Math. Proc. Comb. Phil. Soc. 87 (1980) 199-206.

[Wa] C.T.C. Wall, Surgery on Compact Manifolds, Academic Press.

[Wa2] ___________, Classification of Hermitian Forms VI. Group Rings, Ann. of Math. 103(1976) 1-80.

[We] S. Weinberger, Homologically trivial Group Actions I, II (preprints).